Selected Titles in This Series

48 **Harvey Rubin and David Harlan Wood, Editors,** DNA Based Computers III

47 **Martin Farach-Colton, Fred S. Roberts, Martin Vingron, and Michael Waterman, Editors,** Mathematical Support for Molecular Biology

46 **Peng-Jun Wan, Ding-Zhu Du, and Panos M. Pardalos, Editors,** Multichannel Optical Networks: Theory and Practice

45 **Marios Mavronicolas, Michael Merritt, and Nir Shavit, Editors,** Networks in Distributed Computing

44 **Laura F. Landweber and Eric B. Baum, Editors,** DNA Based Computers II

43 **Panos Pardalos, Sanguthevar Rajasekaran, and José Rolim, Editors,** Randomization Methods in Algorithm Design

42 **Ding-Zhu Du and Frank K. Hwang, Editors,** Advances in Switching Networks

41 **David Aldous and James Propp, Editors,** Microsurveys in Discrete Probability

40 **Panos M. Pardalos and Dingzhu Du, Editors,** Network Design: Connectivity and Facilities Location

39 **Paul W. Beame and Samuel R Buss, Editors,** Proof Complexity and Feasible Arithmetics

38 **Rebecca N. Wright and Peter G. Neumann, Editors,** Network Threats

37 **Boris Mirkin, F. R. McMorris, Fred S. Roberts, and Andrey Rzhetsky, Editors,** Mathematical Hierarchies and Biology

36 **Joseph G. Rosenstein, Deborah S. Franzblau, and Fred S. Roberts, Editors,** Discrete Mathematics in the Schools

35 **Dingzhu Du, Jun Gu, and Panos M. Pardalos, Editors,** Satisfiability Problem: Theory and Applications

34 **Nathaniel Dean, Editor,** African Americans in Mathematics

33 **Ravi B. Boppana and James F. Lynch, Editors,** Logic and Random Structures

32 **Jean-Charles Grégoire, Gerard J. Holzmann, and Doron A. Peled, Editors,** The SPIN Verification System

31 **Neil Immerman and Phokion G. Kolaitis, Editors,** Descriptive Complexity and Finite Models

30 **Sandeep N. Bhatt, Editor,** Parallel Algorithms: Third DIMACS Implementation Challenge

29 **Doron A. Peled, Vaughan R. Pratt, and Gerard J. Holzmann, Editors,** Partial Order Methods in Verification

28 **Larry Finkelstein and William M. Kantor, Editors,** Groups and Computation II

27 **Richard J. Lipton and Eric B. Baum, Editors,** DNA Based Computers

26 **David S. Johnson and Michael A. Trick, Editors,** Cliques, Coloring, and Satisfiability: Second DIMACS Implementation Challenge

25 **Gilbert Baumslag, David Epstein, Robert Gilman, Hamish Short, and Charles Sims, Editors,** Geometric and Computational Perspectives on Infinite Groups

24 **Louis J. Billera, Curtis Greene, Rodica Simion, and Richard P. Stanley, Editors,** Formal Power Series and Algebraic Combinatorics/Séries Formelles et Combinatoire Algébrique, 1994

23 **Panos M. Pardalos, David I. Shalloway, and Guoliang Xue, Editors,** Global Minimization of Nonconvex Energy Functions: Molecular Conformation and Protein Folding

22 **Panos M. Pardalos, Mauricio G. C. Resende, and K. G. Ramakrishnan, Editors,** Parallel Processing of Discrete Optimization Problems

21 **D. Frank Hsu, Arnold L. Rosenberg, and Dominique Sotteau, Editors,** Interconnection Networks and Mapping and Scheduling Parallel Computations

20 **William Cook, László Lovász, and Paul Seymour, Editors,** Combinatorial Optimization

(Continued in the back of this publication)

DNA Based Computers III

DIMACS

Series in Discrete Mathematics
and Theoretical Computer Science

Volume 48

DNA Based Computers III

DIMACS Workshop
June 23–25, 1997

Harvey Rubin
David Harlan Wood
Editors

NSF Science and Technology Center
in Discrete Mathematics and Theoretical Computer Science
A consortium of Rutgers University, Princeton University,
AT&T Labs, Bell Labs, Bellcore, and NEC Research Institute

American Mathematical Society

This DIMACS volume contains papers from the Third DIMACS Workshop on DNA Based Computers. The Workshop was held June 23–25, 1997, at the University of Pennsylvania.

1991 *Mathematics Subject Classification.* Primary 68–06, 92–06, 68Q05, 68Q10, 68Q22, 68Q42, 68Q45, 05C85; Secondary 68M99, 92B05, 94A99.

Library of Congress Cataloging-in-Publication Data

DIMACS Workshop on DNA Computers (3rd : 1997 : university of Pennsylvania)
DNA based computers III : DIMACS Workshop, June 23–25, 1997 / Harvey Rubin, David Harlan Wood, editors.
p. cm. — (DIMACS series in discrete mathematics and theoretical computer science ; v. 48)
"NSF Science and Technology Center in Discrete Mathematics and Theoretical Computer Science, a consortium of Rutgers University, Princeton University, AT&T Labs, Bell Labs, Bellcore and NEC Research Institute."
Includes bibliographical references (p.).
ISBN 0-8218-0842-7
1. Molecular computers—Congresses. I. Rubin, Harvey, 1948– . II. Wood, David Harlan, 1939– . III. NSF Science and Technology Center in Discrete Mathematics. IV. DIMACS (Group) V. Title. VI. Title: DNA based computers 3. VII. Title: DNA based computers three. VIII. Series.
QA76.887.D517 1997
511.3—dc21 98-55491
CIP

∞ The paper used in this book is acid-free and falls within the guidelines
established to ensure permanence and durability.
Visit the AMS home page at URL: http://www.ams.org/

10 9 8 7 6 5 4 3 2 1 04 03 02 01 00 99

Contents

Foreword

The DIMACS Workshop on DNA Based Computers was held on June 23–25, 1997 at the University of Pennsylvania. We would like to express our appreciation to Harvey Rubin, David Wood, Len Adleman, Anne Condon, Richard Lipton and Carter Bancroft for their efforts to organize and plan this successful workshop.

The workshop was part of the Special Year on DNA Computing. We extend our thanks to Richard Lipton and Laura Landweber for their work over many months as special year organizers. The workshop was part of the broader Special Year on Molecular Biology. We extend our thanks to Fred S. Roberts, Joachim Messing, Lawrence Shepp, Michael Waterman, Martin Farach, and Sampath Kannan for their work over many months as special year organizers.

The main focus of the workshop was bringing together researchers from the biological/laboratory community with those from the algorithmic/computability community. The workshop advanced the state of the art by clarifying laboratory capabilities and limitations while simultaneously exploring questions of what might be computed and how.

DIMACS gratefully acknowledges the generous support that makes these programs possible. The National Science Foundation, through its Science and Technology Center program, the New Jersey Commission on Science and Technology, DIMACS' partners at Rutgers, Princeton, AT&T Labs Research, Bell Labs and Bellcore generously supported the special year.

Fred S. Roberts
Director

Bernard Chazelle
Co-Director for Princeton

Preface

Third DIMACS Workshop on DNA Based Computers (June 1997)

The Third DIMACS Workshop on DNA Based Computers was held at The University of Pennsylvania from Monday, June 23 through Wednesday, June 25, 1997. In order to set the stage for the multidisciplinary nature of the conference, two tutorial sessions were held on Sunday, June 22. David Harlan Wood reviewed some basic concepts in computation for the biologists and biochemists and Laura Landweber reviewed ideas and concepts in molecular biology for the computer scientists and mathematicians. A poster session was held during the last half of the conference. An impromptu session was also held, featuring brief presentations of late-breaking research, speculative ideas, and general queries.

Throughout the conference every opportunity was taken to encourage the interaction between the theoreticians and the lab investigators during the coffee breaks, the reception and the conference dinner.

Two sessions addressed "Making it Work in the Lab" were chaired by Len Adleman and Nadrian Seeman. The first paper of Adleman's session was an invited lecture delivered by James Wetmur on the physical chemistry of nucleic acids that discussed in detail the thermodynamics and kinetics of polynucleotide and oligonucleotide association and dissociation and reviewed the mathematical handling of hybridization parameters. This talk was followed by a paper from the MIT group describing a simulator designed to predict the binding specificity of arbitrary deoxy-oligonucleotides. The MIT group then presented a paper on the efficiency of sequence specific separation techniques including solution based annealing of bead bound probes to target molecules followed by magnetic separation of the annealed probe-target complexes from solution. They stressed the need for a standard set of benchmarks for key primitive operations and volunteered to make their probes available to the research community. The morning session was concluded with a lecture from the Univ. of Delaware group presenting a new DNA separation technique according to the DNA substrings based on circular single strands of DNA that become "marked" by virtue of hybridization to a primer followed by treatment with DNA polymerase generating a double strand. Marked and unmarked registers can then be operated upon in a highly efficient fashion using various enzymes with signal or double stranded specificities.

In the session chaired by Nadrian Seeman, the first lecture was an invited talk by Hagiya and Arita from The University of Tokyo on the advantage of the inherent parallelism in DNA computation. They introduced a molecular method for parallel evaluation of multiple μ formulas in which all possible μ formulas are randomly generated and only formulas satisfying the given variable assignments are selected in a single test tube in polynomial time. This was followed by a paper by Laun and

Reddy from Binghamton University on extending the work of Tom Head from "dry" splicing systems which generate a splicing language to "wet" splicing systems. They presented a simple example of such a wet splicing system that generates in vitro a splicing language corresponding to the theory of the dry system. Next, Kaplan, Thaler and Libchaber presented a method to generate a pool of DNA molecules containing paths through a directed graph using a parallel overlap assembly method. The Mount Sinai School of Medicine group then presented work that followed up on their original bit-flipping DNA operations by showing it is possible to run primer extension reactions in parallel. They also showed that the judicious use of restriction sites in the input DNA can result in a robust readout system. This session was rounded out by a presentation from the University of Pennsylvania group in which a DNA based method of bit flipping was discussed. The new twist was presented where the result of the operation is of the same molecular form as the input allowing iterated operations, including the possibility of reversing the computation.

A session devoted to "Proposed Applications" was chaired by Erik Winfree. An invited paper by the Princeton group lead off with a novel idea for using DNA computation to operate on unknown pieces of DNA. This would have applications in DNA sequencing, DNA fingerprinting, detection of mutations in populations. They call this DNA^2DNA computation. The trick to this technique is in the recoding of naturally occurring DNA into DNA that contains a prechosen redundant code. This is carried out without needing to know anything about the sequence(s) of the DNA input. One can then carry out any of the usual techniques associated with DNA computation, but with reduced errors. After this, an approach to DNA computation was presented by Andrew Ellington in which DNA, acting as an allosteric effector can activate ribozyme ligase thereby serving as an operator—transforming one nucleic acid sequence into another sequence—i.e., the ligated product. Eng and Serridge from MIT then presented a surface based algorithm for the minimal set cover problem that has the nice property of not requiring separation steps (which are thought to be error prone). This was followed by a presentation by the University of Tokyo group proposing a solid phase DNA solution to the Hamiltonian Path problem. The session was rounded out by a presentation by Fu and Beigel from Yale and the University of Maryland that described a general technique for constructing molecular based approximation algorithms for NP optimization problems.

A session with the theme of "Proposed Algorithmic Approaches," chaired by Richard Lipton, started with an invited talk by Aviezri Fraenkel from the Weizmann Institute of Science, Israel. He presented a theoretical discussion of the relationship between the protein folding problem and NP complete problems. He related protein folding to finding the ground state of spin glasses. This lead him to suggest that it might be possible to synthesize specific proteins to solve arbitrary instances of the spin glass problem. This was followed by a paper by the University of South Florida group that literally took off into space, suggesting that graphs could be constructed in 3 dimensions using DNA. This could be a more general computational tool than mere linear constructs. The University of Memphis then presented an exploration of the feasibility, reliability and efficiency of DNA based computation by grappling with the ability of DNA to implement nondeterminism as a native mode of computation. Gupta, Parthasarathy and Zaki from the University of Rochester then addressed the theoretical possibility of carrying out arithmetical and logic operations with DNA.

A session devoted to "Models of DNA Computers" was chaired by Lila Kari. John Reif delivered an invited lecture in which he strongly advocated the development of local parallelism (LP) in biomolecular computation as opposed to the more generally understood distributed parallelism (DP). He pointed out that all LP operations could be used in combination with DP methods. This was followed by a paper by Ogihara and Ray entitled DNA-based parallel computation by "counting". Next Andrew Blumberg from MIT presented a theoretical paper on parallel computation using DNA substrates and addressed specifically the problem of sharing information between strands—a key element in applications of parallel computation. He discussed a model which simulates a message routing architecture. Conrad and Zauner from Wayne State University then discussed the design for a DNA conformational processor based on the capacity of DNA to switch from the B form to the Z form in response to methylation of cytosine residues. Eng from MIT then showed, in a theoretical sense, how the self assembly of linear duplexes with hairpins generates sequences that are equivalent to linear context-free languages.

A session entitled "Computability Using DNA" was chaired by John Reif and started with an invited lecture from the multinational group of Freund (Austria), Păun (Romania), Rozenberg (Netherlands) and Salomaa (Finland), delivered by Freund. They related automata theory and biomolecular computation, proposing machines similar to finite automata that handle inputs and outputs in a fashion that ensures universality. This was followed by a paper by Kari, Păun, Thierrin and Yu, delivered by Kari, on characterizing recursively enumerable (RE) languages using insertion/deletion systems. They showed the nexus between insertion and deletion in DNA—a common biological phenomenon—and generating RE languages by inserting and deleting words according to their contexts. Yokomori and Kobayashi from The University of Electro-communications in Japan presented a paper on DNA computation based on a principle of equality checking where the task of equality checking of two memories represented by a string of symbols is isomorphic to a completely hybridized double stranded DNA sequence. The paper by Sakakibara (Tokyo Denki University) and Ferretti (University of Milan) extended the notion of splicing systems on strings to splicing systems on tress and showed how splicing systems on trees with finite sets of axioms and finite sets of rules generates the class of context free languages without the need of multiplicity constraints.

DIMACS Series in Discrete Mathematics
and Theoretical Computer Science
Volume **48**, 1999

Physical Chemistry of Nucleic Acid Hybridization

JAMES G. WETMUR*

Abstract

Thermodynamic and kinetic relationships governing oligonucleotide hybridization are combined with empirical data to permit calculation of hybridization rates, melting temperatures and dissociation temperatures for fully matched and mismatched structures.

Introduction

The goal of this review is to describe a unified approach to all types of nucleic acid hybridization (adapted from Wetmur, 1996; see Wetmur, 1991 and Wetmur and Sninsky, 1995 for additional references). In addition, new data are presented describing the thermodynamics of stacking, dangling and branch migration structures and the advantage of experimental conditions limited by dissociation kinetics rather than equilibria for achieving maximum discrimination against mispaired structures.

Hybrid duplexes containing DNA and/or RNA strands are the products of hybridization reactions between partially or completely complementary single-stranded molecules. Reassociation or renaturation are alternative terms used to describe hybridization of completely complementary DNA or RNA strands. The reverse reaction is called strand separation or dissociation. The structural differences between deoxyribose and ribose in DNA and RNA strands are reflected in different forms of the antiparallel double-helices and in quantitative differences in hybrid stability and the rate of hybrid formation. Nevertheless, they are qualitatively the same, and the basic concepts may be treated together. Most of the illustrations are for DNA:DNA hybridization.

1991 ASM subject classifications: 92C40, 92C45, 80-08.
*We thank Soo-Jung Kim for determining the thermodynamic data illustrated in figures 16 and 20. This work was supported by grant USPH R21 HG01365 from the National Institutes of Health.

Melting and Dissociation Temperatures Defined

Figure 1 contains definitions and a list of dependent variables for three *different* characteristic temperatures associated with hybridization. $\mathbf{T_m^\infty}$ is the melting temperature of a polynucleotide duplex, such as a long DNA molecule. $\mathbf{T_m}$ is the melting temperature where at least one strand of the duplex is an oligonucleotide. $\mathbf{T_d}$ is the dissociation temperature for an oligonucleotide bound to a polynucleotide on a solid support. A fourth temperature, $\mathbf{T_{ss}}$, is the strand separation temperature of a polynucleotide duplex, also called the denaturation temperature in PCR.

Thermodynamics of Polynucleotide Melting

Figure 2 illustrates the definitions of T_m^∞ and T_{ss} for polynucleotide melting. For polynucleotides, but not oligonucleotides, the denaturation occurring at the melting temperature is an intramolecular reaction with molecules containing alternating duplex and denatured regions (loops or ends). For natural nucleic acids, these loops may occur in the same places in many molecules, reflecting regions of lower melting temperature due to compositional heterogeneity. T_m^∞ is the temperature where the fraction of single stranded base pairs, f_{ss}, is 0.5. Because the reaction is intramolecular and equilibrium is rapidly achieved, T_m^∞ is independent of polynucleotide concentration and time, depending only on the composition of the duplex and the composition of the solvent. Strand separation occurs at T_{ss}, a few degrees above T_m^∞, and is not a rapidly reversible reaction. The denaturation temperature in PCR must be greater than T_{ss}. Throughout this review, lower case t will denote °C, whereas upper case T will denote °K.

Figure 3 contains an empirical equation for determining t_m^∞ for polynucleotides of length L containing only G:C and A:T or A:U base pairs as well as the needed constants. This equation includes a salt correction term that is valid at high salt concentrations used for blot hybridization as well as at lower salt concentrations where blots are washed. An additional empirical term has been included to account for divalent ions that are always required in DNA polymerase and ligase buffers. DNA:RNA hybrids adopt the A conformation of RNA:RNA hybrids rather than the B conformation of DNA:DNA hybrids, but are always much less stable than RNA:RNA hybrids. However, DNA:RNA hybrids may be more or less stable than DNA:DNA hybrids, depending on % G+C.

Figure 4 includes worked examples for a 50% G+C DNA:DNA

hybrid of length 250 in a typical PCR buffer with 0.05 M KCl, 0.002 M $MgCl_2$ and long DNA:RNA hybrids versus DNA:DNA of differing GC composition in 1 M NaCl.

Use of different solvents may affect A^* and/or B^*, and hence T_m^∞, with different effects for different polynucleotides. For example, A^* for DNA:DNA melting decreases 0.25°C /% glycerol, which is often included in PCR buffers to stabilize thermostable DNA polymerases, 0.4°C/% ethylene glycol and 0.63°C/% formamide, which is often used for blot hybridizations. The effect of formamide on A^* is much less for DNA:RNA hybrids than DNA duplexes, facilitating R-loop formation, where single-stranded RNA may displace a homologous DNA strand in a DNA duplex. In another example, base composition may be eliminated as a variable for DNA:DNA hybrids ($B^* = 0$) at 3.2 M Me_4NCl ($t_m^\infty = 94$°C) or 2.4 M Et_4NCl ($t_m^\infty = 63$°C) or 5.6 M betaine (Rees et al., 1993). Because unlike Me_4NCl or Et_4NCl, betaine is a zwitterion and does not neutralize the charge on DNA, the melting temperature may be varied from 75°C down to 45°C by varying KCl from 1 M to 0.01 M.

Introduction of modified bases also may affect A^* and/or B^*. Any substitution will affect either AT (or AU) or GC base pairs. If no charge change is introduced, the resulting change in t_m^∞ may be expressed as a change in A^*, ΔA^*, where:

$$\Delta A^* = S\cdot(\%\ G{+}C)\cdot f^M_{AT} \qquad \text{for A or T (U) modifications or}$$

$$\Delta A^* = S\cdot(\%\ G{+}C)\cdot f^M_{GC} \qquad \text{for G or C modifications,}$$

where f^M_{AT} and f^M_{GC} are the fractional substitution of the modified base and S is characteristic of the nature of the substitution. For example, for 5-methyldeoxycytosine substitution for cytosine in duplex DNA, which may also occur in nature, S = 0.12. Thus, for 100% substitution ($f^M_{GC} = 1$) in a 50% G+C DNA, t_m^∞ is increased by 6°C.

Formation of imperfectly matched hybrids also affects A^*, t_m^∞ being reduced approximately 1°C for each percent mismatching. t_{ss} is proportionally reduced. Stringency in washing immobilized targets such as filters or blots (Southern, northern, slot) containing polynucleotide probes, a measure of selective dissociation of imperfectly matched hybrids, is increased as the temperature approaches t_m^∞ in the wash solvent.

Thermodynamics of Oligonucleotide Melting

Figure 5 illustrates the definition of T_m for the all-or-none melting of complementary oligonucleotides, with no denatured (frayed) ends or internal loops. T_m is the temperature at which the fraction of single strands, and hence single stranded base pairs, f_{ss}, is 0.5. All-or-none melting requires that most of the base pairs in each duplex remain paired until dissociation, which should always be true when the melting temperature of the oligonucleotide, T_m, is below the melting temperature, T_m^{∞}, of a polynucleotide containing only AT base pairs (e.g. 77°C in 1 M NaCl). Typically, all-or-none oligonucleotide melting will occur at even higher temperatures if the oligonucleotide duplex ends do not have a high % A+T.

Figure 6 contains an empirical equation for determining T_m. Typically, the reference solvent is 1 M NaCl and the reference temperature, T°, is at 25°C. Because oligonucleotide melting is completely reversible, T_m, unlike T_m^{∞}, depends on oligonucleotide concentration as well as on nucleotide and solvent composition. Thus for the same oligonucleotide in the same solvent, T_m, determined spectrophotometrically at the high oligonucleotide concentration necessary for detection by absorbance, is greater than T_m for PCR priming at a lower oligonucleotide concentration, which is in turn greater than T_m for filter hybridization at an even lower oligonucleotide concentrations. In a spectrophotometric determination of T_m, one oligonucleotide at molar strand concentration C* is often hybridized to a second complementary oligonucleotide at the same concentration. The molar concentration of the single-stranded fraction of the first oligonucleotide at T_m is C*/2, which is equal to C in the calculations below. In the more common experiment, an oligonucleotide probe or primer is present at an excess concentration C* for saturation binding to a complementary region on a polynucleotide. The molar concentration, C, of the single-stranded oligonucleotide at T_m is C*.

Figure 7 contains a nearest neighbor method for obtaining the necessary enthalpies and free energies needed for calculating T_m. The symbols I and nn indicate contributions from initiation of base-pair formation and nearest neighbors base pairs (stacking interactions) as well as other variables such as dangling ends, respectively.

Figure 8 illustrates the data set of Quartin and Wetmur (1989) for calculating enthalpies and free energies. The use of nearest neighbor thermodynamic functions is necessary because oligonucleotides with the same % G+C often have significantly different melting temperatures. Several nearest neighbor data sets are available for DNA:DNA, DNA:RNA and RNA:RNA [Breslauer et al., 1986; Quartin and Wetmur, 1989; SantaLucia et al., 1996; Sugimoto, et al., 1995 (DNA:RNA); Freier, et al, 1986 (RNA:RNA)].

Figure 9 contains a representative calculation using the data set from figure 8 in the equations in figure 7. The calculation agrees with experiment.

Figures 10 depicts dangling ends resulting from binding of an oligonucleotide to a longer strand and lists average values for the contribution of a dangling end to the enthalpy and free energy. Both 3′ and 5′ dangling ends may contribute to helix stability. The magnitude of the contribution depends on the orientation and sequence of a dangling end. Filter hybridization typically involves two dangling ends. During the exponential amplification phase of PCR, one dangling end is formed upon primer-template binding, although PCR primers are typically too long for the dangling end to have a large effect on the melting temperature.

Figure 11 depicts a stacking interaction. Binding of an oligonucleotide adjacent to a duplex generates one additional nearest-neighbor (stacking) interaction. Data are presented below justifying the conclusion that a revised T_m may be calculated by adding the new nearest neighbor contribution to the thermodynamic functions. Both dangling ends and stacking become most important for very short oligonucleotides. Oligonucleotides as short as 8-10 in length may be used for array hybridization (Pease et al., 1994) and may be components of stacked primers for primer extension sequencing (Kieleczawa et al., 1992).

Figure 12 depicts direct and derivative plots of all-or-none oligonucleotide melting. The full width of the direct plot based on the slope at the midpoint, ΔT, as well as the width at half height for the derivative curve, $\Delta T_{1/2}$, are also noted.

Figure 13 presents the calculation of the width of the melting curve, ΔT, where the oligonucleotide is in excess. For the 8-mer in figure 9, ΔT = 12.5 °C. For the derivative curve, the width at half-height, is given by:

$$\Delta T_{1/2} = -\phi \cdot R \cdot T_m^2 / \Delta H^\circ \quad \text{where } \phi = 3.5.$$

Even several degrees above T_m, there is a significant fraction of duplex formed. In PCR, extension by DNA polymerase may be initiated above T_m because extended products no longer dissociate, allowing all templates eventually to be initiated. Thus, discrimination between matched and mismatched primers is a complex phenomenon. Ligase reactions may similarly be driven to completion above the melting temperature.

Figure 14 shows the corresponding results when the complementary sequence is present at the same concentration. The melting curve is not completely symmetric about the melting temperature. Similarly, the derivative curve is not symmetric about the melting temperature and ϕ is increased to 5.22.

Because the nearest neighbor thermodynamic functions have not been determined for modified bases, the best available treatment is to calculate t_m in the absence of the modified bases (N/oligonucleotide) and to correct the result as follows:

$$\Delta t_m = -810000 \cdot N \cdot S / \Delta H^\circ$$

For example, substitution of 5-methyldeoxycytidine of deoxycytidine (S = 0.12) at three positions (N = 3) in the oligonucleotide example above (GGATAACG; $\Delta H^\circ = -58700$ cal/mol) gives $\Delta t_m = 5\,^\circ C$ (Wong et al., 1991). A new and interesting modified base is 3-nitropyrrole (Nichols et al., 1994), which interacts weakly but specifically with all four normal bases in such a way that its inclusion in an oligonucleotide of length L leads to a melting temperature characteristic of an oligonucleotide of length L-1 but with a melting temperature almost independent of which base is pairing with the modified base. Oligonucleotides containing such a universal nucleotide may be quite useful than degenerate oligonucleotides for library screening or cloning by PCR.

Figure 15 depicts a hairpin helix with a 12 nucleotide tail. This type of structure can bind a 12-mer with 1 stacking interaction or 10-mers or 11-mers with either one dangling end or with one dangling end and one stacking interaction. These results may then be compared to binding of the 10-mers, 11-mers or 12-mers to completely complementary oligonucleotides, producing no dangling ends or stacking interactions. Taking into account nearest neighbor contributions, these and similar data form the basis for the conclusions in figures 10 and 11.

Figure 16 presents results from representative optical melting curves obtained using one hairpin sequence and constant oligonucleotide concentrations, clearly demonstrating the importance of stacking interactions.

Figure 17 illustrates a mismatched or mispaired nucleotide corresponding to a heteroduplex formed between a wild-type strand and one containing a transition or transversion mutation. Discrimination against such a defect is a measure of the specificity of hybridization. To estimate the effect of a mismatch, one must first calculate the thermodynamic parameters as if the mismatch had no effect. For example, if a GG mismatch occurs between G and G, one strand will contain GGG, with two GG stacking interactions and the other strand will contain CGC, with a CG and a GC stacking interaction. The entropy and free energy contributions may be calculated by adding these four nearest neighbor contributions and dividing the total by 2. Secondly, one must correct the thermodynamic functions depending on whether one of the mismatched bases is (i) a dG, (ii) a dC, or (iii) neither a dG or a dC. The corrections to add to the enthalpies and free energies are included in the figure. These corrections suggest that destabilization due to a mispair is of the order of the loss of two stacking interactions but with half the change in the temperature coefficient, although mismatches involving G residues are relatively more stable. Average reductions in T_m for internal mismatches are also given in the figure. These calculations are consistent with a decrease in T_m^{∞} for polynucleotides of 1 °C/% mismatching. Consider, however, the case of a mismatch occurring at the end on oligonucleotide, where only one stacking interaction can be lost. The calculated melting temperature for an (L-1)mer is greater than the calculated melting temperature with an internal mismatch. The same may be true for penultimate or even more internal mismatches. The actual melting temperature can be approximated by the larger of the two numbers.

Figure 18 illustrates a bulge defect corresponding to a heteroduplex formed between a wild-type strand and one containing a frameshift mutation. A bulge defect may be estimated by calculating the thermodynamic functions without the bulge nucleotide and adding 2800 cal/mol to the free energy, thus T_m as given in the figure. The position of the bulge defect must again be taken into account.

Figure 19 depicts potential branch migration resulting from overlap (length O) of an oligonucleotide (length L) with a duplex. The total number of nearest neighbor interactions is L-O, including the additional stacking interaction at the junction. The branch destabilizes the structure to the extent of adding 0.8 kcal/mol to the free energy, lowering t_m for a 20, 15 or 10-mer by an average 2.0, 2.6 or 3.6 °C, respectively. If the oligonucleotide overlap sequence is the same as the sequence in the adjacent duplex, branch migration can occur between all O possible structures. The branch migration stabilizes the structure to the extent of subtracting $R \cdot T_m \cdot \ln(O)$ from the free energy.

Figure 20 depicts derivative curves for melting with a single nucleotide overhang. When the nucleotide does not permit branch migration, the melting temperature decreases. When branch migration can occur, in this case of a G overhang, the structure is more stable. The destabilization introduced by the branch is canceled when the overlap, O, is about 4 (data not shown).

Kinetics of Polynucleotide Hybridization

Figure 21 depicts a temperature profile for carrying out a denaturation and hybridization (renaturation, reassociation) reaction with polynucleotides. Denaturation requires a temperature above T_{ss}, the strand separation temperature. The hybridization reaction is initiated (time $t = 0$) by lowering the temperature to T_r, a hybridization (renaturation, reassociation) temperature about 25°C below T_m^{∞}, the polynucleotide melting temperature. Denaturation and renaturation of DNA may also be achieved by increasing the pH to approximately 13 followed by neutralization at T_r. The steps in the hybridization reaction are (i) a reversible nucleation step in which the first 1 (or 2) correct base pairs are formed and (ii) an irreversible zippering reaction, in which the subsequent base pairs are formed (see Wetmur, 1976). Although duplexes containing fewer than 10-20 base pairs are usually unstable at T_r, the rate-determining step is the formation of the second (or third) base pair. The overall hybridization reaction is characterized by the second order rate constant, k_2.

The extent of hybridization may be determined by a variety of methods in solution or involving enzymatic or physical separations techniques (see Wetmur, 1991). Enzymatic separation techniques may employ digestion of single-stranded molecules by RNase or nuclease S1. Physical separation may involve binding of single-stranded DNA to BND-cellulose or double-stranded DNA to hydroxylapatite or the more common Southern and other blotting techniques. Solution measurements away from surfaces include including absorbance (36% hyperchromicity upon melting) and fluorescence depolarization or fluorescence quenching. At a surface, hybridization may be detected by plasmon resonance (Nilsson et al., 1995) and by light scattering or fluorescence using an optical wave guide (Stimpson et al., 1995).

Figure 22 depicts time course plots for the same hybridization reaction, including a C_0t plot for a second order reaction. Logarithmic $C_0 t$ plots permit simultaneous display of reactions differing by orders of magnitude in half-time. Many reactions involve a labeled probe hybridizing to a complementary sequence in a complex target. If the concentrations of complementary probe and target sequences are unequal, the reactant in excess is termed the driver. A second order reaction takes place if the complementary strands are present at equal concentration (e.g. DNA renaturation) or if the driver is double-stranded and capable of reassociation. A pseudo-first order reaction takes place if the driver is single-stranded (e.g. mRNA) or cannot reassociate (e.g. immobilized on a solid support for a Southern blot). For all hybridization reactions, f_{ss} is the fraction of potential duplex remaining single-stranded.

The concentration term used for polynucleotide hybridization is molar nucleotide concentration, C_0, and not strand concentration, C. For example, at 1 μg/ml double-stranded DNA, $C_0 = 3$ μM. If one reactant is in excess, C_0 is the driver concentration. When immobilized DNA is the driver, which is rarely the case, C_0 is difficult to determine, being a measure of available target concentration. Furthermore, because a dilute probe would be used with the immobilized driver target, stirring may be necessary to maintain homogeneity.

Figure 23 depicts reciprocal plots for the same pseudo-first order and second order reactions found in the left panel of Figure 22. Given the nucleotide concentration of the driver, the slopes of these curves yield the hybridization rate constant, k_2 in M^{-1} sec^{-1}.

Figure 24 contains the first and second order equations for the plots in figure 23 and the equations for determining the half-times of these reactions.

Figure 25 contains an empirical equation for predicting k_2 by knowing the values of three variables, (i) the nucleation rate constant, k_N' in M^{-1} sec^{-1}, which depends on properties of the solvent such as temperature, ionic strength and viscosity, (ii) the length of the shorter of the complementary strands participating in duplex formation, L_s in nucleotides, and (iii) the complexity of the driver, N in base pairs, or the total number of base pairs present in non-repeating sequences.

Figure 26 depicts interpretations of N and L. The dependence of k_2 on the square root of L rather than L suggests that nucleation sites become less available as the length of the single strands increases, consistent with an excluded volume effect. An alternative explanation for the square root effect has been suggested by Sikarov and Church, 1991.

Figure 27 presents a model for hybridization where complementary bases with the potential for initiating a productive reaction diffuse into and out of a volume element before forming one base pair, which usually dissociates due to the absence of a stacking interaction. Perhaps one time in 1000, a second base pair forms, producing one stacking interaction, followed by formation of additional base pairs, all of which add one stacking interaction. At typical hybridization temperatures, base pair formation is more rapid than base pair dissociation. Because the rates of base pair formation and dissociation are very fast, partially hybridized intermediates are not observed. The formation and dissociation of the intermediates containing the second and subsequent base pairs may be described by a steady state rate equation.

Figure 28 depicts the temperature dependence of hybridization. The rate of hybridization reactions, and hence k_N', is zero at T_m^∞. k_2 increases to a broad maximum at about 25°C below T_m^∞. At lower temperatures, hybridization rates fall off slowly for homopolymers (dotted line) but precipitously for natural nucleic acids (continuous line) due to intramolecular base-pairing and consequent decreased availability of nucleation sites. At optimal temperatures, k_N' for DNA:RNA hybrid formation is 20 to 50% slower than DNA:DNA hybrid formation, depending on the degree of RNA secondary structure. RNA:RNA hybridization rates are similarly reduced.

Although mismatched base pairs have a significant effect on the melting temperature, mismatching of up to 10% of the bases has little effect on hybridization rates.

k_N' also depends on solvent viscosity and ionic strength. In 1.0 M NaCl at 70°C and other salt solutions, addition of glycerol, ethylene glycol, or formamide reduces k_N' due to increased solvent (microscopic) viscosity compared to water, indicating that the rate determining step for productive base pair formation, including at least one stacking interaction, is viscosity limited. On the other hand, polymers which increase the macroscopic viscosity do not reduce hybridization rates. In fact, hybridization rates may be accelerated by addition of dextran sulfate, which has the effect of concentrating the reactants. Another method for acceleration of hybridization is to concentrate the reactants at the interface of a phenol-water emulsion.

Figure 29 depicts the dependence of k_N' on salt concentration.

Figure 30 presents empirical equations fitting that curve. In typical hybridization buffers approximating 1 M NaCl, $k_N' = 3.5 \cdot 10^5\ M^{-1} sec^{-1}$. Consider as an example the hybridization of a nick-translated 4 kb probe (N = 4000) to a blot. The probe would be the driver. Assume that the nick translation broke the single strands to 625 nucleotides ($L^{1/2} = 25$). Then in 1 M NaCl at 70°C, $k_2 = 2.2 \cdot 10^3\ M^{-1} sec^{-1}$. If the denatured probe is used at 10 ng/ml ($C_o = 30$ nM), $t_{1/2} = 2 / k_2 \cdot C_o = 2 / \{2.2 \cdot 10^3 \cdot 3 \cdot 10^{-8}\} = 3 \cdot 10^4$ sec = 8.5 hours. The half-times for hybridization may vary over many orders of magnitude depending linearly on the concentration and inversely on the complexity of the driver. For example, at 100 μg/ml, $t_{1/2}$ for the same nick-translated probe would be 3 sec, whereas $t_{1/2}$ would be about 40 sec, 1 hour and 30 days for λ (N = 48,500), *E. coli* (N = 4,600,000) and human DNA (N = 3,000,000,000), respectively.

Kinetics of Oligonucleotide Hybridization

Figure 31 depicts the two rate constants governing oligonucleotide hybridization, the hybridization rate constant and the dissociation rate constant. Hybridization with oligonucleotides usually takes place far below T_m^∞ (at or below T_m), so the decrease in k_2 near T_m^∞ is not a factor, and the temperature dependence of k_2 for oligonucleotides is small unless the target has interfering secondary structure. k_2 is reduced only 1.2-fold for every 10°C decrease in hybridization temperature. Oligonucleotide hybridization has the same salt dependence as polynucleotide hybridization. In fact, oligonucleotide hybridization may be described by the same equations. For oligonucleotide-polynucleotide reactions, the oligonucleotide is usually the driver leading to a pseudo-first order reaction characterized by a rate constant with N = L of the oligonucleotide. Oligonucleotide probe reactions are very fast.

Figure 32 gives a worked example for the rate of binding of a PCR primer to its template. Note that $t_{1/2}$ increases proportionally with the degeneracy of mixed oligonucleotides both for blot probes and for PCR primers. In PCR, the change in the rate of primer binding may be more important than a melting temperature depression due to limited mismatching of a primer to its template. The mismatching effect is felt at early cycles, whereas the rate effect persists.

Kinetics of Oligonucleotide Dissociation: The Dissociation Temperature

Figure 33 illustrates an oligonucleotide reaction on a solid support. Capture reactions are simply hybridization reactions. However, when hybridized oligonucleotides on solid supports are washed to eliminate unhybridized oligonucleotides, both T_m and the rate characterized by the rate constant k_2 decrease as C decreases, and the hybrid stability depends entirely on the dissociation rate constant, k_{-1}. The concentration, the enthalpy of melting, $-\Delta H°$, and the melting temperature are sufficient for determining the equilibrium constant for oligonucleotide hybridization, or the ratio of the hybridization and dissociation rate constants. As noted above, k_2 has a small activation energy, approximately 4.0 kcal/mole, primarily due to the dependence of solvent viscosity on temperature. With this information, one may determine k_{-1}. The activation energy for k_{-1}, E_a, is thus 4.0 kcal/mole added to $-\Delta H°$. T_d is the dissociation temperature, which may defined as the temperature at which 50% of correctly matched hybrid is released in a specified length of time, t_{wash}. Because t_d depends on time and not concentration, unlike $t_{1/2}$ for hybridization and t_m, t_d is the same whether using unique or degenerate probes. It is t_d which may be estimated for extended washing of 14 to 20-mers as 2°C for each dA:dT pair plus 4°C for each dG:dC base pair.

Figure 34 explains the calculation of t_d. Note that the dissociation temperature is only weakly (logarithmically) dependent on washing time.

Figure 35 presents worked examples for dissociation temperatures. In the first example, $t_d = 9.5°C$ for 50% probe released with 1 hr (3600 sec) washing. In the second example, $t_d = 48°C$ for 6 hours of washing for the arbitrary 16-mer GATCGTACCAAGGTTC with 2 dangling ends, in agreement with the 2°C for each dA:dT pair plus 4°C for each dG:dC base pair rule. These calculations depend on all the elements of the physical chemistry of hybridization.

Figure 36 contains the output from an Excel spreadsheet of a calculation of matched and mismatched oligonucleotide melting curves at equilibrium. The input variables include the concentration and sequence of the oligonucleotide, the presence or absence of dangling ends, the position and type of mismatch, the monovalent and divalent cation concentrations, and the possible inclusion of glycerol or formamide. The hybridization rate constants and melting temperatures are listed under calculated values. The left panel shows that a single mismatch located near the center of an oligonucleotide leads to significant destabilization. Nevertheless, the melting curves overlap, making it impossible to detect the matched target without residual signal from the mismatched target.

The dissociation temperatures are also listed under calculated values. These calculations required an additional input, the washing time. The right panel depicts a dissociation curve for the same oligonucleotides when they were first bound at low temperature and then depleted of free oligonucleotide. These curves also assume that the dissociated oligonucleotide is continuously removed or that its concentration is too insignificant to permit rebinding. An additional input was needed, the rate of heating in degrees/minute. The apparent curvature of each plot following the last point with a measurable fraction duplex and the next point at complete dissociation simply reflects the absence of intermediate points. Unlike the equilibrium case, there are temperatures at which it is possible to detect the matched target without residual signal from the mismatched target. The same result may be achieved by washing at constant temperature.

References

Breslauer, K.J., Frank, R., Blöcker, H. & Marky, L.A., Predicting DNA duplex stability from the base sequence. *Proc. Natl. Acad. Sci. U. S. A.* **83**, 3746-3750 (1986).

Doktycz, M.J., Morris, M.D., Dormady, S.J., Beattie, K.L. & Jacobson, K.B., Optical melting of 128 octamer DNA duplexes. Effects of base pair location and nearest neighbors on thermal stability. *J. Biol. Chem.* **270**, 8439-8445 (1995).

Freier, S.M., Kierzek, R., Jaeger, J.A., Sugimoto, N., Caruthers, M.H., Neilson, T. & Turner, D.H., Improved free-energy parameters for predictions of RNA duplex stability. *Proc. Natl. Acad. Sci. U. S. A.* **83**, 9373-9377 (1986).

Kieleczawa, J., Dunn, J.J. & Studier, F.W., DNA sequencing by primer walking with strings of contiguous hexamers. *Science* **258**, 1787-1791 (1992).

Nichols, R., Andrews, P.C., Zhang, P. & Bergstrom, D.E., A universal nucleoside for use at ambiguous sites in DNA primers. *Nature* **369**, 492-493 (1994).

Nilsson, P., Persson, B., Uhlén, M. & Nygren, P.-Å., Real-time monitoring of DNA manipulations using biosensor technology. *Anal. Biochem.* **224**, 400-408 (1995).

Pease, A.C., Solas, D., Sullivan, E.J., Cronin, M.T., Holmes, C.P. & Fodor, S.P., Light-generated oligonucleotide arrays for rapid DNA sequence analysis. *Proc. Natl. Acad. Sci. U. S. A.* **91**, 5022-5026 (1994).

Quartin, R.S. & Wetmur, J.G., The effect of ionic strength on the hybridization of oligodeoxynucleotides with reduced charge due to methylphosphonate linkages to unmodified oligodeoxynucleotides containing the complementary sequence. *Biochem.* **28**, 1040-1047 (1989).

Rees, W.A., Yager, T.D., Korte, J. & von-Hippel, P.H., Betaine can eliminate the base pair composition dependence of DNA melting. *Biochemistry* **32**, 137-144 (1993).

SantaLucia, J., Jr., Allawi, H.T. & Seneviratne, P.A., Improved nearest-neighbor parameters for predicting DNA helix stability. *Biochem.* **35**, 3555-3562 (1996).

Sikorav, J.-L. & Church, G.M., Complementary recognition in condensed DNA: accelerated DNA renaturation. *J. Mol. Biol.* **222**,1085-1108 (1991).

Stimpson, D.I., Hoijer, J.V., Hsieh, W.-T., Jou, C., Gordon, J., Theriault, T., Gamble, R & Baldeschwieler, J.D., Real-time detection of DNA hybridization and melting on oligonucleotide arrays by using optical wave guides. *Proc. Natl. Acad. Sci. U. S. A.* **92**, 6379-6383 (1995).

Sugimoto, N., Nakano, S-i, Kotoh, M., Matsumura, A., Nakamuta, H., Ohmichi, T., Yoneyama, M. & Sasaki, M. *Biochem.* **34**, 11211-11216 (1995).

Wetmur, J.G., Hybridization and renaturation kinetics of nucleic acids. *Annu. Rev. Biophys. Bioeng.* **5**, 337-361 (1976).

Wetmur, J.G., DNA probes: Applications of the principles of nucleic acid hybridization. *Crit. Rev. Biochem. Mol. Biol.* **26**, 227-259 (1991).

Wetmur, J.G. Nucleic Acid Hybrids, Formation and Structure of. *In:* Encyclopedia of Molecular Biology and Molecular Medicine, Myers, R.A., ed., VCH Publishers, New York, pp 235-243 (1996).

Wetmur, J.G. & Sninsky, J.J. Nucleic acid hybridization and unconventional bases. *In:* PCR Strategies, Innis, M.A., Gelfand, D.H. and Sninsky, J.J., eds, Academic Press, San Diego, pp 69-83 (1995).

Wong, D.M., Weinstock, P.H. & Wetmur, J.G., Branch capture reactions: Displacers derived from asymmetric PCR. *Nucleic Acids Res.* **19**, 2251-2259 (1991).

DEPARTMENT OF MICROBIOLOGY, BOX 1124,
MOUNT SINAI SCHOOL OF MEDICINE, NEW YORK, NY 10029-6574.
E-MAIL: wetmur@msvax.mssm.edu

Characteristic Temperatures

	Function	Variables
T_m^∞	DNA Melting Temperature	Nucleotide Composition Various Solvent Properties
T_m	Oligonucleotide Melting Temperature	Nucleotide Composition Various Solvent Properties <u>Strand Concentration</u>
T_d	Dissociation Temperature	Nucleotide Composition Various Solvent Properties <u>Time</u>

FIGURE 1

Polynucleotide Melting Temperature

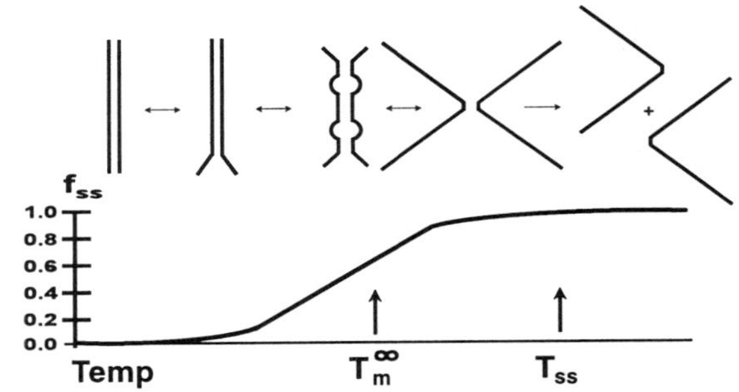

FIGURE 2

DNA-DNA, DNA-RNA and RNA-RNA

- Polynucleotides, length L, containing only G:C and A:T or A:U base pairs.

t_m^∞ = A* + 16.6 $\log_{10}$([salt] / {1.0 + 0.7 [salt] })

+ B* (%G+C) - 500/L where [salt] = $[M^+] + 4\sqrt{[Mg^{++}]}$

Polynucleotides	A*	B*
DNA:DNA	81.5	0.41
RNA:RNA	67	0.8
RNA:RNA	78	0.78

FIGURE 3

Polynucleotide Melting Temperature Calculation

Long Polynucleotides in 1 M NaCl:

Hybrid	DNA:RNA	DNA:DNA
50%G+C	103.2	98.2
30%G+C	87.2	90.0

DNA:DNA, 50%G+C, L = 250,

0.05 M KCl, 0.002 M $MgCl_2$ ([salt] = 0.23)

t_m^∞ = 81.5 - 11.7 + 0.41 x 50 - 2 = 88.3° C.

FIGURE 4

Oligonucleotide Melting Temperature

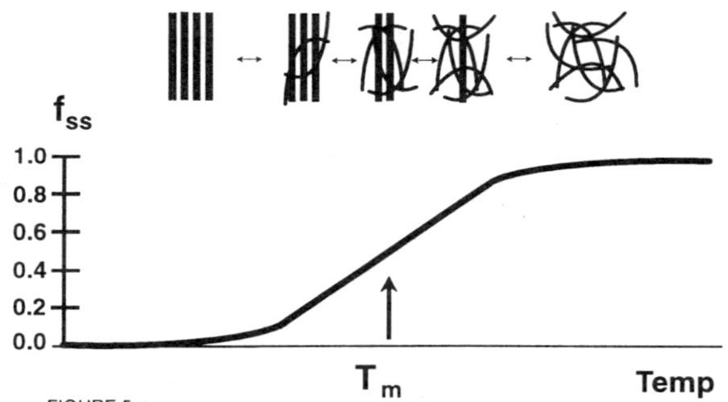

FIGURE 5

Oligonucleotide Thermodynamics

$$\Delta G^{\circ} = \Delta H^{\circ}\ (1 - T^{\circ}/T_m) + RT^{\circ}\ln\ [C/4]$$

where C = total concentration of non-self-complementary oligonucleotides
NOTE: NOT NUCLEOTIDES

$$T_m = T^{\circ}\Delta H^{\circ}\ /\ (\Delta H^{\circ} - \Delta G^{\circ} + RT^{\circ}\ln\ [C/4])$$

$$+\ 16.6\ \log_{10}(\ [salt]\ /\ \{1.0 + 0.7\ [salt]\ \}\ )\ +\ 3.83$$

$$[salt] = [M^{+}] + 4\sqrt{[Mg^{++}]}$$

[C/4] becomes [C] when one oligonucleotide is in excess, as is the case for probe reactions

FIGURE 6

Nearest Neighbors

Enthalpy: $\Delta H^{\circ} = \sum_{nn} (N_{nn}\ \Delta H^{\circ}_{nn})$

Free energy: $\Delta G^{\circ} = \sum_{nn} (N_{nn}\ \Delta G^{\circ}_{nn}) + \Delta G^{\circ}_{i}$

and $T = 298.2^{\circ}$ K

and $R = 0.00199$ kcal/mol° K

FIGURE 7

Thermodynamic Values

For the DNA:DNA nearest neighbor values in the table below

(There is no unique set of thermodynamic values)

ΔG_i = +2.2 kcal/mol AND:

AA TT	AT	TA	CA TG	GT AC	CT AG	GA TC	CG	GC	GG CC
Enthalpies (- kcal/mol):									
9.1	8.6	6.0	5.8	6.5	7.8	5.6	11.9	11.1	11.0
Free Energies (- kcal/mol):									
1.55	1.25	0.85	1.15	1.40	1.45	1.15	3.05	2.70	2.30

FIGURE 8

Oligonucleotide Melting Temperature Calculation

For example, consider excess 1 μM (GGATAACG) binding to (CGTTATCC) at 1 M NaCl:

$-\Delta H^{\circ} = 11+5.6+8.6+6.0+9.1+6.5+11.9 = 58.7$ kcal/mol

$-\Delta G^{\circ} = 2.3+1.15+1.25+0.85+1.55+1.4+3.05 - 2.2$

$= 9.35$ kcal/mol

RT ln [C] = -8.2 kcal/mol

$t_m = 31^{\circ}$C

FIGURE 9

Other Important Structures

• Dangling Ends

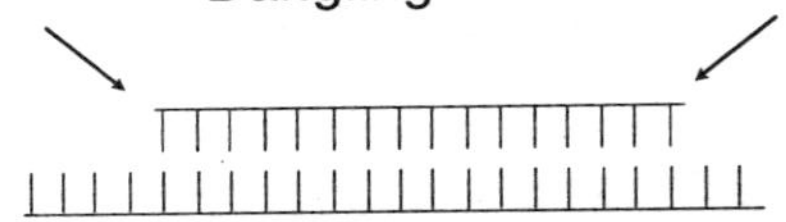

Each dangling end increases the melting temperature approximately 1, 1.5 or 3° C for 20, 15 or 10-mers, respectively.

$-\Delta H_e^{o} = 5.0$ kcal/mol **$-\Delta G_e^{o} = 1.0$ kcal/mol**

FIGURE 10

Other Important Structures

• Stacking

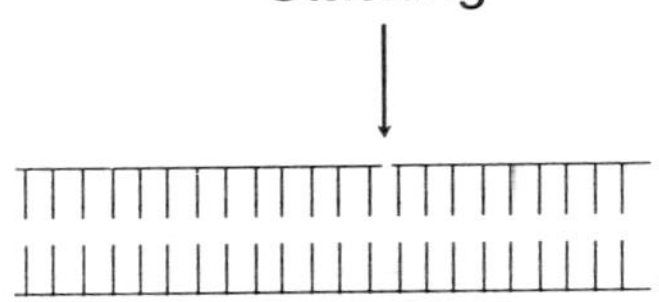

The melting temperature of a L-mer next to a stable duplex region is the same as the melting temperature of an L+1 mer.

FIGURE 11

Derivative Plots

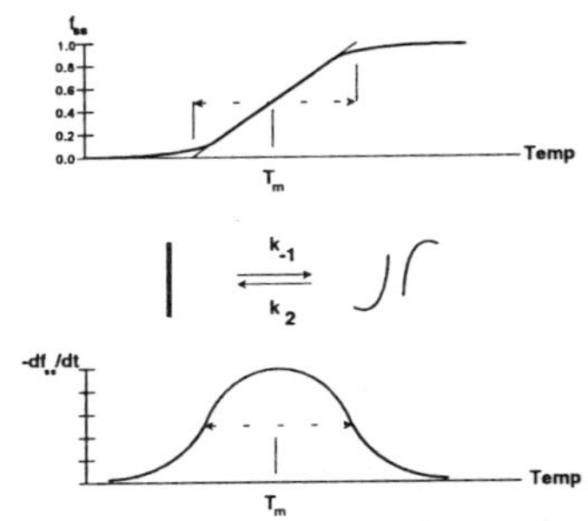

FIGURE 12

Transition Widths - One Oligonucleotide in Excess

- Standard Probe Reaction Conditions

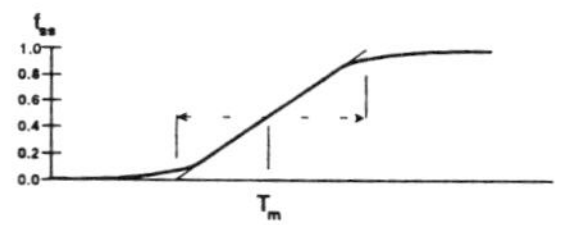

$d \ln K / dT = \Delta H^{o} / RT^{2}$

$f_{ss} = 1 / (1 + KC)$

$df_{ss} / dT = f_{ss}(1 - f_{ss})\, \Delta H^{o} / RT^{2}$

$\Delta T_{(T_m)} = 4\, RT^{2} / \Delta H^{o}$ — 10-mer: $10^{o}C$; 20-mer: $5.5^{o}C$

FIGURE 13

Transition Widths - Equimolar Complementary Oligonucleotides

- Melting in a Spectrophotometer

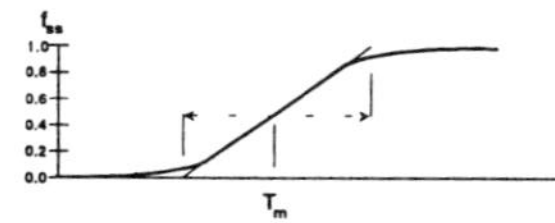

$d \ln K / dT = \Delta H^{o} / RT^{2}$

$df_{ss} / dT = f_{ss}(1 - f_{ss})\, \Delta H^{o} / \{RT^{2}(1 + f_{ss})\}$

$\Delta T_{(T_m)} = 6\, RT^{2} / \Delta H^{o}$ — 10-mer: $15^{o}C$; 20-mer: $8.3^{o}C$

FIGURE 14

Examples on a Hairpin with a 12 Nucleotide Target

```
                         T
CCATCCTGCAGCCATGGCCCGCGC  T
               GTACCGGGCGC T
                          T
```

- 12-mer (Stack, No Dangle)
- 10-mer (Stack and Dangle)
- 10-mer (Dangle Only)

FIGURE 15

Data Plots on a Hairpin with a 12 Nucleotide Target

- 12-mer (Stack, No Dangle)
  $t_m = 65\,^{o}C$
- 10-mer (Stack and Dangle)
  $t_m = 58\,^{o}C$
- 10-mer (Dangle Only)
  $t_m = 46\,^{o}C$

FIGURE 16

Mismatched (Mispaired) Bases

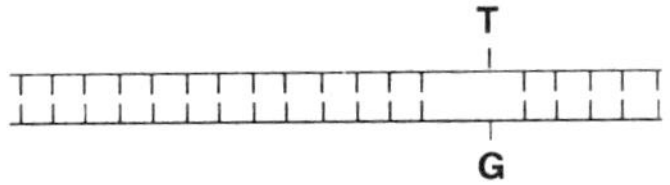

$+\Delta H^{o}$ = 6.7, 8.7 or 9.5 kcal/mol if dG, dC or neither.
$+\Delta G^{o}$ = 2.9, 3.7 or 4.0 kcal/mol if dG, dC or neither.

Each mismatch decreases the melting temperature approximately 5-7, 8-10 or 12-17 ^{o}C for 20, 15 or 10-mers, respectively. The lower values pertain if the mismatch contains a dG.

FIGURE 17

Bulge Loops

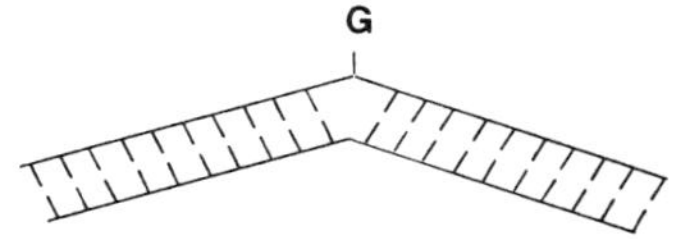

Calculate without bulge. $+\Delta G^{o}$ = 2.8 kcal/mol

Each bulge decreases the melting temperature approximately 7, 9 or 12 ^{o}C for 20, 15 or 10-mers, respectively.

FIGURE 18

Branch Migration

- A Branch Costs 0.8 kcal
- Potential Branch Migration of B Nucleotides Gains RT ln B
- e.g. B = 4

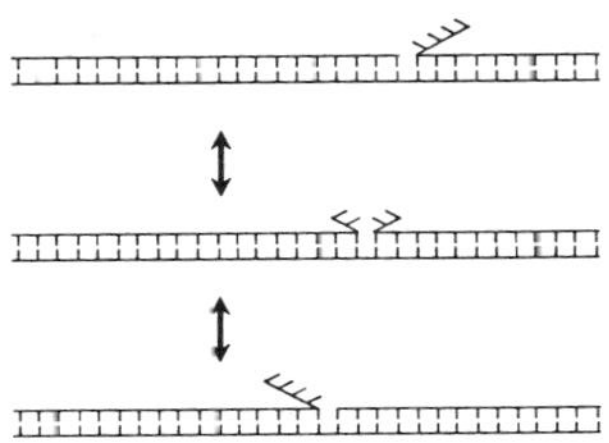

FIGURE 19

Blunt Stack and 1-Base Overhangs

G-Overhang: Branch Migration

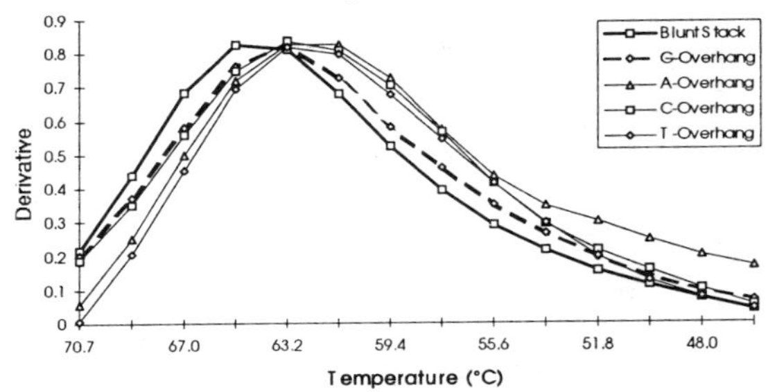

FIGURE 20

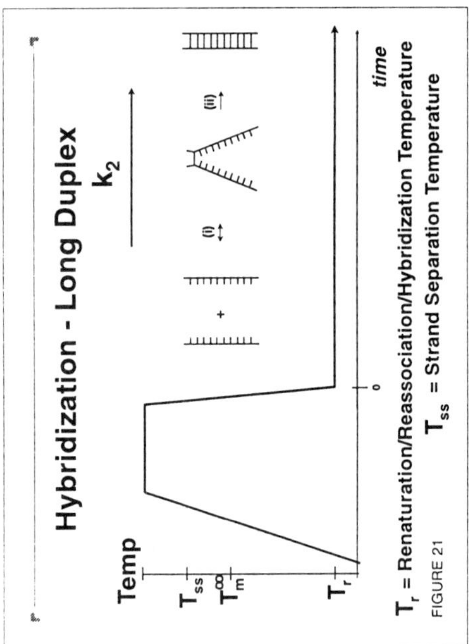

FIGURE 21

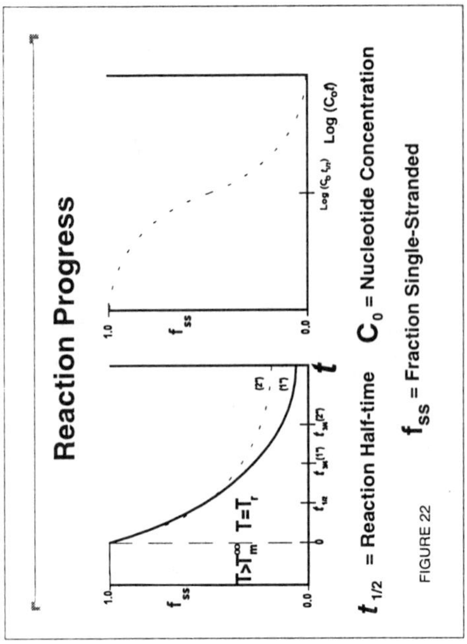

FIGURE 22

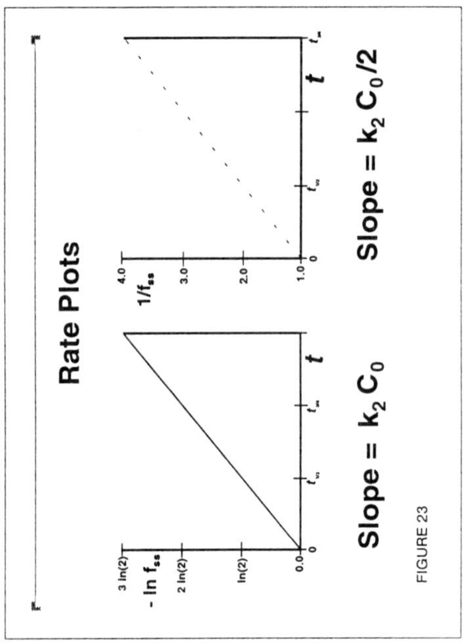

FIGURE 23

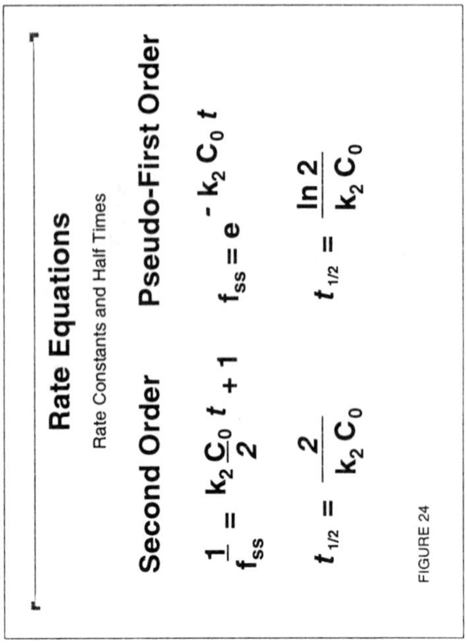

FIGURE 24

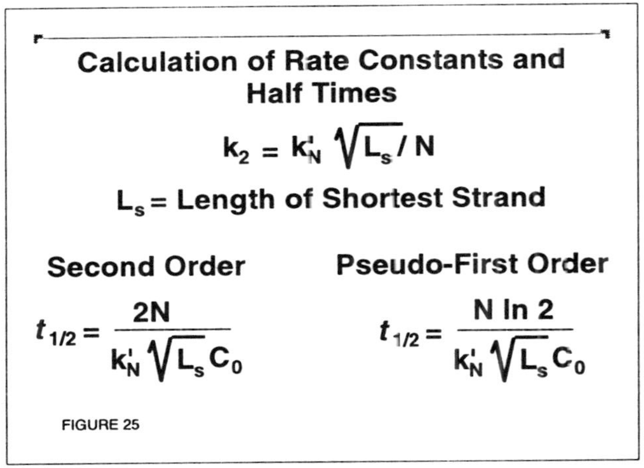

FIGURE 25

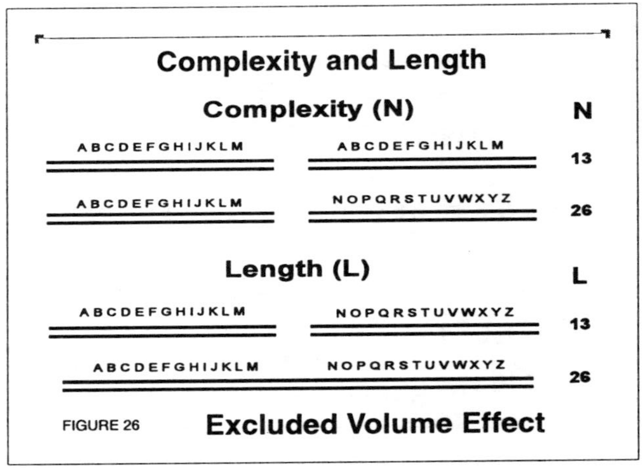

FIGURE 26

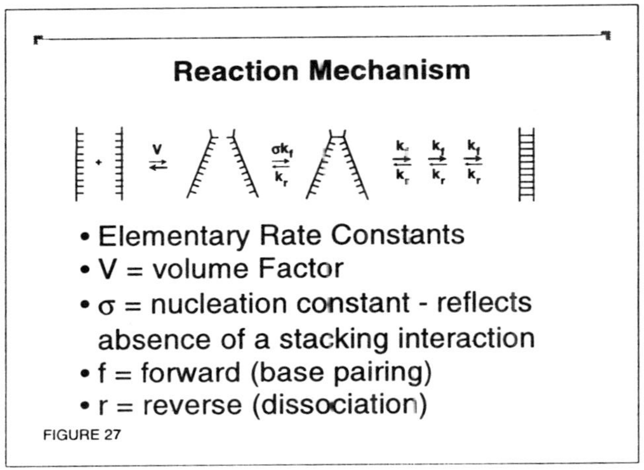

FIGURE 27

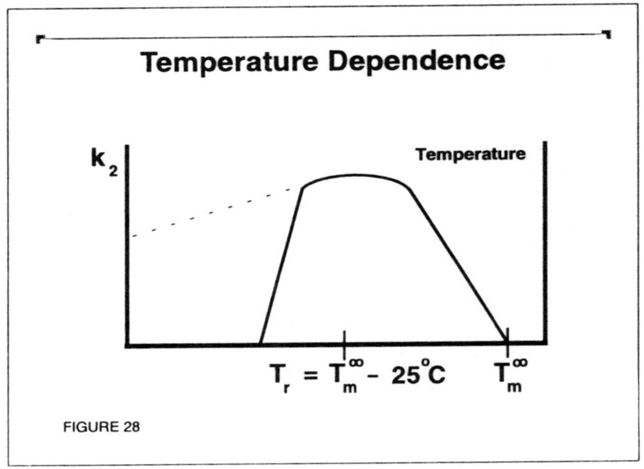

FIGURE 28

Effect of Ionic Strength

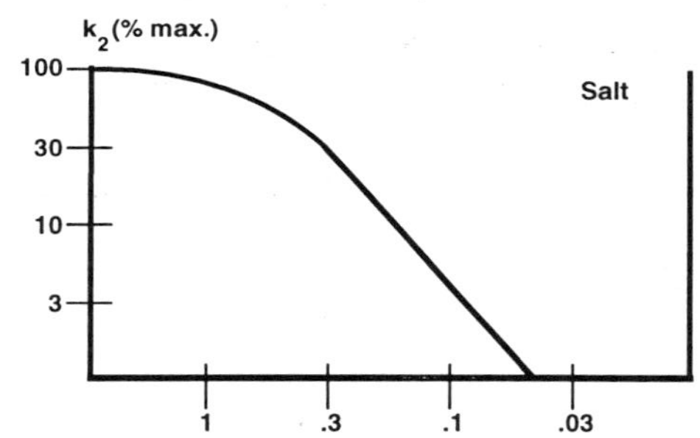

FIGURE 29

Equations for Monovalent and Divalent Cations

For 0.2<[salt]<4.0:

$$k'_N = \{4.35 \log_{10}[\text{salt}] + 3.5\}\ 10^5$$

For [salt]<0.2:

$$k'_N = \{4.35 \log_{10}(1 + 34[\text{salt}]^3\}\ 10^5$$

$$[\text{salt}] = [M^+] + 4\sqrt{[Mg^{++}]}$$

FIGURE 30

Oligonucleotides

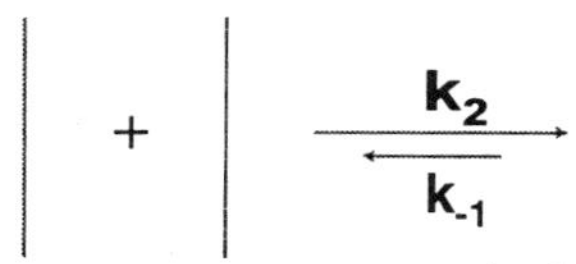

Reactions at $T < T_m^{\infty}$

k_{-1} = First Order Dissociation Rate Constant (s^{-1})

FIGURE 31

PCR Annealing

Pseudo-First Order $t_{1/2} = \dfrac{N \ln 2}{k'_N \sqrt{L_s}\, C_0}$

$$[\text{salt}] = [M^+] + 4\sqrt{[Mg^{++}]}$$

$$[0.25] = [0.05] + 4\sqrt{0.00225}$$

$$k'_N = \{4.35 \log_{10}[\text{salt}] + 3.5\}\ 10^5 = 0.9 \times 10^5$$

$$N = L_s = 25$$

$C_0 = 25 \times 1 \times 10^{-6}$ for 1 μM 25-mer

$$t_{1/2} = \frac{25 \ln 2}{0.9 \times 10^5 \times 5 \times 25 \times 1 \times 10^{-6}} = 1.6 \text{ sec}$$

FIGURE 32

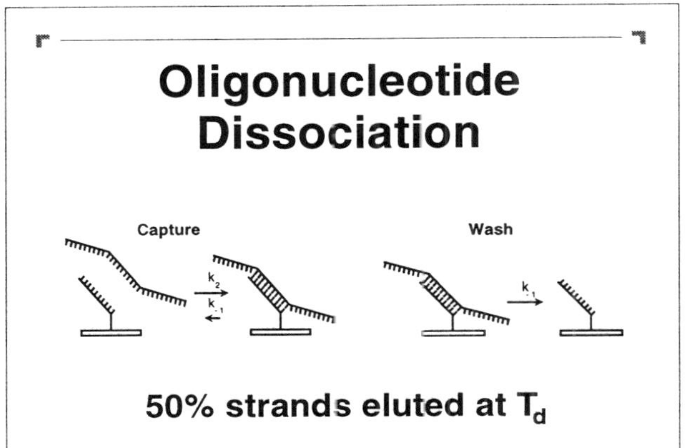

FIGURE 33

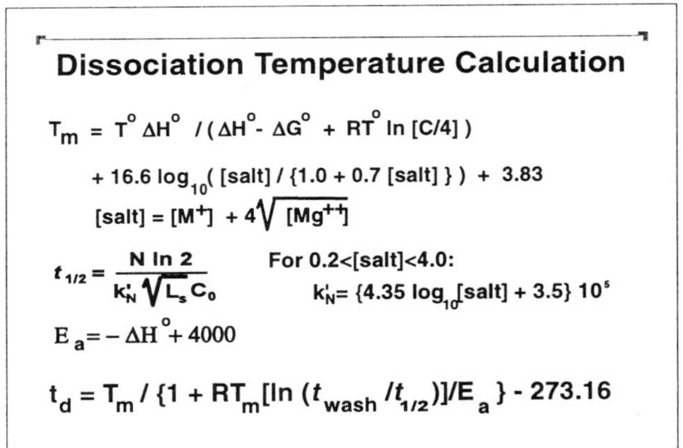

FIGURE 34

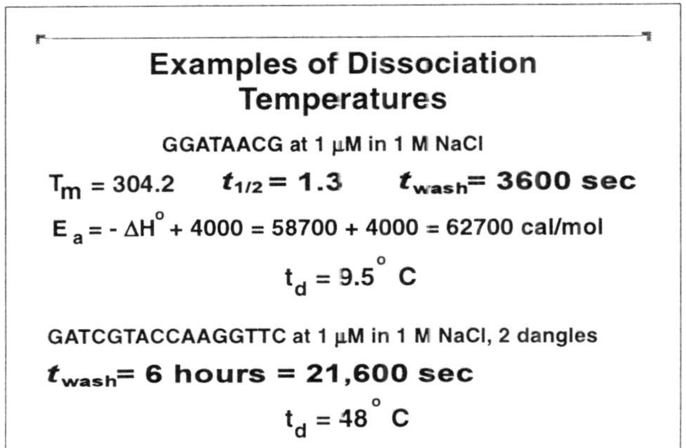

FIGURE 35

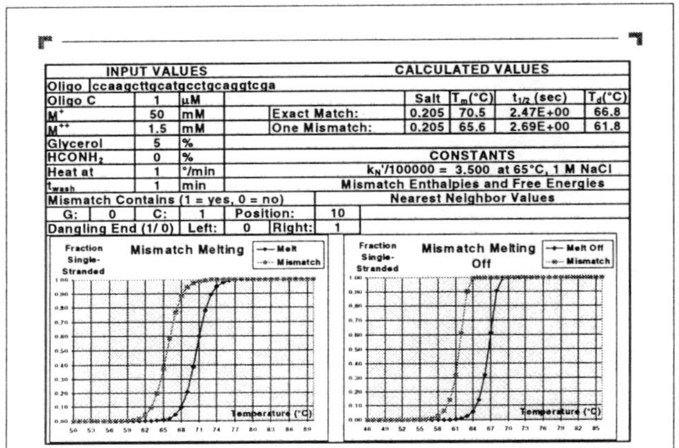

FIGURE 36

DIMACS Series in Discrete Mathematics
and Theoretical Computer Science
Volume **48**, 1999

Thermodynamic Simulation of Deoxyoligonucleotide Hybridization for DNA Computation

Alexander J. Hartemink and David K. Gifford

ABSTRACT. Nearly every model of DNA computation proposed to date depends upon sequence-specific hybridization operations. In order to better predict the binding specificity of arbitrary deoxyoligonucleotides, a simulator named BIND is implemented. BIND operates on a single template DNA sequence and a number of shorter primer sequences. For each primer sequence, BIND calculates a theoretical melting temperature at every position of the primer along the template, yielding a measure of binding specificity between each primer and the template. The simulator differs from previous melting temperature programs in that it is intended to be used with oligonucleotides, is designed to handle mismatched base pairs, makes use of the latest thermodynamic parameters, and provides features with DNA computation expressly in mind. This paper describes how BIND is implemented, provides corroborating evidence as to its accuracy, and offers instances of its usefulness to a range of DNA computing applications.

1. Introduction

Recent activity in the field of DNA computing has generated interest in the dynamics of nucleic acid interaction. By creating another consumer of precise information regarding enzyme activity, thermal and ionic denaturation, binding specificity, and the polymerase chain reaction (PCR), DNA computing has enhanced the already compelling case for deeper fundamental understanding of these processes.

In particular, since many proposed designs for DNA-based computational systems rely on assumptions about the binding specificity between two or more strands of DNA, a thorough investigation of the hybridization process is of paramount importance. To our knowledge, *all* current DNA-based computation proposals make use of annealing steps, whether in constructing large libraries of DNA on which to perform computations, in using PCR to amplify or detect small quantities of DNA containing a specific sequence of nucleotides [**2, 3**], in performing sequence-specific separation operations using magnetic bead or affinity column techniques [**2, 3, 10**], in setting specific bits in sticker-based models [**10**], in marking specific strands in

1991 *Mathematics Subject Classification.* Primary 92-04, 92C40, 80-04; Secondary 68U20.

Hartemink is partially funded by a National Science Foundation Graduate Research Fellowship.

surface-based models [**6**], or in performing sequence-specific mutagenic operations in programmed mutagenesis models. Because these operations all rely on site-specific nucleic acid hybridization, it is critical that significant progress be made in more precisely characterizing the way the temperature, ionic environment, and degree of mismatch influence the hybridization process.

By evaluating the current state of DNA-based computation, and by looking in the direction in which the field will need to expand in the next few years, it becomes clear that many of the proposed computational systems will need to move "from the blackboard into the laboratory" in order to iron out the subtle details associated with their implementation. Before particular problems can be investigated, however, the algorithms (and the input to those algorithms) will need to be translated into concrete DNA sequences, which is a daunting task, especially in light of the necessary binding specificity demands made by the systems' specifications. With this in mind, a simulator capable of quickly evaluating the binding specificity of oligonucleotides serving as PCR primers, stickers, markers, probes, or mutagenic rewrite rules should prove helpful to the DNA computing community, as well as to the molecular biological community at large.

Programs which compute DNA melting temperatures do exist, but most are intended only for polynucleotide hybridization, do not handle mismatched nucleotide binding or multiple binding possibilities, or use outdated thermodynamic parameters. Moreover, to our knowledge, none of these programs is designed with DNA computation in mind. Therefore, there appears to be a niche for an accurate simulator that uses the latest thermodynamic parameters, is capable of handling nucleotide mismatches, and is designed to provide needed functionality to scientists investigating DNA computation.

Overview of this Paper. The preliminary implementation of such a simulator, evaluations of its accuracy, and demonstrations of its general applicability provide the basis of this paper. The simulator described here, named BIND, has been calibrated using melting temperature data reported in the literature and has already begun to provide useful information to our group's ongoing work in the area of programmed mutagenesis. In the following section, we briefly describe programmed mutagenesis in order to provide a contextual motivation for our interest in such a simulator. Then, in the next section, we proceed to discuss the BIND simulator itself and derive the thermodynamic equations used in its melting temperature calculations. Then we turn to an evaluation of BIND's accuracy by comparing it with melting temperature data determined in the laboratory and reported in the literature. This is followed by a discussion of the simulator's uses and some of the results obtained with the aid of the simulator. In the last section, we conclude by presenting a number of possible extensions to BIND which we are in the process of developing.

2. Programmed Mutagenesis

Programmed mutagenesis is a technique for programmatically rewriting DNA sequences by incorporating highly sequence-specific oligonucleotides into newly manufactured strands of DNA. There are several advantages to using programmed mutagenesis for DNA-based computation:

- The pool of oligonucleotides can be designed to cause sequence-specific programmed changes to occur, including the propagation of programmed changes up and down a DNA molecule and the evolution of a programmed sequence of changes over the course of future replication events. Thus, sequential computations with programmatically evolving state can be carried out, resulting in *constructive computation*, as contrasted with *selective computation* which requires all possible solutions to a problem to be present *ab initio*.

- The sequence specificity of the oligonucleotides permits a set of oligonucleotides to be present at each step of the reaction, with only a fraction of them being active during each cycle. This reduces human effort since it allows the computation to be carried forward by thermocycling the reactants in the presence of thermostable polymerase and ligase. Ideally, there would be no need for human (or robotic) intervention between computation steps.
- Although input and output of data from programmed mutagenesis reactions can be accomplished by the direct synthesis of input molecules and sequencing of output molecules, other more indirect methods exist which lend this technique greater flexibility and modularity. For example, input molecules could be created by ligating computational DNA subunits together, with the subunits chosen to encode specific problems.
- Programmed mutagenesis can be used in conjunction with other proposed computational systems by creating output strands to be used as input to other systems. In this manner, input strands for other systems could be generated in a systematic fashion. Because of this "intelligent design" property, the need to have all possible solutions present *ab initio* may be obviated.
- All the components necessary to implement programmed mutagenesis are present *in vivo*. Therefore it may eventually be possible to harness the internal workings of the cell for computation, thereby capitalizing on the cell's homeostatic capabilities to ensure that the computation takes place in a stable chemical environment.

We have begun to build programmed mutagenic machines in the laboratory and believe the techniques to be generally extensible. In the subsection below, we describe in detail how we have applied the technique of programmed mutagenesis to build a simple unary counter.

2.1. The Unary Counter Machine. To demonstrate the function of programmed mutagenesis, we implemented a counter that will in theory be incremented every time DNA is replicated. We encode the value of the counter in a DNA molecule as the number of **X** and **Y** symbols it contains, where **X**, **Y**, and **Z** symbols are shorthand representations of specific 12-nucleotide sequences, designed so that both **X** and **Y** differ from **Z** at two base positions, but from each other at four. The initial template molecule includes the sequence **ZZZZZX**, which encodes the number one. The mutagenic oligonucleotides which implement the counting mechanism are $\mathbf{X'Y'}$ and **XY**, denoted ρ_1 and ρ_2, respectively.

This counter construction presumes that 24-mer oligonucleotides which anneal to the template with two mismatches can be extended and successfully ligated to other polynucleotides, while oligonucleotides with a greater number of mismatches cannot be effectively incorporated into the new strand. Note that we do not need as many oligonucleotides as counter values; in this case, two oligonucleotides are sufficient to allow the counter to advance to an arbitrary value.

Each cycle, the counter uses as its input template the product of the immediately preceding cycle, and the product of cycle N encodes the number $N+1$. The counter is implemented by thermocycling a reaction that begins with the initial template strand, ρ_1, ρ_2, thermostable polymerase and ligase, and outside primers LP and RP (needed to produce the full-length product on each cycle). Each thermal cycle consists of a high temperature step, which denatures the double-stranded DNA and prepares it for the polymerization-ligation step; and a low temperature step, which is permissive for primer annealing, polymerization, and ligation. Ideal values for these temperatures were chosen on the basis of data from laboratory experiments. In a later section of this paper, the BIND simulator is used to demonstrate how such choices might be made more easily, circumventing the need for detailed laboratory experiments.

2.2. Summary. The salient point regarding programmed mutagenesis is that it relies on the binding specificity of its rewrite rules to ensure that the strand of DNA is being rewritten in a systematic way. For example, if rewrite rule ρ_i is meant to be applied to a strand of DNA representing state σ_i, producing a strand representing state σ_{i+1}, and rewrite rule ρ_{i+1} is subsequently meant to be applied to the strand of DNA representing state σ_{i+1} to produce a strand representing σ_{i+2}, it should be the case that ρ_{i+1} cannot be applied to σ_i and ρ_i cannot be applied to σ_{i+1}. If this condition is satisfiable, then both of the rewrite rules can be present in the reaction and yet the system can only evolve from the state representing σ_i to the state representing σ_{i+2} by first passing through σ_{i+1}, with each rewrite rule being applied in sequence, thereby capturing the notion of programmatic computation.

3. Simulator Design

In order to provide a means for rapidly testing binding specificity hypotheses, a simulator called BIND was developed. After being provided a single template sequence and any number of primer sequences, the BIND simulator considers individually how each primer can bind to the template strand, at the specified ionic and reactant concentrations. It does this by calculating, for each possible position of the given primer along the template strand, a theoretical melting temperature for the primer-template duplex at that position. A straightforward loop generates melting temperatures for all such positions along the template. This procedure is repeated for each primer. The resulting data can be used to generate plots revealing the most stable binding sites along the template for each of the primers.

Note that if the template is the same length as the primer, BIND just calculates the melting temperature between the two strands at a single position, the position where the two are perfectly aligned. Therefore, calculation of the melting temperature of a simple double-stranded segment of DNA is easily handled. Consequently, we can test BIND's melting temperature calculation by comparing its output with data reported in the literature for a number of oligonucleotide duplexes.

3.1. The Calculation of Theoretical Melting Temperature. Theoretical melting temperature is typically calculated assuming that the coil-helix transition is two-state, which is a justifiable assumption for small oligonucleotides.[1] SantaLucia, *et al.* [**11**] suggest that the two-state model is capable of providing a reasonable approximation of melting temperature for duplexes with non-two-state transitions,

[1]Usually defined in this context as fewer than 15-25 bases.

but the applicability of the assumption obviously decreases as the size of the duplex under consideration increases. For a two-state model of the

$$\text{single-strand} + \text{single-strand} \rightleftharpoons \text{double-strand}$$

transition between two distinct oligonucleotides in equimolar concentration, the equilibrium constant is given by

$$K = \frac{2f}{(1-f)^2[C_T]} \tag{3.1}$$

where f is the fraction of strands in the double-stranded state and $[C_T]$ is the total molar strand concentration.

But the equilibrium constant can also be expressed in thermodynamic terms as

$$K = \exp(-\frac{\Delta G^\circ}{RT}) = \exp(-\frac{\Delta H^\circ - T\Delta S^\circ}{RT}) \tag{3.2}$$

where R is Boltzmann's constant and T is the temperature in Kelvin. Since the melting temperature, T_m, is defined as the temperature at which half of the strands are in the double-stranded state, it follows that $f = 1/2$ when $T = T_m$. Setting $f = 1/2$ and setting the two expressions for K in equations 3.1 and 3.2 equal to one another, we get an equation relating the melting temperature, T_m, to the total molar strand concentration and the enthalpy and entropy of the forward state transition:

$$T_m = \frac{\Delta H^\circ}{\Delta S^\circ + R\ln([C_T]/4)} \tag{3.3}$$

In order to adjust for $[\mathrm{Na}^+]$ concentrations different from 1M, the salt adjustment term[2] introduced by Wetmur [**13**] is appended to the right hand side of equation 3.3, yielding equation 3.4:

$$T_m = \frac{\Delta H^\circ}{\Delta S^\circ + R\ln([C_T]/4)} + 16.6\log_{10}\left(\frac{[\mathrm{Na}^+]}{1+0.7[\mathrm{Na}^+]}\right) - 269.3 \tag{3.4}$$

The nearest neighbor stacking model allows the total enthalpy and entropy of the transition to be expressed as

$$\Delta H^\circ = \Delta H^\circ_{ends} + \Delta H^\circ_{init} + \sum_{k\in\{stacks\}} \Delta H^\circ_k \tag{3.5}$$

$$\Delta S^\circ = \Delta S^\circ_{ends} + \Delta S^\circ_{init} + \sum_{k\in\{stacks\}} \Delta S^\circ_k \tag{3.6}$$

Equations 3.4, 3.5, and 3.6 are similar to the ones reported in Wetmur [**13**], but they make use of the entropy, ΔS°, rather than the free energy, ΔG°, which is temperature dependent. Since temperature is the variable of interest in this calculation, the expression for melting temperature which does not include free energy terms is preferred.

---

[2]Note that when $[\mathrm{Na}^+]$ is 1M, the term drops out.

If divalent cations like Mg^{++} are present in solution, their effect on duplex stability can be expressed in terms of an $[Na^+]$ equivalent using the conversion [**14**]:

$$[\mathrm{Na}^+]_{equiv} = 4\sqrt{[\mathrm{Mg}^{++}]} \tag{3.7}$$

For self-complementary oligonucleotides, the $[C_T]/4$ term is replaced by $[C_T]/2$: the equilibrium constant K changes by a factor of 4, but because of symmetric entropic considerations, another factor of 2 must be introduced in the other direction, partially offsetting the first factor [**7, 14**]. If the strands are not in equimolar concentration, but one strand is present in gross excess over the other, the $[C_T]/4$ term becomes $[C_T]$. Intermediate cases can also be handled by modifying this term appropriately.

The thermodynamic parameters used in calculating the enthalpy and entropy values in equations 3.5 and 3.6 are taken from the literature. The BIND simulator's modular design enables it to use multiple thermodynamic basis sets, so it was tested on two different basis sets. The enthalpy and entropy values for nearest neighbor stacks were taken first from Quartin and Wetmur [**9**] and then from SantaLucia, Allawi, and Seneviratne [**11**]. The former paper uses enthalpy values previously determined by Breslauer, *et al.* [**5**] but provides new entropy values.

Neither of these basis sets accounts for mismatched base pairs, so mismatches were handled by treating them as "virtual stacks", as described by Werntges, *et al.* [**12**]. The simulator uses the thermodynamic parameters associated with virtual stacks reported in their paper, but a correction is incorporated to compensate for the specific nucleotide context in which their mismatches occurred. Since it is possible for multiple mismatched base pairs to occur consecutively and produce a large internal loop, BIND treats such a loop as a single large virtual stack, adding the average of all the mismatch enthalpies and entropies for the bases in the loop, rather than adding the mismatch enthalpy and entropy for each.

3.2. Calibration. In order to calibrate the BIND simulator, the melting temperatures for ninety-three oligonucleotide sequences were calculated and compared with experimental melting temperatures reported in the literature. Each sequence and concentration was provided as input to the simulator which computed the melting temperatures for all the sequences using each of the two thermodynamic parameter basis sets, with the primary intention of ensuring BIND's accuracy and with the secondary intention of determining which basis set's parameters yielded better predictions of the oligonucleotide duplex melting temperature.

The calibration process was relatively *ad hoc* in that it was only used to verify that the parameter basis sets were reasonable predictors of melting temperature for a vast range of oligonucleotides. The results of this process should be taken with a grain of salt. For example, there is a dependence issue arising from the fact that forty-four of the oligonucleotide sequences tested were also used by SantaLucia, *et al.* [**11**] in the determination of their thermodynamic parameters so one would expect their basis set to perform quite well on those sequences. Similarly, the Quartin and Wetmur [**9**] thermodynamic parameter basis set was calculated using some of the other test sequences, again invalidating independence assumptions. Consequently, these results should not be misinterpreted as conclusive proof that one basis set is superior to the other.

TABLE 1. Results of BIND calibration tests

		Quartin & Wetmur		SantaLucia		Weighted 3:8	
Test	Number	*av* ΔT	*av* $\|\Delta T\|$	*av* ΔT	*av* $\|\Delta T\|$	*av* ΔT	*av* $\|\Delta T\|$
Quartin & Wetmur	12	-1.0	2.1	4.2	4.2	2.7	2.8
SantaLucia	44	-2.5	4.5	0.3	1.8	-0.5	1.9
Aboul-ela Mismatches	16	-1.4	3.3	-2.1	3.4	-1.9	3.3
Werntges Mismatches	16	1.6	1.8	1.8	1.9	1.8	1.8
Salt Concentration	5	-5.2	5.2	0.5	0.9	-1.1	1.1
Total	93	-1.6	3.5	0.6	2.3	0.0	2.2

av ΔT indicates the average difference between the calculated T_m and the experimental T_m.
av $|\Delta T|$ indicates the average absolute value of the difference between the calculated T_m and the experimental T_m.

Thirty-two mismatched sequences from work by Aboul-ela, *et al.* [**1**] and Werntges, *et al.* [**12**] were also tested to verify that the simulator is predicting melting temperatures for sequences with mismatched base pairs reasonably well. Additionally, a few sequences from Quartin and Wetmur [**9**] were used to test the simulator's ability to handle sequences with dangling ends. Finally, a single sequence from work by Braunlin and Bloomfield [**4**] was tested at five different salt concentrations to verify that the melting temperature is being predicted correctly as the ionic environment changes.

A summary of the results of these test runs is presented in table 1. As expected, the Quartin and Wetmur basis set outperformed the SantaLucia basis set on its own sequences and *vice versa*. Both sets of parameters performed equally well on the sequences containing mismatches. On the whole, the average absolute deviation from experimentally determined melting temperature was 3.5° C for the Quartin and Wetmur basis set and 2.3° C for the SantaLucia basis set. The former tended to underpredict T_m by 1.6° C, while the latter tended to overpredict T_m by 0.6° C. When the predicted temperatures from the two basis sets were averaged with weights of 3 and 8 to compensate for the under- and overprediction, the average absolute deviation fell to 2.2° C.

It should be emphasized that no outlying data points were discarded, the test sequences were not culled for their simplicity, a single algorithm is being used for all the sequences, whether they contain mismatches or dangling ends or exist in an environment with low salt concentration, and no changes were made to the algorithm or underlying parameters on the basis of these test sequences. In fact, absolutely no optimization of the algorithm was performed *ex post facto*. Therefore, it is expected that this performance could be improved somewhat if a new set of thermodynamic parameters were determined (especially parameters describing the thermodynamics of base pair mismatches) or a few adjustments were made to the algorithm in light of these test results.

In conclusion, neither basis set is definitively superior to the other, especially in light of the dependence issues mentioned earlier. Nevertheless, because the SantaLucia basis set slightly outperformed the Quartin and Wetmur basis set, melting temperatures for the remainder of this paper will be calculated using the SantaLucia basis set.

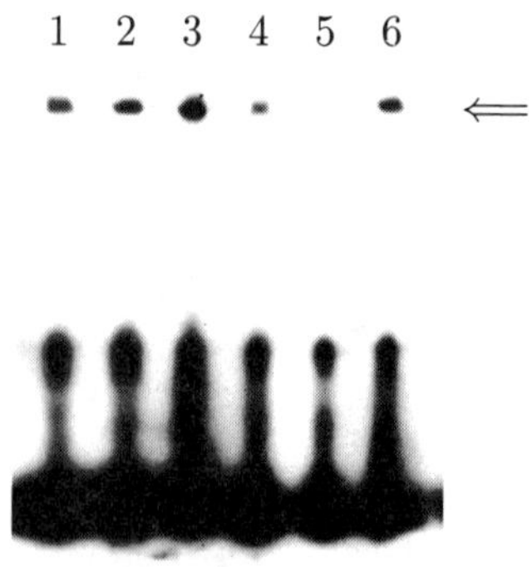

FIGURE 1. Experiment 45 gel, lanes 1-6: The six lanes are indicated by the numerals at the top, while the band of interest is indicated by the arrow at right. Descriptions of the reactions in each lane appear in the text.

3.3. Laboratory Confirmation. In addition to calibrating the simulator using melting temperature data reported in the literature, two experiments were carried out in our own laboratory to solidify the conclusion that BIND is useful in providing accurate melting temperature predictions.

In the first experiment, three distinct primers with differing degrees of mismatch were individually mixed with a single template strand at two different temperatures and given time to be extended by polymerase. Lanes 1-3 in the gel shown in figure 1 correspond to the three primer reactions run at 45° C, while lanes 4-6 in the gel correspond to the three primer reactions run at 55° C. The gel clearly indicates that while all three primers are successfully extended at 45° C, only the first and third primer are successfully extended at 55° C. This would suggest that the melting temperature of the primer-template duplex at the optimal binding position[3] along the template for the first and third primers would be above 55° C, but would be between 45° C and 55° C for the second primer. In fact, BIND predicts the melting temperatures to be 60.2° C, 47.7° C, and 65.6° C, respectively.

In the second experiment, two distinct primer template combinations were given time to be extended by polymerase at four different temperatures. Both 24-mer primers possessed two mismatches in the optimal binding position along their respective templates. Lanes 1-4 in the gel shown in figure 2 correspond to the first reaction being run at 45° C, 50° C, 55° C, and 60° C, respectively. Lanes 5-8 correspond to the second reaction being run at the same four temperatures. The gel indicates that while the first reaction allows polymerase extension at all four temperatures, the second reaction seems to falter between 55° C and 60° C. Indeed, BIND predicts the melting temperature of the primer template duplex to be 60.4° C for the first reaction and 57.1° C for the second, in close agreement with the conclusions drawn in the laboratory.

Additional confirmation of BIND's ability to find the optimal binding location along the template strand is given by the fact that in each of these experiments involving mismatched oligonucleotides, the optimal binding position reported by

[3]Optimal binding position simply means the position with the highest melting temperature.

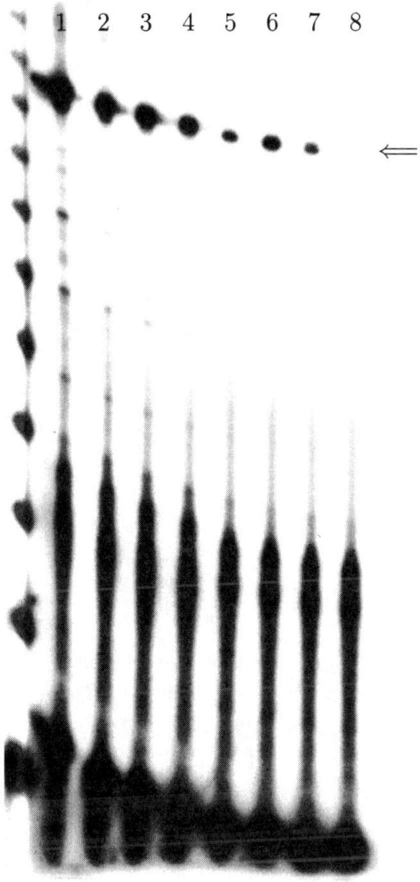

FIGURE 2. Experiment 74 gel, lanes 1-8: The eight lanes are indicated by the numerals at the top, while the band of interest is indicated by the arrow at right. Descriptions of the reactions in each lane appear in the text. This gel is slightly skewed, with the bands sloping gradually downwards moving from left to right.

BIND corresponded precisely with the optimal binding position as determined by the length of the polymerase-extended product appearing in the gels. Thus it seems likely that BIND is successful in locating optimal primer binding position along template strands, even in the presence of base pair mismatch.

4. Using the Simulator

Now we turn to a discussion of some examples of how BIND can be used in designing DNA-based computational systems, with reference to assumptions about binding specificity. In the following subsections, we consider a few steps in the design of a simple unary counter, much like the one being constructed in our laboratory. Throughout these subsections, the template under consideration consists

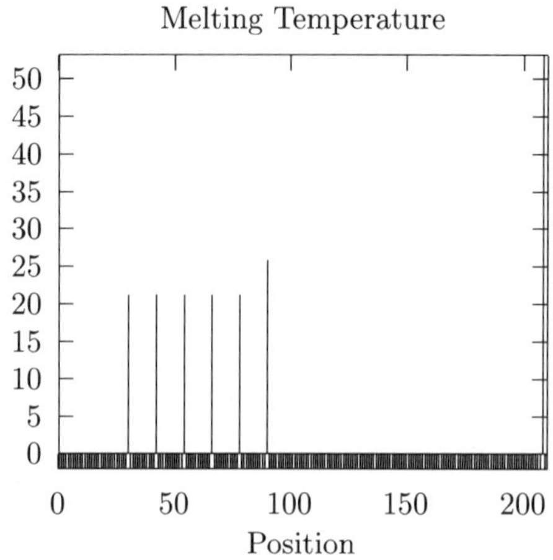

FIGURE 3. Primer binding: Melting temperature as a function of position for an outside primer with respect to the cycle 1 template. The primer should bind specifically at the end of the template but not interfere anywhere else.

of 224 nucleotides and contains within it the sequence **ZZZZZX**, as mentioned before. This sequence will henceforth be referred to as the *active site* of the template. Also as before, the first cycle rewrite rule, denoted ρ_1, is $\mathbf{X'Y'}$ and the second cycle rewrite rule, denoted ρ_2, is **XY**.

4.1. Primer Specificity. As an example of BIND's usefulness, consider the choice of an outside primer for the template used in the unary counter machine. After selecting a site away from the active site for a primer to bind, the outside primer is chosen to bind perfectly to the selected site. For this primer to be safe to use, however, it must be the case that the primer does not interfere with any of the other reactions which may be taking place in the test tube simultaneously. In particular, the primer should not be able to bind in the active site of the template.

Once chosen, the template and primer are passed to BIND as input, whereupon BIND returns the information displayed in figure 3. In the rightmost position (corresponding to the $3'$ end of the template), there is a perfect match between the template and the selected 17-mer primer, yielding a melting temperature around 53° C. The primer has very unfavorable binding specificity for most of the length of the template strand, but some potential interference can be detected near the active site for temperatures in the 20-25° C range. Although this is unlikely to cause a problem for high-temperature reaction cycles, BIND has indicated a possible source of interference between the outside primer and the active site, leading a machine designer to ensure that temperatures remain well above 25° C.

4.2. Rewrite Rule Incorporation and Sequential Computation. Programmed mutagenesis relies on the sequence specificity of its rewrite rules in order to guarantee that certain steps are being performed before others. Recall that if rewrite rule ρ_i is meant to be applied to a strand of DNA representing state σ_i, producing a strand representing state σ_{i+1}, and rewrite rule ρ_{i+1} is subsequently meant to be applied to the strand of DNA representing state σ_{i+1} to produce a

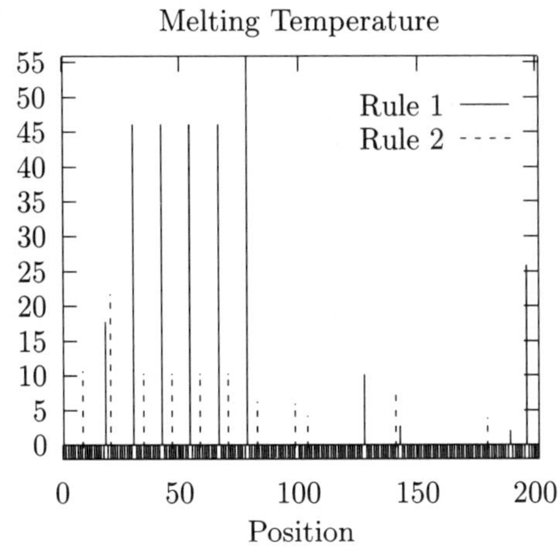

FIGURE 4. Cycle 1 binding: Melting temperature as a function of position for rewrite rules 1 and 2 with respect to the cycle 1 template. In this cycle, rule 1 should bind specifically while rule 2 remains uninvolved.

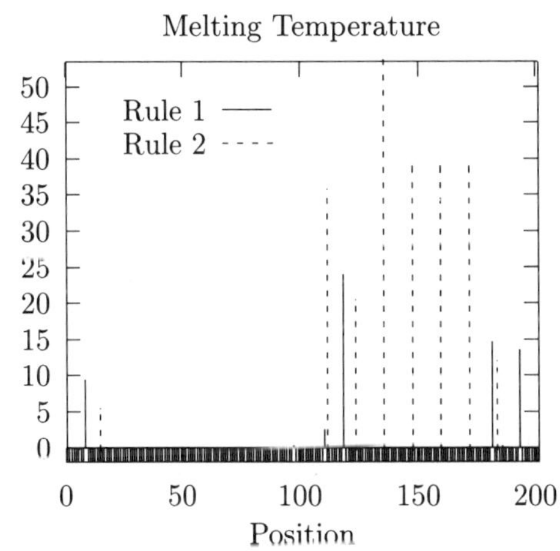

FIGURE 5. Cycle 2 binding: Melting temperature as a function of position for rewrite rules 1 and 2 with respect to the cycle 2 template. In this cycle, rule 2 should bind specifically while rule 1 remains uninvolved.

strand representing σ_{i+2}, it should be the case that ρ_{i+1} cannot be applied to σ_i and ρ_i cannot be applied to σ_{i+1}. In the context of the unary counter, we can use BIND to test that ρ_1 can bind to the template in the first cycle but not in the second, while ρ_2 can bind in the second cycle but not in the first. Additionally, we would like to be able to determine a temperature which would guarantee this binding specificity condition is met.

To test this, the same template as above can be passed to BIND, along with both rewrite rules, ρ_1 and ρ_2. The results for this first cycle reaction are shown in figure 4. Then the template which is the product of the first cycle can be passed to BIND, again with both rewrite rules. Figure 5 displays the results for this second cycle reaction. As the figures indicate, the correct rewrite rule binds to the template in the correct cycle, with very little interference from the opposing rewrite rule. As in the case of the outside primer, if the temperature of the reaction is maintained well above 25° C, no cross-rule interference should be observed.

However, these two figures reveal that it is possible for the rewrite rules themselves to inappropriately bind to the template elsewhere in the active site, annealing to any **ZZ** subsequence in the template. In order to prevent this, reactions should be run somewhere in the 47-52° C range. In fact, these same observations were made in the laboratory before BIND was developed.

4.3. Sticker Testing. As a final sample application of BIND, an experiment was conducted to measure the likelihood of sticker interference in the sticker model proposed by Roweis, *et al.* [**10**]. A template strand of length 10,000 was generated at random. Then five stickers were chosen, of lengths 10, 15, 20, 25, and 30, respectively. Each sticker was guaranteed to match the template in at least one location because the stickers were chosen to be complementary to a subsection of the original template.

The 10,000-mer template was passed to BIND along with each sticker. The sticker of length 10 bound to its complementary subsection at a temperature of 53° C, at another location at 46° C, and at a third at 34° C. No other locations had melting temperatures above 34° C. For stickers of length 15 and 20, no "false" binding position was found with a melting temperature within 30° C of that of the optimal binding location, while for stickers of length 25 and 30, no "false" binding position was found with a melting temperature within 40° C of that of the optimal binding location. Of course, one strand constructed haphazardly does not a proof of specificity make, but it is reassuring that BIND is able to confirm that the randomly chosen stickers of length at least 15 anneal to the template strand in this example in a highly specific fashion.

This simple-minded demonstration is intended to illustrate how BIND can be used in other DNA computation contexts, such as the verification of sticker choices. Although sticker interference is unlikely at sufficiently long lengths, BIND can be used to conclusively rule out such interference or point out where such interference might be occurring. It also provides valuable information about the temperature at which sticker operations should be carried out in order to maximize binding specificity.

For this reason, the simulator seems to hold a great deal of promise, not only for use in programmed mutagenesis contexts but also in the context of many other proposed models of computation. In fact, it should prove valuable to any research effort involving site-specific annealing, whether in biological computation or molecular biology at large.

5. Extensions

The BIND simulator is being extended in a number of directions. First, because polymerization plays a large role in some proposed DNA-based computational systems, it would be helpful if BIND categorized various binding sites by the stability of their open 3′ end. That is, BIND should verify that a primer which is intended to be extended by polymerase does in fact bind to the template strand with sufficient stability at the primer's 3′ end to allow for successful polymerization.

Another extension involves explicitly incorporating a time element into the simulator's calculations, in order to be able to handle more complex reaction types where more than one binding site is possible at the prevailing temperature or competition between primers arises. This dynamic version of BIND would also provide information on how long reactions would need to be run in order to achieve certain predetermined levels of binding and/or primer extension.

Of course, better thermodynamic enthalpy and entropy values would be extremely beneficial in making a more precise version of BIND. In particular, accurate parameters describing mismatched binding would be of great use. Additionally, more elaborate models of the influence of the ionic and chemical environments on deoxyoligonucleotide hybridization would allow BIND to provide accurate melting temperature predictions over a wider range of possible reactions.

Acknowledgements. The authors would like to thank Julia Khodor for her generous assistance in the provision of laboratory gels and experiment protocols. In addition, Hartemink gratefully acknowledges the support of the National Science Foundation and wishes to thank James Wetmur and John SantaLucia, Jr. for their helpful communication.

References

[1] Aboul-ela, F., Koh, D. & Tinoco, I. (1985) *Nucleic Acids Res.* **13**, 4811-4824.

[2] Adleman, L. (1994) *Science* **266**, 1021-1024.

[3] ——— (1995) On Constructing a Molecular Computer. Univ. of Southern California, draft, January 8.

[4] Braunlin, W. & Bloomfield, V. (1991) *Biochemistry* **30**, 754-758.

[5] Breslauer, K., Frank, R., Blöcker, H. & Marky, L. (1986) *Proc. Natl. Acad. Sci. USA* **83**, 3746-3750.

[6] Cai, W., Condon, A., Corn, R., Glaser, E., Fei, Z., Frutos, T., Guo, Z., Lagally, M., Liu, Q., Smith. L. & Thiel, A. (1996) The Power of Surface-Based DNA Computation. Univ. of Wisconsin, draft, August 13. Available by anonymous ftp from *corninfo.chem.wisc.edu* as */Papers/powerDNA.ps.*

[7] Cantor, C. & Schimmel, P. (1980) *Biophysical Chemistry, Part III: The Behavior of Biological Macromolecules.* W. H. Freeman, San Francisco.

[8] Doktycz, M., Paner, T., Amaratunga, M. & Benight, A. (1990) *Biopolymers* **30**, 829-845.

[9] Quartin, R. & Wetmur, J. (1989) *Biochemistry* **28**, 1040-1047.

[10] Roweis, S., Winfree, E., Burgoyne, R., Chelyapov, N. V., Goodman, M. F., Rothemund, P. W. K. & Adleman, L. (1996) A Sticker Based Model for DNA Computation. *Proceedings of a DIMACS Workshop on DNA Based Computers*, May.

[11] SantaLucia, J., Allawi, H. & Seneviratne, P. A. (1996) *Biochemistry* **35**, 3555-3562.

[12] Werntges, H., Steger, G., Riesner, D. & Fritz, H. (1986) *Nucleic Acids Res.* **14**, 3773-3790.

[13] Wetmur, J. (1991) *Crit. Rev. in Biochem. and Mol. Biol.* **26**, 227-259.

[14] ——— (1997) Personal communication.

Massachusetts Institute of Technology, Laboratory for Computer Science, 545 Technology Square, Cambridge, Massachusetts 02139

E-mail address: `amink@mit.edu`

Massachusetts Institute of Technology, Laboratory for Computer Science, 545 Technology Square, Cambridge, Massachusetts 02139

E-mail address: `gifford@mit.edu`

DIMACS Series in Discrete Mathematics
and Theoretical Computer Science
Volume **48**, 1999

The Efficiency of Sequence-Specific Separation of DNA Mixtures for Biological Computing

Julia Khodor and David K. Gifford
Laboratory for Computer Science,
Massachusetts Institute of Technology

Abstract

We report a series of experimental observations on the efficiency and fidelity of sequence-specific DNA extraction operations. We examined the solution-based annealing of bead-bound probes to target molecules followed by magnetic separation of the annealed probe-target complexes from solution. Our experiments measured how efficiently a 20-mer probe could be used to purify a 40-mer radiolabeled single-stranded target with varying target concentrations. Our results suggest that with perfectly homologous probes and targets, recovery rates of 3-28% can be expected. When the target is not homologous to the probe less than 1% of the target is recovered. The high variance of our recovery rate is consistent with the PNA-DNA results of Orum, et al.[5]. Based on our experience we propose that the biological computing community develop a standard set of benchmarks for key primitive operations. To this end, we are making our sequences and oligonucleotides available to other interested research groups.

1 Introduction

A central primitive operation for DNA computing is the ability to extract molecules that contain a desired DNA sequence from solution. One application of an extract operation is to impose constraints on a mixture of molecules that represent possible answers to a problem in order to find one or more answer molecules that meet the desired constraints.

Two well known sequence based affinity purification techniques are affinity column separation, where probe DNA is immobilized on solid support, and magnetic bead separation, where probe DNA is free to circulate in solution until a magnetic field is applied, drawing the magnetic beads and probe-target complexes attached to the beads to one side of the test tube.

Most laboratory work in DNA computing[1, 4] has used magnetic bead separation technology. In this previous work, the product of the final separation step was amplified by PCR before proceeding to the next computational step. Later theoretical work, such as the sticker model and the refinery model[6], proposed the use of separation without the subsequent use of PCR. The feasibility of using separation techniques without intervening amplification depends on whether the separation techniques satisfy certain assumptions, such as 1) high efficiency of separation (low false negative rate); 2) high fidelity of separation (low false positive rate); 3) insignificant degradation of efficiency and fidelity of separation with the change in

1991 *Mathematics Subject Classification.* Primary 68M99, 94A99; Secondary 92B05.

concentration of the target molecule; and 4) negligible loss of materials during the separations. The last two assumptions are particularly important in the context of the refinery model. While no claims on the separation efficiency have been made by the manufacturers of the magnetic beads, the DNA computing literature has assumed 90 to 95% efficiency. To the best of our knowledge, there exists no published quantitative study of the efficiency of the magnetic bead separation using DNA-DNA hybrids of single stranded molecules. At the same time, there exists a body of quantitative work on various other sequence-specific separation techniques. Particularly impressive is the work by Orum, et al.[5], which used PNA-controlled hybrid selection to obtain 72%-92% efficiency of purification.

In this paper we investigate the efficiency of magnetic bead separation under a number of experimental conditions. We discuss our choice of reaction conditions and provide a short summary of our experimental protocol (Section 2), present the results of the experiments (Section 3), discuss the implications of our experimental results to existing computing paradigms (Section 4), and conclude that amplification steps may be needed in certain DNA computation systems if separation technology is not improved beyond our experimental observations (Section 5).

2 Materials and Methods

In the experiments described below, we used Dynabeads M-280 streptavidin beads and a Magnetic Particle Concentrator (MPC) from Dynal, Inc. All oligonucleotides were manufactured by Gibco BRL.

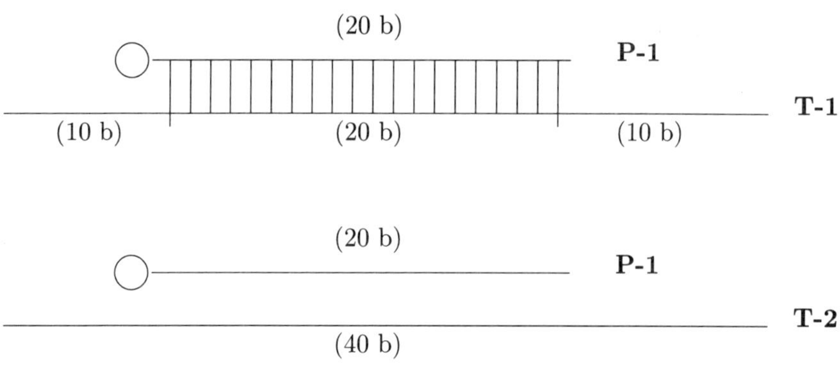

Figure 1: Match patterns between the bead-attached probe P-1 and free-floating targets T-1 and T-2.

Our goal in designing these experiments was to create an ideal set of conditions, so as to discover upper bounds on the efficiency of magnetic bead separation. To that end we designed a 20-mer probe P-1, biotinylated on the 5' end. To minimize effects of steric hindrance, we designed our target molecules to be only 40 bases long. As shown in Figure 1, T-1 is a 40-mer, whose middle 20 bases are an exact match for P-1, while T-2 is a 40-mer with no appreciable matches with P-1. We also manufactured complements of T-1 and T-2, called respectively T-1C and T-2C. To ensure secondary structure played no role in the behavior of the system, all five oligonucleotides were designed to have no appreciable secondary structure, and were tested for inappropriate intramolecular interactions.

We chose reaction conditions similar to those encountered by Kaplan[4]. We used bead-bound P-1 at 10-8 molar, and target molecules at 10-10 molar and 10-12 molar.

As previously described in the literature[1, 4], we did not insert a spacer region to place the probe sequence as far as possible from the beads on the P-1 molecule. Such a spacer could reduce the probability of bead/probe complex formation, but might increase the probability of probe/target complex formation.

Other conditions, such as the number of washes, temperature and times of hybridization and washes, buffers, and the vigor with which the tubes were mixed, were those recommended by Dynal, Inc. Note that the buffer in which the experiments were performed is 6X SSC, as opposed to the 0.5X SSC buffer used by both Adleman[1] and Kaplan[4]. Higher salt concentration in our buffer should stabilize the DNA bonds, and therefore increase the yields of the protocol.

In preparation for the experiment itself, we followed the protocol provided by Dynal, Inc for binding the biotinylated oligonucleotide to magnetic beads. Dynal has determined that 1 milligram of Dynabeads M-280 Streptavidin binds up to 200pmoles of biotinylated single-stranded oligonucleotides. The beads themselves were washed twice prior to binding. To control the amount of P-1 bound to the beads, we took OD readings of the oligonucleotide in solution before binding, and of the supernatant removed from the magnetic beads after binding. The readings indicated that virtually all of P-1 remained with the magnetic beads. Moreover, to ensure that no P-1 remained in solution while not attached to the beads (as well as to remove poorly attached probes), we performed 3 washes subsequent to the binding. All washes up to this point were done in the Binding and Washing buffer, made according to the recipe provided by Dynal, Inc. Final resuspension was done in 6X SSC buffer.

The experiments consisted of five steps that were performed according the instructions supplied by Dynal, Inc. We first mix solutions of P-labeled (*) T-1 or T-2 with P-1-coated magnetic beads in 6X SSC buffer. We allow 30 minutes for the reactions to hybridize. Second, we apply a magnetic field, and remove the supernatant into a "-" (negative) scintillation counter (SC) vial. Third, we wash the pellet three times in 6X SSC, each time removing the supernatant into the "-" SC vial. Fourth, after adding water and heating at 67C for 5 minutes, we apply a magnetic field again, this time removing the supernatant into a "+" (positive) SC vial. The fifth and final step is to resuspend the pellet in water again and to move the resuspension into an "L" (loss) SC vial.

In the case where we use *T-1, we can measure the rate of false negatives by measuring the amount of *T-1 eluted into the "-" vial. In case where we use *T-2, we can measure the rate of false positives by measuring the amount of *T-2 eluted into the "+" vial. In both cases we can measure the rate of loss by measuring the amount of radiolabeled target which ended up in the "L" vial. Since the probe is in molar excess of at least 100x over the target in all experiments, we expect our single-target experiments to provide error rates reasonably close to multiple-target experiments. Since we know single-stranded DNA to be rather non-specific, in that it prefers a bad match to no match, we also performed a series of experiments where labeled target was supplemented with an equal concentration of its nonradiolabeled complement.

We also determined control readings, which are the SC readings of the same amount of radioactive target as was put in the corresponding reaction. Notice that it is impossible to count for control the same radioactivity as went into each particular reaction, since that radioactivity remains physically divided between the "+," "-," and "L" vials.

3 Results

We performed two sets of experiments. In the first "general survey" set, we report on 13 separation experiments. Experiments designed to determine false negative rates were performed, as discussed in Section 2, for two different concentrations with *T-1 by itself (repeated twice for the higher concentration and three times for the lower), and with only the higher concentration in the presence of T-1C molecules (repeated three times). Those periments aimed at determining the false positive rates were performed only with the higher concentration of the target, both with (repeated three times) and without (repeated twice) T-2C present. The value of the baseline SC counts was obtained by averaging the values of 46 runs of blank samples (samples that only contain SC fluid). The second "closer look" set consisted of 10 separations of *T-1 by itself, at the concentration of 10E-10M. Control value in this set was determined by averaging the counting five samples of the same amount of radioactive target as was put in each of the reactions. The value of the baseline SC counts was obtained by averaging the values of 25 runs of blank samples.

Table 1 shows the raw SC counts from the experiments, while Table 2 shows data adjusted for the base counts. Notice that the Loss counts for tube 4 in Table 1 are below the background. Since radioactive decay is a Poisson process, its error function is the square root of its value[3]. The amount by which Loss counts in tube 4 are below the background is within the square root error and, indeed, within the 10% error commonly considered acceptable. We therefore replace the negative number that would have appeared in the tube 4 Loss space in Table 2 with 0.

Finally, Table 3 shows the calculated efficiencies and error rates for the above experiments. It is convenient to define several terms at this point. In a separation reaction, let p be the probability a match is identified as such; q the probability mismatch is identified as such; e- false negative rate; e+ false positive rate.

We observe from the experimental data that p, q, e-, and e+ all change with the variation in the target concentration. Adding the complement of the target to the reaction seems to have a small effect on the values of interest. More experiments are needed to precisely characterize the effect of adding the complement, or any other DNA strand, on the behavior of a magnetic bead separation reaction.

4 Discussion

Our experimental data indicates that the efficiency and fidelity of the magnetic bead separation reactions are dependent on the initial concentration of target molecules. Furthermore, the experimental data indicates that while the fidelity (q) of the protocol is over 99%, the efficiency (p) is between 3 and 28%.

Low separation efficiency, p, may be troublesome in the context of models such as the sticker model[6] which rely on performing a large number of sequential separations without intervening DNA amplification steps. While the fidelity of separation,

Target	Tube	Conc. (M)	Positive	Negative	Loss	Control	Baseline
*T-1	1	1.00E-10	118267.28	22393.66	4626.97	944111.06	616.61
	3		173159.12	22236.70	5253.00		
	4	1.00E-12	1668.57	1088.29	608.15	9076.52	
	5		1797.89	918.35	634.90		
	6		2155.94	931.33	632.50		
*T-2	7	1.00E-10	1231.79	43897.46	1225.59	1010867.81	
	8		4702.60	44970.43	960.18		
	9		1611.17	43466.70	822.66		
*T-1 & T-1C	13	1.00E-10	107101.15	19034.06	6652.16	944111.06	
	14		96559.67	11390.36	4836.40		
	15		175445.78	20507.44	12232.26		
*T-2 & T-2C		19	1.00E-10	5862.52	38908.08	2242.21	1010867.81
	20		3327.71	38727.52	668.99		
*T-1	1	1.00E-10	169424.64	69871.06	11951.56	1013588.36	475.2
	2		62546.65	71251.21	4861.43		
	3		115777.37	76927.08	3890.32		
	4		491907.03	84555.29	12524.65		
	5		237071.45	44173.49	35254.1		
	6		496687.09	61019.99	12758.81		
	7		144433.06	80012.48	14505.28		
	8		830919	85334.97	25678.02		
	9		294438.59	102059.18	12588.29		
	10		96123.9	118005.95	5198.34		

Table 1: Raw data from the separation experiments. Values in columns "Positive," "Negative", and "Loss" are the SC counter readings of the corresponding vials (see Section 2). Values in the Control column are the SC counter readings of the same amount of radioactive target as was put in the corresponding reaction. Notice that values in the "Negative" column represent only 1/25th of the content of the "-" vial.

q, is high, sequential separations to increase fidelity will result in substantial target losses.

For example, if it is desired to have more probe-homologous "good" target molecules after the separation than non-homologous "bad" target molecules then multiple separation steps may be required. Multiple separation steps will be required when the starting ratio of "good"/"bad" targets is less then (1-q)/p. Suppose that original mixture consists of M molecules, G good, and B bad. Then after the separation, the tube labeled "good" will contain $M' = pG + (1-q)B$ molecules, of which pG will be good, and (1-q)B will be bad. As we have assumed above that $G/B < (1-q)/p$, then M' will consist of $< (1-q)B$ good molecules, and (1-q)B bad molecules. This would necessitate the use of sequential separations which would further decrease the yield of the protocol, necessitating the use of DNA amplification.

In Section 1 we outlined some of the basic assumptions made by the models which propose to employ separations with no intervening DNA amplification steps. Our results suggest that 1) p has a moderate value, which may require DNA amplification between separation steps; 2) q is relatively high, but can still present problems due to moderate p; 3) p and q appear concentration-dependent; 4) the loss of target molecules is appreciable. The last point is important, since it restricts the number

Target	Tube	Conc. (M)	Pos.	Neg.	Loss	Control	Total Count Pos+Neg +Loss	Accounted for
*T-1	1	1.00E-10	117650.67	544426.25	4010.36	943494.45	666087.28	70.60%
	3		172542.51	540502.25	4636.39		717681.15	76.07%
	4	1.00E-12	1051.96	11792.00	0.00	8459.91	12843.96	151.82%
	5		1181.28	7543.50	18.29		8743.07	103.35%
	6		1539.33	7868.00	15.89		9423.22	111.39%
*T-2	7	1.00E-10	615.18	1082021.25	608.98	1010251.20	1083245.41	107.23%
	8		4085.99	1108845.50	343.57		1113275.06	110.20%
	9		994.56	1071252.25	206.05		1072452.86	106.16%
*T-1 & T-1C	13	1.00E-10	106484.54	460436.25	6035.55	943494.45	572956.34	60.73%
	14		95943.06	269343.75	4219.79		369506.60	39.16%
	15		174829.17	497270.75	11615.65		683715.57	72.47%
*T-2 & T-2C	19	1.00E-10	5245.91	957286.75	1625.60	1010251.20	964158.26	95.44%
	20		2711.10	952772.75	52.38		955536.23	94.58%
*T-1	1	1.00E-10	168949.44	1734896.50	11476.36	1013113.16	1915322.30	189.05%
	2		62071.45	1769400.25	4386.23		1835857.93	181.21%
	3		115302.17	1911297.00	3415.12		2030014.29	200.37%
	4		491431.83	2102002.25	12049.45		2605483.53	257.18%
	5		236596.25	1092457.25	34778.90		1363832.40	134.62%
	6		496211.89	1513619.75	12283.61		2022115.25	199.59%
	7		173957.86	1988432.00	14030.08		2146419.94	211.86%
	8		830443.80	2121494.25	25202.82		2977140.87	293.86%
	9		293963.39	2539599.50	12113.09		2845675.98	280.88&
	10		95648.70	2938268.75	4723.14		3038640.59	299.93%

Table 2: Adjusted data from the separation experiments. Data from Table 1 has been adjusted for background (and scaled in the case of the "Negative" column). Total counts (the sum of adjusted and scaled "Positive," "Negative," and "Loss" values), and the percentage of control values they represent have also been calculated.

of cascaded separations, and, in the case of the refinery model[6], the width of the compound separator unit.

The refinery model will need to be analyzed further if it is based on magnetic bead separation[3, 6]. This is due to the apparent dependency of efficiency, p, and fidelity, q, of separation on the concentration of the target molecules. This dependency, combined with the moderate value of p (which would force the drops in the concentration of target), would necessitate the use of the generating functions in the probability calculations of the refinery model. Also, the fact that p and q are vastly different would require the probability calculation formulas in the refinery model to be expanded to reflect that difference.

As discussed in Amos, et al.[2], some authors that analyze the complexity of DNA computing do not take into account all of the steps that are required to perform a computation, and thus their analyses underestimate expected completion times. Based on our experimental data, complexity measures of DNA computations may need to account for additional steps of DNA amplification if a sequence of separations is employed.

Target	Tube	Concen-tration (M)	Recovery, p Pos/Total	Negative, q Neg/Total	Negative error Neg/Total	Positive error Pos/Total	Loss Loss/Total	False negative, e- N-error+Loss	False positive, e+ P-error+Loss
*T-1	1	1.00E-10	17.66%		81.73%		0.60%	82.34%	
	3		24.04%		75.31%		0.65%	75.96%	
	4	1.00E-12	8.19%		91.81%		0.00%	91.81%	
	5		13.51%		86.28%		0.21%	86.49%	
	6		16.34%		83.50%		0.17%	83.66%	
*T-2	7	1.00E-10		99.89%		0.06%	0.06%		0.11%
	8			99.60%		0.37%	0.03%		0.40%
	9			99.89%		0.09%	0.02%		0.11%
*T-1 & T-1C	13	1.00E-10	18.59%		80.36%		1.05%	81.41%	
	14		25.97%		72.89%		1.14%	74.03%	
	15		25.57%		72.73%		1.70%	74.43%	
*T-2 & T-2C	19	1.00E-10		99.29%		0.54%	0.17%		0.71%
	20			99.71%		0.28%	0.01%		0.29%
*T-1	1	1.00E-10	8.82%		90.58%		0.60%	91.18%	
	2		3.38%		96.38%		0.24%	96.62%	
	3		5.68%		94.15%		0.17%	94.32%	
	4		18.86%		80.68%		0.46%	81.14%	
	5		17.35%		80.10%		2.55%	82.65%	
	6		24.54%		74.85%		0.61%	75.46%	
	7		6.71%		92.64%		0.65%	93.29%	
	8		27.89%		71.26%		0.85%	72.11%	
	9		10.33%		89.24%		0.43%	89.67%	
	10		3.15%		96.70%		0.16%	96.86%	

Table 3: Efficiency, fidelity, and error rates of magnetic bead separation have been calculated from the experimental data in Table 2.

5 Conclusions

We presented experimental data that show that in magnetic bead separation one can expect efficiency of 3 to 28%, and fidelity of more than 99%. We have observed that the efficiency and fidelity of the separation depend on target concentrations. Based on our results, it appears that DNA amplification, such as PCR, may be required in certain types of DNA computing applications.

We performed our experiments under carefully controlled conditions. The results we obtained have noticeably high variance, which is consistent with the PNA-DNA results of Orum, et al[5]. We therefore propose that further investigation of separation techniques to find the origin of this high variance is warranted. We also propose that the biological computing community develop a standard set of benchmarks for key primitive operations, including sequence-specific separations. To this end, we are making our sequences and oligonucleotides available to other interested research groups.

6 Acknowledgments

This paper has its origins in a class project for Computer Science and Molecular Biology, taught at MIT by Prof. David K. Gifford and Prof. Bonnie Berger. We would like to thank the rest of the team which worked on the class project: Alexander Hartemink, Brad Lichtenstein, Mariya Minkova, and Peter Szilagyi, all of Massachusetts Institute of Technology. The authors would also like to thank Dr. Brenton R. Graveley of Harvard University for his expert advice on magnetic bead separation, and Prof. Douglas Melton of Harvard University for providing laboratory space and resources for performing experiments described in this paper. In addition, Khodor gratefully acknowledges the supported of Genomics Training Grant.

References

[1] Leonard Adleman. Molecular computation of solutions to combinatorial problems. Science, 266:1021-1024, 1994.

[2] Martyn Amos, Alan Gibbons, and Paul E. Dunne. The complexity and viability of DNA computations (Extended draft).

[3] William Feller. An Introduction to Probability Theory and Its Applications, Volume I, 3rd edition. John Wiley & Sons, 1968.

[4] Peter Kaplan, Guillermo Cecchi, Albert Libchaber. Molecular computation: Adleman's experiment repeated, 1995.

[5] H. Orum, P.E. Nielsen, M. Jorgensen, C. Larsson, C. Staneley. Sequence-specific purification of nucleic-acids by PNA-controlled hybrid selection. Biotechniques, v19 (3): 472-480, 1995.

[6] Sam Roweis, Erik Winfree, Richard Burgoyne, Nickolas V. Chelyapov, Myron F. Goodman, Paul W. K. Rothemund, Leonard M. Adleman. A sticker based model for DNA computation, 1996.

DIMACS Series in Discrete Mathematics
and Theoretical Computer Science
Volume **48**, 1999

A New DNA Separation Technique with a Low Error Rate

Junghuei Chen and David Harlan Wood

ABSTRACT. A new method for separation of DNA according to its substrings is developed and demonstrated. Our method is for circular single strands of DNA. These strands are assumed to be designed to minimize unwanted hybridization (sticking to each other).

The new method is enzyme-based and notable for its simplicity. Errors are below the level of detectablity in gels that exhibit the desired, separated strands. This purity is a significant improvement over the usual "affinity purification" separation method. Additional more sensitive instrumentation will be required to precisely measure the error rate of the new method.

1. Separation, And Why New Methods Are Needed

A fundamental need in DNA laboratory technology is to isolate fragments or structures of DNA that contain a given sequence or sequences from other DNA. This laboratory operation is generally referred to as "separation." However, the objective is often not to obtain two separated sets of DNA, but only to isolate the one of interest. This version of the process is often referred to as extraction or as purification. In any case, we will pursue methods that can *separate* a collection of DNA into two parts: one part where the DNA has some distinguishable property encoded, and a second part where the DNA does not.

The purities obtained by present separation techniques remain the subject of controversy. However, it appears that even 95% accuracy may be unobtainable with conventional techniques.

Our Purity Goals. Our technique aims for much higher accuracy than that obtained by present laboratory techniques. In the experiments reported in this paper, the errors are *below the level of detectablity* in gels that exhibit the desired, separated strands.

The anticipated sources of error in our method are discussed in Section 4, Limitations and Difficulties. These sources of error are orders of magnitude lower than the resolution of the gel comparisons we exhibit. The error rate of our new method is, as expected, apparently so low that more sensitive instrumentation will be required to precisely measure the error rate.

1991 *Mathematics Subject Classification.* Primary 68Q40, 68U99, 68Q15.

Purity is our objective, while yield is a secondary consideration. Our yields are typically 20% to 50%, but we assume that separated products can be readily increased by using the Polymerase Chain Reaction (PCR). Indeed, our specifications provide for incorporating short "universal" sequences that facilitate PCR.

Applications To DNA Computer Logic Register Design. The generic idea of a DNA computer "logic register" is simply a structure fabricated out of DNA that encodes information in DNA sequences we refer to as "variables." In DNA computing, one often wants to answer logic operations, such as "Which registers contain the variable x or the variable y?" The fundamental issue is to separate registers in the laboratory on the basis of whether or not a logical statement is TRUE or FALSE for these registers.

It is notoriously difficult to achieve separation in the laboratory with the high purity needed for DNA computation (which generally requires many sequential steps without human intervention). Even a 95% accuracy degenerates after 100 steps to less than 1% accuracy.

We vigorously exploit a few fundamental assumptions about the properties of *DNA logic registers structured to specifically facilitate separation.* These registers are not linear strands of DNA. We will assume each register has circular structure that facilitates laboratory detection of variables and separation.

2. Approach Used In The New Techniques

Our approach begins in the expected way. "Variables" (DNA sequences encoding information) are to be detected in "logic registers" made of DNA. Single strands of DNA (called primers) that are Watson-Crick complementary to the variables are individually available.

Throughout our design process we make the idealized assumptions listed below. These assumptions are examined more closely in Section 4, as are the sources of experimental errors. Aside from the listed assumptions, "logic registers" and "variables" are otherwise unrestricted.

1. *Assume no primer will hybridize (stick) anywhere except with its corresponding variable.*
2. *Assume logic registers are circles of single-stranded DNA, and that they do not hybridize with each other or with themselves.*
3. *Assume all registers contain exactly one copy of the "universal" variable whose coding includes a restriction enzyme site in its center.*

Basic Techniques for Implementing Separation. We now describe the basic techniques that will be used. Figure 1 and Figure 2 illustrate some of these techniques applied to a simple example.

- Marking registers. Primers are added to a solution that contains the logic registers. The primers hybridize with the logic registers that contain the corresponding variable(s). The enzyme polymerase is then added. It extends each primer in a unique ($5' \rightarrow 3'$) direction, continuing along forming a double strand of DNA, until it encounters another primer (or the other end of itself). It does not, however, connect to the primer that it encounters. A logic register that has become double stranded is said to be "marked." A logic register that does not contain any variables corresponding to the primers is unchanged, and is said to be "unmarked."

- Recovering marked registers. To recover the logic registers that have been marked, the enzyme *S1* Nuclease is used to digest all single-stranded DNA, namely the unmarked registers and the excess primers. Then the enzyme *Exonuclease III* is used which digests the remaining DNA *except the closed circles* constituting the logic registers that were marked. Here it is important to recall that the extended primer(s) constituting the second strands of the marked registers are digested *because they were never closed.*
- Recovering unmarked registers. To recover the logic registers that were unmarked, the appropriate restriction enzyme is added that digests (severs) the registers at the restriction site in the universal variable — but only the marked registers, because a restriction enzyme *cuts only double stranded DNA* at its restriction site. Now *Exonuclease III* is used to digest the marked registers (they are no longer closed) and the surplus primers. All that is left are the closed unmarked logic registers that we wanted to extract.
- Separating marked from unmarked registers. To separate the logic registers that were marked from those that were unmarked, divide the solution into two equal parts and perform the two above extractions on the two parts. Naturally, yields are reduced by half.
- Amplification by PCR. Yield is a secondary concern for us. The structure of logic registers allows us to easily increase their quantity. To a solution containing any mixture of registers, add the primer complementary to the universal variable which is present in every register. Restriction enzyme is used to cut all registers at the universal restriction site; these strands have known sequences at each end. Use PCR to increase quantity as desired. By adding a excess of the universal primer, hybridization and ligation will close large quantities of registers. Digest all leftover (non-closed) materials with *Exonuclease III.*

Construction and Separation of Circular Logic Registers. In Figure 1 left we illustrate the routine method of construction of circles of DNA, which we consider to be logic registers, each containing several variables. In Figure 1 right we illustrate how a mixture of two different logic registers can be separated according to the presence or absence of a specific variable.

Laboratory Demonstrations Using Our New Techniques. Our laboratory results are in two categories. In our first set of laboratory results (see Figure 2 left) we show successful construction of circles of DNA, which we consider to be logic registers. In our second set of laboratory results (see Figure 2 right) we demonstrate the separation of different logic registers

Figure 2: Left. Lane 1 contains a 100 base ladder as calabration. Lanes 2, 3 and 4 contain the starting materials used to synthesize one 147 mer single-stranded circle. These materials are 42, 71, and 76 nucleotides long, respectively. Corresponding to Figure 1 left part (I), Lane 5 contains the products of the first ligation reaction using T4 DNA ligase. A 147 mer linear molecule is formed from the 71 and 76 mers. A 42 mer labeled U was used as the complementary strand for the ligation reaction. Corresponding to Figure 1 left part a (II), Lane 6 contains the rsults of the second ligation, which converts the 147 mer linear molecule to a circular molecule. Here, a second 42 mer was used as a complementery strand to bring together the ends of the 147 mer linear molecule. Lane 7 contains the product obtained after the material shown in Lane 6 was treated with *Exonuclease III*, which

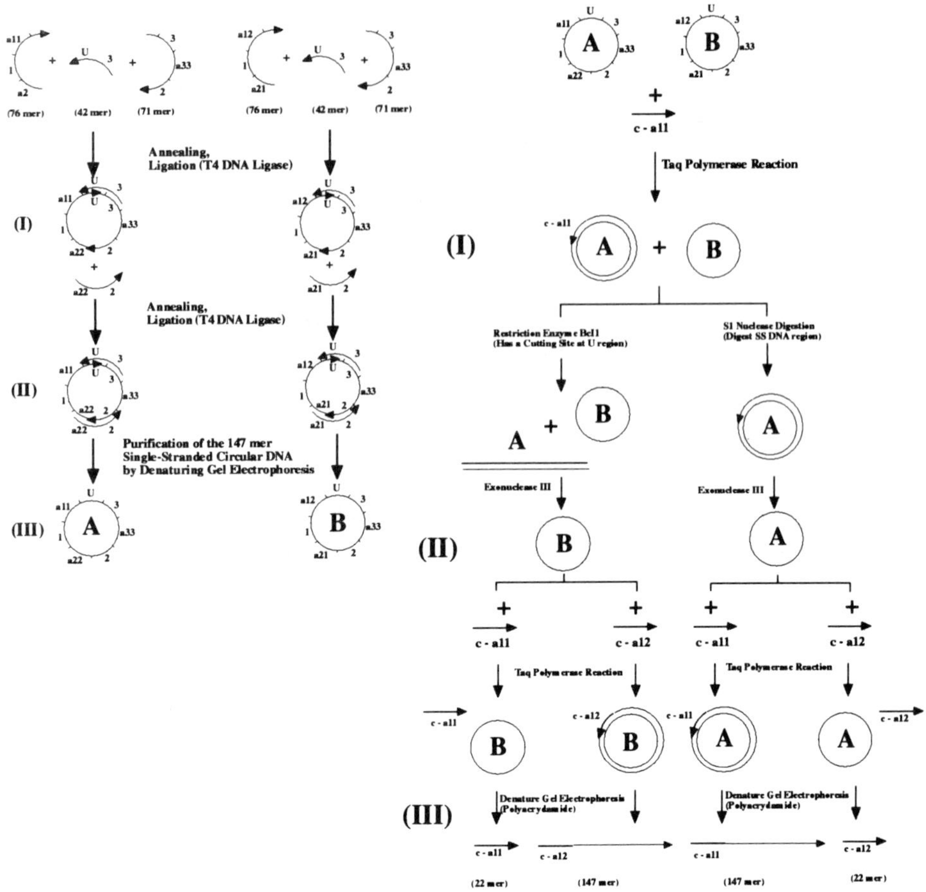

FIGURE 1. Left: Diagram illustrating the construction of the DNA circles A and B. Right: Diagram illustrating circles A and B can be separated from a mixture of A and B. See the text of Section 2.

digests all the linear DNA, leaving only the closed circular single-stranded DNA. In this lane, the upper band resists the *Exo III* digestion, indicating it is the 147 mer single-stranded closed circle. Lane 8 contains 147 mer single-stranded circular DNA which was purified by excising it from a gel prepared in the same way as Lane 7. The gel-purified circular DNA was used for the experiments in Figure 1 right to better quantize the error rates in those experiments.

Figure 2: Right. A mixture of circles is separated using the criterion of whether or not each circle contains the substring a11. Since, in this case, A contains a11 while B does not, the experiment separates the circles A from the circles B. The identities and purity of the separated materials are confirmed in the final steps. Lane 1 contains the unpurified reaction synthesizing the circular 147 mer template, circle A (as shown in Figure 2 left, Lane 6). It is included in this gel to establish calibration. Lane 2 contains a mixture of purified single stranded circular DNAs, labeled A and B in Figure 1 right. Lane 3 illustrates the products of *Taq* polymerase reaction after 30 thermal cycles through the reannealing temperature of 55° C. Corresponding to Figure 1 right (I), the 22 mer c-a11 was added. The c-a11 reanneals with circle A and is extended by the polmerase, but circle B is unaffected.

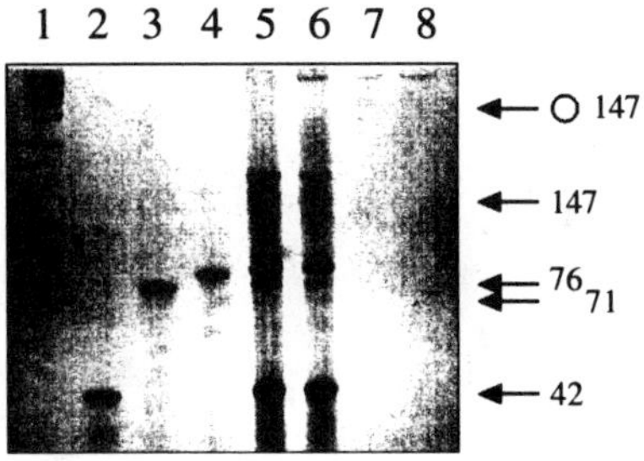

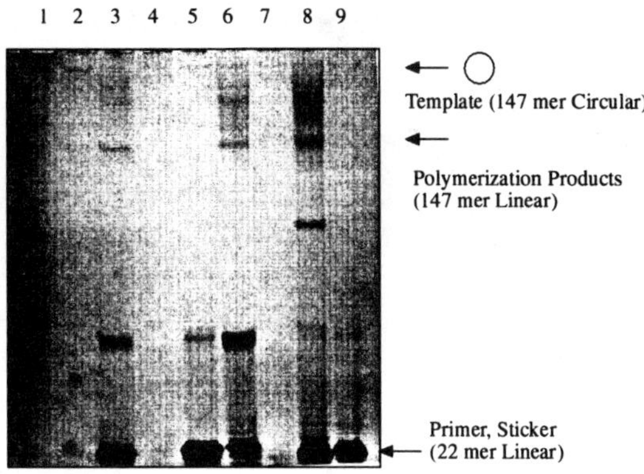

FIGURE 2. Left: This is a 12% polyacrylamide denaturing gel (8 M urea). This gel illustrates intermediate and final results in the construction of a closed circle of DNA. This gel corresponds to the construction diagramed on the left half of Figure 1. Right: This is a 12% polyacrylamide denaturing gel (8 M urea). This gel illustrates intermediate and final results in separating circles A and B being separated from a mixture of A and B. These results correspond to the right half of Figure 1.

Corresponding to the left half of Figure 1 right (II), Lane 4 illustrates the products resulting when the material sampled in Lane 3 was digested with the restriction enzyme *BcI 1*. That enzyme digests the double stranded circle (Circle A) into a linear duplex strand of DNA with a nick. Subsequently, the sample was treated with *Exonuclease III* which digested the linear duplex, but not the closed single strand circular DNA. Thus, B has been extracted from a mixture of A and B because it does not contain a11 (while A does). Corresponding to the right half of Figure 1 right (II), Lane 7 illustrates the products resulting when a sample of the material exhibited in Lane 3 was treated with S1 nuclease. That enzyme digests only single stranded DNA regions; in this case, circle B. Subsequently the sample was treated with *Exonuclease III* to digest all remaining DNA except the closed single-stranded DNA, circle A. Thus, A has been extracted from a mixture of A and B because it is the only circle that contains a11. The remaining four lanes show a method for determining the contents of the separations sampled in Lane 4 and Lane 7. This method attempts, on one hand, to anneal and extend c-a11 (which is present only in circle A) and, on the other hand, to anneal and extend c-a12 (which is present only in circle B). Corresponding to the left half of Figure 1 right (III), Lanes 5 and 6 display the reactions in which the separated material shown in lane 4 (which happens to have only circle B) was used. In Lane 5, the primer that is unique to circle A, c-a11, was used. This lane indicates the absence of Circle A. In Lane 6, the primer that is unique to circle B, c-a12, was used. This lane indicates the presence of Circle B. Corresponding to the right half of Figure 1 right (III), Lanes 8 and 9 display the reactions in which the separated material shown in lane 7 (which happens to have only circle A) was used. In lane 8, the primer that is unique to circle A, c-a11, was used. This lane indicates the presence of Circle A. In lane 9, the primer that is unique to circle B, c-a12, was used. This lane indicates the absence of Circle B.

3. Existing Separation Techniques And Their Drawbacks

We describe the three most commonly mentioned separation methods, not all possible techniques. There are rumors that some DNA computing groups have tried ten or more methods! Further, we do not consider the direct use of separation by length using gel electrophoresis because our concern is the presence or absence of certain sequences (variables).

General methods are known which can improve the purity of any separation process to a arbitrary degree by a repeating or cascading the process [**17**]. This is paid for in the increased number of time steps and / or increased laboratory processes and equipment.

Fishing: Biotin-Streptavidin Bead Separation. In this method, Biotinylated primers complementary to the variable we seek are bonded to the surface of beads coated with Streptavidin which are packed in a column. As the mixture containing registers is poured through the column, those registers with the desired variables are retained, while the others pass through. Afterward, the column can be flushed with warm buffer to denature (melt) the DNA bonds, releasing the captured registers. Regrettably, this method, which was used in Adleman's historic experiment, delivers quite low purity. The specificity of DNA hybridization is reduced by the process of bonding DNA to the beads. As a result, only about 20% of the DNA retained in the column is correct.

Eating: Digesting Unwanted DNA With Enzymes. Linear single strands of DNA can be used as logic registers with this approach. As usual, primers are allowed to hybridize with registers. One uses restriction enzymes to modify lengths that can be separated by gel electrophoresis. Unmarked registers can be recovered from the gel. Marked registers have their parts recovered from the gel and have to be reconstituted by annealing with primers followed by ligation. As with circular DNA registers, one can avoid requiring the encoding of each variable to contain a restriction site. A universal variable with a restriction site is built in to the $5'$ end of each register. If any primer hybridizes, polymerase can be used to create a double strand through the restriction site. The restriction enzyme results in shortening the marked registers. With knowledge of the distribution of register lengths, the shortened ones may be separated in a denaturing gel, and reconstituted after hybridizing with the universal primer. The primary concern is that the kilogram quantities of DNA envisioned for DNA computing are a million times the capacity of current gel electrophoresis equipment.

Exaggerating: PCR Amplifying Desired DNA Strands. At last, a high-yield method! Repeated doubling of desired linear registers overwhelms anything else, giving a form of purification and extraction. The problem is that the entirety of the desired register is not obtained unless the primers hybridize only at both of the two extreme ends. *Thus, the only variables that can be detected are those located at the ends of a register.* No PCR method is known for overcoming this drawback.

4. Limitations And Difficulties

Three laboratory process are used in our separation methods. We discuss them in the order of least to most troublesome: digestion (by enzymes), polymerization (extension of primers), and hybridization (sticking). We advocate the use of digestion and polymerization because of their low error rates, relatively high yield, and suitability for scaling up to DNA computers which might use kilogram quantities of DNA. We avoid all but secondary use of gel purification, out of concern about scale-up. Current laboratory gel electrophoresis equipment would require a 1,000,000 fold increase in capacity to handle kilogram quantities of DNA.

Enzyme Error Rates And Their Effects. In the laboratory, enzymes are most predictable when present in excess. The expected result is that enzyme digestion is essentially complete, and error consists of overdigestion, which is the digestion of materials that are supposedly resistant to the enzyme. Thus, digestion enzymes are favorably biased for our purposes, in that errors marginally reduce yields, rather than reduce purity. This is consistent with our goal of achieving high purity, while tolerating a moderate yield.

Hundreds of enzymes are commercially available, and vary greatly in their specific activity. We use only a few enzymes, all of which have high specific activity and cause little very error.

Closed DNA structures, such as circles, resist *Exonuclease III*. But overdigestion reduces the number of surviving circles, marginally decreasing yield [**16**].

On the left side of Figure 1 right step II, overdigestion by the enzyme *BcI* again reduces the yield of the result.

On the right side of Figure 1 right step II, overdigestion by *S1* Nuclease, which is not supposed to digest double stranded DNA, causes reduction in yield. *S1* is reported [**18**] to digest single stranded DNA 75,000 times faster than double stranded DNA. Unwanted overdigestion of double circles A produces a slight reduction in yield, which can be neglected. Recall the circles A are double stranded except for a "nick." *S1* is reported [**19**] to digest so slowly at a nick that it only negligibly reduces yield. It can happen that extension by polymerization is incomplete, leaving a single-stranded gap rather than a nick. Digestion in a gap would lower yield, but gaps are negligibly infrequent (see next section).

Polymerase Error Rates And Their Effects. *Taq* polymerase occasionally fails to correctly complement each base in a template strand of DNA. But this error rate is only 2×10^{-4} to 7×10^{-5}, according to two studies cited in [**7**, page 45, Table 1]. Further, even when these rare errors occur, they seldom affect the hybridization of primer extensions with our circle templates. Thus, *Taq* errors in fidelity are negligible in our context.

A single polymerase reaction does not fully extend all hybridized primers. The two studies involving *Taq* which were cited in the previous paragraph quote 36% and 88% production of completed primer extensions (under two different conditions). We cope with this deficiency by repeated thermal cycling. Assuming only 36% efficiency, the 30 cycles we use would theoretically fail to completely extend primers only at the insignificant rate of 1.5×10^{-6}, which we disregard.

Hybridization Difficulties And Their Effects. Information is encoded in pieces of DNA with the intention that the pieces will hybridize with their Watson-Crick complementary locations *and will not hybridize where they are not wanted.* This can be difficult to achieve, because exact complementary pairing is not required for the pieces of DNA to hybridize. This can give rise to errors in experiments

[**10, 9**]. The encoding problem has been addressed by Baum [**4**] and the research group at Memphis University [**6**].

For DNA computing, the ultimate level of hybridization difficulties is not yet completely clear. However, the purity of our experimental results in Figure 2 is very encouraging, especially compared to roughly similar published experiments [**8, 9, 10, 6, 2**]. In our gels exhibiting the separated strands, the errors are below the level of detectablity.

Encoding problems appear to be significant in laboratory experiments [**10, 9**] involving DNA computing. Similar problems have been widely discussed in the context of PCR error rates. But in contrast to PCR, DNA computing has an important advantage: *one is able to control the DNA encodings to reduce hybridization errors.*

Hybridization is to be used in all proposed DNA computer designs that we are aware of. All are vulnerable to the difficulties posed by the hybridization "coding problem." Our separation method is not exempt. Just consider the uncontrolled errors that would result in Figure 1 right Part I if a badly encoded primer c-a11 could sometimes hybridize with circle B!

However, our laboratory results shown in Figure 2 right already indicate low rates for errors due to unwanted hybridization. Precise error measurements need to be made. Our low error rates owe much to the encodings kindly provided to us by the DNA computing group at Memphis University. These encodings are based on their research [**6**] and are designed to minimize cross-hybridization, hairpins, and other problems. We are grateful for their help.

Even with advantageous encodings at our disposal, we at first failed to notice the importance of choosing subgroups of encodings having closely matched annealing temperatures, T_m. Figure 3 indicates the extra effort that can be required to compensate for this oversight.

Figure 3. Part (a) is a diagram showing the two single-stranded circular templates used in our experiment, circles labeled A and B. Each of these circle has three 22 nucleotide regions that can anneal with three distinct 22 mer primers having complementary sequences. As shown in the figure, c-a11, c-a22 and c-a33 can all anneal with circle A, while c-a12, c-a21 and c-a33 can all anneal with circle B.

Parts (b), (c) and (d) are 12% polyacrylamide denaturing gels (8 M urea). These gels show the results of the reannealing and extension of six primers with the two templates at different reannealing temperatures during 30 thermal cycles attained in the presence of *Taq* polymerase. For calibration, Lane 1 in each gel contains a full length polymerase reaction product of the length we wish to produce in all the other lanes. Each desired product is a 147 mer linear single strand of DNA. Lanes 2, 3 and 4 are the results of using circle A as the template with c-a11, c-a22, and c-a33, respectively as the primers. Lanes 5, 6, and 7 are the results of using circle B as the template with c-a12, c-a21, and c-a33 as the primers, in that order.

In gel (b) the reannealing temperature was cycled through a minimum of 50^oC — obviously not the best choice. In gel (c) the reannealing temperature was cycled through 55^oC, with erratic results. To achieve the improved results shown in gel (d), each reaction was individually reannealed at a different temperature. For c-a11 and c-a12 the reannealing temperature was 55^oC, while for c-a22 it was 60^oC, and for c-a21 and c-a33 was it was 47^oC.

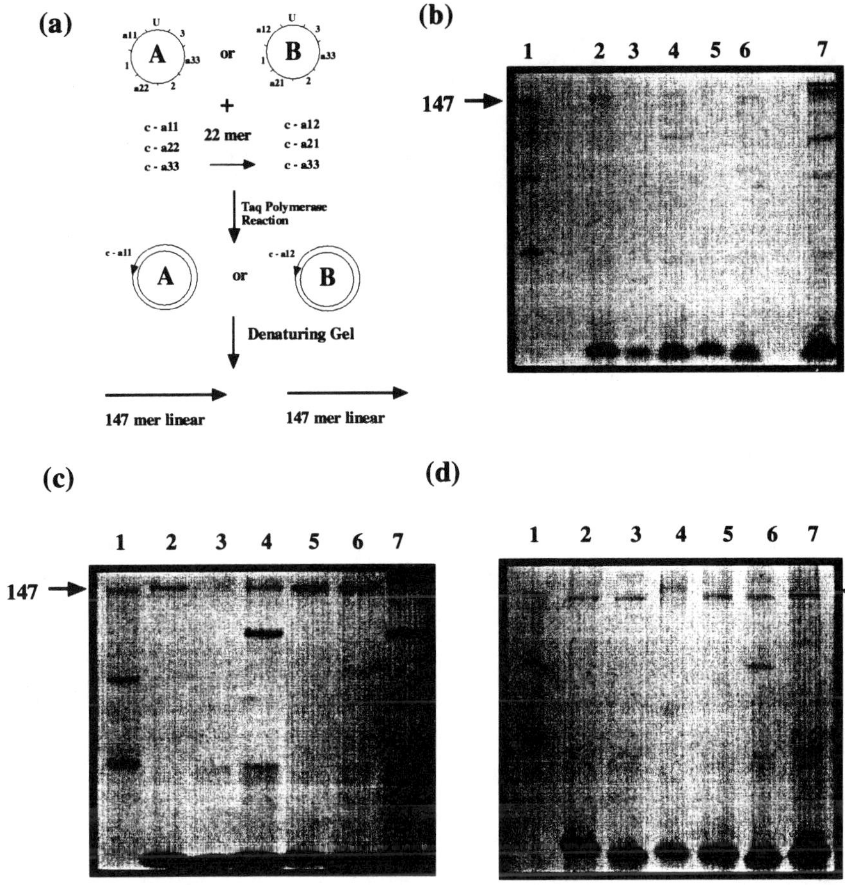

FIGURE 3. Experimental demonstration of hybridization sensitivity to melting temperatures, T_m.

Ultimately, we adjusted our thermal cycling to produce the results shown in Figure 2 left.

Possible Applications To Molecular Biology And Nanotechnology. It would be very nice to be able to point out problems in biochemistry and nanotechnology likely to remain intractable until high-purity separation becomes available. We are not aware of any such problems, even though some may come to light when better separation techniques are known.

However, biochemistry and nanotechnology may well be affected. In these disciplines, laboratory practice frequently follows the slogan, "Put yield before purity." Our methods emphasizing "Put purity before yield" will offer contrasting tradeoffs that could alter conventional laboratory practice in these disciplines.

The materials to be separated in biochemistry and nanotechnology generally do not conform to the assumptions on structure that expedite our high-purity separation. Adaptions and modifications can help but laboratory practices reflecting reduced purity will likely remain the rule, not the exception.

References

[1] L. M. Adleman. On constructing a molecular computer. University of Southern California Department of Computer Science Draft.

[2] Leonard M. Adleman. Molecular computation of solutions to combinatorial problems. *Science*, 266:1021–1024, November 1994.

[3] Leonard M. Adleman. On constructing a molecular computer. In Lipton and Baum [**15**]. Based on Manuscript, Computer Science Department, University of Southern California, January 11, 1995.

[4] Eric B. Baum. DNA sequences useful for computation. In Landweber and Lipton [**12**].

[5] Dan Boneh and Richard Lipton. Making DNA computers error resistant. In Landweber and Lipton [**12**]. Previously: Princeton University TR-491-95, May 1995 fixmeStatus?

[6] R. Deaton, R.C. Murphy, M. Garzon, D.R. Franceschetti, and S.E. Stevens, Jr. Good encodings for DNA-based solutions to combinatorial problems. In Landweber and Lipton [**12**].

[7] Carl W. Dieffenbach and Gabriela S. Dveksler. *PCR Primer: A Laboratory Manual.* Cold Spring Harbor Laboratory Press, Plainview, New York, 1995.

[8] Peter Kaplan, Guillermo Cecchi, and Albert Libchaber. DNA based molecular computation: Template-template interactions in PCR. In Landweber and Lipton [**12**].

[9] Peter D. Kaplan, Guillermo Cecchi, and Albert Libchaber. Molecular computation: Adleman's experiment repeated. Technical report, NEC Research Institute, 1995.

[10] Peter D. Kaplan, Guillermo Cecchi, and Albert Libchaber. DNA based molecular computation: template-template interactions in PCR. In Landweber and Lipton [**12**].

[11] Richard M. Karp, Claire Kenyon, and Orli Waarts. Error-resilient DNA computation. In *Proceedings of the Seventh Annual ACM-SIAM Symposium on Discrete Algorithms*, pages 458–467. SIAM, January28–30 1996.

[12] Laura F. Landweber and Richard J. Lipton, editors. *Proceedings of the Second Annual Meeting on DNA Based Computers, held at Princeton University, June 10-12, 1996*, DIMACS series in discrete mathematics and theoretical computer science, Providence, to appear. American Mathematical Society.

[13] Richard J. Lipton. Speeding up computations via molecular biology. Unpublished manuscript Dec. 9, 1994,, December 11, 1994.

[14] Richard J. Lipton. Speeding up computations via molecular biology. In Lipton and Baum [**15**].

[15] Richard J. Lipton and Eric B. Baum, editors. *DNA Based Computers: Proceedings of a DIACS Workshop, April 4, 1995, Princeton University*, volume 27 of *DIMACS series in discrete mathematics and theoretical computer science*, Providence, 1996. American Mathematical Society.

[16] S. G. Rogers and B. Weiss. Exonuclease III of *escherichia coli* k-12. *Meth. Enzymol*, 65:201–211, 1980.

[17] Sam Roweis, Erik Winfree, Richard Burgoyne, Nickolas V. Chelyapov, Myron F. Goodman, Paul W. K. Rothemund, and Leonard M. Adleman. A sticker based architecture for DNA computation. In Landweber and Lipton [**12**].

[18] V. M. Vogt. Purification and properties of S1 Nuclease from *aspergillus. Meth. Enzymol*, 65:248–254, 1980.

[19] Roger C. Weigand, G. Nigel Godson, and Charles M. Radding. Specificity of the S1 Nuclease from *aspergillus oryzae. J. Biological Chemistry*, 250(22):8848–8855, 1975.

(J. Chen) Dept. of Chemistry & Biochemistry, University of Delaware Newark, DE
E-mail address: `junghuei@udel.edu`
URL: `http://www.udel.edu/chem/chen/chen.html`

(D. Wood) Department of Computer Science, University of Delaware Newark, DE
E-mail address: `wood@cis.udel.edu`
URL: `http://www.cis.udel.edu/~wood`

DIMACS Series in Discrete Mathematics
and Theoretical Computer Science
Volume **48**, 1999

Towards Parallel Evaluation and Learning of Boolean μ-Formulas with Molecules

Masami Hagiya, Masanori Arita, Daisuke Kiga, Kensaku Sakamoto, and Shigeyuki Yokoyama

ABSTRACT. A μ-formula is a Boolean formula in which each variable occurs at most once. The paper treats its molecular representation including its queries using techniques in molecular biology. The novelty is that this method can evaluate a Boolean formula in a single tube within a short time. The preliminary experimental result suggests the possibility of parallel evaluation and learning of general μ-formulas. This method is essentially a simulation of state transitions and can be used to simulate a decision tree or a state machine.

1. Introduction

From the beginning of molecular computation, its parallelism is what theoretical analyses are concerned with and what researchers have tried to fully utilize in lab experiments. Despite the initial optimism, however, researchers came to know pros and cons of molecular operations [**11**]. The most unfavorable is the instability of lab operations susceptible to temperature, concentration and enzyme activity. Another hurdle is the difficulty in realizing complex procedures and data structures in terms of primitive operations on molecules. Although DNA molecules can be concatenated or rotated by lab operations [**2**], the set of operations on DNA molecules is still far behind what can be called programmable. A molecular Random Access Machine (RAM) is only theoretically examined, and has not yet been realized [**10**]. Even with these obstacles, however, the parallelism inherent in molecular computation is hard to abandon. When fully utilized, this parallelism could be a great tool for a wide range of computation.

With standard molecular biological techniques, the best way to utilize the molecular parallelism is to execute what can be called *generate and search* as was initially performed by Adleman [**1**]. He first generated at random all the candidate solutions for an instance of Hamiltonian path problem, then chose the correct solution from them in standard lab operations. Lipton's method for solving a 3-SAT problem is also along the line of *generate and search* [**5**].

However, the power of *generate and search* with molecules is currently far from sufficient for solving actual problems because the available operations on molecules are severely limited. In order to make molecular computation practically feasible,

1991 *Mathematics Subject Classification.* Primary 68Q10, 68Q22; Secondary 68Q05.

it is vital to prepare molecular operations that allow complex procedures and data structures.

In this paper, a molecular method for parallel evaluation of multiple μ–formulas is introduced. A μ–formula is a Boolean formula in which each variable occurs at most once. Our ultimate goal is to learn μ–formulas from examples utilizing the parallelism of molecules. Pitt and Valiant showed that μ–formulas are not learnable from examples alone in polynomial time [**9**]. In our method, however, all the possible μ–formulas can be randomly generated, and then only formulas satisfying the given variable assignments can be selected in a single tube in polynomial time.

Unlike in other approaches towards evaluation of Boolean circuits with molecules [**5, 4, 7**], we represent both Boolean formulas and their variable assignments with DNA molecules. This is a significant progress from the existing paradigm for molecular computation. Since the DNA molecules representing Boolean formulas can be regarded as programs, and molecules for variable assignments can be regarded as inputs, our approach allows multiple programs with different inputs to run in parallel (MPMD or MIMD). On the other hand, in previous approaches [**5, 4**], programs are implemented as a sequence of operations that are uniformly applied on molecules in a tube (SIMD). A method for parallel evaluation of a single circuit was also proposed [**7**] but it is not directly related with our goal.

In our approach, a variable assignment (or an input), a Boolean formula (or a program), and the current state of evaluation (or a head in Turing machine's sense) are encoded in one single–stranded DNA in this order (Fig.1). Evaluation of the Boolean formula with its variable assignment is performed by state transitions implemented by molecular operations. In each transition, the head reads information from the Boolean formula as well as from the assingment to change the state. More precisely, single–stranded DNA forms a hairpin structure where the current state hybridizes to an appropriate position in the formula or the assignment, then the 3' end of the state is extended by one symbol to obtain the next state using polymerization.

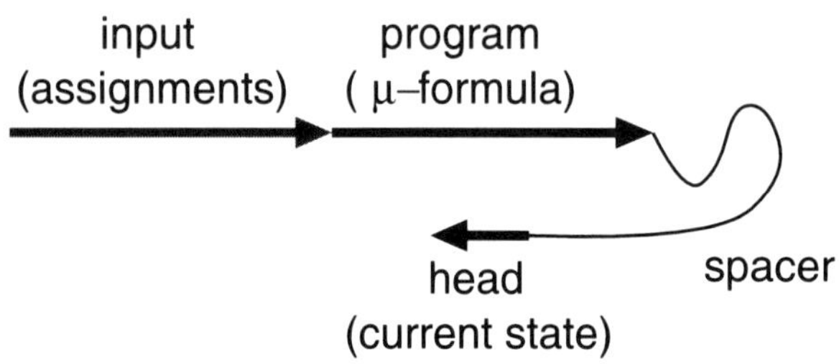

FIGURE 1. Our implementation

The contribution of this work is twofold: in theory and in experimental technique. In theory, we implement a successive state transitions of multiple μ-formulas in parallel as a single–tube reaction. This method is, to the authors' knowledge, the first molecular computation in which both programs and data are implemented and evaluated in parallel. The transitions are realized as a succession of thermal cycles, achieving parallel evaluation of Boolean formulas in a single tube.

As a new experimental technique, we also introduce a method to stop polymerization exactly after the head copies one symbol. For this purpose, we use a stopper sequence consisting of a deoxyribonucleotide whose complement is missing in the polymerization buffer. We call this technique *polymerization stop.*

This paper also reports the result of two lab experiments that have been done for showing the feasibility of the above method. One is that of a single transition and the other checked two successive transitions.

Utilizing the molecular operations for state transitions, we want to solve computationally hard combinatorial problems by generate and search. As is said before, we propose a method for learning Boolean μ–formulas from examples. Since we can randomly generate and then evaluate Boolean formulas in parallel, it is possible to select those formulas that behave as specified by examples.

The rest of this paper is organized as follows. Section 2 explains how μ–formulas are represented with molecules, and how they are generated and selected. Section 3 introduces experimental results for the basic operations required. The last section is for discussions, including that of implementing state machines with molecules in general.

2. Molecular representation

A μ–formula is inductively defined as follows.

- A Boolean variable is a μ–formula.
- If e is a μ–formula, then its negation $\neg e$ is also a μ–formula.
- If e_1 and e_2 are both μ–formulas, then so are their conjunction $e_1 \wedge e_2$ and disjunction $e_1 \vee e_2$, provided that e_1 and e_2 do not share a common variable.

In our method, the number of Boolean variables that can appear in μ–formulas is fixed, say n. They are denoted by n symbols, `var1`, `var2`, $\cdots$, `var`n. For short, we also write v_1, v_2, $\cdots$, v_n, respectively.

A Boolean vector whose length is the same as the number of Boolean variables is called an *input vector*, because it determines an assignment of true or false to each Boolean variable. The value of a μ–formula with respect to an input vector is defined as usual.

The symbols in Fig.2 are used to represent μ–formulas and input vectors. Each symbol will be encoded into a single–stranded DNA sequence.

Without loss of generality, the first variable in each μ–formula can be fixed. In the following, we assume that the first variable in a μ–formula is always `var1`, or v_1 for short.

2.1. Input vectors. Each input vector is represented by the sequence of symbols as in Fig.3. The 5'–end symbol denotes whether the input is positive or negative; i.e., the sequence begins with `input+` (`in+` hereafter) if a μ–formula should take the value true for this input, and it begins with `input-` (`in-`) if the value is expected to be false. To the 3' downstream of this symbol, the assignment of true or false for each of the n Boolean variables v_1, v_2, $\cdots$, v_n is listed. If true is assigned to the variable v_i, it is preceded by v_i+. If false, it is preceded by v_i-. An example, the positive input vector `10`, is shown in Fig.3.

2.2. μ–formula. Each μ–formula is represented by a sequence of $4 \times n$ symbols of the form in Fig.4. A μ–formula is translated into the above form in such a way that its subsequence is interpreted as follows:

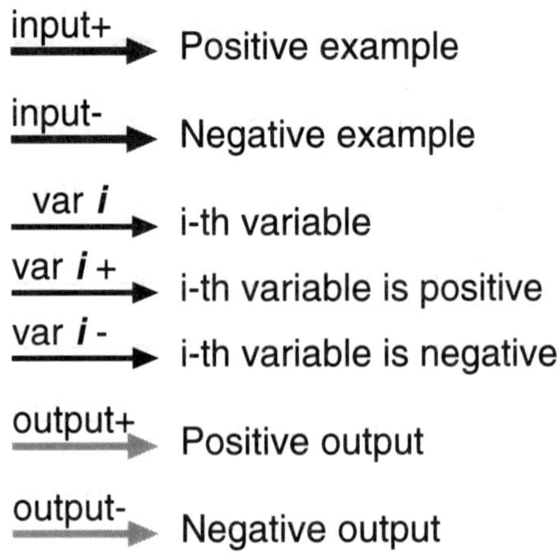

FIGURE 2. Symbols

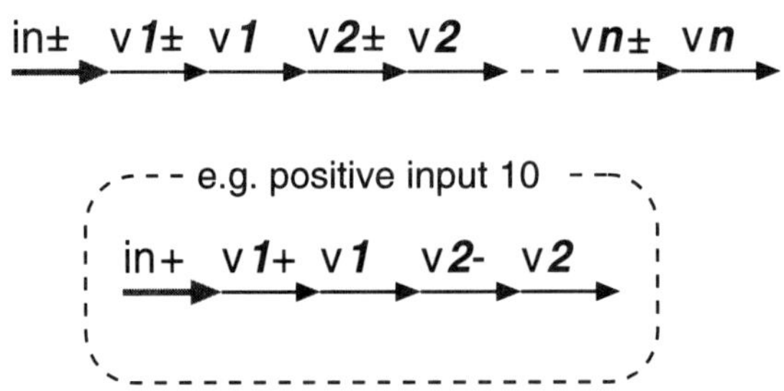

FIGURE 3. Input vectors

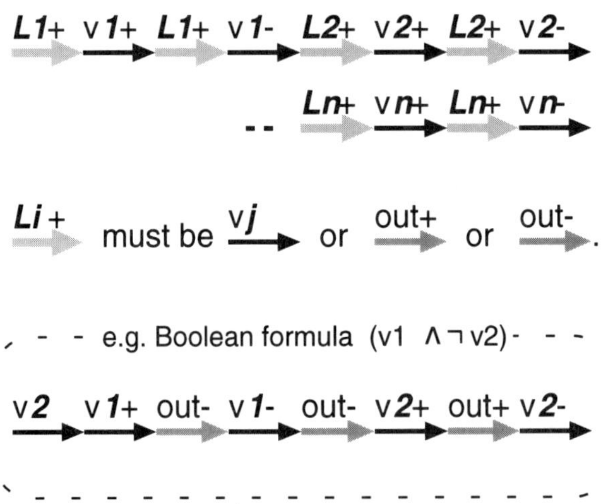

FIGURE 4. μ–formula

$$\mathtt{trans}(v_i, x, y) = \xrightarrow{x}\xrightarrow{v_i+}\xrightarrow{y}\xrightarrow{v_i-}$$
$$\mathtt{trans}(\neg e, x, y) = \mathtt{trans}(e, y, x)$$
$$\mathtt{trans}(e_1 \wedge e_2, x, y) = \mathtt{concatenate}(\mathtt{trans}(e_1, \mathtt{first}(e_2), y), \mathtt{trans}(e_2, x, y))$$
$$\mathtt{trans}(e_1 \vee e_2, x, y) = \mathtt{concatenate}(\mathtt{trans}(e_1, x, \mathtt{first}(e_2)), \mathtt{trans}(e_2, x, y))$$
$$\mathtt{first}(v_i) = v_i$$
$$\mathtt{first}(\neg e) = \mathtt{first}(e)$$
$$\mathtt{first}(e_1 \wedge e_2) = \mathtt{first}(e_1)$$
$$\mathtt{first}(e_1 \vee e_2) = \mathtt{first}(e_1)$$

TABLE 1. Translation of μ-formulas

$\xrightarrow{v_j}\xrightarrow{v_i+}$ :: If the value of the i-th variable is true, then jump to the place to check the j-th variable.

$\xrightarrow{v_j}\xrightarrow{v_i-}$ :: If the value of the i-th variable is false, then jump to the place to check the j-th variable.

$\xrightarrow{\mathtt{out+}}\xrightarrow{v_i+}$ :: If the value of the i-th variable is true, then terminate with the output `1` (true).

$\xrightarrow{\mathtt{out+}}\xrightarrow{v_i-}$ :: If the value of the i-th variable is false, then terminate with the output `1` (true).

$\xrightarrow{\mathtt{out-}}\xrightarrow{v_i\pm}$ :: Similarly, produce the output `0` (false).

The algorithm for translating μ–formulas to their representations is trivial. It is described by the function $\mathtt{trans}(e, x, y)$ in Table.1. One can obtain the representation of μ–formula e by computing $\mathtt{trans}(e, \mathtt{out+}, \mathtt{out-})$.

A μ–formula is equivalent to a decision tree in this translation. Note that a variable to be evaluated next can be uniquely determined in μ–formula. A general logical formula does not have this property.

2.3. Evaluation. Given an input vector and a μ–formula, its value can be evaluated using the concatenated sequence in Fig.5. The 3' end is a Watson–Crick

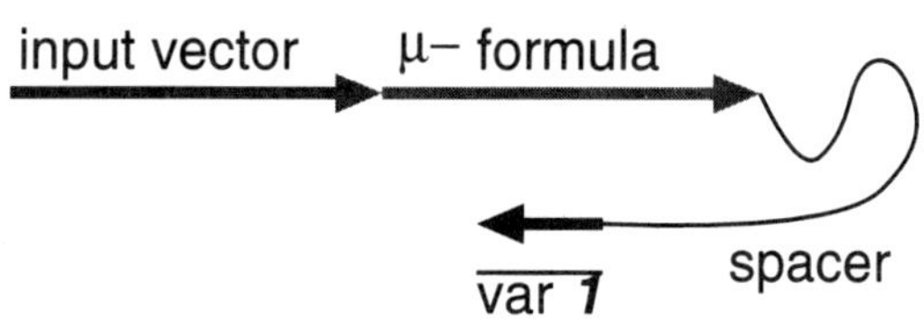

FIGURE 5. Evaluation

complement of the symbol v_1 bound by a spacer sequence to the input vector and the μ–formula. The entire sequence is assumed to be single–stranded.

The evaluation of the μ–formula consists of the repetition of the following three steps (Fig.6):

1. Anneal the 3' end symbol of the sequence to its complementary part of the same sequence.

2. Add exactly one symbol to the 3' end by extending it according to the result of annealing.
3. Denature the 3' end, and go to the first step.

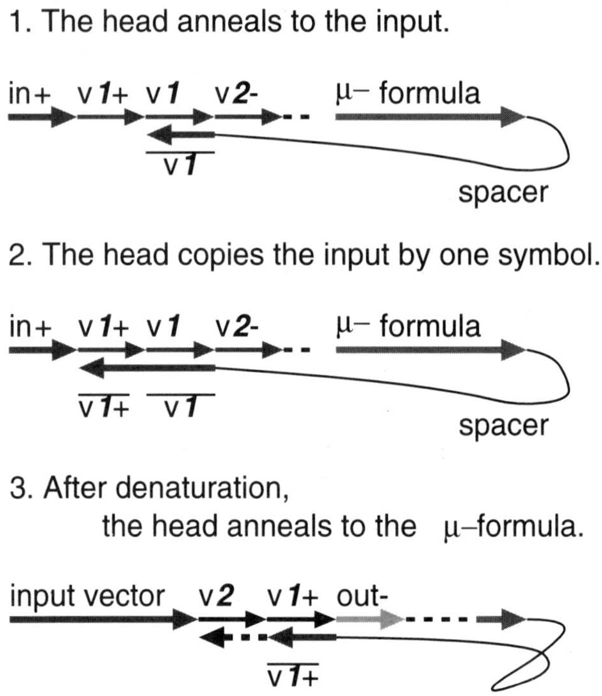

FIGURE 6. Evaluation steps

In order to stop the extension exactly after one symbol is added to the 3' end, both input vectors and μ–formulas contain a stopper sequence (denoted by `stop`) in between symbols. The actual implementation of the stopper sequence is described in Sec. 3.

2.4. Selection. After repeating the evaluation steps, the 3' end of a sequence eventually becomes `out+` or `out-`, meaning that the value of the μ–formula is true or false for the given input vector. Therefore, if the sign of `out`$\pm$ is identical to that of `in`$\pm$, the μ–formula has been evaluated as expected. This can be detected by PCR with `in`$\pm$ and `out`$\pm$ as primers.

Given a series of input vectors, it is possible to select only those μ–formulas that produce the expected result for all the input vectors from a pool of candidates in a test tube. By the process of evaluation and selection described in the last section, only valid μ–formulas are successively chosen for each input vector as follows:

1. Concatenate the copy of an input vector to the 5' end of each μ–formula in the tube. Also concatenate a spacer sequence and the symbol $\overline{v_1}$ to its 3' end.
2. Apply the process of evaluation and selection described in the last section.

3. Remove the copy of an input vector and the spacer with extended symbols from each μ–formula, so that the next input vector can be processed.

3. Experimental results

This section describes the experiments that have been done for showing the feasibility of the approach proposed in this paper. We have done two experiments: one is that of a single transition by extension of a hairpin structure, and the other is that of two successive transitions, where the DNA fragment produced in the first transition is denatured and the second transition is performed in the same tube.

In both experiments, symbols are assumed to be encoded with three out of the four common deoxyribonucleotides (dATP, dGTP, dCTP and dTTP). If a polymerization buffer contains only these three deoxyribonucleotides, a repetition of the missing deoxyribonucleotide works as a stopper sequence in polymerization. In the actual experiments, we used a triplet "ggg" or "aaa" as the stopper.

3.1. Single transition.

Sequence design. The triplet "ggg" is used as a stopper sequence in Table 2. All the sequences are single–stranded if not specially mentioned. A sequence with a bar indicates its Watson–Crick complement. For example, $\overline{\texttt{s1}}$ represents the complementary sequence of `s1`. A hyphen between sequences indicates their concatenation, i.e., phosphodiester bonding between the sequences. For example, `s1-s2` represents the concatenated sequence of `s1` and `s2` in this order.

The sequence `ini-s1-s2-s3-`$\overline{\texttt{s2}}$ is chemically synthesized in this experiment. Note that the subsequence `ini` contains the stopper `ggg`. Also synthesized are `ini-s1-s2-s3-`$\overline{\texttt{s2}}$`-`$\overline{\texttt{s1}}$ and `ini-s1-s2-s3-`$\overline{\texttt{s2}}$`-`$\overline{\texttt{s1}}$`-`$\overline{\texttt{ini}}$ as markers for electrophoresis.

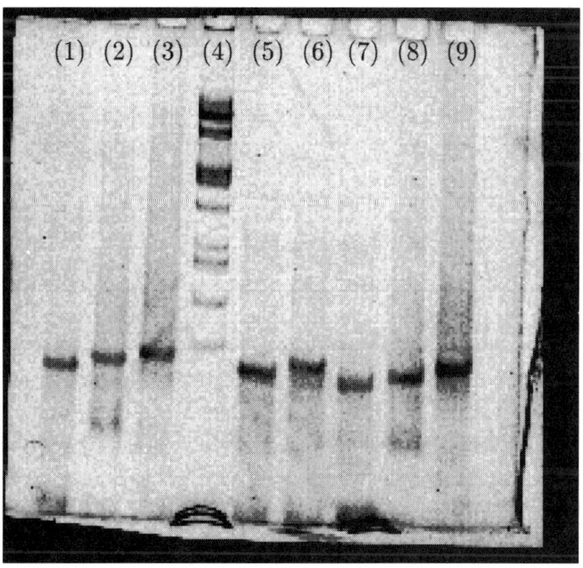

FIGURE 7. Electrophoresis without urea

s1	tccattctaa
s2	ctcacccttatacat
s3	cgtttcag
ini	cgacaagcttggg

TABLE 2. DNA sequences 1

Transition. The buffer in this experiment lacks dCTP. In 20 μl scale, the buffer contains 2 μg of DNA, 250 μM of dATP, dGTP and dTTP, and 2 units of AmpliTaq DNA polymerase (Perkin Elmer). Other conditions are the same as given by the supplier.

For comparison, we also prepared the buffer containing 250 μM of dCTP together with other three deoxyribonucleotides. In this case, it is expected that polymerization does not stop and produces ini-s1-s2-s3-$\overline{\text{s2}}$-$\overline{\text{s1}}$-$\overline{\text{ini}}$.

Transition was performed by initial denaturaturing at 90 ℃ for 3 min, followed by an extension step, 68 ℃ for 10 min, using Perkin Elmer DNA Thermal Cycler 480. Finally, the reaction mixture was cooled down on ice.

Electrophoresis. Samples were analyzed by electrophoresis on 12% polyacrylamide gels with or without 8 M urea. The band of the transition product was compared with those of the two markers: one is ini-s1-s2-s3-$\overline{\text{s2}}$-$\overline{\text{s1}}$ and the other is ini-s1-s2-s3-$\overline{\text{s2}}$-$\overline{\text{s1}}$-$\overline{\text{ini}}$.

The result of electrophoresis without urea is shown in Fig.7. From the left, the lanes correspond to:

(1) the initial DNA ini-s1-s2-s3-$\overline{\text{s2}}$,
(2) the marker ini-s1-s2-s3-$\overline{\text{s2}}$-$\overline{\text{s1}}$,
(3) the marker ini-s1-s2-s3-$\overline{\text{s2}}$-$\overline{\text{s1}}$-$\overline{\text{ini}}$,
(4) some known standard markers,
(5) the polymerization product with three deoxyribonucleotides,
(6) the polymerization product with four deoxyribonucleotides,
(7) the initial DNA ini-s1-s2-s3-$\overline{\text{s2}}$,
(8) the marker ini-s1-s2-s3-$\overline{\text{s2}}$-$\overline{\text{s1}}$, and
(9) the marker ini-s1-s2-s3-$\overline{\text{s2}}$-$\overline{\text{s1}}$-$\overline{\text{ini}}$.

In lanes (1), (2) and (3), DNA molecules were dissolved in water and renatured (cooled down on ice). Lanes (7), (8) and (9) contain the same molecules as in (1), (2) and (3), respectively, but they were dissolved in the polymerization buffer without AmpliTaq DNA polymerase; they were also renatured before electrophoresis.

The DNA molecules in these lanes each produced a single sharp band, indicating that they each formed a single stable structure, probably a hairpin structure. In contrast, the DNA molecules which were not renatured showed a smeared band as well as another band unmoved during the electrophoresis. These bands probably indicate the formation of various aggregates (data not shown).

In lane (5), the band of the transition product corresponds to that of the marker ini-s1-s2-s3-$\overline{\text{s2}}$-$\overline{\text{s1}}$ in (8), while there is no band corresponding to that of ini-s1-s2-s3-$\overline{\text{s2}}$-$\overline{\text{s1}}$-$\overline{\text{ini}}$ in (9). This means that hairpin structures were properly formed and their extension was stopped after $\overline{\text{s1}}$ was copied. On the other

s1	gcacgatctaggaaa
s2	tcccgttctgggcct
s3	gtggtttgcgctcgt
ini	ccaaa

TABLE 3. DNA sequences 2

hand, the extension with all four deoxyribonucleotides produced a band in lane (6), corresponding to the marker in lane (9).

Denaturing gel electrophoresis with 8 M urea exhibited a pattern similar in correspondence between the transition products and the markers to Fig.7 (data not shown).

3.2. Two successive transitions.

Sequence design. In this experiment, we used "aaa" as a stopper sequence.

The sequence used is `ini-s2-s1-s3-s2-`$\overline{\texttt{s1}}$ as defined in Table 3. The expected extension is shown in Fig.8.

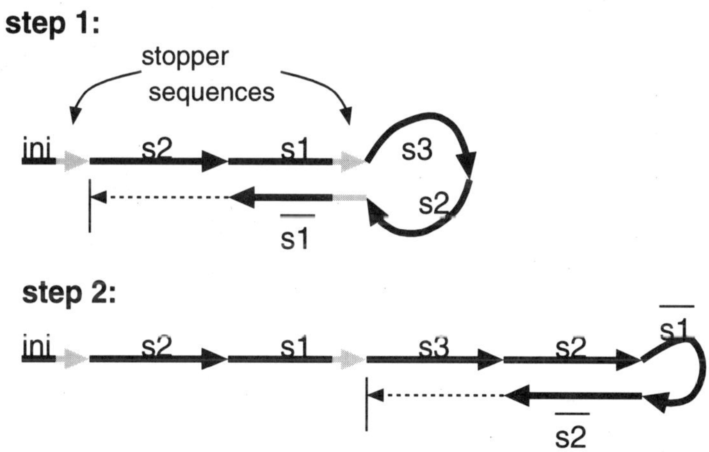

FIGURE 8. Expected extension

Transition. The buffer used in this experiment lacks dTTP. In 20 μl scale, the buffer contains 1 μg of DNA, 250 μM of dATP, dGTP and dCTP, and 2 units of AmpliTaq DNA polymerase (Perkin Elmer). Other conditions are the same as given by the supplier.

We programmed Perkin Elmer DNA Thermal Cycler 480 as follows: initial denaturation step, 90 ℃ for 1 min; extension step, 68 ℃ for 1 min; four cycles with denaturation step at 90 ℃ for 1 min, cooling down on ice for 1 min, incubation at 40 ℃ for 30 sec, and extension at 68 ℃ for 30 sec. The cooling step is for forming hairpin structures before extension.[1]

[1]When cooled down, tubes were temporarily detatched from the thermal cycler and put on ice.

Primer 1	gctcgtcgac gcacgatctaggaaa gtgg
Primer 2	gacggaattc aaaaaaaaaaaaaaaaaaa gtgg

TABLE 4. Primers for cloning

Electrophoresis. Samples were analyzed by electrophoresis on 12% polyacrylamide gels containing 8 M urea. The bands of the transition products were compared with those of the markers, which had been chemically synthesized.

The product of the first extension step exibited two bands in electrophoresis. One is that of ini-s2-s1-s2-s3-$\overline{\texttt{s2}}$. The other is considered that of ini-s2-s1-s2-s3-$\overline{\texttt{s2}}$-$\overline{\texttt{s1}}$.

After the four cycles of denaturation, annealing, and cooling, the above two bands disappeared and the band corresponding to that of the marker ini-s2-s1-s3-s2-$\overline{\texttt{s1}}$-$\overline{\texttt{s2}}$-$\overline{\texttt{s3}}$ emerged.

Sequencing of Transition Product. The DNA molecules that have undergone the successive transitions was processed by terminal deoxynucleotidyl transferase, and thus a poly-T sequence was added to the tail of each DNA molecule. It was then amplified by PCR with the primers of the sequences shown in Table 4. Primer 1 includes s1 and Primer 2 has a poly-A sequence. The product of PCR was digested with EcoRI and SalI, and was ligated to the corresponding sites in vector pUC119. The sequences of the clones were determined, confirming that the successive transitions certainly occurred as we expected.

4. Discussions

In this section, we first discuss some crucial points for the implementation of state transitions. Then we introduce how to implement a finite automaton using our method, and compare its effectiveness with yet another implementation of state transition using PCR.

4.1. Experimental techniques. As crucial points, we consider another implementation of polymerization stop, and some conditions for more efficient state transitions.

Polymerization stop. In the previous section, we described the technique of polymerization stop using triplets as stopper sequences. This technique, however, requires encoding using only three deoxyribonucleotides.

We plan to use artificial nucleosides such as iso-guanosine or iso-cytosine [**8**] for stopper sequences. As a result, all four common deoxyribonucleotides can be used for encoding. This allows more information to be encoded in sequences of the same length. Another advantage is that it is easier to adjust the gc-contents at an appropriate level.

Intra-molecular reaction. In our approach, since a program and its input reside on a single DNA strand, parallel execution of multiple programs requires that intra-molecular reaction occur exclusively.

In the above experiments, we cooled down single-stranded DNA molecules on ice to facilitate intra–molecular reaction, i.e., to prevent two or more molecules from interacting with one another.

Facilitating intra–molecular reaction with a sufficient certainty is an important issue to be settled down. There is a report of the stable formation of a DNA hairpin *in vitro* [**3**], and the adjustment in experimental conditions will greatly improve the efficiency and the accuracy of extension. So far, we propose the following two simple methods:

- Keep the concentration of DNA molecules sufficiently low.
- Quickly cool down DNA molecules during annealing.

The effectiveness of these methods should be verified in further experiments.

We also plan to utilize surface chemistry [**2, 4**]. By binding DNA molecules sparsely on a surface, it is possible to avoid inter–molecular interaction, thus facilitating intra–molecular reaction.

Successive hairpin formation. To perform two successive transitions, as in our last experiment, thermal cycles are repeated so that a hairpin structure once created and extended is denatured, and a new hairpin is formed for the second transition. However, it is highly probable that the first structure is formed again and no successive transition occurs (Fig.9).

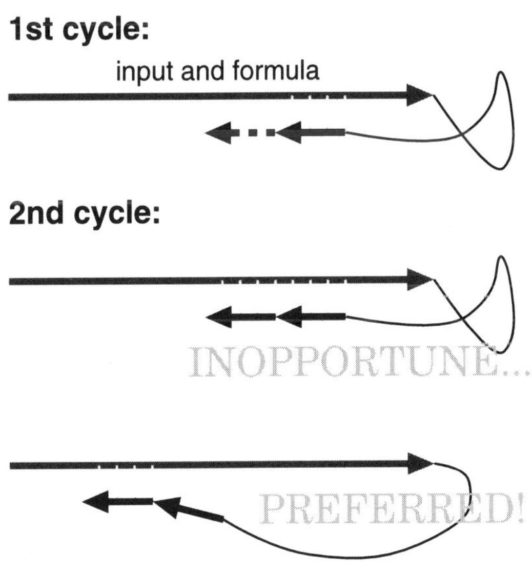

FIGURE 9. Forming the first structure again

This problem does not occur if the probability of the second structure is reasonably large compared with (not necessarily larger than) the first one, because by repeating thermal cycles, the second structure will be eventually formed. Note that the formation of the first structure produces no harmful results, because further extension from this structure is prohibited by the stopper sequence.

The stability of a hairpin structure is affected by the length of its loop. Sequences used in our experiment were designed so that the second structure has higher stability. More experiments and theoretical analyses are required to settle down this issue.

4.2. Finite automata. The method introduced in this paper is essentially that for simulating deterministic or non–deterministic state transitions. Therefore, it can be applied to implement state machines in general, and we have already considered how to implement finite automata with molecules.

The transition table of a finite (deterministic or non–deterministic) automaton is a sequence of triples of the form s'–c–s, where s denotes the current state and s' the next state corresponds to the input symbol c. For example, the table of the automaton M in Fig.10 is represented by the sequence `P-0-S-Q-1-S-Q-1-P-R-0-P`.

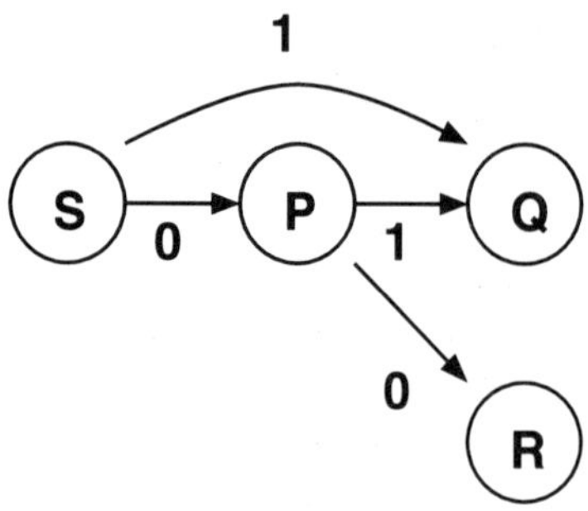

FIGURE 10. Automaton M

To realize state transitions, we concatenate the complement $\overline{s}$ of the current state s of the automaton as in Fig.11, where a spacer sequence with an appropriate length is inserted in between the table and the state.

The state transition corresponding to the input symbol c is performed as follows. First, $\overline{c}$ is attached to the end of the sequence using a method for concatenating DNA molecules as in [**6**]. The end consists of the subsequence $\overline{s}$–$\overline{c}$, which can hybridize to the triple s'–c–s in the transition table in the hairpin structure. By polymerization, the next state $\overline{s'}$ is copied to the end.

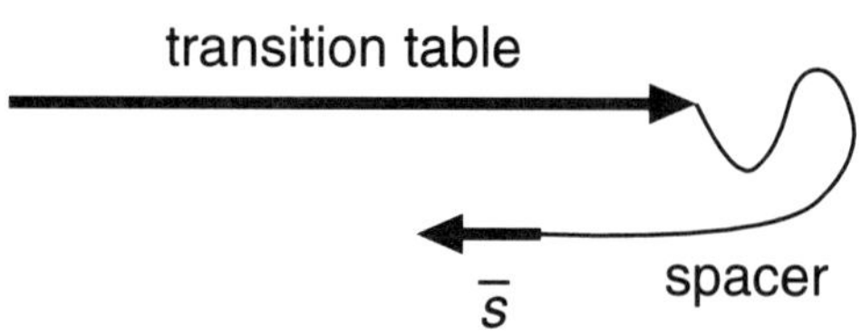

FIGURE 11. Current state of an automaton

To stop polymeration right after $\overline{s'}$ is copied, we again use the technique of polymerization stop.

By the method described above, it is possible to run finite automata in parallel with respect to an input sequence of symbols.

4.3. Yet another representation. State transitions can be realized still in other ways. We introduce another method in which one PCR step corresponds to one transition step, and compare its effetiveness with our original method.

We briefly explain the new method using the example automaton M in Fig.10.

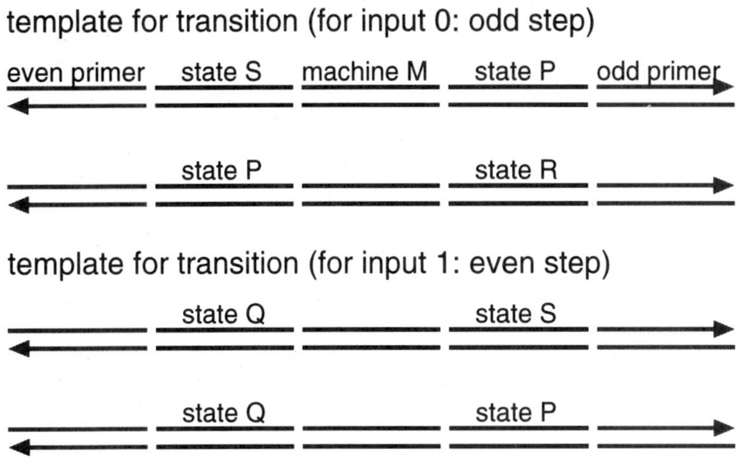

FIGURE 12. Another representation

As in Fig.12, each transition is represented by a double–stranded sequence of symbols used later as a PCR template. A different set of templates is prepared for each input (in this case `0` or `1`), i.e., the possible transitions for the input in M is represented by a set of PCR templates. The current state is represented by a primer so that only templates containing the next state is amplified in PCR. The primer is changed by turns (`odd` or `even`) to proceed the transition. A sample transition of the automaton M for the input `01` is shown in Fig.13. The corresponding molecular operation is shown in Fig.14.

This method can be easily parallelized by changing the name of an automaton in the middle of templates. The amount of each PCR template required for an amplification is very small, so many automata can be simulated in parallel.

This method (*PCR transition* for short) uses PCR, a standard operation when compared to our single–tube polymerization. Our method, however, is considered better for several reasons.

Random generation of data. In general, synthesis of molecular data is a costly process. In a generate and search approach, it is important to randomly generate all and only valid data. DNA for the single–tube polymerization can be randomly generated by a stepwise extension of symbols as in [**6**], but DNA for the PCR transition is hard to randomly generate. The reason is that PCR templates representing the same automaton must have the same machine name at their middle part. With this restriction, it is almost impossible to randomly generate all the possible automaton of the given number of nodes.

Time in experiment. PCR or gel electrophoresis requires more than an hour to perform, while enzymatic modification (e.g. ligation or restriction) requires much shorter time. (e.g. 10 min. The time depends on temperature and concentration

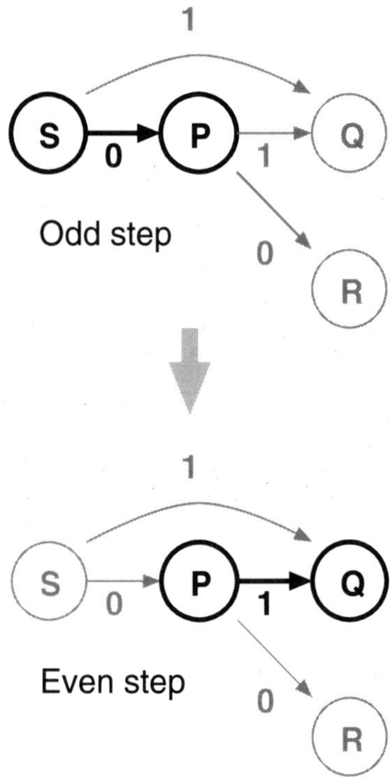

FIGURE 13. State transition

of enzymes.) The single–tube polymerization can repeat the transition steps using only these quicker operations without changing buffers, a great advantage which suggests the automation of the repetitive process like PCR in the future.

Amount of molecules. To preserve the intermediate variety of solutions, each solution should not be stored in a large population. PCR is certainly a reliable operation for amplification, but at the same time, it runs the risk of extermination of solutions of a small population. As long as the total population does not become too small, PCR should be avoided in a generate and search approach.

Acknowledgments

The second and the third authors are supported by Research Fellowships of the Japan Society for the Promotion of Science for Young Scientists. This work is also supported by the Japan Society for the Promotion of Science "Research for the Future" Program (JSPS-RFTF 96I00101).

References

[1] L. Adleman, *Molecular computation of solutions to combinatorial problems*, Science, **266** (1994), pp.1021–1024.

[2] M. Arita, M. Hagiya and A Suyama, *Joining and rotating data with molecules*, Proceedings of 1997 IEEE International Conference on Evolutionary Computation (ICEC'97), 1997, pp.243-248.

[3] R.R. Breaker and G.F. Joyce, *Emergence of a replicating species from an* in vitro *RNA evolution reaction*, Proc. Natl. Acad. Sci. USA, **91** (1994), pp.6093–6097.

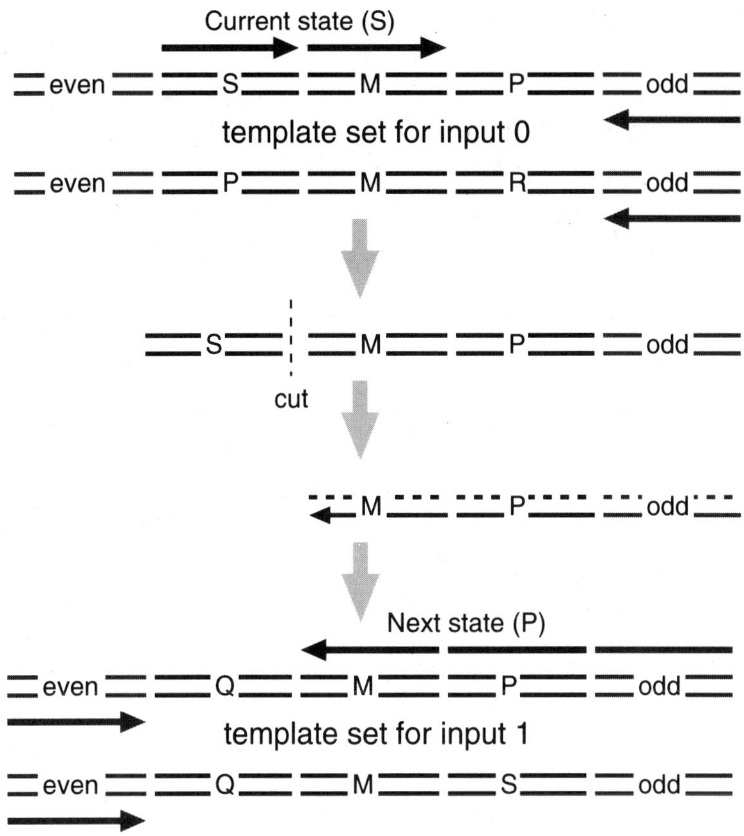

FIGURE 14. DNA for state transition

[4] W. Cai, A.E. Condon, and R.M. Corn *et al.*, *The power of surface-based DNA computation*, Proc. RECOMB'97, Assoc. Comput. Mach., 1997, pp.67–74.

[5] R. Lipton, *DNA solution of hard computational problems*, Science, **268** (1995), pp.542–544.

[6] N. Morimoto, M. Arita, and A. Suyama, *Stepwise generation of hamiltonian path with molecules*, Proc. Biocomputing and Emergent Computation (BCEC'97), World Scientific, 1997, pp.184–192.

[7] M. Ogihara and A. Ray, *Simulating boolean circuits on a DNA computer*, Technical Report **631**, University of Rochester, 1996.

[8] J.A. Piccirilli *et al.*, *Enzymatic incorporation of a new base pair into DNA and RNA extends the genetic alphabet*, Nature, **343** (1990), pp.33–37.

[9] L. Pitt and G.L. Valiant, *Computational limitations on learning from examples*, J. Assoc. Comput. Mach., vol.**35**, no.**4** (1988), pp.965–984.

[10] E. Roweis, S. Winfree and B. Richard *et al.*, *A sticker based architecture for DNA computation*, Proc. 2nd meeting on DNA based computers, 1996, pp.1–27.

[11] N.C. Seeman, N. Wang and B. Liu *et al.*, *The perils of polynucleotides: the experimental gap between the design and assembly of unusual DNA structures*, Proc. 2nd meeting on DNA based computers, 1996, pp.191–205.

Department of Information Science, Graduate School of Science, University of Tokyo, Bunkyo-ku Hongo 7-3-1, 113 Tokyo, Japan
E-mail address: hagiya@is.s.u-tokyo.ac.jp
URL: http://nicosia.is.s.u-tokyo.ac.jp/members/hagiya.html

Department of Information Science, Graduate School of Science, University of Tokyo, Bunkyo-ku Hongo 7-3-1, 113 Tokyo, Japan
E-mail address: ari@ims.u-tokyo.ac.jp

Department of Biophysics and Biochemistry, Graduate School of Science, University of Tokyo, Bunkyo-ku Hongo 7-3-1, 113 Tokyo, Japan
E-mail address: kiga@y-sun.biochem.s.u-tokyo.ac.jp

Department of Biophysics and Biochemistry, Graduate School of Science, University of Tokyo, Bunkyo-ku Hongo 7-3-1, 113 Tokyo, Japan
E-mail address: sakamoto@y-sun.biochem.s.u-tokyo.ac.jp

Department of Biophysics and Biochemistry, Graduate School of Science, University of Tokyo, Bunkyo-ku Hongo 7-3-1, 113 Tokyo, Japan
E-mail address: yokoyama@y-sun.biochem.s.u-tokyo.ac.jp

DIMACS Series in Discrete Mathematics
and Theoretical Computer Science
Volume **48**, 1999

Wet Splicing Systems

Elizabeth Laun and Kalluru J. Reddy

This work is dedicated to Tom Head.

ABSTRACT. Splicing systems were originally developed as a mathematical or *dry* model of the generative capacity of DNA molecules in the presence of appropriate enzymes. The accuracy with which the model predicts the behavior of the corresponding biological or *wet* splicing system is investigated. A simple example of a wet splicing system is shown to generate *in vitro* the splicing language predicted by the corresponding dry splicing system.

1. Introduction

In 1987 Tom Head introduced a mathematical model of the generative capacity of linear DNA molecules in the presence of restriction enzymes and a ligase, which he called *splicing systems*. According to the original model, a splicing system consists of a finite initial set of strings over an alphabet, and a finite set of rules by which the strings can be spliced together to form new strings in addition to the initial strings. The closure of the initial set under the splicing operation generates a *splicing language*. In his original work Head posed the problem of characterizing the splicing languages. In the same work, Head showed the equivalence of the class of strictly locally testable languages to the class of languages which can be generated by splicing using a restricted, uniform set of rules.

Since renamed H-systems, splicing systems have been extensively studied [**7, 8, 12, 13, 21, 20, 22, 24, 26**]. The generalization of H-systems to include systems having infinite sets of initial strings and rules has proved fruitful with regard to the characterization problem. In particular, G. Paun demonstrated in 1995 that extended H-systems can generate arbitrary recursively enumerable languages [**21**]. Other interesting and important closure results have been formulated, [**7, 22**] as have results pertaining to splicing with circular strings in addition to linear strings, as discussed by D. Pixton in [**23**].

1991 *Mathematics Subject Classification.* Primary 68Q99, 92B99; Secondary 00A99.
Research supported by the NSF under grant number CCR9509831.

The scientific community has recently taken great interest in biomolecular models of computation. In particular, Leonard Adleman's seminal 1994 work [**3**] inspired a surge of research focused on exploring the possibility of using DNA or other biomolecules to solve mathematical problems which are computationally hard [**5, 18**]. In light of such developments, it seemed worthwhile to demonstrate the viability of Head's model using experimental methods in the laboratory. It was with this aim that an investigation of the feasibility of performing the splicing operation as originally defined was undertaken.

Results of our study indicate that the splicing operation can be easily performed in a single-step laboratory procedure. In this procedure, restriction digestion and religation of linear double-stranded DNA (dsDNA) take place simultaneously in a single buffer, producing the new strings which are predicted by the model. While the significance of this finding is yet to be realized completely, the importance of experimentally verifying the predictive capability of the H-system model of splicing is clear.

We would like to thank T. Head for suggesting this problem. We would also like to thank Zhen Tian for her invaluable technical assistance in the laboratory.

2. Definitions and Examples

Let A be a finite set of alphabet symbols. The free monoid over A is denoted by A*, and the identity element is the empty string, denoted by 1. The free semigroup over A is denoted A^+, as usual.

A splicing system S = (A, I, B, C) consists of the finite alphabet A, a finite set I of initial strings in A*, and finite sets B and C of triples (c, x, d) with c, x, and d in A*. Each such triple is called a *pattern*. For the triple (c, x, d), the string cxd is called a *site*, and the string x is called a *crossing*. For the purposes of this discussion it will be helpful to consider the initial set to be linear strings of dsDNA over an alphabet consisting of the nucleotide pairs formed by hydrogen bonding between the bases adenine, guanine, cytosine, and thymine. These nucleotide pairs can be represented by the set {[A/T], [T/A], [C/G], [G/C]}, which we shorten to {a, t, c, g}. Single stranded DNA are denoted by the capitalized alphabet {A, T, C, G}. It is also helpful to think of a site as the recognition site of a restriction enzyme, and the crossing as the single-stranded overhang which is formed when the enzyme cleaves dsDNA containing the site. Patterns in B are called left-handed, and correspond to recognition sites of restriction enzymes which produce either 5' or blunt end overhangs. Patterns in C are called right-handed, and correspond to the sites of enzymes which produce 3' overhangs.

Strings are spliced in L according to the following algorithm: If there are strings ucxdv and pexfq in L, and patterns (c, x, d) and (e, x, f) of the same hand, then the strings ucxfq and pexdv are formed by splicing ucxdv and pexfq together. Again, it is helpful to consider patterns of the same hand which contain the same crossing as corresponding to restriction enzyme sites which produce the same overhangs, i.e. overhangs which allow subsequent religation to occur.

The language L = L(S) generated by S consists of the strings in I and all strings in the closure of I under the operation of splicing, i.e. strings which can be generated by splicing in L are adjoined to L. Strings in L(S) which cannot be used for splicing are called the *adult* strings in L(S).

A simple example of a splicing language is the language L=(a)*. This infinite language can be easily seen to be generated by the system S = (A, I, B, C) where A = {a}, I is the initial set { 1, aa }, B = {(1, a, 1) } and C = $\emptyset$. Notice that there are infinitely many splicing systems which can generate (a)*. It is also worth noting that not every regular language is a splicing language. For example, as noted by Gatterdam in [**11**], the language L= (aa)* cannot be generated by any splicing system having a finite initial set and finite set of rules. The language a*ta*ta* is another such example.

We would like to emphasize again that, although the generalizations of the concept of a splicing system by Paun, Pixton, Rozenberg, Salomaa and others permit arbitrary languages as initial sets and arbitrary rule sets, we shall confine ourselves to the original definition of an H-system in which both the set of rules and initial strings are finite. Furthermore, we use the original notation as presented by Head in 1987, rather than the notation almost always used in the literature devoted to the mathematical generaliztion of the splicing concept. See [**13**] for a more complete explanation of the development of the splicing notation.

2.1. The Primary Example. A simple biological example of a splicing system is S = (A, I, B, C) where A is the set {a, g, c, t} introduced above, and I = {αgccgcaccggcβ, γcaccacgtgδ }, with α, β, γ, and δ in {a,c,t,g}* = A*. These initial strings are sequences which appear in the genome of the bacteriophage lambda. The strings are assumed to be dephosphorylated on their 5' ends in order to prevent blunt end ligation. Set B is empty in this example. The substrings gccgcaccggc and caccacgtg appearing in the initial strings are recognition sites for the restriction enzymes *Bgl*I and *Dra*III , respectively. The patterns (gccg, cac, cggc) and (cac, cac, gtg) encode specific actions of these enzymes in the presence of T4 DNA ligase, and together they are the elements in set C. Notice that there are other sites which can be encoded for both of these enzymes. For the purposes of this example, however, only the two patterns above are required. Both restriction sites, when cut, leave 3' CAC single-stranded overhangs which allow religation. There are no other recognition sites for these enzymes in either initial string, so the splicing language L(S) for this example is the set I $\cup$ {αgccgcacgtgδ, γcaccaccggcβ}. We are motivated to verify experimentally that this language is in fact generated *in vitro* as predicted by the splicing model.

3. The Experimental Design

The motivation for this experiment was to produce laboratory verification of Head's theoretical model of splicing systems. The experiment was designed to test the hypothesis that a particular splicing system will converge to a fixed set of strings. The initial set was taken to be two distinct sequences of linear dsDNA, each with dephosphorylated 5' ends, one having a *Bgl*I restriction site and the other having a *Dra*III restriction site. The action of iterated cleavage and religation was predicted to result in a dynamical splicing system which would converge to a particular set of adult strings. The experiment was designed so that this progression would be clearly apparent if the molecular (also termed *wet*) splicing system behaved as predicted by the mathematical (or *dry*) model.

There were two major considerations in the experimental design. First, appropriate molecules were chosen such that the predicted outcome could be easily verified by standard laboratory techniques. Second, optimal conditions for simultaneous restriction digestion and ligation of dsDNA using the chosen enzymes had to be determined by preliminary experimental investigation.

The two substrings of phage lambda DNA which constitute the initial set in the primary example described above were chosen for the following properties: String 1, approximately 1.6 kbp long, was chosen because it contains exactly one *Bgl*I site, about 1.45 kbp from one end. String 2, which we shall refer to as the 2.1 kbp string, contains exactly one *Dra*III site, approximately 1.3 kbp from one end. The 2.1 kbp string contains no *Bgl*I restriction site and the 1.6 kbp string contains no *Dra*III site.

The first string represents the 1,597 bp fragment of phage lambda DNA between the genome locations 16,196 and 17,793, and the second string represents the 2,137 bp fragment of the phage lambda DNA between locations 40,754 and 42,891. The substrings gccgcaccggc and caccacgtg appear approximately at locations 17,638 and 41,579 in the genome, respectively [**25**]. Both sites, when cut by the appropriate enzyme, leave 3' overhangs of the sequence CAC which allow subsequent ligation of fragments to occur. Further, if two fragments originating from the two different strings are ligated together, neither restriction site is present in the product string, and no subsequent cleavage of these molecules is possible, i.e. these are the adult strings in the splicing language. Two such adult strings can be formed in this system. No blunt end ligation could occur at all, since the initial molecules were dephosphorylated on their 5' ends. In theory therefore, repeated cleavage and ligation steps should reduce the quantity of the original 2.1 and 1.6 kbp molecules, and increase the quantity of the adult molecules, which are approximately 0.98 and 2.7 kbp in length. Such strands would be readily detectable and easily differentiated by agarose gel electrophoresis and subsequent UV analysis [**4**]. The initial molecules and the adult product molecules expected after cleavage and religation are shown in Figure 1.

It was predicted that an analysis of the dynamical bahavior of this splicing system in its wet form would show a time-related decrease in the initial strings, and a corresponding increase in the adult strings. Intermediate fragments having 3' CAC overhangs were expected as well.

The simultaneous restriction digestion and ligation reaction was performed using *Bgl*I, *Dra*III and T4 DNA ligase in the ligase buffer. The addition of ATP and BSA to enhance the action of T4 DNA ligase and *Dra*III was necessary. Optimization of the relative quantities of the different reagents was explored in a series of preliminary experiments the results of which are omitted here.

4. Laboratory Procedure

The initial two kinds of dsDNA molecules were generated by PCR amplification using phage lambda genomic DNA as the template. The primers were twenty base pair exact complements of the ends of the chosen phage lambda DNA substrings of lengths 1.6 and 2.1 kbp, and they were dephosphorylated on their 5' ends to prevent blunt end ligation of the PCR product molecules during the later reaction phase of the experiment. The primers were purchased from Gibco BRL (Gaithersburg, MD). The PCR reactions were run as 100 μl aliquots, each containing 0.8 mMdNTPs (0.2 mM each dATP, dCTP, dGTP, dTTP) from Pharmicia, 1.5 mM $MgCl_2$, 1 μM each

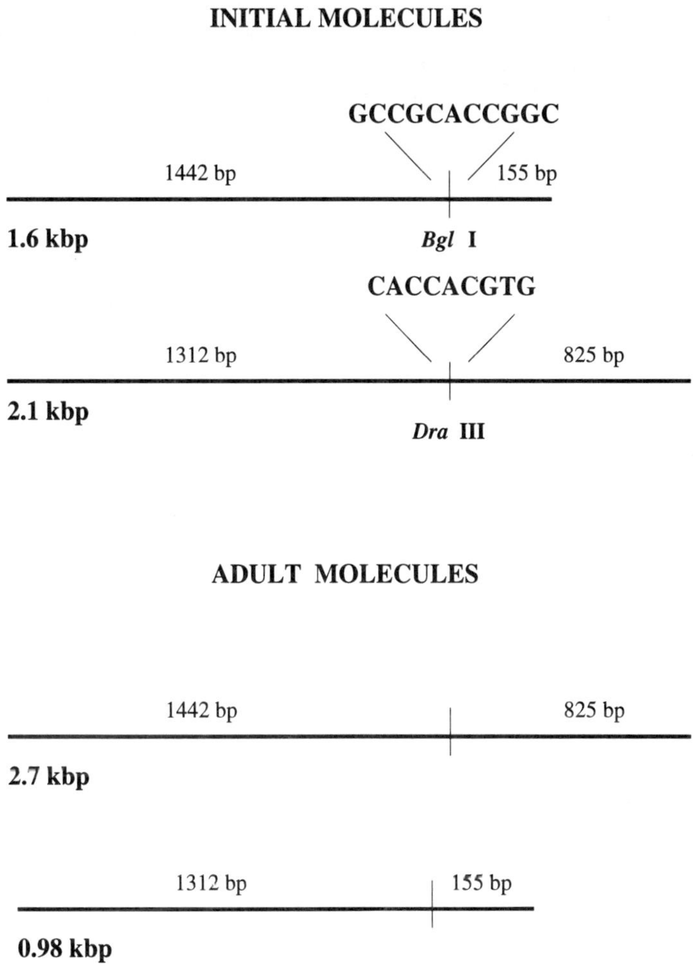

FIGURE 1. The initial molecules and the adult molecules which are the expected products in the primary example.

forward and reverse primer, 50 ng target lambda DNA from New England Biolabs (Beverly, MA), 0.5 ul Taq polymerase (5 U/ul) from Gibco BRL in 1xPCR buffer. The reactions were cycled 40 times at 94°C for 1 minute, 55°C for 1 minute and 72°C for two minutes, ending with 5 minutes at 72°. The Peltier Thermal Cycler (MJ Research, Inc., Watertown, MA) was used to perform the PCR. The PCR products were purified from 0.7% agarose gel using the QIAEX Gel Extraction Kit by Qiagen. Each QIAquick purification column was used to purify two 100 μl PCR reactions. The final purified product from each column was eluted in 50 μl elution buffer (see Appendix). The purified PCR products were quantitated using a UV-visible spectrophotometer. The concentration of the 1.6 kbp DNA was approximately 36 ng/μl in purified form, and the concentration of the 2.1 kbp DNA was approximately 34 ng/μl. A volumetric ratio (1.6 kbp purified PCR DNA/2.1 kbp purified PCR DNA) of 0.7196 was calculated to yield approximately equal molarity of the two types of initial molecules. In preliminary experiments, however, a volumetric ratio of 0.625 of the two PCR products yielded good results.

A preliminary run of the simultaneous restriction digestion and ligation experiment was performed, and the results of this reaction are shown in Figure 2. The reaction contained 15 μl purified 1.6 kbp DNA, 24 μl purified 2.1 kbp DNA, 4 μl *Bgl*I restriction enzyme (8U/μl), 10 μl *Dra*III restriction enzyme (3U/μl), 10 μl T4 DNA ligase, 0.5 μl Bovine Serum Albumin (10$\mu g/\mu l$), 20 μl 10mM rATP, 12 μl 10x T4 Ligase Buffer and 24.5 μl sterile water. The enzymes were obtained from Strategene, and the BSA came from NEBiolabs. This reaction was incubated at 22°C overnight. The reaction product was then purified according to the purification protocol in the Appendix. Approximately half (11 μl) of the purified reaction product was then run on a 1.0% agarose gel stained with ethidium bromide. See Figure 2.

A time sequence study of the primary simultaneous restriction digestion and ligation reaction, called Reaction T, was then performed. The results of the experiment appear in Figure 3. Reaction T was a 600 μl solution containing 105 μl of purified 1.6 kbp DNA, 168 μl of purified 2.1 kbp DNA, 28 μl *Bgl*I restriction enzyme (8U/μl), 70 μl *Dra*III restriction enzyme (3U/μl), 45 μl T4 DNA ligase (all enzymes from Strategene), 2.5 μl Bovine Serum Albumin (10$\mu g/\mu l$)from NEBiolabs, 80 μl 10mM rATP, 60 μl 10x T4 DNA ligase buffer and 41.5 μl sterile water. The reaction was initially incubated at room temperature (22°C). Aliquots of 145 μl were removed from the reaction mixture after 5, 15, and 60 minutes, respectively. The first three samples were immediately stored at −70°C. The remaining 145 μl sample was incubated approximately 20 additional hours at 16°C. All the samples were then heated to 65°C for 15 minutes and purified according to the purification protocol in the Appendix. Slightly more than half (about 14 μl) of each purified sample was then run on a 1.0% agarose gel stained with ethidium bromide. The gel photograph of the time sequence reaction appears in Figure 3.

All photographs were taken using Polaroid 55 Professional film and a Fotodyne Polaroid MP-4 Land Camera apparatus.

5. Results and Discussion

The restriction enzymes chosen apparently work with good efficiency at the concentrations used and in the T4 DNA ligase buffer. The *Bgl*I and *Dra*III began to cleave the initial strands of DNA in T4 DNA ligase buffer within 5 minutes after Reaction T was begun, as seen in Lane 3 of the gel in Figure 3. Digestion of the 2.1 and 1.6 kbp fragments apparently continued throughout the reaction, although the majority of both initial fragments were cleaved within an hour at room temperature, as apparent from Lane 5. The resulting DNA pieces with overhangs appeared within 5 minutes, as seen in Lane 3. The ligase worked well at 22°C, as expected, and the fragments with overhangs began to religate immediately. The 5 minute aliquot in Lane 3 of the gel appearing in Figure 3 showed some apparent 2.7 and 0.98 kbp adult products of religation. Within 60 minutes the two adult DNA strands of lengths 2.7 and 0.98 kbp were clearly visible (Lane 5). The complete

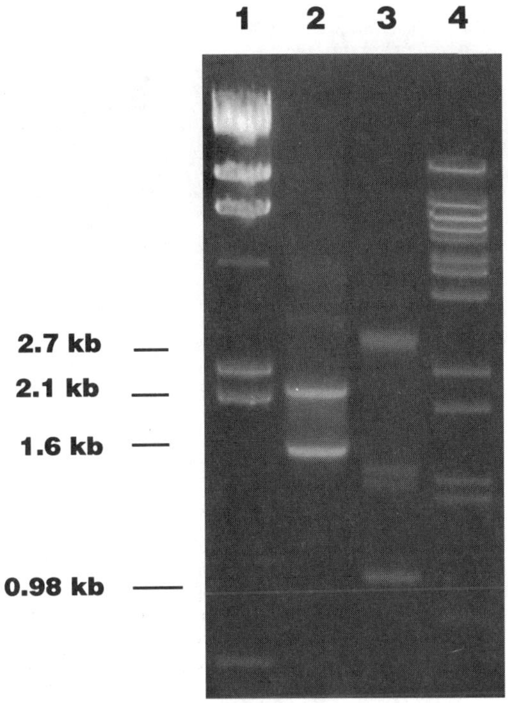

FIGURE 2. Preliminary Reaction: Simultaneous restriction digestion and ligation of 1.6 kbp and 2.1 kbp dsDNA fragments. Lanes 1 and 4 are the standard molecular weight markers (phage lambda DNA cut with *Hind*III and *BstE*II, respectively). Lane 2 contains the initial dsDNA strings of lengths 2.1 and 1.6 kbp. Lane 3 contains the products after overnight incubation at room temperature. Note that the expected adult strings of lengths 2.7 and .98 kbp appear. The faint bands appearing in Lane 3 are unreligated fragments from restriction digestion of the initial strings.

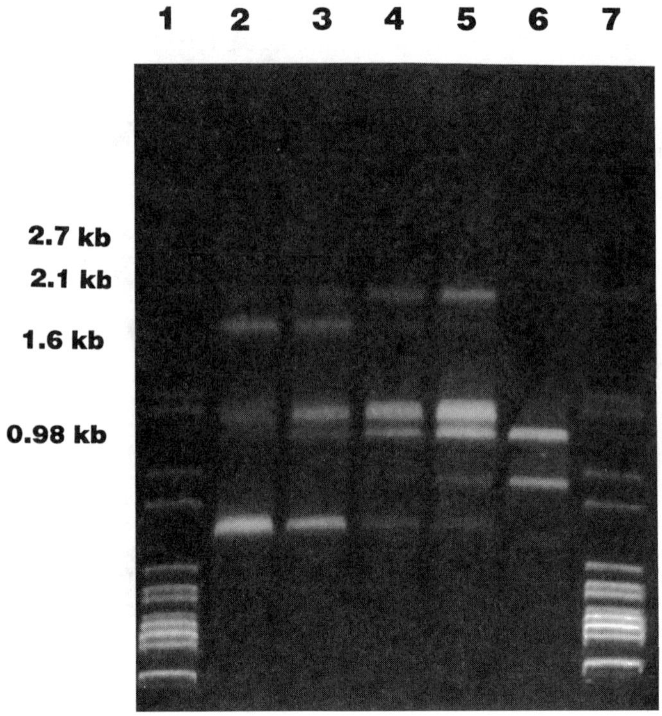

FIGURE 3. Reaction T: Time sequence study of simultaneous restriction digestion and ligation of 1.6 kbp and 2.1 kbp dsDNA fragments. Lanes 1 and 7 are standard molecular weight markers (phage lambda DNA cut with *Hind*III and *BstE*II, respectively). Lane 2 contains the initial dsDNA strings of lengths 2.1 and 1.6 kbp. Lanes 3, 4 and 5 contain the products after incubation at room temperature for 5, 15 and 60 minutes, respectively. Lane 6 contains the product strings after overnight incubation at 16°C. Note that the bands containing the expected adult strings of lengths 2.7 and .98 kbp appear increasingly bright as time passes, while the bands of both the original dsDNA and the unreligated fragments grow fainter with time.

time sequence indicates a progressive decrease in the quantities of the initial 2.1 and 1.6 kbp strings, and a progressive increase in the quantities of the adult strings, as expected.

Repetitions of this experiment have produced results which indicate that the general dynamical behavior of such a system is quite predictable. The quantities of the initial molecules decreases over time, until virtually all initial molecules disappear. The intermediate fragments with overhangs first increase in apparent quantity as the initial molecules are cleaved by the restriction enzymes during the first minutes of the reaction, and then the quantities of these intermediate fragments decrease as they are used up during the religation process, forming adult molecules which cannot be recleaved. The adult molecules show a steady increase in quantity throughout the reaction, and are apparently not involved in further interactions with other molecules or enzymes. Somewhat surprising was the speed with which the restriction digestion occurred at room temperature. After only 5 minutes, cleavage of the initial molecules began to occur. The high concentrations of these enzymes relative to the amount of DNA in the reaction may account for this unexpectedly favorable result.

Variations of the experiment, however, have yielded slight variations in the speed with which ligation appears to occur. Factors such as ligase concentration in the reaction, and temperature seem to be important. Different results might be observed with other restriction enzymes, particularly those which cleave optimally in different salt solutions. In all cases, the ligation reaction required the addition of ATP to the reaction mixture, as evident from preliminary work not presented here.

This splicing system behaved as predicted by the dry mathematical model proposed by Head. The intial set consisting of the two dsDNA strings was dynamically transformed by the process of cleavage and religation into the predicted set of adult strings of dsDNA. The adult strands showed no evidence of further cleavage, and the initial strings gradually disappeared from the reaction mixture over time. The intermediate fragments were visible as quickly as five minutes after the reaction was initiated. Their concentrations initially appeared to steadily increase until they reached a maximum concentration in the reaction, after which time the quantities of the intermediate fragments in the reaction mixture appeared to monotonically diminish. The apparent quantities of the adult strings steadily increased throughout the duration of the reaction.

The general dynamical behavior of the system suggested by these results was verified by several repetitions of the experiment, the details of which we do not include here.

6. Conclusions and Suggestions for Further Research

The results of these experiments verify that Head's mathematical model accurately predicts the splicing language generated by the wet splicing system for the example we investigated. The splicing language predicted by the dry splicing model was generated *in vitro*. Although no unexpected strings were generated, we would like to find conditions under which complete religation of the intermediate fagments readily occurs.

Further studies using other wet splicing systems are necessary. In particular, we expect that finding compatible buffer solutions may be a non-trivial matter, depending upon the restriction enzymes chosen. We propose to attempt this type of experiment using two restriction enzymes having different restriction sites producing identical palindromic overhangs, in which case we would expect to see four intermediate fragments which have sticky ends, 10 intermediate dsDNA with no sticky ends, and four adult molecules which could not be recleaved by either restriction enzyme. An example of such a pair of restriction enzymes is *BamH*I and *Bgl*II.

Experiments utilizing chemically synthesized dsDNA should also be performed, since the possibility of using the splicing operation as a computational tool may necessarily depend on the ability to use arbitrary intial sets of strings. We also propose to verify by experiment that the splicing model can predict the behavior of a splicing system which is designed to produce an infinite set of adult strings. Again, such a step seems reasonable and necessary for an ongoing investigation of the splicing system as a computational tool.

In conclusion, we have initial results which seem promising. Further research must be done in the areas suggested above, as well as in areas which will broaden the scope of the investigation of the splicing system model.

7. Appendix

•**Elution Buffer** : 10mM Tris-HCl, 0.1mM EDTA

•**Standard TE Buffer** : 10mM Tris-HCl, 1.0mM EDTA

Purification Protocol:

Qiagen Buffer QX1 was added to the reaction product in a 5:1 (vol:vol) ratio. This mixture was loaded into a QIAquick column and spun in a standard table-top centrifuge for 1 minute at 13,000 rpm. The eluent was poured off, and .75 ml of Qiagen Bufer PE was added. The sample was allowed to stand 2-3 minutes, then the column was spun 1 minute at 13,000rpm, the eluent discarded, and the column was respun for another minute. The elution buffer was added (25 μl), and the column was allowed to sit 5 minutes before the DNA was eluted by spinning the column at 14,000rpm for 5 minutes.

References

[1] *IEEE International Conference on Evolutionary Computation*, Indiana University, Purdue University, Indianapolis, Illinois, April13–16, 1997, Special Session on DNA-based Computation.

[2] *Pacific Symposium on Biocomputing*, 1997.

[3] Leonard M. Adleman, *Molecular computation of solutions to combinatorial problems*, Science **266** (1994), 1021–1024.

[4] A. Ausubel, R. Brent, R.E. Kingston, D.D. Moore, J.G. Seidman, J.A. Smith, and K. Struhl, *Current Protocals in Molecular Biology*, Greene Publishing Associates and Wiley-Interscience, 1994.

[5] Dan Boneh, Christopher Dunworth, and Richard J. Lipton, *Breaking DES using a molecular computer*, Tech. Report CS-TR-489-95, Princeton University, May 1995.

[6] Erzsébet Csuhaj-Varjú, Rudolf Freund, Lila Kari, and Gheorghe Păun, *DNA computation based on splicing: universality results*, In Hunter and Klein [**14**].

[7] Karel Culik II and Tero Harju, *The regularity of splicing systems and DNA*, 16th International Colloquium on Automata, Languages and Programming (Berlin), Lecture Notes in Computer Science, vol. 372, Springer Verlag, 1989, pp. 222–233.

[8] K. L. Denninghoff and R. W. Gatterdam, *On the undecidability of splicing systems*, Internat. J. Comput. Math. **27** (1989), 133–145.

[9] Rudolf Freund, Erzsébet Csuhaj-Varjú, and Franz Wachtler, *Test tube systems with cutting/recombination operations*, In *Pacific Symposium on Biocomputing* [**2**].

[10] Rudolf Freund, Lila Kari, and Gheorghe Păun, *DNA computation based on splicing: The existence of universal computers*, Journal of the ACM, To appear. Also Technical Report 185-2/FR-2/95, TU Wien, 1995.

[11] R. W. Gatterdam, *Splicing systems and regularity*, Internat. J. Comput. Math. **31** (1989), 63–67.

[12] Tom Head, *Splicing systems and DNA*, Lindenmayer Systems: Impacts on Theoretical Computer Science, Computer Graphics and Developmental Biology (Grzegorz Rozenberg and Arto Salomaa, eds.), Springer Verlag, Berlin, Heidelberg, New York, 1992, pp. 371–383.

[13] Tom Head, Gheorghe Păun, and Dennis Pixton, *Generative mechanisms suggested by DNA recombination*, Handbook of Formal Languages (Grzegorz Rozenberg and Arto Salomaa, eds.), vol. 2, Springer Verlag, Berlin, Heidelberg, New York, October 1996.

[14] Lawrence Hunter and Teri Klein (eds.), *Biocomputing: Proceedings of the 1996 pacific symposium*, Maui, HI, World Scientific Publishing Co., Singapore, January 1996.

[15] IEEE Computer Society Technical Committee on Pattern Recognition and Machine Intelligence (PAMI), *Proceedings of the First International Symposium on Intelligence in Neural and Biological Systems*, Herndon, VA, IEEE Computer Society Press, May 1995.

[16] Nataša Jonoska and Steven A. Karl, *Ligation Experiments in Computing with DNA*, In *IEEE International Conference on Evolutionary Computation* [**1**], Special Session on DNA-based Computation.

[17] Lila Kari, *DNA computing: Arrival of biological mathematics*, Math. Intelligencer **19** (1997), no. 2, 9–22.

[18] Richard J. Lipton, *DNA solution of hard computational problems*, Science **268** (1995), 542–545.

[19] A. Mateescu, Gheorghe Păun, Grzegorz Rozenberg, and Arto Salomaa, *Simple splicing systems*, Discrete Applied Mathematics (1996).

[20] Gheorghe Păun, *The splicing as an operation on formal languages*, In *Proceedings of the First International Symposium on Intelligence in Neural and Biological Systems* [**15**], pp. 176–180.

[21] ______, *Regular extended H systems are computationally universal*, Journal of Automata, Languages, Combinatorics **1** (1996), no. 1, 27–36.

[22] Dennis Pixton, *Splicing in abstract families of languages*, in preparation.

[23] ______, *Linear and circular splicing systems*, In *Proceedings of the First International Symposium on Intelligence in Neural and Biological Systems* [**15**], pp. 181–188.

[24] ______, *Regularity of splicing languages*, Discrete Appl. Math. **69** (1996), no. 1–2, 99–122.

[25] F.W. Stahl R.W. Hendrix, J.W. Roberts and R.A. Weisberg, *Lambda II*, Cold Spring Harbor Laboratory.

[26] Rani Siromoney, K.G. Subramanian, and V. Rajkumar Dare, *Circular DNA and splicing systems*, Proceedings of Parallel Image Analysis, 2nd International Conference ICPIA '92, Ube, Japan, 21-23 Dec 1992. (Ube, Japan) (A. Nakamura, M. Nivat, A. Saoudi, P.S.P. Wang, and K. Inoue, eds.), Lecture Notes in Computer Science, no. 654, Springer Verlag, Berlin, Heidelberg, New York, 1992, pp. 260–273.

Department of Mathematics, Binghamton University, Binghamton, NY 13905
E-mail address: `liz@math.binghamton.edu`

Department of Biological Sciences, Binghamton University, Binghamton, NY 13905
E-mail address: `kjreddy@binghamton.edu`

DIMACS Series in Discrete Mathematics
and Theoretical Computer Science
Volume **48**, 1999

Parallel Operations in DNA-Based Computation

[1,2] Frank Guarnieri, [2]Martin Orlian, and [2]Carter Bancroft

ABSTRACT

The new field of molecular computation holds out exciting and challenging prospects for both miniaturizing and massively parallelizing computing. Although molecular computing is still in its early stages of development, it is already apparent that future progress will depend upon combining theory and experiment. We have previously developed and executed a DNA-based algorithm for adding binary numbers, which is based upon a linked series of primer extension reactons, collectively termed a horizontal chain reaction (HCR). More recently, we have begun to apply the HCR concept to the problem of parallel operations. We describe here the theoretical and experimental design of an algorithm for employing parallel, intersecting HCR's to solve a simple search problem. We also describe some of the experimental problems we have encountered and solved in executing this algorithm, as well as experimental investigations implying that the HCR approach is sufficiently robust to permit its application to more complex search problems.

1. INTRODUCTION

The promise of molecular computation is its potential to perform

----

1991 *Mathematics Subject Classification.* Primary 68-04; Secondary 68-U99.

Supported by DARPA/NSF Grant CCR-9724012. The second author was supported by NIDA Training Grant DA07135.

massively parallel calculations using molecular scale operations. Leonard Adleman introduced DNA as a computational tool by associating unique DNA sequences with every vertex and edge of a small graph, employingcontrolled ligase reactions, and isolating a strand of DNA corresponding to the unique Hamiltonian path of the system (1). This seminal work has inspired several interesting new theories of DNA computation. Since Adleman's work, however, there has been only one new theoretical approach which has also been carried out experimentally (2). Given both the limited number of experimental results currently available, and the recent discussions highlighting the practical difficulties associated with DNA-based computations, it is clear that purely theoretical formulations of molecular computing are inadequate. For example, Baum (3), Mir (4), Deaton *et al* (5), and Kaplan *et al* (6) discuss sequence-dependent phenomena such as inadvertent complementarity, intramolecular annealing, and heteroduplex formation as being potential problems. These and many other experimental pitfalls of nucleic acid chemistry are discussed by Seeman *et al* (7). In this paper, we discuss our current work in DNA computation. Specifically, we discuss our attempts to integrate theory and algorithms with experiment and some of the experimental difficulties that we have encountered.

We recently generated and applied a paradigm for making DNA add any two rational nonnegative binary numbers, demonstrating that DNA can be used to both encode binary digits and carry out the dynamical binary operations necessary to perform addition (2). The addition reaction employs a linear series of primer extension reactions, collectively called the "horizontal chain reaction", in which input DNA sequences serve as successive templates for extension of a result strand. Our algorithm is quite general: there is no theoretical limit to the size of the numbers that can be represented with DNA, there is no theoretical limit to the number of addition operations, and the addition operations may be performed in any base (8). The major problem with our algorithm, however, is that binary operations are carried out irreversibly, thus iterative addition cannot be performed. Since a major goal of molecular computation is parallel processing, we have begun to investigate whether algorithms can be developed and executed that employ a series of horizontal chain reactions run in parallel. In this paper, we describe the algorithm we have developed to do this, and the experimental difficulties we have faced in the experimental execution of this protocol.

2. THEORETICAL AND EXPERIMENTAL CONSIDERATIONS

To test the utility of the horizontal chain reaction for parallel processing, we have generalized the procedure to perform searching operations. These searching operations have been applied to solving Adleman's Hamiltonian path problem (**Figure 1, left**). In the design of our algorithm, we made use of the fact that DNA primer extension reactions proceed in a 5' to 3' direction. We therefore encoded edges instead of vertices as our fundamental DNA unit, thus giving our algorithm directionality.

To examine whether the horizontal chain reaction could be applied to searching operations, we began with a greatly simplified path example, corresponding to the single path through Adleman's graph that corresponds to a Hamiltonian solution (**Fig. 1, right**). As illustrated in **Fig. 2**, the input DNA for traversing this simple path consists of a single biotin-linked starting primer, plus five template DNA strands. As discussed further below, the starting primer DNA contains a 5'-terminal biotin moiety to permit attachment to streptavidin-coated paramagnetic beads, followed by a DNA linker region to permit the PCR amplifcation of the result stands as well as to prevent steric interference by the beads. The 5' terminal element of this primer encodes the edge 01. The templates each contain, starting at the 3' end, a junk (J) sequence that, as described previously (2) prevents template extension, followed by elements encoding the DNA complements (indicated by a bar) of the DNA encoding two successive connected edges of the path (e.g., the complements of the sequences encoding edges 01 and 12). Each element is 20 bases long, and the 3' element represents the complement of an edge already traversed, while the next element is the complement of a new potential path which has not yet been traversed. When these DNA molecules are combined in a test tube, and subjected to primer extension, the horizontal chain reaction illustrated in **Fig. 3** should occur. In the first reaction, the primer strand hybridizes specifically with the template containing the complements of the edges 01,12, and primer extension extends the primer to encode the edges 01,12. In four similar successive reactions, each involving specific hybridization with the extended primer to a particular template DNA, a result DNA strand is produced that encodes the entire path 0123456.

The same principles are employed to encode DNA strands that, upon primer extension, will traverse the entire graph (**Fig. 4**). The input consists of two biotinylated starting primers, corresponding to the two starting paths 01 and 03 (the 06 path is not encoded, since it clearly cannot produce a Hamiltonian solution to the graph problem), plus 21 different template strands corresponding to the 21 different connected edges of the graph. The expected result of multiple primer extension reactions performed on these DNA strands in a test tube is illustrated in **Fig. 5**. To permit immobilization of the starting primers, they are attached via their biotin moiety to streptavidin-coated paramagnetic beads. Following multiple rounds of primer extension, the primers are extended to yield various result DNA strands, each encoding a path through the graph. Following incubation of the test tube under alkali denaturation conditions, the result DNA strands remain bound, but the templates employed are solubilized. Thereafter, a magnet is employed to pellet the bound result DNA strands (**Fig. 6,left**), and the template strands are removed with a pipet (**Fig. 6, right**). Following several washes of the pellet with buffer, the templates can be completely removed. This is an important separation step, since the template DNA elements are complementary to result strand DNA elements, and the templates would thus interfere with a PCR-based readout of the result.

3. EXPERIMENTAL RESULTS AND DISCUSSION

We have recently described (9) the application both of the approach outlined above to traverse Adleman's graph, and of readout protocols based upon either PCR or restriction enzyme analysis to determine the Hamiltonian path through the graph. We describe here some of the experimental difficulties we have encountered in the execution of this algorithm, and how we have gone about overcoming them. In addition, we present experimental results implying that it should be possible to apply the horizontal chain reaction approach experimentally to more complex search procedures.

We first investigated whether successive primer extension reactions could be employed experimentally to traverse the single path through Adleman's graph that corresponds to a Hamiltonian solution (illustrated in Fig. 1, right). AmpliTaq (Perkin-Elmer) was employed to catalyze multiple rounds of primer extension, employing the DNA strands illustrated in Fig. 2, and with 50 pmole of each template present. Following removal of the templates as illustrated in Fig. 6, PCR with appropriate amplification primers was employed to amplify immobilized result DNA strands that begin and terminate at, respectively cities 0 and 6. As shown in **Fig. 7**, this yielded a single amplified product 170 bp in size, corresponding to the size expected for a result DNA strand encoding the path 0123456. This demonstrated that performance of the five successive primer extension reactions illustrated in Fig. 3 is experimentally feasible.

We next attempted to carry out the full "graph reaction", by using AmpliTaq to catalyze multiple rounds of primer extension employing the DNA strands illustrated in Fig. 4, followed by PCR as above to amplify immobilized result DNA strands that begin and terminate at, respectively, cities 0 and 6. By inspection of the graph in Fig. 1, it can be seen that the 01 and 03 starting primers should yield the following PCR-amplified bands ≤170 bp in size: 01 primer, 170 bp (path 0123456) and 150 bp (path 013456); 03 primer, 130 bp (path 013456). However, experimental execution of this algorithm yielded a "template concentration" problem (**Fig. 8A**). When equal inputs of the 01 and 03 primers were employed with a high (40 pmole/template) input of the 21 template strands, only the 03-initiated 130 bp band was observed (**lane 2**), implying that, under these conditions, 03-initiated paths completely predominate over 01-initiated paths. Even when the ratio of 01:03 starting primer was increased to 50:1, only faint traces of the 01-initiated 170 and 150 bp bands were detected (**lane 3**). This predominance of the 03-initiated paths was partly alleviated when template input was lowered to 20 pmole/template, since a 50:1 ratio of 01:03 starting primers yielded approximately equal levels of all three bands (**lane 5**). Finally, lowering the template input to 5 pmole/template yielded clearly detectable levels of all three bands, even when equal inputs of the 01 and 03 starting primers was employed (**lane 6**). In more recent experiments, we have found that when a "hot start" with AmpliTaq Gold (Perkin-Elmer) (which minimizes inappropriate hybridizations) is used instead of AmpliTaq, a 5:1 ratio of 01:03 input primers yields roughly equal levels of the 170, 150, and 130 PCR-amplified bands (**Fig. 8B, lane 3**) and that template inputs can be further reduced to 1 pmole/template (unpublished observations). It may be possible to lower templates even further, although we have not yet examined this possibility. These results imply that inequalities between the efficiencies of different primer-initiated horizontal chain reactions can be alleviated to a great extent simply by significantly decreasing the level of template input.

The ability to use very low template inputs (≤1 pmole/template) to execute an algorithm employing parallel, intersecting horizontal chain reactions bodes well for scaling up this type of algorithm for the solution of more extensive search problems. However, scale-up will clearly require the ability to run a number of successive primer extension reactions larger than the six chained reactions required to solve the graph problem investigated in our studies to date. The ability to do this might be limited by the cumulative effects of inefficiences occurring at each primer extension step, related either to properties of the reagents used (i.e., the DNA polymerase, the quality of the oligonucleotides used, etc.) or to some other properties inherent to the reactions themselves. We have performed a preliminary investigation to determine the number of successive primer extension reactions that can be carried out with our present experimental protocol. The result DNA strand products of the entire graph reaction were PCR-amplified with amplimers designed to amplify DNA strands that begin and terminate at, respectively, cities 0 and 6, as in **Fig. 8**. These PCR products were then reamplified using a partially nested proximal primer along with a nested distal primer in order to facilitate the subsequent analysis of the putative result stand. It is seen in **Fig. 9** that faint bands $\geq$400 bp in band were amplified, implying that at least 20 successive primer extension reactions had occurred, under conditions where the graph reaction had been run for 30 cycles. Thus, the horizontal chain reaction as presently performed can apparently be extended to at least 20 successive primer extensions. This number can probably be further increased both by increasing the number of graph reaction cycles, and by appropriate modifications in the experimental conditions employed. Again, this bodes well for scaling up the horizontal chain reaction to more complex search problems.

4. CONCLUSIONS

We have described here an experimental protocol for using a horizontal chain reaction-based algorithm that employs parallel operations to carry out a search procedure, and the experimental execution of this protocol. We have found that numerous successive primer extension reactions can occur in parallel, thus potentially exploring all possible paths simultaneoously. The apparent efficiency of this process suggests that both the size and parallelism of calculations employing the horizontal chain reaction can be increased.

During the course of this work, we encountered and overcame significant experimental problems. We anticipated a cumulative diminution in the overall efficiency of the process at each successive cycle. Hence, for our initial experiments we employed large quantities of input DNA (both primers and templates). However, we have found that, under these conditions, the relative efficiencies with which different paths run differs significantly, even though the oligonucleotides used were chosen at random and subjected to several algorithms designed to detect aberrant and incompatible sequences. Lowering the template DNA concentration and "hot starting" the graph reaction to minimize inappropriate hybridizations significantly reduced the disparate efficiencies. Furthermore, we found that the concentrations of path-initiating primers used in these experiments could be reduced significantly as well. The problems we have encountered may be due at least in part to our particular experimental design. However, if these problems are indeed inherent to reactions where concentrated DNA solutions are used, this may point to a fundamental lower limit on the overall concentration of DNA that can be used, and hence to an upper limit on the number of computations that can effectively be carried out in parallel. On the other hand, it may be possible to develop second generation algorithms that permit the isolation and/or analysis of intermediate products. These products could then be subjected to additional manipulations, thus effectively extending the practical limits of DNA-based computers.

REFERENCES

1. L.M. Adleman, *Science* **226**, 1021 (1994).
2. F. Guarnieri, M. Fliss and C. Bancroft, *Science* **273**, 220 (1996)
3. E. Baum, in *Proceedings of the Second Annual Meeting on DNA Based Computers,* Princeton, N.J., 10 to 12 June 1996 (American Mathematical Society, Providence, RI, in press).
4. K.U. Mir, in *Proceedings of the Second Annual Meeting on DNA Based Computers,* Princeton, N.J., 10 to 12 June 1996 (American Mathematical Society, Providence, RI, in press).
5. R. Deaton, R.C. Murphy, M. Garzon, D.R. Franceschetti and S.E. Stevens, Jr., in *Proceedings of the Second Annual Meeting on DNA Based Computers,* Princeton, N.J., 10 to 12 June 1996 (American Mathematical Society, Providence, RI, in press).
6. P.D. Kaplan, G. Cecchi and A. Libchaber, in *Proceedings of the Second Annual Meeting on DNA Based Computers,* Princeton, N.J., 10 to 12 June 1996 (American Mathematical Society, Providence, RI, in press).
7. N.C. Seeman, H. Wang, B. Liu, J. Qi, X. Li, X. Yang, F. Liu, W. Sun, Z. Shen, R. Sha, C. Mao, Y. Wang, S. Zhang, T-J. Fu and J. Chen, in *Proceedings of the Second Annual Meeting on DNA Based Computers,* Princeton, N.J., 10 to 12 June 1996 (American Mathematical Society, Providence, RI, in press).
8. F. Guarnieri and C. Bancroft, in *Proceedings of the Second Annual Meeting on DNA Based Computers,* Princeton, N.J., 10 to 12 June 1996 (American Mathematical Society, Providence, RI, in press).
9. M. Orlian, F. Guarnieri and C. Bancroft. Use of Linked, Intersecting DNA Primer Reactions to Solve a Computationally Difficult Graph Problem. Submitted for publication.

[1]Sarnoff Corporation, 201 Washington Road, Princeton, NJ 08543
[2]Department of Physiology & Biophysics, The Mount Sinai School of Medicine New York, NY 10029.

Figure Legends

Figure 1. Graphs to Be Solved. Shown are the directed path problem originally encoded in DNA and solved by Adleman (1994) (**left**), and a simple test graph consisting of the seven-city path that represents the Hamiltonian solution to this graph (**right**).

Figure 2. Input DNA for Traversal of the Simple Test Graph. The required 5'-biotinylated 01 starting primer plus five connected edge templates are illustrated.

Figure 3. Expected Horizontal Chain Reaction Resulting in Traversal of the Simple Graph. Illustrated are the five successive primer extension reactions that are expected to occur, yielding finally a result DNA strand encoding the path 0123456.

Figure 4. Input DNA for Adleman's Graph. Twenty-four different DNA strands are combined in one test tube. These strands consist of 5'-biotinylated starting primers corresponding to either the 01 and 03 edge, shown at the top; plus 21 connected edge templates with the general structure shown in the middle, and containing one of the 21 sets of connected edges listed at the bottom.

Figure 5. Use of Immobilized Primers for the Horizontal Chain Reaction Employed to Traverse Adleman's Graph. The two types of biotinylated starting primers are bound to separate strepavidin-coated paramagnetic beads, to permit adjustment of the input concentration of each starting primer. Following addition of the bead-bound primers to a reaction mixture containing all 21 connected edge templates, multiple rounds of primer extension are carried out. Alkali denaturation is then employed to release the connected edge template DNA's from the newly synthesized result DNA strands.

Figure 6. Separation of Result DNA Strands from Connected Edge Template DNA. (**Left**) After performance of the graph reaction followed by denaturation as illustrated in Fig. 5, a magnetic particle concentrator is used to immobilize the bead-bound result DNA strands. **(Right)** The solubilized connected edge template DNA's are then removed by aspiration with a micropipettor.

Figure 7. PCR-Amplified Products of the Simple Graph Reaction. The simple graph reaction was run, and an aliquot subjected to PCR amplification of paths beginning and ending at, respectively, cities 0 and 6, all as described in the text. PCR products were then subjected to agarose gel electrophoresis (lane 2). Lane 1 shows an Msp I digest of pBR322 used as a molecular weight marker in all gels. Bands were visualized by ethidium bromide staining.

Figure 8. Investigation of Conditions Required to Equalize the Apparent Efficiencies of Traversal of Paths Initiated by Different Primers.

(Panel A) Reduction of Template Concentration Significantly Reduces the Predominance of 03-Initiated Path Products. Different template concentrations were employed to run the complete graph reaction illustrated in Fig. 5, and the result DNA strands were then subjected to PCR amplification as in Fig. 7, followed by agarose gel electrophoresis and ethidium bromide staining. Shown are the PCR-amplified products of complete graph reactions that had been run employing the indicated amounts of each template ("pmole/Template") and of the starting ratios of 01 and 03 starting primers ("01/03").

(Panel B) Use of AmpliTaq Gold to "Hot Start" Both the Graph Reaction and Subsequent Amplification Reactions Further Reduces the Predominance of the 03-Initiated Path Products. The complete graph reaction was run and analyzed as in Panel A, except that AmpliTaq Gold and a "Hot Start" was substituted for AmpliTaq in both graph reactions and PCR amplifications. Shown are the PCR-amplified products of complete graph reactions that had been run employing template inputs of 5 pmole/template and the indicated ratios of 01 and 03 starting primers ("01/03 Input").

Figure 9. The Horizontal Chain Reaction Can Be Employed to Perform at Least Twenty Chained Primer Extension Reactions. The complete graph reaction was run and the result DNA strands PCR-amplified as in Fig. 8. Aliquots of these PCR-amplified products were then reamplified, employing a common partially nested proximal primer plus a nested distal primer corresponding to the complement of one of the indicated edges. These PCR products were then subjected to agarose gel electrophoresis and ethidium bromide staining.

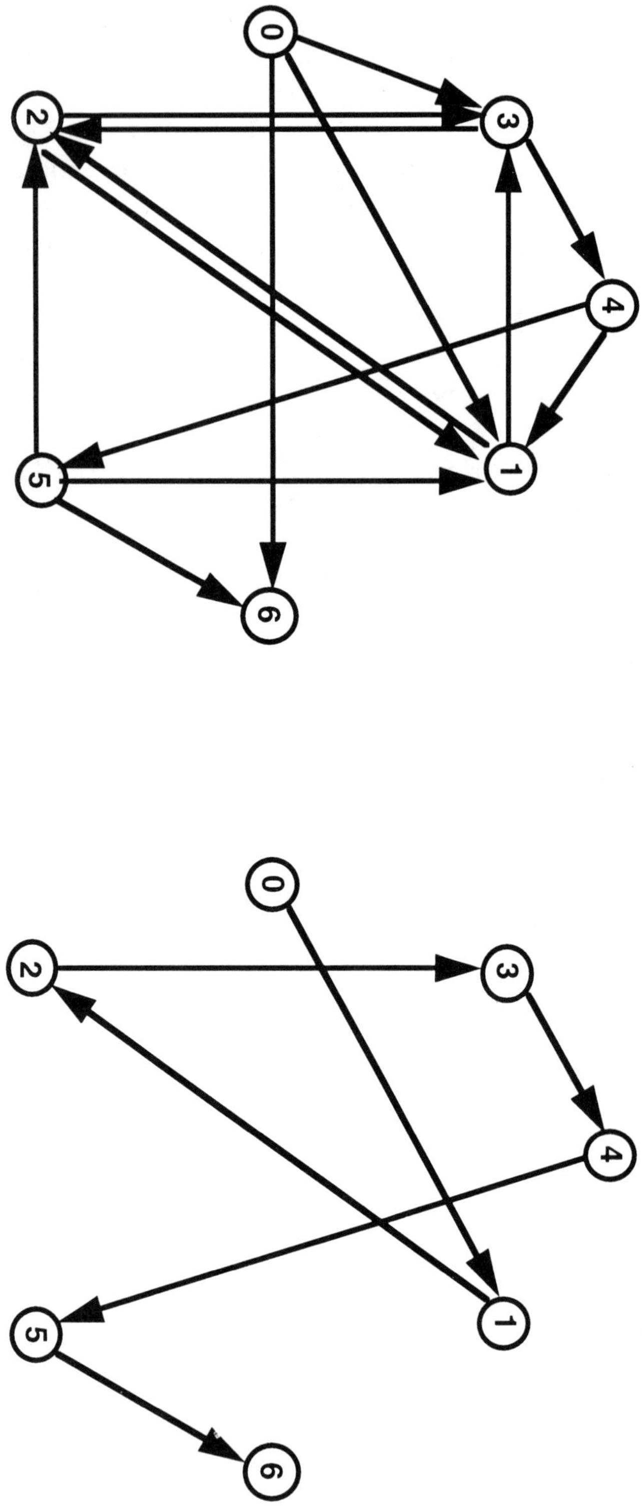

Figure 1

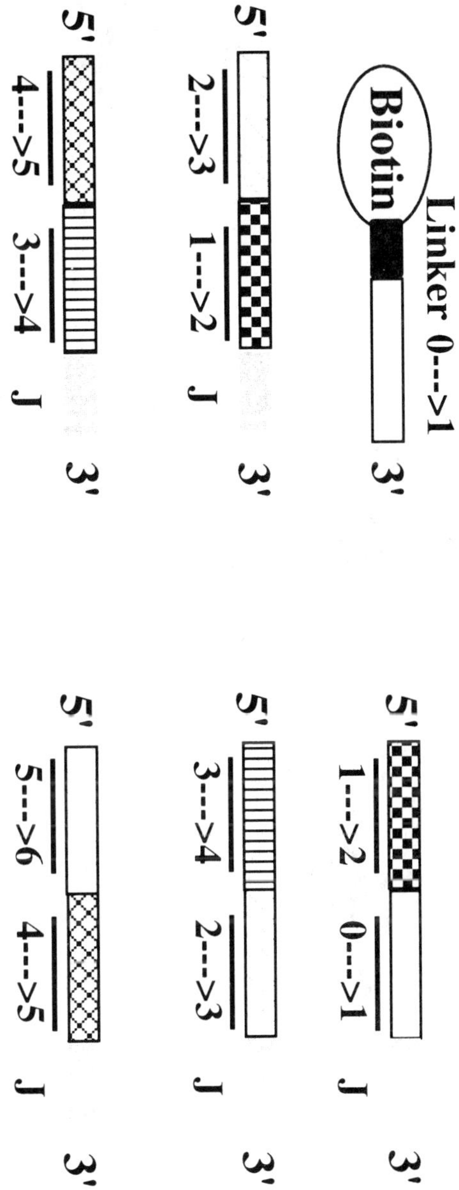
Biotin
Linker 0--->1
5'
3'
5'
2--->3
1--->2
J
3'
5'
4--->5
3--->4
J
3'
5'
1--->2
0--->1
J
3'
5'
3--->4
2--->3
J
3'
5'
5--->6
4--->5
J
3'

Figure 2

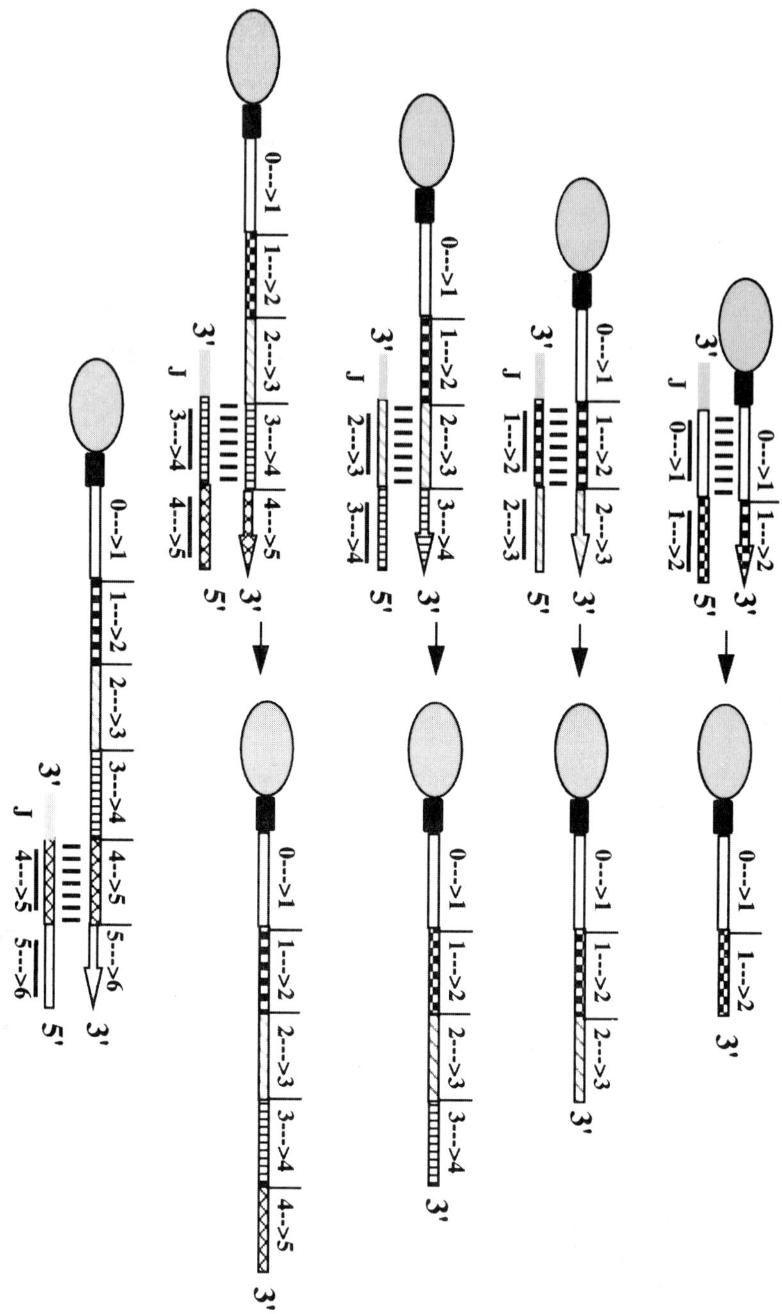

Figure 3

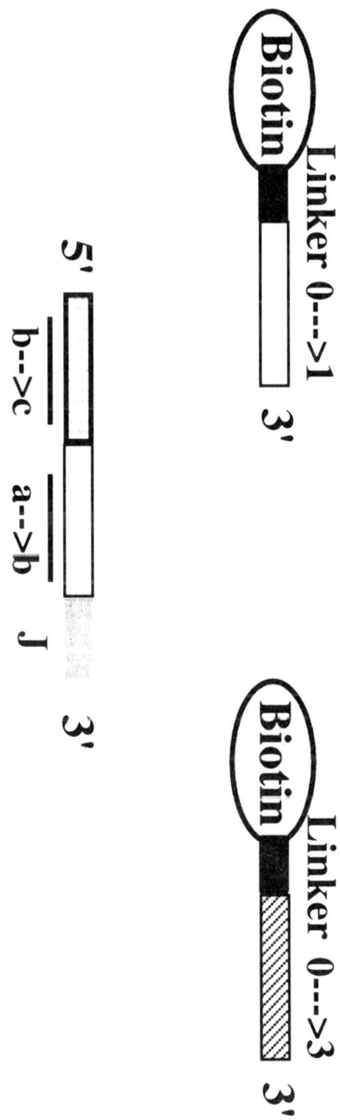

a b c = 012, 013, 213, 412, 413, 512, 513, 123, 321, 521, 523, 032, 034, 234, 132, 134, 341, 345, 451, 452, 456

Figure 4

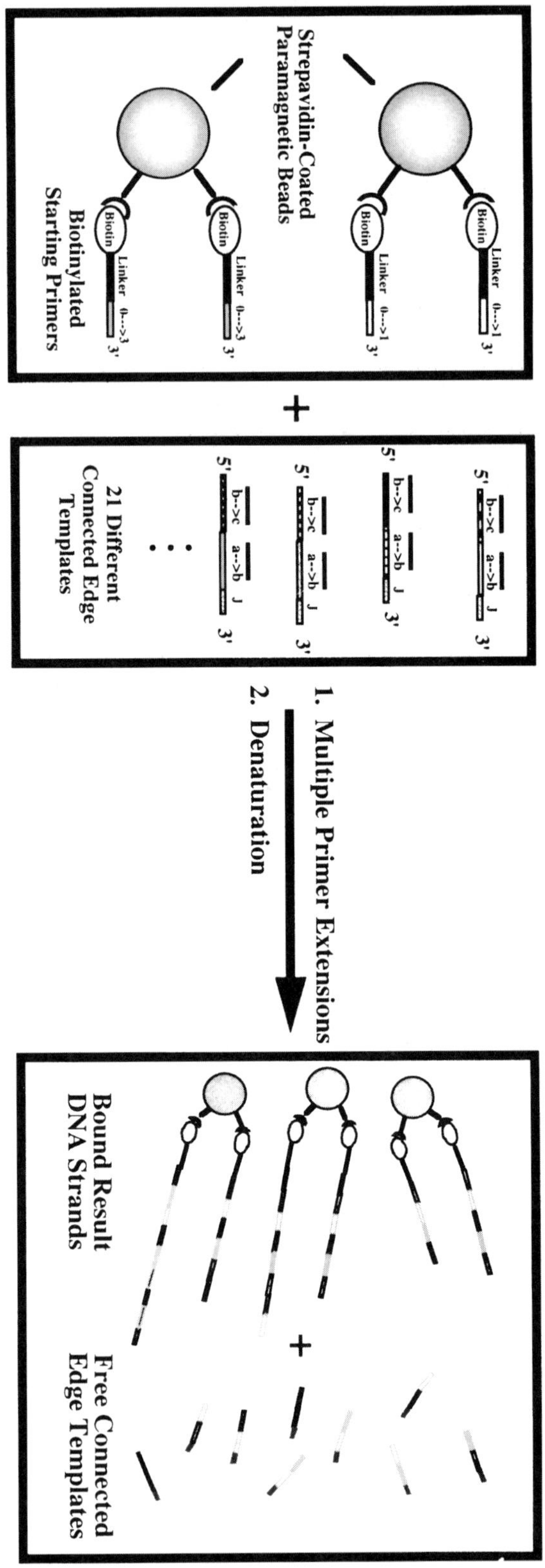
Strepavidin-Coated
Paramagnetic Beads
Biotin
Linker
0-->1
0-->3
3'
Biotinylated
Starting Primers
+
5'
b-->c
a-->b
J
3'
21 Different
Connected Edge
Templates
1. Multiple Primer Extensions
2. Denaturation
Bound Result
DNA Strands
Free Connected
Edge Templates

Figure 5

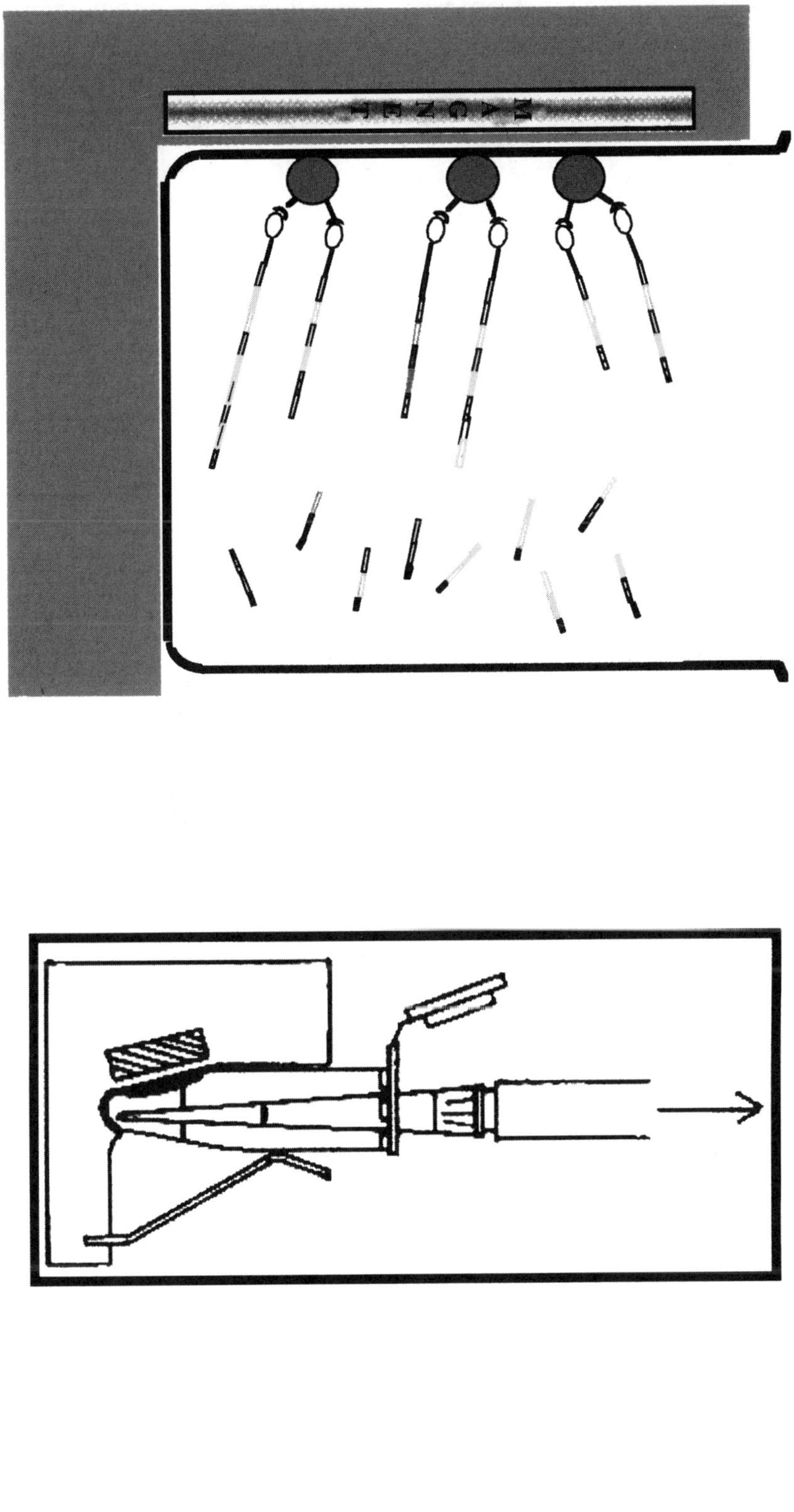

Figure 6

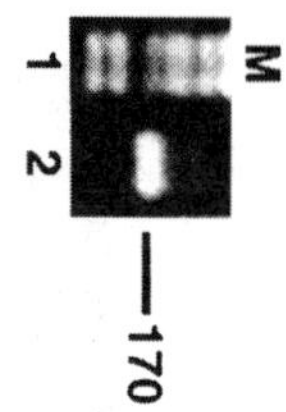

Figure 7

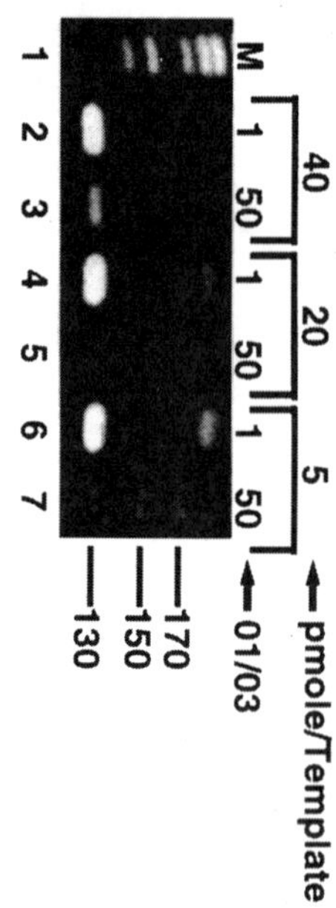

Figure 8A

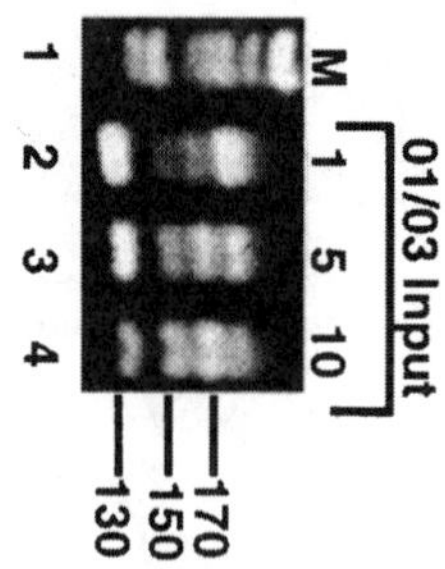

Figure 8B

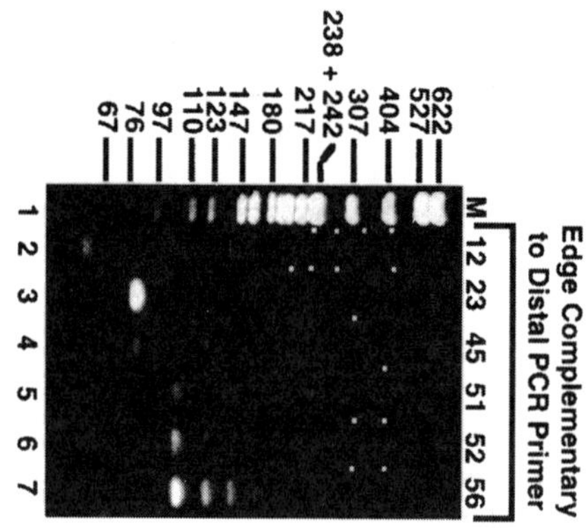

Figure 9

DIMACS Series in Discrete Mathematics
and Theoretical Computer Science
Volume **48**, 1999

Protein Folding, Spin Glass and Computational Complexity[1]

Aviezri S. Fraenkel

Abstract. A reduction from "Ground State of Spin Glass" in statistical mechanics to a minimum-energy model of protein folding is made, which shows that the latter is NP-complete (high complexity). The reduction approximates true folding of a protein. The method also enables to show that even if the backbone of the protein is fixed, the folding of the side-chains is NP-complete. In a separate second part, the possibility of synthesizing proteins to solve arbitrary instances of the spin glass problem is speculated upon.

1. Introduction

The motivation for this work is the speculation of exploiting nature's capability of protein folding to solve computationally intractable problems. One way of investigating this idea is to encode known NP-complete problems in terms of protein folding. The main content of this paper is to do this for the spin glass problem. We construct a protein that achieves the encoding, i.e., the folded protein provides a solution to spin glass. More precisely, albeit incidentally, we prove that a minimum energy model of protein folding is NP-complete.

In [**Fra1990**] the NP-completeness of a minimum energy model of protein folding was announced, and the proof appeared in [**Fra1993**]. Independent proofs were given by [**NgM1992**], [**UnM1993**] and [**HaI1996b**]. See also [**HaI1997**]. All of these proofs were rather artificial, and didn't involve true folding of a protein. In [**PaP1996**], an embedding of a chain in a grid was shown to be NP-complete, where the chain is over an unbounded alphabet, and the score is the number of adjacent matching string symbols. In the present paper we come a little closer to true protein folding, in that the number of amino acids is only three, the amino acids adjacent along the chain are also adjacent on the lattice, and a process of folding is involved which approximates that of true proteins. The problem "Ground State of a Spin Glass" which models phase transitions in statistical mechanics, magnetic fields, etc., bears a similarity to protein folding. In both, the steady state is a

1991 *Mathematics Subject Classification.* Primary 92-08; Secondary 92D20

[1]Expanded form of invited paper presented at Third Annual DIMACS Workshop on DNA Based Computers, University of Pennsylvania, June 23-25, 1997.

free energy conformation. We also mention that in [**Hal1996a**], [**Hal1996b**] and [**Hal≥1997**] good approximations to optimal folding in a minimal energy model were given.

We chose the spin glass problem, whose general case is also NP-complete (see [**Joh1983**]), to reduce from to protein folding, in the hope of encoding an NP-complete problem more realistically in terms of protein folding. This is of importance if we hope to use protein folding to solve computational problems, especially NP-complete problems, very fast. Of course the present reduction provides, in addition, another proof of the NP-completeness of protein folding. The protein we generate is still rather restricted in its allowed movements. All rotations are restricted to be in a fixed plane, but some rotations are at least unrestricted in the angle of rotation, subject only to the lattice restriction, i.e., the vertices have to be at integer coordinates. This situation is actually somewhat similar to that of real proteins. See e.g., [**NMK1994**] §1: "Although the bond lengths and angles in a protein can be predicted easily since they cannot vary much, torsion angles (rotations about bonds) are not easily predicted. Rotatable torsions tend to have three preferred values, none of which can be ruled out a priori." A similar statement appears in [**Ree1988**]. The angle restrictions in real proteins stem from spatial constraints, since folded proteins are densely packed, so distinct atoms cannot occupy the same location.

We now begin at the beginning.

A protein is a sequence of amino acids, created as an essentially linear sequence from the well understood genetic code inherent in a DNA chain, with semi-rigid bonds (peptide bonds) between pairs of amino acids which are adjacent in the linear sequence, and other, less rigid bonds, between some other amino acids. The protein then folds into a complex 3-dimensional *native conformation*, which determines its biological functions, such as the geometry of receptor sites, enzymatic properties, etc. The folding mechanism is not known, but it is believed that the native folded conformation of a protein is its lowest free energy state [**Anf1973**]. Typically a protein consists of 1000–20,000 atoms, and has a "diameter" of 35–100Å (1Å=10^{-8} cm).

One approach to model protein folding is to consider the protein to be a collection of hard impenetrable spheres (atoms) held together by elastic strings (covalent bonds). The atoms have electric charges that obey Coulomb's law [**LeL1969**]. The electromagnetic force between any two atoms diminishes as d^{-2}, and the potential energy as d^{-1}, where d is the distance between a pair of interacting atoms. The interaction is really between every atom of the protein and every atom in the universe! Yet within about one second the protein attains its final native 3-dimensional conformation. Some chemical physicists simulate this system, where they will typically neglect forces between atoms at distance >6Å. Under this simplifying assumption, simulation of one nanosecond (10^{-9} sec) of the protein folding process still takes some 150 hours on a modern main-frame computer [**LeS1988**]. For more information about protein folding see e.g., [**GiK1990**] and [**NMK1994**].

In §2 we give some background on computational complexity and NP-completeness, and in §3 we present a 3-dimensional model of protein folding and prove it to be NP-complete. In the final §4 we speculate about possible ramifications of this and other NP-complete models of fragments of nature, leading to a brief discussion of how this proof, or rather modifications thereof, might lead to a fast computation of the spin glass problem.

2. Computational Complexity and NP-Completeness

The computational complexity of a problem is usually measured in terms of the number of "steps", i.e., the "time" required to solve it, as a function of the problem's input size. A problem is called *tractable* if this function is polynomial; *intractable* otherwise. Thus sorting n integers is tractable: it can be done in $O(n \log n)$ comparison steps. The following reasons motivate this convention:

1. Normally, only tractable problems can be solved on a computer in reasonable time. Suppose that each of the problems π_1, π_2, π_3 has input size n and that the best algorithms (= "lower bounds") for solving them need n, n^2 and 2^n steps respectively. If the rate of our machine is 10^6 steps/second, then for $n = 60$, π_1 requires 0.00006 sec for execution, π_2 0.0036 sec and π_3 366 centuries!

2. The input size of a problem that can be solved in a reasonable fixed time is of practical value only for tractable problems. Thus for, say, five hours of uninterrupted computation, π_1 can have input size 18×10^9; π_2 has size 13.4×10^4 and π_3 only 34. Moreover, any 10-fold gain in speed that technological advances may yield increases the size of a problem with an $O(n^k)$ algorithm that may be solved in a fixed time by the *factor* $10^{1/k}$, whereas the size of a problem with an $O(c^n)$ algorithm is increased only by an *additive* amount of $\log_c 10$ ($c > 1$ a constant).

3. The most simplistic approach to solving a problem is to explore its entire "search tree", i.e., searching through all possibilities. Except for trivial problems, this search constitutes an exponential algorithm. Thus a problem whose best algorithm is exponential has often no essentially better algorithm than to search through all or most possibilities.

(In reality the world is not so simple; mostly in the pessimistic direction: there are problems which are polynomial and still appear to be intuitively intractable in two different senses! This is implied by the recent Robertson-Seymour theory in graph minors. See e.g., [**RoS1996**].)

Many problems can be shown to be tractable, simply by producing a polynomial algorithm for them. Very few problems can be proved to be intractable at present, however. For the bulk of interesting problems, both tractability and intractability appear to be rather difficult to establish at this time. For a large subset of them we can do a next best thing, which is to establish *completeness*, such as NP-completeness. For the moment we restrict attention to *decision problems*, i.e., problems for which the answer is YES or NO.

A decision problem π is *NP-complete* if:

(i) Given any solution for π, its validity can be *verified* in polynomial time (but there may not be a deterministic polynomial algorithm for *finding* a solution; we say that the problem has a "nondeterministic" algorithm). So the P of NP stands for Polynomial, N for Nondeterministic.

(ii) If π can be shown to be tractable, then all NP-complete problems are tractable; if π can be shown to be intractable, then all NP-complete problems are intractable.

Since the best known algorithms for any NP-complete problem are at present non-polynomial, all NP-complete problems are presently "practically intractable", or "conditionally intractable". For a thorough treatment of NP-completeness see [**GaJ1979**].

A common way to prove that a problem π is NP-complete consists of three phases:

(a) *NP-Membership.* Show (i) directly. This is usually, but not always, the easy part of the proof.

(b) *Construction.* Select an NP-complete problem π', consider an arbitrary generic *instance* x of π' and select a function f such that $f(x)$ is some instance of π and $f(x)$ is constructed in time which is polynomial in the size of x. This phase is normally called the polynomial *construction*.

(c) *YES-Equivalence.* Show that the answer to x is YES if and only if the answer to $f(x)$ is YES.

Phases (b) and (c) together are called a *reduction* of π' to π. Notation: $\pi' \propto \pi$.

The intuitive meaning of (a) and (b) should be clear: given an arbitrary instance x of π', it is transformed polynomially into a particular instance $f(x)$ of π (Fig. 0). If π is polynomial, then, in particular, $f(x)$ can be decided in polynomial time; and the answer to $f(x)$ is YES if and only if the answer to x is YES. Thus the polynomial algorithm for solving π and the polynomial f constitute a polynomial algorithm for solving π'. Equivalently, π' intractable implies π intractable.

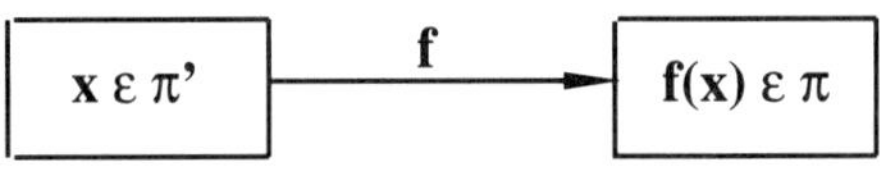

FIGURE 0. THE INTUITIVE MEANING OF A REDUCTION.

Within the so-called *Turing machine* model (see e.g. [**GaJ1979**]), it is customary to denote by P the set of all tractable problems, and by NP the set of all problems whose solutions can be verified in polynomial time. Then clearly $P \subseteq NP$. A major unsolved problem in Theoretical Computer Science is whether $P = NP$ or not. It is customarily conjectured that $P \neq NP$. Any NP-complete problem belongs to the hardest problems in NP, in the sense that if $P \neq NP$, then the NP-complete problems are intractable.

Consider the problem

GROUND STATE OF A SPIN GLASS (GSG). Given positive integers M and N, the 3-dimensional grid graph $\Gamma = (T, E)$, where the set T of vertices consists of the integer-coordinate points (x, y, z) satisfying $1 \le x \le M$, $1 \le y \le N$, $1 \le z \le 2$, and whose edges connect each pair of vertices that are adjacent in one of the three directions, an *interaction weight* $J(e)$ with values in $\{-1, 0, 1\}$ for each edge $e \in E$, and a negative integer H. Is there an assignment of a *spin* $s(u) \in \{-1, 1\}$ for all $u \in T$, such that the "ground state spin energy" is at most H, i.e., such that

$$\sum_{(u,v)\in E} J(u,v)s(u)s(v) \le H? \tag{1}$$

This problem is known to be NP-complete; see [**Joh1983**].

3. NP-Completeness of Protein Folding

We consider the following 2-dimensional model for protein folding.

MINIMUM FREE ENERGY CONFORMATION OF PROTEIN (MEP). Given a graph $G = (V, A)$ (with vertex set V and edge set A), $V \subset \mathbb{Z} \times \mathbb{Z}$ (V consists of lattice points in the plane), $V = V_1 \cup V_2$, $V_1 \cap V_2 = \emptyset$, $A = A_1 \cup A_2$, $A_1 \cap A_2 = \emptyset$, a function $C: V \to \{-1, 0, 1\}$ (the charges), $K \in \mathbb{Z}^-$ (energy bound) and $L \in \mathbb{Z}^+$ (maximum distance). Specified subgraphs may rotate about vertices of V_2 (but not V_1) in restricted specified ways. Is there a rearrangement V' of V with $V' \subset \mathbb{Z} \times \mathbb{Z}$ which preserves $d(u, v)$ for all $(u, v) \in A_1$, such that the total *free energy* is at most K, i.e., such that

$$\sum_{(u,v)\in A} C(u)C(v)/d(u,v) \le K, \tag{2}$$

and $d \le L$, where d is the discretized Euclidean distance $\left\lceil \left((x_2 - x_1)^2 + (y_2 - y_1)^2\right)^{1/2} \right\rceil$?

We also consider the following restricted protein folding problem.

SIDE CHAINS OF PROTEINS (SCOP). Given a graph $G = (V, A)$, $V \subset \mathbb{Z} \times \mathbb{Z}$, $V = V_1 \cup V_2$, $V_1 \cap V_2 = \emptyset$, $A = A_1 \cup A_2$, $A_1 \cap A_2 = \emptyset$, a function $C: V \to \{-1, 0, 1\}$ (the charges), $K \in \mathbb{Z}^-$ (energy bound) and $L \in \mathbb{Z}^+$ (maximum distance), where V_2 is restricted to the middle vertices of subgraphs called "side-chains", permitting rotation by 180° into one of two stable states in a plane parallel to $x = 0$. No rotation is permitted for V_1 (the induced subgraph of which is the "backbone"). Is there a rearrangement V' of V with $V' \subset \mathbb{Z} \times \mathbb{Z}$ which preserves $d(u, v)$ for all $(u, v) \in A_1$, such that the total *free energy* is at most K, i.e., such that $\sum_{(u,v)\in A} C(u)C(v)/d(u, v) \le K$, and $d \le L$, where d is the discretized Euclidean distance $\left\lceil \left((x_2 - x_1)^2 + (y_2 - y_1)^2\right)^{1/2} \right\rceil$?

Remarks.

(i) For both MEP and SCOP we will chose $L = 1$, so only atoms at the minimal distance of 1 interact. This is actually a better approximation for the interaction between amino acids, where the interaction diminishes as the distance to the 6-th power at least (see e.g., [**DDL1994**]), than for atoms. Thus MEP and SCOP can and will be considered to be models for protein folding with 3 amino acids.

(ii) The subset A_1 models peptide bonds, which are rigid, whereas A_2 resembles the other, less rigid bonds. Similarly, the subsets V_1 and V_2 model amino acids which do not permit and permit rotation, respectively.

Our purpose is to show that the decision problem MEP is NP-complete. The corresponding optimization problem — where we ask for $\min E$ rather than only $E \le K$ — is then clearly not any easier than MEP. In general, if a decision problem π_1 is NP-complete, then its corresponding optimization problem π_2 is said to be NP-*hard*. We note that NP-hard problems are, generally, not any easier to solve than the NP-complete problems they correspond to. We also show that even the simpler problem SCOP is NP-complete.

The NP-membership of MEP or SCOP follows from the observation that, given any solution to MEP or SCOP, we have only to check that $d(u,v)$ is preserved for all $(u,v) \in A_1$ and that $E \le K$, both of which can be done in time which is a polynomial in the input size of G.

We show GSG $\propto$ MEP. Before doing so, it is convenient to prove the following auxiliary result.

Lemma 1. *Let $x_1, \ldots, x_k$ be any real numbers, and $(\pi(x_1), \ldots, \pi(x_k))$ any permutation of $(x_1, \ldots, x_k)$. Then $\sum_{i=1}^k x_i \pi(x_i) \le \sum_{i=1}^k x_i^2$.*

Proof. Consider the two k-dimensional vectors $\mathbb{V} = (x_1, \ldots, x_k)$ and $\mathbb{V}_\pi = (\pi(x_1), \ldots, \pi(x_k))$. As usual, the inner product of vectors $\mathbb{V}$ and $\mathbb{V}' = (y_1 \ldots, y_k)$ is defined by $(\mathbb{V}, \mathbb{V}') = \sum_{i=1}^k x_i y_i$. The well-known Cauchy-Schwarz inequality asserts that $(\mathbb{V}, \mathbb{V}') \le \| \mathbb{V} \| \| \mathbb{V}' \|$, where $\| \mathbb{V} \| = (\mathbb{V}, \mathbb{V})^{1/2}$ is the *norm* or *length* of $\mathbb{V}$. Since clearly $\| \mathbb{V} \| = \| \mathbb{V}_\pi \|$, we have, $(\mathbb{V}, \mathbb{V}_\pi) \le \| \mathbb{V} \|^2 = (\mathbb{V}, \mathbb{V})$, which is the desired result. ■

Theorem 1. *The problem* MEP *is* NP-*complete.*

Proof. Given an arbitrary instance $\Gamma = (T, E)$ of GSG, we construct an instance $G = (V, A)$ of MEP. Put $L = 1$. In all our figures we denote by $\bullet$ a vertex of V with charge +1, by $\circ$ a vertex with charge -1, by $\times$ a vertex with charge 0, and by $\star$ a "rotation axis", i.e., a vertex in V_2, about which an attached subgraph has restricted rotation capability, as specified for each figure with rotation axes. Every rotation axis has charge 0. Informally, the expected native (folded) conformation of G will be a 3-dimensional box, as is Γ, but each vertex of Γ will transform into a small subgraph in G, which will constitute the "side chains".

Our basic building blocks are *strands*. A typical strand has a central vertex called *center*, from which emanate rays of length 1 or 2 in each of the three directions, possibly in both the negative and positive orientation for the x- and y-axes, but in only one orientation for z. Thus a center has degree 3, 4 or 5, where the *degree* of a vertex is the number of edges impinging on it. Edges joining vertices in the figures all belong to A_1; they have length 1 unless otherwise specified. Vertices u, v at distance 1 of each other with no edge drawn between them form an edge $(u, v) \in E_2$. In a *charged* strand, the center has charge -1 or 1, and adjacent vertices have alternating charges; and in a *neutral* strand, the center and other vertices have charge 0, except that the *ends* (vertices of degree 1), all have degree -1 or 0; or all have degree 1 or 0 (Fig. 1). The centers of the strands simulate the location and degrees of the vertices of the original instance of Γ in a sense to become clear in the sequel.

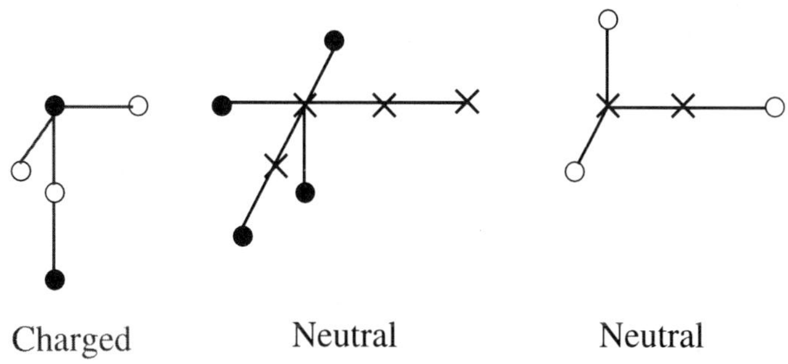

FIGURE 1. EXAMPLES OF STRANDS.

For a charged strand S with vertical ray pointing in the negative z-direction, there is a "mirror-image" called *complement* strand $\overline{S}$ two units "above" S, and parallel to it in the y-axis, antiparallel in the x- and z-axes. If the vertical ray of S points in the positive z-direction, then the complement, analogous to the above, is two units "below" S (Fig. 2). In either case, the orientations of the rays in the x- and z-axes are reversed in the complement, and all the corresponding charges are *complemented*, where 1 and -1 are complements, and the complement of 0 is 0. The two strands are joined via rotation vertices (with charge 0) in the vertical plane between S and $\overline{S}$. The joining network, together with S and $\overline{S}$ constitute a *frame*.

Analogously to the frames there are the *shells*: Four units above or below a neutral strand S there is its complement $\overline{S}$, parallel in the y-axis, antiparallel in the x- and z-axes, and complemented charges (Fig. 3). The two middle vertices of the joining network are rotation vertices, and the four others are vertices with charge 0 which cannot rotate about themselves. A frame or shell is also referred to as a *side-chain.*

It's time to explore the global aspects of the construction. We illustrate with an instance of GSG with $M = 3$, $N = 4$, complete with its J-weights, as depicted in Fig. 4. It will aid us in shaping our view of the overall construction. An important feature of Γ is that it is *bipartite*, i.e., its vertex set T can be partitioned into two disjoint subsets T_0 and T_1 such that all edges of Γ join vertices between T_0 and T_1.

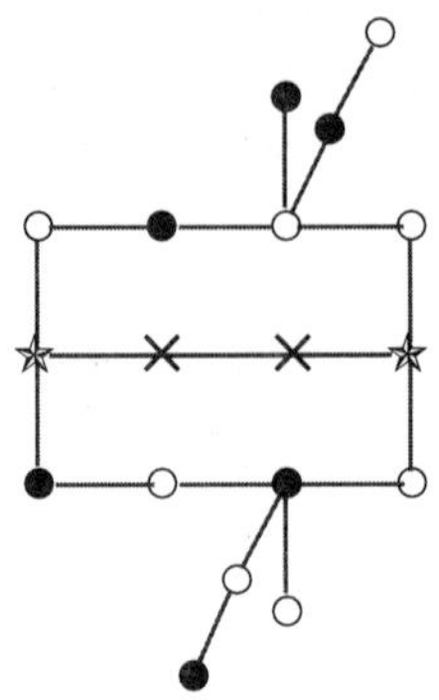

FIGURE 2. EXAMPLE OF A FRAME.

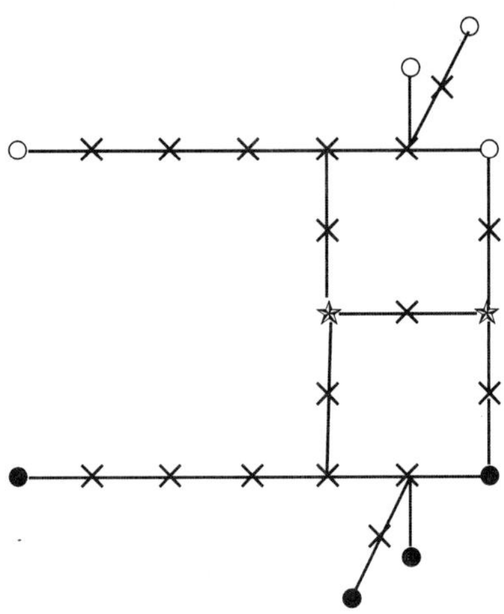

FIGURE 3. A SHELL.

This follows from the fact that all cycles in Γ are evidently even. Without loss of generality we may assume that the "origin" (vertex at (1,1,1)) is in T_0.

To fix the construction, view each shell as simulating a vertex in T_0, and each frame as simulating a vertex in T_1. Frames and shells are, in fact, adjacent. There are three types of bonds for tying together shells and frames. The first involves *glue*, to be defined below. For the moment it suffices to say that the twin *strings* with oppositely charged vertices facing each other at distance 1, on top of Fig. 5, constitute glue. At their bottoms, the two strings impinge on rotation vertices in V_2. The rotation of each string can be by any amount in a plane which, in Fig. 5, is parallel to $y = 0$. Only in the position shown in Fig. 5 the charges are at distance 1, and the strings "stabilize". Notice that the glue is in a plane parallel to the plane of the rest of the figure, 1 unit "behind" it.

The second type of bond (Fig. 6) is by means of the *backbone*, which impinges

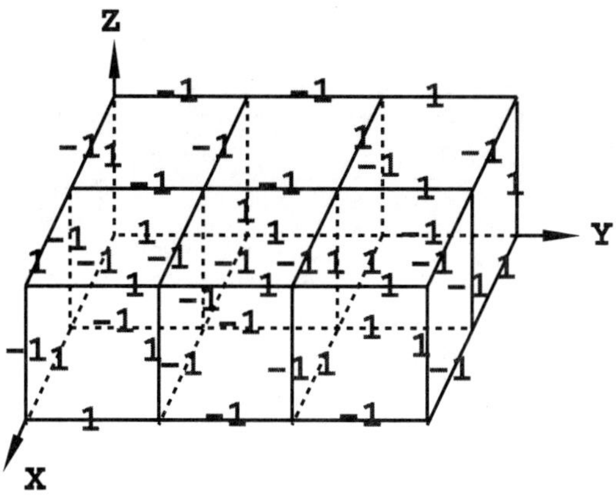

FIGURE 4. INSTANCE OF GSG WITH M=3, N=4.

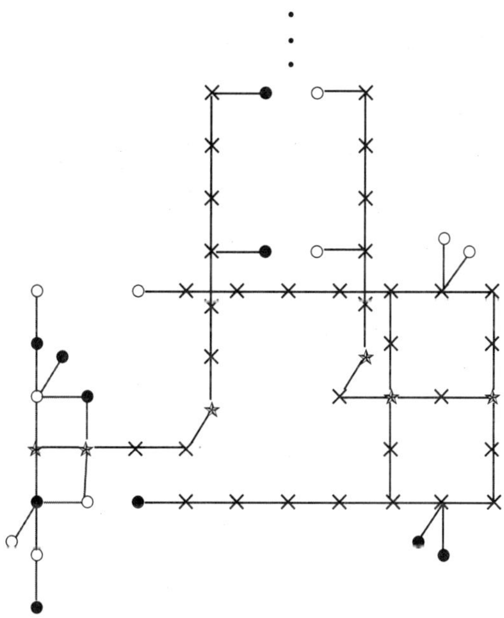

FIGURE 5. INTERCONNECTION OF FRAMES AND SHELLS: I. GLUE

on every frame and shell (as seen in Fig. 8). The third bond is *implicit* (Fig. 7). It looks as if there is no bond at all, but the frame and shell are held together by the backbone, which intersects them in a different plane.

Whatever the bonding of a shell and frame is, each can rotate, independently of each other, by 180° about two rotation axes (Figs. 5, 6, 7); so the two stable states for a side-chain are both inside a plane parallel to $x = 0$, with either S above and $\overline{S}$

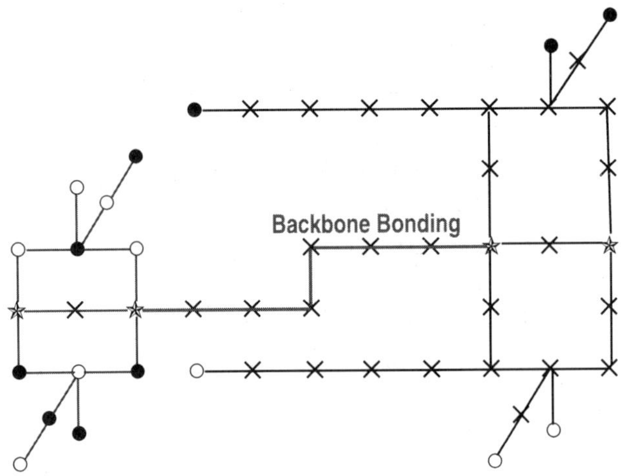

FIGURE 6. INTERCONNECTION OF FRAMES AND SHELLS: II. BACKBONE

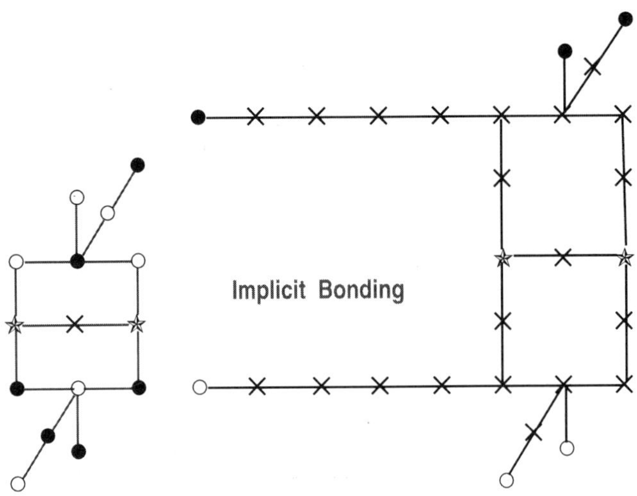

FIGURE 7. INTERCONNECTION OF FRAMES AND SHELLS: III. IMPLICIT

below, or the other way around. The aggregate of the joining edges, together with the *glue*, to be specified below, constitutes the *main chain*, or simply the *chain*. Note that there are MN side-chains intersecting the plane $z = 1$, and that many intersecting the plane $z = 2$.

The chain is expected to fold into the native conformation, portrayed schematically in Fig. 8, which is also known as the *backbone*. The chain giving rise to the backbone is depicted in Fig. 9. It is the glue which is likely to cause the chain to fold into the backbone.

The glue g_i ($i \in \{1, \ldots, 2M\}$) consists of a run of $i+7$ negatively charged vertices

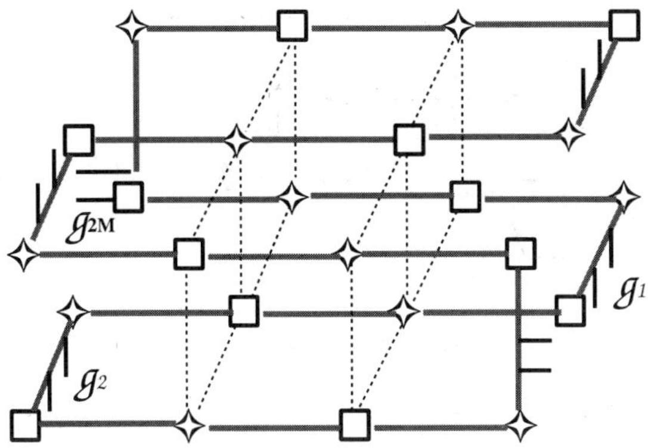

FIGURE 8. SCHEMA OF EXPECTED BACKBONE. OVALS DENOTE FRAMES, SQUARES—SHELLS. THE iTH GLUE IS g_i. (THE DOTTED LINES ARE NOT PART OF THE BACKBONE.)

FIGURE 9. SCHEMA OF CHAIN: FRAMES (OVALS), SHELLS (SQUARES) AND GLUE (g).

spaced at distances 3, and another run of $i + 7$ positively charged vertices spaced at distances 3, arranged on edges of adjacent vertices with charge 0, which are called strings. Each vertex of a glue is encased on 5 sides with vertices of charge 0 at distance 1, and only the side expected to interact with the matching vertex of opposite charge is unblocked. (In Fig. 5, the encasing of the glue is not shown, to avoid cluttering.) Glue strings with positive charges impinge on a subset of frames, and strings with negative charges—on a subset of shells. The rotation vertices of the glue permit rotation of the glue strings by any angle, but in one plane only. The unblocked interaction direction of the oppositely charged vertices of g_i is designed to make the expected folding into the backbone (Fig. 8) possible. Fig. 10 portrays two vertices of g_i, The expected folding of this glue is shown in Fig. 11.

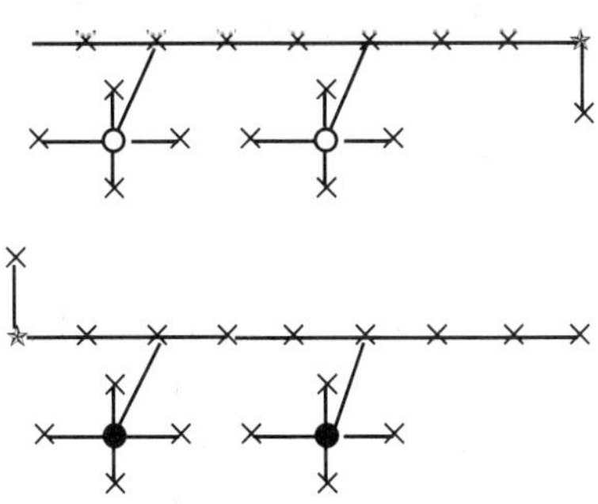

FIGURE 10. TWO VERTICES OF GLUE OF POSITIVE AND NEGATIVE CHARGE.

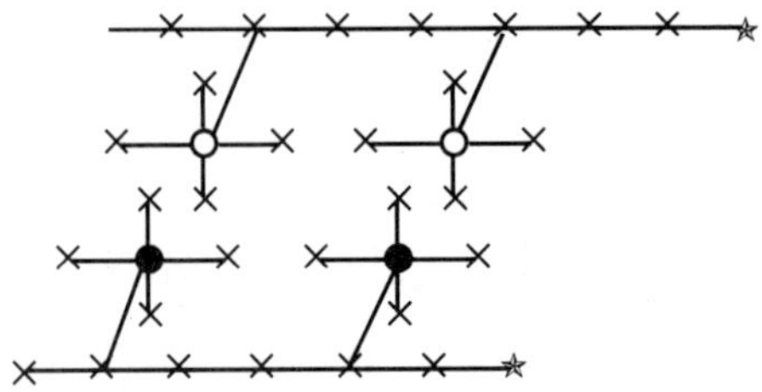

FIGURE 11. EXPECTED FOLDED FORM OF GLUE OF PREVIOUS FIGURE.

The glue bonds $2M$ shells and frames, as shown in Fig. 8, which also shows that in the expected folding, the beginning and end of the chain attach to each other, by virtue of g_{2m}.

In Fig. 12 we see the side chains in their expected native conformation. Only one complete frame and one "compressed" shell with a small part of the chain hugging them is shown, to avoid cluttering the figure.

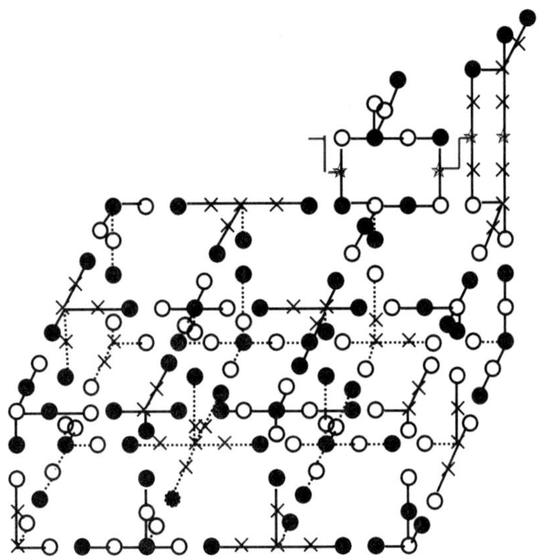

FIGURE 12. ESSENTIAL PART OF THE EXPECTED NATIVE CONFORMATION.

At edges of Γ where $J = -1$ or $J = 0$, the strands of the frames have length 1, and at edges where $J = 1$, the strands of the frames have length 2. The situation for the strands of the shells is in a way reversed: they have length 6 for $J = -1$ or $J = 0$, and length 5 for $J = 1$. (The larger length of the shell-strands enables to accommodate the glue—see Fig. 5.) This is illustrated in Fig. 12 for the values of J given for the instance of GSG of Fig. 4, except that the shell-strands are shortened to length 3 or 4, to enable portraying the essential features of the construction in one figure. Note that in the expected native conformation, each edge of the emerging grid G has an end-vertex of a frame-ray at distance 1 from the end-vertex of a shell-ray: they *dock* with each other, forming an edge in E_2. The two docking rays constitute the *docking site.* The two end vertices at a docking site are the *docking vertices.*

The end vertex of a shell-ray has charge 0 if and only if it corresponds to an edge of Γ with $J = 0$ (for S and $\overline{S}$). All other end vertices of a shell-ray have charge 1 in S, and -1 in $\overline{S}$. The construction is completed by putting $K = H + 2m + 2m_1 + M(2M + 15)$, where $m = |E|$ and m_1 denotes the number of edges of Γ with weight $J = 1$.

Any instance of Γ has 8 vertices of degree 3, $4((M-2)+(N-2))$ vertices of degree 4 and $2(M-2)(N-2)$ vertices of degree 5. Thus,

$$m = 3 \times 4 + 8(M + N - 4) + 5(M-2)(N-2) = 5MN - 2(M+N),$$

which is 46 for the instance of GSG depicted in Fig. 4. Since $m_1 \leq m$, the construction is polynomial.

Suppose now that the given instance of GSG has a spin assignment $s(u) \in \{-1, 1\}$ for all $u \in T$, such that (1) holds. We show that there exists a folding of G satisfying (2).

The charged strands, of all the frames, contribute $-2(m_{-1}+m_0)-4m_1 = -2m-2m_1$ (-67 for Fig. 12), where $m_{-1}+m_0 = m - m_1$, and m_{-1} is the number of edges of Γ with weight $J = -1$; m_0 the number of edges with $J = 0$. We claim that the folding of G such that (2) holds is the expected one, where the positive and negative charges of g_i match each other precisely for all i, as per Fig. 11. The total contribution of the glue is then $-\sum_{i=1}^{2M}(i+7) = -M(2M+15)$. It thus suffices to show that there exists a folding of the side-chains such that the docking vertices contribute energy $\leq H$, i.e.,

$$\sum_{(w,z)\in D} C(w)C(z)/d(w,z) \leq H, \tag{3}$$

where $D \subset A$ is the subset of docking vertices.

Let (u, v) be any edge of Γ. Without loss of generality we may assume $u \in T_0$, $v \in T_1$. If $s(u) = -1$, rotate the shell which simulates u such that its end vertices with charge -1 or 0 are at the docking site. If $s(u) = 1$, rotate it such that its end vertices with charge 1 or 0 are the docking vertices. If $s(v) = -1$, rotate the frame which simulates v such that its center vertex r has charge -1, and if $s(v) = 1$, rotate it such that r has charge 1. Thus $C(r) = s(v)$.

Denote by w, z, the two docking vertices in G which simulate the edge $(u, v) \in \Gamma$, where w is at the end of a frame-ray with center r simulating v, and z is at the end of a shell-ray simulating u. Note that the specified folding of the shell implies that $C(z) = s(u)$ if $J(u, v) \neq 0$; and $C(z) = 0$ if $J(u, v) = 0$. The folding of the frame shows that $C(w) = C(r) = s(v)$ if $J(u, v) = 1$, because then the frame-ray has length 2; and $C(w) = -C(r) = -s(v)$ if $J(u, v) \in \{-1, 0\}$. This information is summarized in Table 1. Since its middle-column and last column turn out to have identical entries, we conclude that $J(u,v)s(u)s(v) = C(w)C(z)/d(w,z)$ for all cases. It follows that (1) implies (3).

TABLE 1.

$J(u,v)$	$s(u)$	$s(v)$	$Js(u)s(v)$	$C(w)$	$C(z)$	$C(w)C(z)/d(w,z)$
1	−1	−1	1	−1	−1	1
1	−1	1	−1	1	−1	−1
1	1	−1	−1	−1	1	−1
1	1	1	1	1	1	1
−1	−1	−1	−1	1	−1	−1
−1	−1	1	1	−1	−1	1
−1	1	−1	1	1	1	1
−1	1	1	−1	−1	1	−1
0	−1	−1	0	1	0	0
0	−1	1	0	−1	0	0
0	1	−1	0	1	0	0
0	1	1	0	−1	0	0

Conversely, suppose that there exists a folding of G which satisfies (2). We have to show that there exists a spin assignment $s(u) \in \{-1,1\}$ for all $u \in T$ satisfying (1). Since the frames contribute $-2m - 2m_1$ to K, we have actually

$$\sum_{(w,z)\in P} C(w)C(z)/d(w,z) \le H + M(2M+15),$$

where $P = D \cup Q$ and Q is the set of edges each of which intersects the glue. Thus a vertex u of the glue which folds to within distance 1 from a vertex v of a side-chain contributes an edge $(u,v) \in Q$.

Consider first the case where the glue forms a closed set, i.e., glue vertices are adjacent to each other only. Since glue vertices are blocked by 0-charges on 5 sides, at most one pair of vertices of glue can interact with each other. Putting $x_i = i + 7$ in Lemma 1, we see that the smallest energy contribution is achieved when all the glues are matched up perfectly; in other words, any glue association contributes $\ge -M(2M+15)$. Thus (3) holds a fortiori for any closed glue folding.

Secondly, consider the case where the glue interacts with the side-chains. Note that the rotation vertices keep the glue strings and the strands in directions parallel to the $x-$, $y-$ and $z-$axes only. The smallest glue run has length 8, and from Fig. 12 we see that the side-chains contain at most 7 consecutive vertices at distance 1 with charge $\neq 0$. Note further that each glue-vertex is encased by 0-charges at distance 1 from it from 5 sides. Thus no clusters of nonzero charges can be formed. It follows that any interaction between glue and side-chains has a larger contribution than the perfect glue match-up, i.e., we see that the contribution for this case is also $\geq -M(2M+15)$. Thus (3) holds in either case.

So it turns out, incidentally, that the folding may possibly *not* be the expected one, which is, in a way, analogous to what seems to happen in the prion protein, where α-helices are transformed into β-sheets, precipitating a diseased folded state, which is claimed to induce the mad-cow syndrome [**HBDPC**]. In any case (3) holds, and this suffices to assign spins in $\{-1,1\}$ to the vertices of Γ to satisfy (1), as is shown now.

We use again the notation $(u,v) \in E$ with $u \in T_0$, $v \in T_1$; and z is at the end of a shell-ray simulating u; w at the end of a frame-ray with center r which simulates v. If $C(z) = -1$, then put $s(u) = -1$, and if $C(z) = 1$, put $s(u) = 1$. If $C(z) = 0$, examine the charges at the other ends of the same strand z belongs to. If any of its charges is 1, put $s(u) = 1$; if one of them is -1, put $s(u) = -1$; if all of them are 0, $s(u)$ can be assigned either -1 or 1. Further, put $s(v) = C(w)$ if w is at distance 2 from r (then $J(u,v) = 1$); and $s(v) = -C(w)$ if w is at distance 1 from r (then $J(u,v) \in \{-1,0\}$). It then follows from Table 1 that $J(u,v)s(u)s(v) = C(w)C(z)$ in all cases, so (3) implies (1). ■

Corollary 1. SCOP *is* NP-*complete.*

Proof. We use a small modification of the proof of Theorem 1, namely we replace the glue with its vertices in V_2 by backbone bonding as in Fig. 6, and put $K = H + 2m + 2m_1$. Everything else remains the same. ■

Remark The protein constructed in the proof of Theorem 1 can be modified in many ways, such as in the z-axis more than two layers can be used. We used only two in order to simplify the construction.

4. Speculations about Implications

Analogously to [**Fra1990**] and [**Fra1993**], we now indulge in some unscientific speculations about nature. In the present context it is aimed at the possible design of experiments for synthesizing proteins for fast computing. We first present two opposing views:

A. Nature can solve NP-complete problems in polynomial time.

The following is but a small fraction of the supporting evidence.

Each amino acid in a protein can adopt, on the average, 8 different conformations [**Pri1979**]. A relatively small protein, consisting of 100 amino acids, can thus potentially assume 8^{100} conformations. Yet nature attains the native conformation in about one second. (Note that the claim that nature assumes the global minimum free energy conformation in one second is *not* equivalent to saying that it explores all the 8^{100} potential conformations in one second: when we sort 100 distinct integers in ascending order, we don't have to test all the $100! > 100^{100}$ different permutations!)

The double helices of DNA become knotted and linked in the course of biological processes such as replication. There are methods for realizing given knots and links on DNA chains [**WaC1986**]. Moreover, the untying mechanisms follow topological transformations, so we can tell at the end of the process whether the unknot is a knot or not. The complexity of this question is a major unsolved problem in mathematics; see e.g.[**Wel1993**]. In statistical mechanics, many models have been studied to explain phase transitions [**Bax1982**]. Among them is the spin glass model GSG which we used in the proof of Theorem 1. This problem was shown to be NP-complete in [**Bar1982**]. Special cases are polynomial; see [**BMRU1980**], [**Bar1982**], [**BMRU1982**]. Nature usually manages to accomplish the phase transitions without a hitch and very fast.

Deutsch [**Deu1985**], [**Deu1989**] argues that "quantum computers" (computers based on quantum physics theory) can be constructed that can carry out in polynomial time computations which require exponential time. Bennett and Brassard [**BeB1989**] proposed and implemented practical protocols —realizing an idea of Wiesner [**Wie1983**] — and constructed a quantum counter for a certain public key cryptography application. See also [**Bra1988**], [**BrC1990**] and [**BBBSS1992**] [**BBCS1992**].

In the cryptography quantum computer, polarized photons are used to transmit digital information. Single photons (with high probability) are transmitted over a communication channel in one of 4 polarizations. There is no complexity gain in this device, yet it makes a dent in the Turing machine model: a Turing machine M can, by eavesdropping, learn the secrets being exchanged by two interactive Turing machines M_1 and M_2, without M_1 and M_2 learning that M has learned their secrets. But no machine can eavesdrop on two interactive quantum machines without being detected with high probability, assuming Heisenberg's uncertainty principle, which implies that eavesdropping to single photon transmission is tantamount to tampering with it.

Because of reasons of this type, it was suggested in [**Fra1990**], to try and reverse our usual scientific endeavor: in *addition* to modeling nature, studying the models, solving them or proving them to be NP-complete, etc., try to use devices of nature, such as proteins, quantum apparatus, DNA-chains etc. as black boxes, to which we input instances of NP-complete problems and output solutions in reasonable time. One of the problems to be faced here is the design of efficient input/output interfaces to proteins and to other devices of nature. This is a problem even for the large DNA chains (containing some 10^9 nucleotides).

The opposite view is:

B. Nature functions within the Turing machine model.

How then can nature's apparent capability of solving NP-complete problems be explained without being forced to conclude $P = NP$? There are various answers to this question.

(i) Nature does not necessarily achieve *global* optimization. If only a local energy minimum is achieved, say in protein folding, even if it is at bounded distance from the global minimum, it may not produce a near-optimal solution for an NP-complete problem encoded in terms of protein folding: NP-complete problems are equivalent and can be encoded in terms of each other; yet some of them have good approximation properties, and some don't; see e.g., [**ArL1997**].

(ii) NP-completeness is an asymptotic property, whereas the universe seems to be finite. Moreover, it seems that once a protein is sufficiently large, it is subdivided by nature into units of smaller size, say up to 200 amino acids per unit, which fold independently! See e.g. [**Pri1982**] and [**JaW1983**].

(iii) A problem π is polynomial if it is *universally* polynomial, i.e., if *all* its instances can be solved in polynomial time; it is NP-complete if *some* of its instances are NP-complete, though some of them may be solvable in polynomial time. In fact, NP-completeness reflects worst case behavior, but the average case behavior, under the assumption of some probability distribution, may be polynomial. Processes of nature are not necessarily universal. Thus perhaps the natural selection of nature may help to preserve proteins with polynomial folding mechanisms, and reject the others. Indeed, experiments show that some synthesized proteins may fail to fold into a stable conformation. Similarly, perhaps nature creates only very few knot types in DNA helices.

(iv) The protein folding mechanism may be encoded in the protein's amino acid sequence, analogously to the genetic code of a DNA chain, but the code is still unknown. If this is the case, then folding is not a search process, and there is no issue of complexity.

It may be difficult to decide between the views A and B at present. Some discussions on this point have been made. See e.g., [**Tra1996**]. But perhaps we can design experiments to create artificial proteins for exploiting fast computations even if the second view will turn out to be the valid one. For recent progress in protein design see e.g., [**BBD1995**], [**CDS1996**]. We don't understand the folding of proteins. Ken Dill put it thus: "Protein folding is a problem for us, but not for the proteins" (as quoted in [**Joh1997**]). However, mankind has exploited fire, dams, electricity, wireless communication, even before understanding the chemistry of fire, the hydraulics of water streams, the theory of electricity and waves. There is the possibility, if highly speculative, that we may be able to exploit protein folding for fast computing, even though we don't quite understand how nature goes about folding proteins.

It seems that the protein constructed in the proof of Theorem 1 shares a few properties of real proteins. It can fold, if unlike the folding of true proteins. The subsets A_1 and A_2 of edges resemble peptide bonds and other, less strong couplings respectively. The subset V_2 of vertices enables restricted rotations of specified subgraphs. The glue is possibly analogous to the subset of five generic nonpolar amino acids, any of which can be placed at strategic locations on the chain, which produce a similar pattern to an entire family of synthetic and natural proteins [**KSXBH1993**], [**WeH1995**].

For the purpose of illustration, to solve an arbitrary instance of GSG with given J-values on the edges, using the unrealistic protein constructed above, we might design a family of such synthetic proteins with the same glue, but random side chains of the frame/shell type above. Then a protein which will have frames whose strands have length 1 corresponding to edges of Γ where $J = -1$ or $J = 0$, length 2 for edges where $J = 1$, and matching lengths for the shells, will fold such that its minimum energy conformation enables to read off the spin values for the given instance of GSG, as specified in the last part of the proof of Theorem 1.

One approach for making the protein and ensuing experiments more realistic, would be to initiate protein-cooperation between physical or structural chemists, biologists and computer scientists or mathematicians, similarly to the DNA-cooperation that already exists in the area of the human genome and DNA-computing.

Conclusion

We have encoded the NP-complete spin glass problem in terms of a minimum energy model of protein folding, such that an instance of the former is a YES-instance if and only if its encoding is a YES-instance of the latter. We have also seen that even the problem of binary side chain foldings of proteins is NP-complete. The protein constructed in the proof contains only three amino acids and approximates true folding.

It would be of interest to encode other NP-complete problems in terms of protein folding. It would also be interesting to do some experimental work with real proteins, aimed at using them for speeding up computations. Whereas each element in DNA computations is well-understood and involves a simple slow computation, but the high parallelism of their great quantity is exploited for simulating the operation of a nondeterministic Turing machine, each protein is a "black box" computer which may have higher computational power than a DNA strand.

Acknowledgment. I thank Ron Unger, whose criticism on an early draft led me to an improved protein construction.

References

1. [Anf1973] C.B. Anfinsen [1973], Principles that govern the folding of protein chains, *Science* **181**, 223–230.

2. [ArL1997] S. Arora and C. Lund [1997], Hardness of Approximations, Ch. 10 in: *Approximation Algorithms for NP-Hard Problems* (D.S. Hochbaum, ed.), PWS Publishing Company, Boston, pp. 399–446.

3. [Bar1982] F. Barahona [1982], On the computational complexity of Ising spin glass models, *J. Phys. A: Math. Gen.* **15**, 3241–3253.

4. [BMRU1982] F. Barahona, R. Maynard, R. Rammal and J.P. Uhry [1982], Morphology of ground states of two-dimensional frustration model, *J. Phys. A: Math. Gen.* **15**, 673–699.

5. [Bax1982] R.J. Baxter [1982], *Exactly Solved Models in Statistical Mechanics*, Academic Press, London.

6. [BBBSS1992] C.H. Bennett, F. Bessette, G. Brassard, L. Salvail and J. Smolin [1992], Experimental quantum cryptography, *J. Crypto.* **5**, 3–28.

7. [BB1989] C.H. Bennett and G. Brassard [1989], The dawn of a new era for quantum cryptography: the experimental prototype is working!, *Assoc. Comput. Mach. SIGACT News* **20**(4), 78–82.

8. [BBCS1991] C.H. Bennett, G. Brassard, C. Crépeau and M.-H. Skubiszewska [1991], Practical quantum oblivious transfer, *Proc. Crypto.* **91**, 351–366.

9. [BBM1992] C.H. Bennett, G. Brassard and N.D. Mermin [1992], Quantum cryptography without Bell's theorem, *Phys. Rev. Lett.* **68**, 557–559.

10. [BBD1995] S.F. Betz, J.W. Bryson and W.F. DeGrado [1995], Native-like and structurally characterized designed α-helical bundles, *Current Opinion in Structural Biology* **5**, 457–463.

11. [BMRU1980] I. Bieche, R. Maynard, R. Rammal and J.P. Uhry [1980], On the ground states of the frustration model of a spin glass by a matching method of graph theory, *J. Phys. A: Math. Gen.* **13**, 2553–2567.

12. [Bra1988] G. Brassard [1988], *Modern Cryptology*, Lecture Notes in Computer Science, Springer-Verlag, Vol. 325, New York.

13. [BC1990] G. Brassard and C. Crépeau [1990], Quantum bit commitment and coin tossing protocols, *Proc. Crypto.* **90**, 49–61.

14. [CDS1996] M.H.J. Cordes, A.R. Davidson and R.T. Sauer [1996], Sequence space, folding and protein design, *Current Opinion in Structural Biology* **6**, 3–10.

15. [DDL1994] J. Desmet, M. De Maeyer and I. Lasters [1994], The "dead-end elimination" theorem: a new approach to the side-chain packing problem, in: *The Protein Folding Problem and Tertiary Structure Prediction* (K. Merz, Jr. and S. Le Grand, eds.), Birkhäuser, Boston, 1994, Ch. 10, pp. 308–337.

16. [Deu1985] D. Deutsch [1985], Quantum theory, the Church-Turing principle and the universal quantum computer, *Proc. Roy. Soc. London A* **400**, 97–117.

17. [Deu1989] D. Deutsch [1989], Quantum computational networks, *Proc. Roy. Soc. London A* **425**, 73–90.

18. [Fra1990] A.S. Fraenkel [1990], Deexponentializing complex computational mathematical problems using physical or biological systems, Tech. Report CS90–30, Dept. of Appl. Math. and Computer Science, Weizmann Inst. of Science.

19. [Fra1993] A.S. Fraenkel [1993], Complexity of protein folding, *Bull. Math. Biol.* **55**, 1199–1210.

20. [GaJ1979] M.R. Garey and D.S. Johnson [1979], *Computers and Intractability: A Guide to the Theory of NP-Completeness*, Freeman, San Francisco, CA.

21. [GiK1990] L.M. Gierasch and J. King (eds.) [1990], *Protein Folding: Deciphering the Second Half of the Genetic Code*, Amer. Assoc. Adv. Sci., Washington, D.C.

22. [HaI1996a] W.E. Hart and S. Istrail [1996], Fast protein folding in the hydrophobic-hydrophilic model within three-eights of optimal, *J. Computational Biology* **3**, 53–96.

23. [HaI1996b] W.E. Hart and S. Istrail [1996], Invariant patterns in crystal lattices: implications for protein folding algorithms (Extended Abstract), Proc. 1996 Combinatorial Pattern Matching, Lecture Notes in Computer Science No.

1075, pp. 288-303, Springer Verlag 1996.

24. [HaI1997] W.E. Hart and S. Istrail [1997], Robust proofs of NP-hardness for protein folding: general lattices and energy potentials, *J. Computational Biology* **4**, 1–22.

25. [HaI≥1997] W.E. Hart and S. Istrail [≥1997], Lattice and off-lattice side chain models of protein folding: linear time structure prediction better than 86% of optimal, *J. Computational Biology*, to appear.

26. [HBDPC1997] P.M. Harrison, P. Bamborough, V. Daggett, S.P. Prusiner and F.E. Cohen [1997], The prion folding problem, *Current Opinion in Structural Biology* **7**, 53–59.

27. [JaW1983] J. Janin and S.J. Wodak [1983], Structural domains in proteins and their role in the dynamics of protein function, *Prog. Biophys. Mol. Biol.* **42**, 21–78.

28. [Joh1983] D.S. Johnson [1983], The NP-completeness column: an ongoing guide, *J. Algorithms* **4**, 87–100.

29. [Joh1997] G. Johnson [1997], Designing life: proteins 1, computers 0, The New York Times (Science Times), March 25, 1997.

30. [KSXBH1993] S. Kamtekar, J.M. Schiffer, H. Xiong, J.M. Babik and M.H. Hecht [1993], Protein design by binary patterning of polar and nonpolar amino acids, *Science* **262**, 1680–1685.

31. [LeL1969] M. Levitt and S. Lifson [1969], Refinement of protein conformations using a macromolecular energy minimization procedure, *J. Mol. Biol.* **46**, 269–279.

32. [LeS1988] M. Levitt and R. Sharon [1988], Accurate simulation of protein dynamics in solution, *Proc. Natl. Acad. Sci. USA* **85**, 7557–7561.

33. [NgM1992] J.T. Ngo and J. Marks [1992], Computational complexity of a problem in molecular-structure prediction, *Protein Engineering* **5**, 313–321.

34. [NMK1994] J.T. Ngo, J. Marks and M. Karplus [1994], Computational complexity, protein structure prediction, and the Levinthal paradox, in: *The Protein Folding Problem and Tertiary Structure Prediction* (K. Merz, Jr. and S. Le Grand, eds.), Birkhäuser, Boston, 1994, Ch. 14, pp. 433–506.

35. [PaP1996] M. Paterson and T. Przytycka [1996], On the complexity of string folding, *Discr. Appl. Math.* **71**, 217–230.

36. [Pri1979] P.L. Privalov [1979], Stability of proteins, *Adv. Protein Chem.* **33**, 167–241.

37. [Pri1982] P.O. Privalov [1982], Stability of proteins: proteins which do not present a single cooperative system, in *Advances in Protein Chemistry* **35**, 1–104.

38. [Ree1988] G.N. Reeke, Jr. [1988], Protein folding: computational approaches to an exponential time problem, in: *Annual Reviews of Computer Science* **3**, 59–84, Annual Reviews, Inc.

39. [RoS1996] N. Robertson and P.D. Seymour [1996], Graph minors. XV. Giant steps, *J. Combin. Theory* Ser.B **68**,112–148, 1996.

40. [Tra1996] J. Traub [1996], On reality and models, in: *Boundaries and Barriers: On the Limits to Scientific Knowledge* (J.l. Casti and A. Karlqvist, eds.), Addison-Wesley, 1996, pp. 238-254.

41. [UnM1993] R. Unger and J. Moult [1993], Finding the lowest free energy conformation of a protein is a NP-hard problem: proof and implications, *Bull. Math. Biol.* **55**, 1183–1198.

42. [WaC1986] S.A. Wasserman and N.R. Cozzarelli [1986], Biochemical topology: applications to DNA recombination and replication, *Science* **232**, 951–960.

43. [Wel1993] D.J.A. Welsh [1993], The complexity of knots, *Ann. Disc. Math.* **55**, 159–171.

44. [WeH1995] M.W. West and M.H. Hecht [1995], Binary patterning of polar and nonpolar amino acids in the sequences and structures of native proteins, *Protein Science* **4**, 2032–2039.

45. [Wie1983] S. Wiesner [1983], Conjugate coding, *Assoc. Comput. Mach. SIGACT News* **15**(1), 78–88 (manuscript written about 1970).

Department of Applied Mathematics and Computer Science, Weizmann Institute of Science, Rehovot 76100, Israel

fraenkel@wisdom.weizmann.ac.il

`http://www.wisdom.weizmann.ac.il/~fraenkel`

DIMACS Series in Discrete Mathematics
and Theoretical Computer Science
Volume **48**, 1999

Creating 3-Dimensional Graph Structures with DNA

Nataša Jonoska*, Stephen A. Karl, and Masahico Saito*

ABSTRACT. We propose solving computational problems with DNA molecules by physically constructing 3-dimensional graph structures. Building blocks consisting of intertwined strands of DNA are used to represent graph edges and vertices. Different blocks would be combined to form all possible 3-dimensional structures representing a graph. The solution to the Hamiltonian cycle problem provided requires a constant number of steps regardless of the number of vertices. If a solution to the graph problem exists, then a fully closed circular molecule would be formed and can be isolated. This paper introduces a method of using 3D structures in computing which might significantly improve the efficiency of computations with DNA.

1. Introduction

In 1994, Adleman [**1**] described a laboratory experiment involving DNA in which an example of a Directed Hamiltonian Path Problem was solved. This paper opened the field of practical DNA computing. Lipton [**5**] demonstrated that a large class of NP-complete problems also could be solved by encoding the problem in DNA molecules. Shortly after, several authors have suggested applications of DNA methodology for computational purposes (see for example [**4**] and [**9**]).

Generally, DNA molecules have been treated as linear strings where much of the information content is encoded in the order of nucleotides that make up the DNA. These algorithms require polynomial increases in the number of steps necessary to identify a solution with increasing size of the problem (e.g., the number of vertices in a graph in [**1**] and the number of variables in a formula in [**5**]). Recently reported research [**11**] however, has demonstrated that it is possible to form higher order, three dimensional (3D) structures with DNA molecules. In this paper we explore the possibility of the use of 3D DNA structures. We show that, theoretically, the use of 3D DNA structures could significantly reduce the time and steps needed to identify a solution. In fact, 3D structures allow the Hamiltonian cycle problem to be solved with a constant number of steps, regardless of the number of vertices in the graph.

1991 *Mathematics Subject Classification.* Primary 68Q99, 68R10; Secondary 92E10 .

* Supported, in part, by the University of South Florida Research and Creative Scholarship Grant Program under grants number 1249933RO and 1249932RO.

In [**13**] a method was proposed to solve the Hamiltonian Path Problem in a constant number of steps by self-assembling double-crossover molecules into larger structures such that if a solution exists, then there is a closed DNA molecule which encodes the answers and which can be isolated. Self-assembling molecules build complex DNA structures that can be viewed as lying on a 2D surface, and the computation does not make use of 3D DNA structures. Our approach is different in that we propose use of 3D flexible structures of DNA for solving the problem geometrically.

In this paper we start with a review of the Hamiltonian cycle problem and describe our approach through a simple example. A general case also is considered and we present the algorithm for arbitrary graphs. Finally, we discuss the general feasibility of the method.

Acknowledgement. We are grateful to N.C. Seeman and E. Winfree for valuable comments and discussions.

2. The Hamiltonian Cycle Problem

2.1. Definition. Let G be a finite graph with $V(G)$ the set of vertices and $E(G)$ the set of edges. A *Hamiltonian cycle* c of G is a cycle (a loop on the graph when we regard G as a 1-complex, without self intersection points) such that it goes through every vertex exactly once. The *Hamiltonian cycle problem (HCP)* asks whether a given graph G has a Hamiltonian cycle.

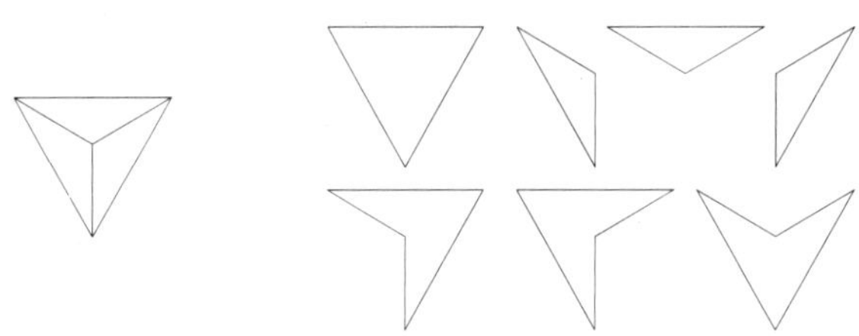

FIGURE 1. A simple four vertex graph and all individual cycles

2.2. Example. An example of a graph is depicted in Fig. 1 (left). This is the complete graph with 4 vertices (i.e., C_4). Individual cycles through the graph are illustrated to the right of the whole graph. Since none of the four cycles in the right top of the figure pass through all vertices, they are non-Hamiltonian cycles. The three cycles illustrated in the right bottom of the figure are Hamiltonian cycles.

The HCP is a well known member of a larger class of problems known as NP-complete. NP-complete problems have posed a significant challenge to computer scientists since all conventional algorithms require (in the worst case) an exponential increase in the number of steps (or time) relative to the size of the problem. Identifying a solution to even a moderately simple problem therefore, can require a prohibitively large number of steps. In fact, the excitement over DNA as a computational tool rests in the ability to perform in parallel a large number of the computational steps. Thus, the number of steps to identify a solution does not increase exponentially with increases in the size of the graph.

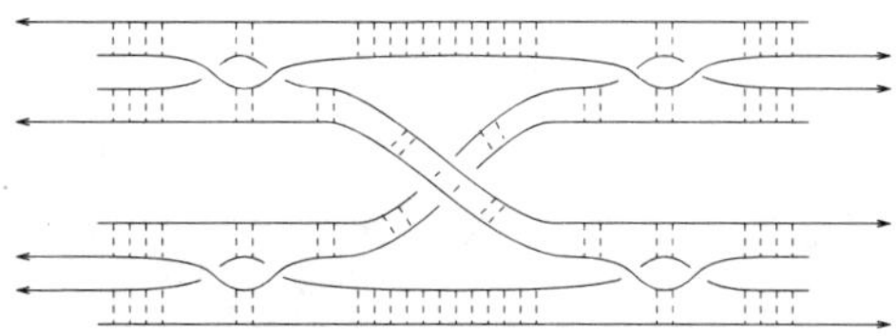

FIGURE 2. Type I edge building block

2.3. Review of Adleman's Algorithm. Adleman's solution involves 4 basic steps:

1. Form random DNA paths in the graph.
2. Extract paths that go through exactly n (possibly repeating) vertices of the graph, where n is the number of vertices.
3. For each vertex i ($i = 1, \ldots, n$) extract the paths that go through vertex i (i.e. extract paths that visit every vertex).
4. If a path remains, a Hamiltonian cycle exists.

The first step presumably generates an exponential number of paths that contain the result. Steps 2 and 3 are used to isolate and detect the result generated by step 1. Step 3 of this algorithm must be repeated n times for a graph of n vertices. The laboratory procedure suggested for this step uses biotin label oligos that are complementary in DNA sequence to the vertices. Paths containing a specific vertex are removed from the mix when hybridized to the biotin labeled oligo that is conjugated to paramagnetic beads. The resulting mixture is sequentially treated in a similar way with each vertex specific oligo. The extraction procedure is undoubtedly less then 100% effective [**7**]. With repeated use, errors will tend to accumulate and could result in false positives (i.e., concluding that a Hamiltonian cycle exists when in fact one does not). For this reason, other extraction techniques such as restriction endonuclease digestion have been suggested [**2**]. Although these methods may be more efficient than biotin beads, the repetitive use of the extraction technique is not avoided and errors still can accumulate.

2.4. Constructing 3D graphs with DNA. We have chosen a simple four vertex Hamiltonian cycle problem (Fig. 1) to illustrate how to construct 3D graphs with DNA. Each edge of the graph is represented by the 3D DNA structure shown in Fig. 2. Orientation of the strands of DNA are indicated with arrowheads being placed at the $3'$ end. Hydrogen bonds between the anti-parallel, complementary Watson-Crick (WC) bonds are depicted as dotted segments between the strands. As can be seen in the figure, each edge is a complex of intertwined single and double stranded segments of DNA. The double stranded sections function to hold the 3D structure of the edge and can vary from 25 to 35 nucleotides long depending on double-stranded stability of the sequence. For simplicity, the double helix of each of the molecules is not presented in the figure. This DNA structure is referred to as a *type I edge block*. Different such blocks are needed for each edge of the graph.

The $3'$ ends of the DNA strands in the edge block end with single stranded segments of 20 to 30 nucleotides length that are vertex-edge specific. They have complementary sequence tails such that compatible edge blocks can form WC paired double-stranded DNA segments and be joined together. Figure 3 shows a formation of a vertex in the graph. For example, the $3'$ ends labeled 1 of the bottom edge

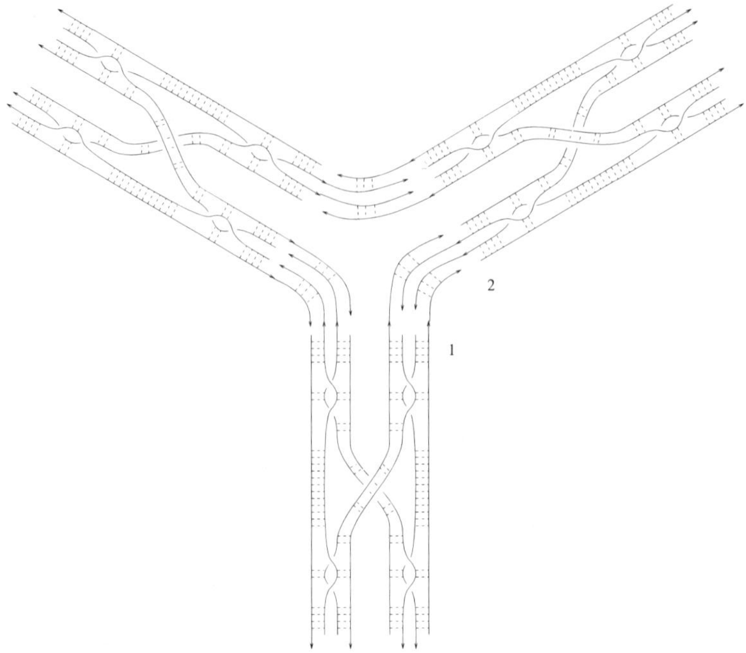

FIGURE 3. Construction of a trivalent vertex

are anti-parallel and complementary to the $3'$ ends labeled 2 of the right top edge. To form the graph, all edge blocks are combined and their compatible ends are allowed to from double-stranded DNA. Once formed, the edges are locked together by sealing all open "nicks" in the DNA strands with DNA ligase. Three-dimensional DNA structures that do not contain open ends are referred to as *graph structures.*

2.5. Solving the Hamiltonian path problem: specific case.

1. Combine the edge blocks in a single mix, allow them to hybridize and then ligate them together.
2. Remove partially formed 3D DNA structures with open ends that have not been matched. Biotin labeled oligos complementary to the junction points of all of the edge blocks can be used to remove incomplete structures intact. By this process paths that visit a vertex more than once are removed except those that are part of a covering space of the graph. See Section 4.3 for discussions on covering graphs and how to remove them.
3. Choose a vertex in the graph (e.g., v_0; Fig. 4). Primers specific to this vertex are used in polymerase chain reaction (PCR) amplification of the graph. Primers for each of the six double strands of DNA that pass through v_0 are used and all circular strings corresponding to cycles passing through v_0 are amplified and isolated. Other circular strings not passing through v_0 (and hence not Hamiltonian cycles) will remain in the mixture at negligible concentrations.
4. Test for the presence of a Hamiltonian cycle. The lengths of DNA fragments after PCR amplification are determined by gel electrophoresis. If any of the fragments are of the appropriate length for a Hamiltonian cycle (i.e.,

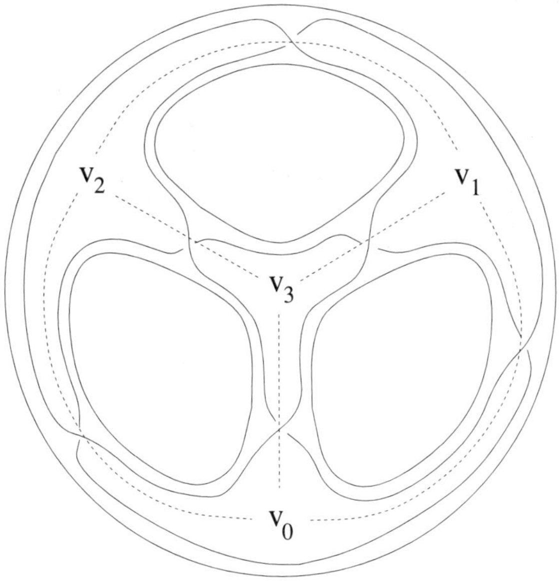

FIGURE 4. Cycles formed by DNA

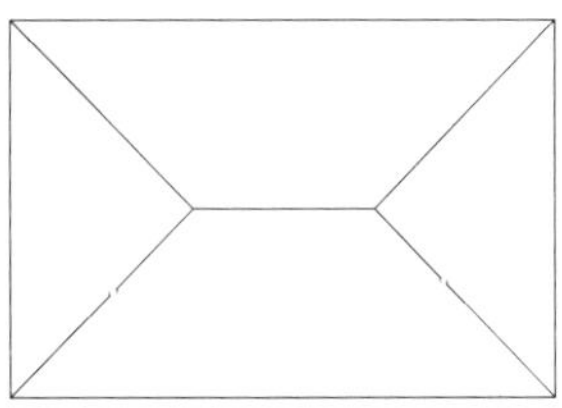

FIGURE 5. A more complex graph requiring type I and type II blocks

the number of vertices times the length of each building block), then a Hamiltonian cycle exists for the graph. If, for example, the strands that make a building block of an edge are 50 base pairs long and the number of vertices of the graph is 4, a Hamiltonian cycle is 200 bp in length. No further assays are needed.

Figure 4 illustrates the cycles formed by DNA. To simplify the figure each line represents a double stranded DNA molecule. The graph C_4 is presented with thin dotted lines and vertices are labeled. In this example, every Hamiltonian cycle is realized in the graph. By following the above procedure, we could conclude if this graph has a Hamiltonian cycle. In general, however, type I edge blocks cannot form every possible Hamiltonian cycle. Our proposed procedure gives a solution for this particular graph because each vertex belongs to exactly 6 cycles (and there are 6 double strands "passing" through a vertex) and all possible cycles in the graph are produced. Since this may not always be the case, a more generalized procedure is needed.

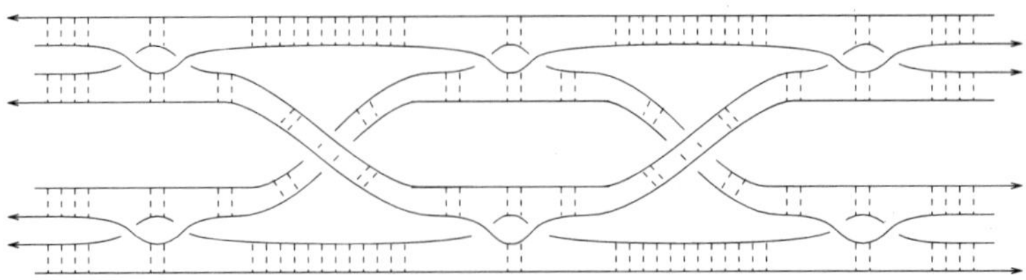

FIGURE 6. Type II edge building block

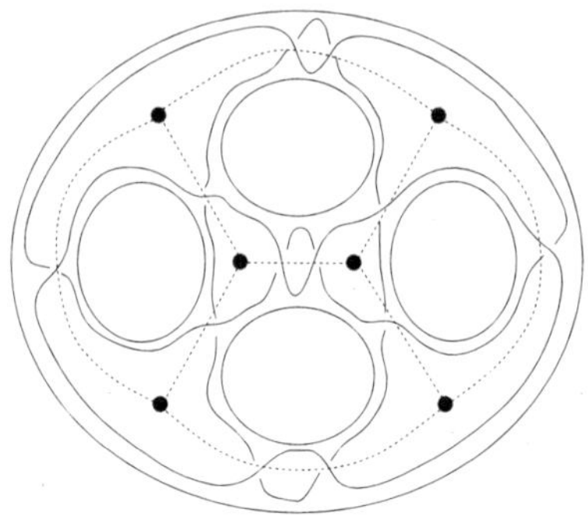

FIGURE 7. One of the DNA patterns

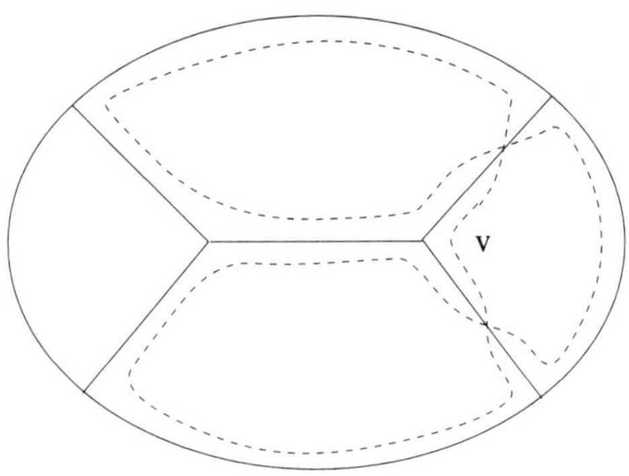

FIGURE 8. A DNA strand passing through the vertex v three times

3. Solving the Hamiltonian cycle problem: general case

In this section we present a general solution to the Hamiltonian cycle problem. When considering the graph presented in Fig. 5, one can easily see that it is impossible to produce all cycles in the graph using only type I edge blocks. To obtain all cycles of this graph some of the internal cycles must contain non-crossover DNA segments. In order to circumvent this problem, we propose a *type II edge block* as shown in Fig. 6. By using both type I and type II blocks all possible cycles of the graph can be generated. A specific pattern of cycles obtained by these blocks is presented in Fig. 7. The vertices of the graph are represented by black dots and the edges are represented by thin dotted segments. In this pattern Hamiltonian cycles are realized.

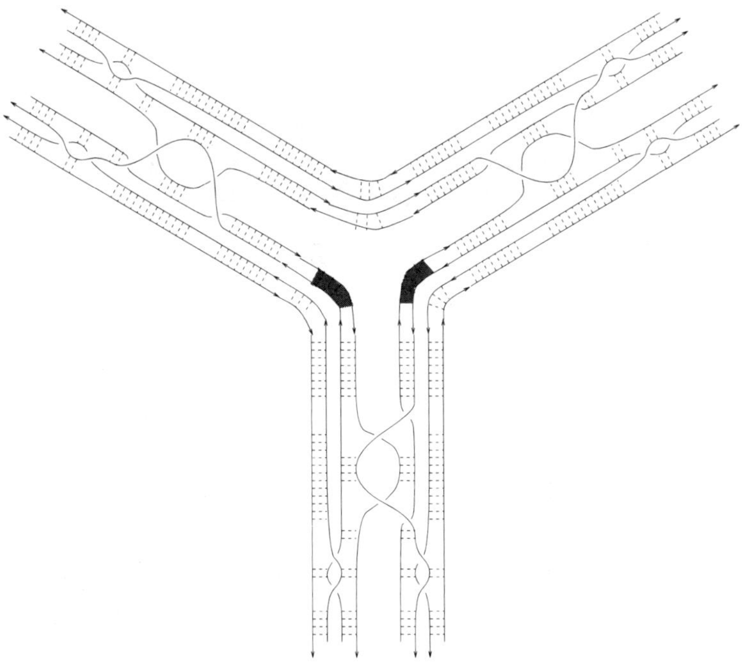

FIGURE 9. A building block for vertices

With this example, however, another problem arises. Although all cycles through the graph can be produced by using both types of edge blocks, circular molecules that pass through a vertex up to three times (Fig. 8) also are possible. Each of the inner three double strands presented in Fig. 3 can be a part of the same circular double stranded molecule. We therefore need to include an additional DNA structure; a *vertex building block*.

A vertex building block of degree 3 is shown in Fig. 9. Shaded areas in the double stranded molecules of the block indicate segments that encode sequences for sites that can be cleaved by restriction endonuclease enzymes. Both sites are encoded with the same recognition site and can be cleaved simultaneously. Only the "inner" strands of the blocks are important to the graph solution. The "outer" strands are needed only as support for the inner. Three blocks are constructed for every vertex combination such that all cyclically rotated versions are represented. This allows a configuration where every pair of inner double strands of a vertex block can be cut by a restriction endonuclease.

Building blocks for higher degree vertices are constructed in a manner similar to that shown in Fig. 9. For a given vertex v (with degree ≥ 3), a separate building block for every combination of two edges e_i, e_j, that pass through the vertex is needed. For example, if a vertex has degree 5 with edges $\{e_1, \ldots, e_5\}$ passing through v, then a building block of the form presented in Fig. 10 is needed for each $\{i, j\}$, $(i \neq j)$ where $\{k, k', k''\} = \{1, \ldots, 5\} - \{i, j\}$. Figure 10 represents the inner double strands of the building block and the actual structure is essentially similar to the one presented in Fig. 9. The shaded areas in Fig. 10 again represent restriction endonuclease recognition sites. The nonshaded areas of the block are

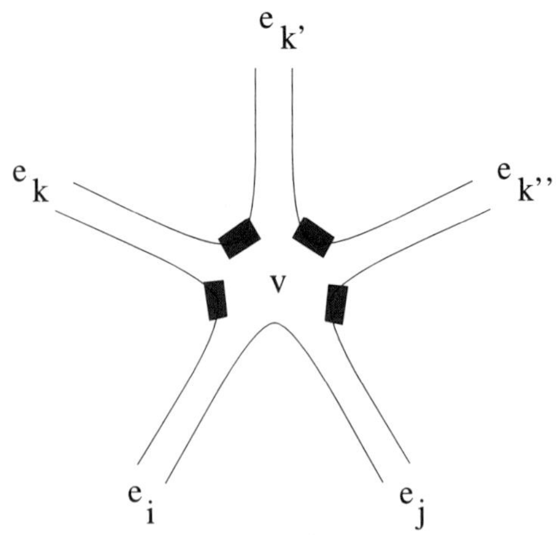

FIGURE 10. A five degree vertex v building block

double stranded molecules encoding path e_ie_j (or e_je_i) that could be part of a cycle passing through vertex v. This block is referred to as *compatible* with e_i, e_j and is denoted by $B_v(e_i, e_j)$.

Note that if a graph has a degree 1 vertex, then there is no Hamiltonian cycle, and if a vertex has degree 2 shared by edges e_1 and e_2, then the path e_1e_2 can be substituted by an edge block.

3.1. Remark. If the degree of a vertex v is $k_v \geq 3$ then the number of blocks needed is:

$$\sum_{i=1}^{n} \binom{k_v}{2} + 2m$$

where m is the number of edges. This expression is bounded by a cubic polynomial on the number of vertices (n).

3.2. Proposition. Let $C = \{G_1, \ldots, G_m\}$ be the set of graph structures G_i that are obtained by adding all building block molecules in a tube and allowing them to join and be ligated. Assume that C contains all possible graph structures. Let $c = v_0v_1 \cdots v_kv_0$ be a cycle in the graph G. Then there is $G_j \in C$ that contains a circular double stranded DNA molecule that encodes c and does not contain a restriction site.

Proof: Let e_i be the edge from v_i to v_{i+1} ($i+1$ is taken modulo $k+1$) for $i = 0 \ldots, k$. If a vertex v_i has degree 2 then there are edge building blocks for the path $e_{i-1}e_i$. Thus we can assume that all vertices have degree at least 3. We form a graph structure G_j containing vertex blocks $B_0, \ldots, B_k$ where $B_i = B_{v_i}(e_{i-1}, e_i)$ ($i-1$ is modulo $k+1$) and edge blocks for edges $e_0, \ldots, e_k$. If G has more than $k+1$ vertices, then for G_j we use arbitrary blocks for vertices other than $v_0, \ldots, v_k$. A building block for e_i connects blocks B_i and B_{i+1}. The inner double strands of the two edge building blocks (type I and II) of e_i give four possibilities of paths to be constructed: $e_{i-1}e_ie_{i+1}, e_{i-1}e_ie', e''e_ie_{i+1}$ and $e''e_ie'$ where e', e'' are some other edges incident to vertices v_i and v_{i+1} respectively. Each type (I or II) of building blocks for e_i allows exactly two of these paths to be encoded. The structure G_j is then constructed such that the building block for e_i always assures a DNA segment encoding a path $e_{i-1}e_ie_{i+1}$ from blocks B_i to B_{i+1} for $i = 0, \cdots, k-1$.

Hence, the structure G_j contains a double stranded circular DNA molecule encoding the cycle c and by the specific choice of $B_0, \ldots, B_k$ (each B_i is compatible with $e_{i-1}e_i$), c does not contain a restriction endonuclease recognition site. □

3.3. Procedure. The following procedure describes the DNA based algorithm for solving the Hamiltonian cycle problem:

1. Combine all building blocks molecules in a tube and allow them to hybridize and then be ligated.
2. Remove molecules that are partially annealed or are not part of a graph structure.
3. Remove larger "covering graphs" by gel electrophoresis. Covering graphs are discussed below, (see Section 4.3).
4. Digest the DNA with a restriction endonuclease. This cleaves cycles that pass through a vertex more than once.
5. PCR amplify using primers specific to a chosen vertex v_0.
6. Size sort the DNA fragments by gel electrophoresis to determine if a Hamiltonian cycle is present.

In step 1 it is assumed that every possible graph structure is obtained. By Proposition 3.2, every cycle in the graph is represented by a circular, double-stranded DNA molecule in one of the graph structures obtained. In step 2 we remove the partially formed structures. This is done with biotin beads, as before. In step 3 all closed paths that pass through a vertex more than once are linearized. Any circular molecules that remain represent a cycle in the graph. The Hamiltonian cycle is obtained by PCR amplification in the same way as previously described.

It is important to note that none of the steps performed depends on the number of vertices or edges in the graph. This procedure, therefore, describes a constant time algorithm for solving the NP-complete HCP.

4. Discussion of the feasibility of the method

4.1. Complexity of the building blocks. The idea of using different building blocks to depict the entire graph structure was inspired by the work of N. Seeman and his research group [**11**] where they describe the construction of different 3D DNA structures. The building blocks proposed here are somewhat different, but contain double parallel crossover molecules which have been previously discussed (see references in [**11**]).

Currently, a major limitation to our approach is that the algorithm presented assumes that desired building blocks already are available. This is likely to be the most difficult and complex part of the procedure and specific laboratory experiments are necessary to determine the degree of difficulty involved in forming these building blocks.

Seeman [**10**] suggested two physical models for exploring 3D DNA structures. As a support for the plausibility of our construction, we built the proposed structure (type I edge building block) using one of the models described in [**10**], as shown in Fig. 11.

It could be simpler to encode building blocks with single stranded molecules which would carry the information of the double stranded molecules employed in our building blocks. Figure 12 shows an example of a possible single-stranded

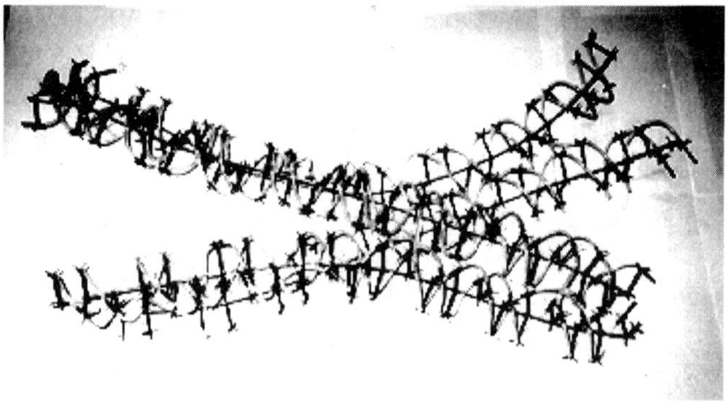

FIGURE 11. A physical model of type I edge block

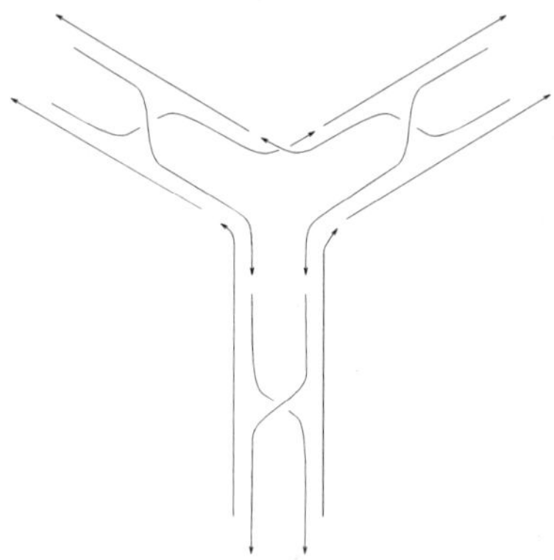

FIGURE 12. Incompatible orientations

structures forming a degree 3 vertex. Since the "inner" and "outer" segments are switched at the top, the argument of Proposition 3.2 can not be applied. In order to apply the argument of Proposition 3.2 additional building blocks for the vertices that would realize all the possible extra permuatations between the "inner" and the "outer" segments are needed. Furthermore, a difficulty arises in avoiding circular single stranded molecules that pass through a vertex more than once. This difficulty could be overcome by adding a hairpin like structure to the single strands which would contain a restriction site (Fig. 13). This additional structure was suggested to us by Seeman. The shaded part indicates the site that can be cleaved by restriction endonuclease enzymes and can be used to eliminate circular molecules that pass through a vertex more than once. This restriction site corresponds to the restriction site (shaded area of Fig. 9) of the vertex block. With this alterations of the single-stranded building blocks, the Proposition 3.2 holds and the algorithm suggested with Procedure 3.3 solves the HCP in constant time.

Another possible approach, suggested by Winfree [**14**], is a use of k-armed junction molecules with sticky ends as a vertex building blocks. These molecules have been described recently in [**11**] and [**13**], (see also references therein). Such k-armed junction molecules could be formed from partially phosphorylated oligos such that after formation of the graph structure by hybridization, the ligation process

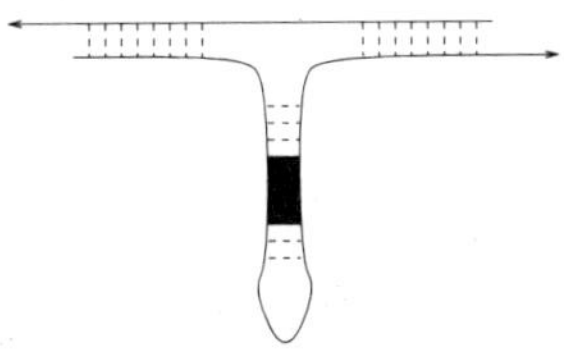

FIGURE 13. T-shaped structure

generates only circular molecules that encode cycles (paths that visit a vertex at most once). This simple design of vertex building blocks somewhat increases the number of building blocks needed, but its simplicity provides a more feasible way of using 3D DNA structures and solving the HCP in constant time.

4.2. Generating graphs. Step 1 of the procedure described in the previous section assumes that all possible combinations of graph structures are generated. This step is similar to the first step in many DNA based algorithms as well as Adleman's algorithm described in section 2. Studies have shown, however, that even in a fairly simple mixture, appropriate reaction conditions for annealing are important and difficult to determine [**6**]. To date, we are unaware of laboratory experiments of this type being performed on 3D DNA structures. It is impossible, therefore, to assess the feasibility of generating all possible graph structures.

4.3. Covering graphs. All of the graph constructions proposed here are local in the sense that we start from vertex and edge building blocks that are assembled to form graphs. Graphs other than the ones intended, however, can be obtained (Fig. 14) and may consists of two (or more) copies of the given graph with a pair of edges crossing each other to form a new larger graph. The graph in Fig. 14 is a double covering space of C_4 when the graphs are considered as topological spaces (1-dimensional complex). Theoretically any covering graph of the intended graph is possible.

4.4. Proposition. Let $G = (V, E)$ be a graph. Let $\hat{E}$ be a set of building blocks obtained by DNA molecules that encode the edges of G and $\hat{V}$ a set of building blocks obtained by DNA molecules that encode the vertices of G. Assume that the building blocks have single stranded $3'$ ends such that the building blocks of each vertex v in V have Watson-Crick compatible regions with the edges that are incident to v. If multiple copies of building blocks $\hat{V}$ and $\hat{E}$ are allowed to hybridize and ligate in a tube, then the resulting graph structures are, topologically, covering spaces of G.

This proposition follows from the topological definition of a covering space and the fact that the structure that is formed is obtained locally (see for example [**8**] (Section 8.3)).

Since the structures representing a covering space of a graph are much larger than the structures representing the original graph, it may be possible to remove them by gel electrophoresis. Avoiding the formation of a covering space graphs all together, however, may be impossible.

We end this section with a note that, with the exception of forming the building blocks, many of the laboratory techniques suggested are fairly standard and

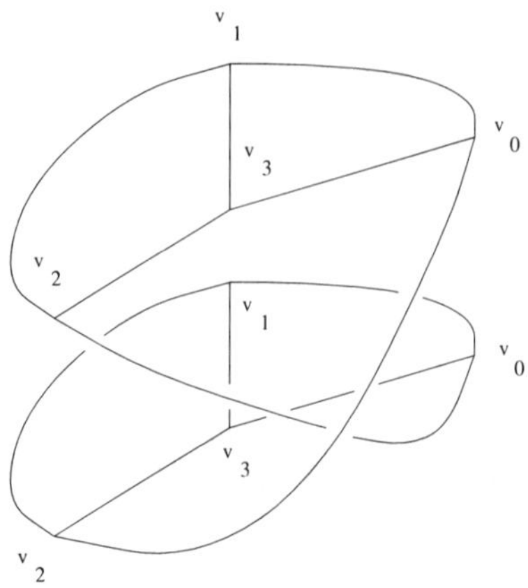

FIGURE 14. A double cover of a graph

their detailed description can be found in [**3**]. Their application to complex DNA structures might, however, require extensive experimentation before they can be used as proposed.

4.5. Conclusion. The procedure and the algorithm presented are only an attempt to demonstrate that using higher dimensional structures in DNA based algorithms might provide simpler and faster solutions to many problems. Electronic computers use information only in a linear (string) form. By using DNA, we can take advantage of 3D structures that can contain considerably more information than the sequence of nucleotides from which they are made.

At the moment, however, obtaining 3D DNA structures is difficult. The algorithms suggested by Adleman [**1**] and Lipton are not of constant time, but the laboratory techniques involved are fairly standard and their general characteristics are known. To the contrary, our approach requires a constant number of steps (regardless of the size of the problem) but the laboratory techniques are generally uncharacterized as to their difficulty and specificity. It is our belief, however, that as polymer technology develops, constructing the vertex and edge building blocks needed for this approach should become easier. Even if building blocks were readily available, the uncontrolled formation of covering space structures remains a formidable obstacle. Nonetheless, the covering spaces themselves carry a substantial amount of information and may be useful in other computational problems.

References

[1] L. Adleman, *Molecular computation of solutions of combinatorial problems*, Science **266** (1994), 1021-1024.

[2] M. Amos, A. Gibbons, D. Hodgson, *Error-resistant implementation of DNA computations*, in [**9**]

[3] F.M. Ausubel, R. Brent, R.E. Kingston, D.D. Moore, J.G. Seidman, J.A. Smith, K. Struhl, P. Wang-Iverson and S.G. Bonitz. *Current Protocols in Molecular Biology*, Greene Publishing Associates and Wiley-Interscience, New York, NY (1993).

[4] *DNA Based Computers*, edited by E. Baum, R. Lipton, *DIMACS series* **vol. 27** AMS 1996.

[5] R. Lipton, *DNA solution of hard computational problems*, Science **268** (1995), 542-545.

[6] N. Jonoska, S.A. Karl, *Ligation Experiments in DNA Computations* Proceedings of 1997 IEEE International Conference on Evolutionary Computation (ICEC'97), April 13-16, 1997, 261-265.
[7] J. Khodor, D.K. Gifford, *The efficiency of sequence-specific separation of DNA mixtures for biological computing,* in Preliminary Proceedings, 3rd DIMACS Workshop om DNA Based Computers, June 23rd – June 25th, 1997, 26–34.
[8] J. R. Munkres, *Topology, a first course,* Prentice-Hall, 1975.
[9] Proceedings of the Second Annual Meeting on *DNA Based Computers*, DIMACS Workshop, Princeton, NJ, June 10-12, 1996, (in press).
[10] N.C. Seeman, *Physical Models for Exploring DNA Topology.* Journal of Biomolecular Structure and Dynamics 5, 997-1004 (1988).
[11] N. C. Seeman et al. *The Perils of Polynucleotides: The Experimental gap Between the Design and Assembly of Unusual DNA Structures* in [**9**].
[12] N.C. Seeman, private communication.
[13] E. Winfree, X. Yang, N.C. Seeman, *Universal Computation via Self-assembry of DNA: Some Theory and Experiments.* in [**9**].
[14] E. Winfree, private communication.

Department of Mathematics, University of South Florida, Tampa, Florida 33620
E-mail address: jonoska@math.usf.edu

Department of Biology, University of South Florida, Tampa, Florida 33620
E-mail address: karl@chuma.cas.usf.edu

Department of Mathematics, University of South Florida, Tampa, Florida 33620
E-mail address: saito@math.usf.edu

DIMACS Series in Discrete Mathematics
and Theoretical Computer Science
Volume **48**, 1999

DNA Implementation of Nondeterminism

The Molecular Computing Group
Y. Gao, M. Garzon, R. C. Murphy, J. A. Rose, R. Deaton,
D. R. Franceschetti, and S. E. Stevens, Jr.

ABSTRACT. We explore the capabilities of DNA reactions to implement nondeterministic computations at levels lower than the HAMILTONIAN PATH problem. We provide a technique for programmable fault-tolerant implementation of nondeterministic finite-state machines that enforces the basic conditions in the subset constructions that permit efficient computation. The implementation can be extended to arbitrary nondeterministic Turing machines of a moderate size in practice since they are basically finite state machines with a large state set. It also suggests an experiment to provide evidence for the open analogous questions about the tradeoff between nondeterminism and computational resources.

1. Introduction

Adleman [**1**] showed how the actual mechanisms underlying DNA recombination and separation can perform computations meaningful to human endeavors, such as solving instances of the HAMILTONIAN PATH problem. A lot of the work generated has concentrated on determining the fundamental information processing capabilities of DNA based reactions in two areas. On the one hand, they have been explored at the upper end of the computability spectrum in several places. Bennet [**5**] used imaginary enzymes to show how to simulate the transitions of arbitrary Turing machines. Wilhem and Rothemund [**22**] show the same with commercially available enzymes. Smith [**19**] starts with a similar result, but further provides a somewhat negative assessment of the feasibility of DNA based computing. On the other hand, DNA capabilities have also been explored for problems on the fringe of what is theoretically solvable on VLSI computers by Lipton [**14**] (all **NP** problems) and Beaver [**3**] (all **PSPACE** problems). [**19**] and [**22**], however, question the feasibility of such approaches, and whether that would be the best way to approach the relationship between biology and computation. They point out that, although DNA computing allows implementation of nondeterminism and massive parallelism, unbridled use of molecules can quickly offset the potential gain offered by these advantages and/or introduce errors than render the computation infeasible or unreliable.

Perhaps more important from a practical point of view is therefore to understand the range of feasible tasks that DNA computing can perform efficiently and

---

1991 *Mathematics Subject Classification.* Primary 68Q40, 68U99, 68Q15.

reliably under realistic assumptions on the chemical environments in which DNA computations take place. Several DNA-based algorithms such as binary addition [**10**], real-valued multiplication [**17**], breadth-first search and dynamic programming [**2**] have been suggested. More recently, attention has been focused on models which, although not as powerful from a theoretical point of view, hold an appeal in terms of practical applications, such as finite-state machines (fsm) [**18**]. Two implementations of the deterministic model have been given in [**8**], one based on ligation and another one based on hybridization without ligation. The ligation-based approach allows input of arbitrary length and can be readily implemented with current biotechnology, but requires sequential input feed and different molecules for different machines. In a second implementation not based on ligation, transitions are represented by reusable molecules, and the input, coded as a molecule, can be introduced at once. Both implementations allow optical extraction of the status of the machine.

This paper explores feasible, reliable and efficient ways to implement controlled nondeterminism on restricted devices such as finite-state machines. Nondeterminism is at the core of the difficulties that Adleman's original experiment was designed to overcome. On the other hand, nondeterminism is supposed to be well understood in the context of finite-state machines, specifically, a so-called *subset construction* produces a deterministic equivalent of a given nondeterministic fsm. It is conceivable that a better idea about the key difficulties of effectively using DNA computing may be gained by looking for ways to implement nondeterminism efficiently in DNA. The intent is not to have DNA implement a deterministic equivalent (which was done in [**8**]) or just nondeterminism (which was done in several places [**3, 22, 19**]), but rather to understand the ability of DNA to implement nondeterminism as a native mode of computation in a fault-tolerant and efficient way. The class **NP**, for example, is defined as the class of algorithmic problems whose solutions that can be *verified*, although not necessarily *found*, efficiently in time polynomial in the size of the input instance. From the computational point of view, given the importance of problems in the class **NP** that can be efficiently solved in nondeterministic mode, this is perhaps the most remarkable contribution in Adleman's original insight. We show that efficient nondeterministic computations can be performed by DNA, not only of finite-state machines, but perhaps even of more powerful low level Turing machines, at least to the extent that experimental evidence can be gathered for open questions about the tradeoff between nondeterminism and time and memory resources.

2. Definitions

In DNA-based computation, the instances of a problem are encoded in oligonucleotides, or strands, of DNA. The encoding alphabet is the set of nucleic acid bases `A`, `T`, `G`, `C`, which bind according to the Watson-Crick complement condition, $\overline{\mathtt{A}} = \mathtt{T}$ and $\overline{\mathtt{C}} \equiv \mathtt{G}$, and *vice versa*. Oligonucleotides bind in an antiparallel way with respect to the chemically distinct ends, $5'$ and $3'$, of the DNA molecule. Unlike the ordinary convention, oligonucleotides x are always written in this paper from the $3'$ to the $5'$ end throughout, unless we are describing double strands, in which case the lower oligonucleotide is always directed from $5'$ to $3'$ as a Watson-Crick (WC) complement. *Hybridization* is a chemical process that joins two complementary single strands into a double strand. *Ligation* is a chemical process whereby two double strands are joined into one double strand. A *restriction enzyme* (such as SmaI or EcoRI) is a protein characterized by the double strand DNA sequence which it recognizes (called a *site*, such as $\begin{array}{c}\texttt{CCC'GGG}\\ \texttt{GGG'CCC}\end{array}$ for SmaI, and $\begin{array}{c}\texttt{CTTAA'G}\\ \texttt{G'AATTC}\end{array}$ for EcoRI) and cuts into two segments (ending in $\begin{array}{c}\texttt{CCC}\\ \texttt{GGG}\end{array}$ and $\begin{array}{c}\texttt{GGG}\\ \texttt{CCC}\end{array}$ for SmaI, and $\begin{array}{l}\texttt{CTTAA}\\ \texttt{G}\end{array}$ and $\begin{array}{r}\texttt{G}\\ \texttt{AATTC}\end{array}$ for EcoRI). See [**20, 21**] for further background in molecular biology.

The basic ingredients in an information processing device are:

(a) a mechanism for reading and encoding input information from the external world;

(b) a mechanism for processing information.

(c) a mechanism for returning output information back to the environment.

(The word 'mechanism' is used here in a generic sense and is not meant to imply an electronic device. In particular, it could be a piece of bioware.) In a finite-state machine (fsm), (a) is given by words over an input alphabet written over a one-way read-only tape. (b) is a finite control reading problem instances on the input tape (in the form of strings of symbols) and computing according to a prewired program δ that drives the machine from one memory (source) state to the next (target) state depending on the symbols just read on the tape. We assume that the reader is familiar with finite-state machines (see [**11**] for background). An example is shown in Fig. 1 that checks divisibility by 2 or 3 in a binary string, which will be used throughout for the implementation in this section. A nondeterministic fsm performs computations in *runs*. A run is a sequence of valid state transitions $q_0, q_1, \cdots, q_m$ according to the transition table δ, i.e., so that each $q_t \in \delta(q_{t-1}, x_t)$, where $x := x_1 x_2 \cdots x_m$ is the input string and q_0 is the start state (incoming arrow without source). In general, there is more than one legal run for a nondeterministic machine. The nfsm accepts its input x iff *some* run successfully arrives at a so-called final state (double circles in Fig. 1; the final states are part of the machine's specification). Alternatively, one can view a nondeterministic fsm as an ordinary deterministic fsm with the capability of making copies of itself when making a nondeterministic move (more than one next state possible in the transition diagram), with each one of the copies covering for each of the choices in the move. This latter view is particularly suited for DNA implementation due to the massive parallelism in DNA computations.

3. Implementing Nondeterministic Finite-state machines

The method used below to implement nondeterminism in DNA computation consists of three basic steps: *encoding* that maps the problem onto DNA strands, *hybridization/ligation* that performs the basic core processing, and *extraction* that makes the results visible to the naked eye. The implementations are based on DNA hybridization, that is, under proper conditions, complementary single-stranded DNA will hybridize to form a double-stranded helix. Error preventing strategies, which are critical in practice for the prevention of unwanted hybridizations, have been discussed elsewhere [**6, 7, 9**] and will be assumed but not addressed here.

Since the input to the fsm is represented as binary strings, for part (a) we need to encode them into DNA strands. We use *adapters*, i.e., double-stranded

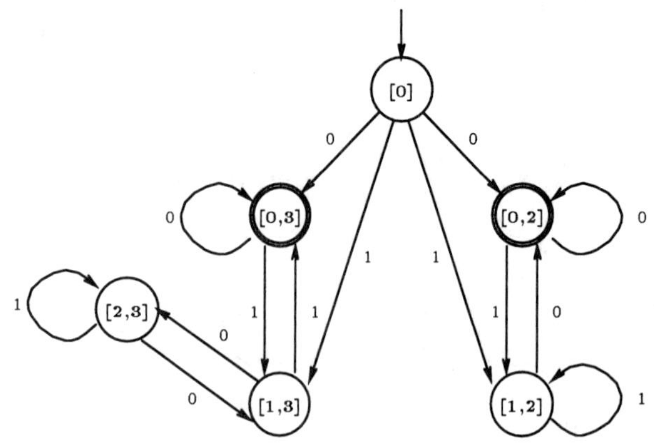

FIGURE 1. A nondeterministic finite-state machine for divisibility by 2 or 3.

segments of DNA, usually with a single-stranded overhang, and name them after the corresponding input symbols. The double-stranded part codes for the previous input symbol (`GAG` is the nucleotide encoding for binary input 0 and `GCA` encodes for input 1; there is a special start molecule), and the overhang codes for current state information. The dynamic molecule representing the finite control is a double strand with a segment encoding for the current state and another segment encoding the last symbol read that led to the current state.

Specifically, the state of the finite state machine is encoded by nucleotide sequences containing discriminating bases (lower case 4-mers) given by

state [0]: `atat`
state [0,3]: `ttat`
state [1,3]: `gctg`
state [2,3]: `ctca`
state [0,2]: `tatt`
state [1,2]: `gtcg`

The start molecule is:
```
GGGGAGATCatatCTTAA
CCCCTCTAG
```

Other state representations (adapters of fairly different lengths) may be chosen if extraction demands it (see section 3.2).

For proper simulation of the transitions, each of the states needs to be embedded within a ring in as many molecules as there are transitions emanating from it as follows. (A ring is a double strand of DNA with an unhybridized single-stranded oligo in the form of a loop. The loop size and the size of the nucleotide sequence at the end will vary, depending on the size of the fsm being simulated). In the example, let's call the states as they are labeled in Fig. 1, i.e., **[0]**, **[0,3]**, **[1,3]**, **[2,3]**, **[0,2]**, and **[1,2]**, and the input symbols 0 and 1, so that we have 12 combinations **[0]**0, **[0]**1, **[0,3]**0, **[0,3]**1, **[1,3]**0, **[1,3]**1, **[2,3]**1, etc. Each input d (0 or 1) is represented by the seven molecules representing the transitions $\delta(\mathbf{p}, d)$ labeled by d, where $\mathbf{p}$ is a state, as shown next:

```
                                   TCt
                                 A       t
[0]0:               CCCGGGGAG              atCTTAAGAG c   , and additionally
          tataGAATTGGGCCCCTC      -        TAGAATTCTC
                             TCt
                           T       a
          CCCGGGGAG                ttCTTAAGAG c
tataGAATTGGGCCCCTC      -          AAGAATTCTC
```

Note that the second molecule is obtained by replacing GAGATCttat on the first ring by GAGTTCtatt .

```
                                   GCg
                                 C       t
[0]1:               CCCGGGGCA              cgCTTAAGAG c   , and additionally
          tataGAATTGGGCCCCGT      -        GCGAATTCTC
                             GCg
                           T       c
          CCCGGGGCA                tgCTTAAGAG c
tataGAATTGGGCCCCGT      -          ACGAATTCTC
```

Note that these molecules are obtained by replacing GAGATCtatt and GAGTTCtatt on the first two rings by GCACGCgtcg and GCATGCgctg , respectively.

Likewise:

```
                                   TCt
                                 A       t
[0,3]0:             CCCGGGGAG              atCTTAAGAG c
          aataGAATTGGGCCCCTC      -        TAGAATTCTC
                                   GCg
                                 T       c
[0,3]1:             CCCGGGGCA              tgCTTAAGAG c
          aataGAATTGGGCCCCGT      -        ACGAATTCTC
```

Note that this strand is obtained by replacing GAGATCttat on the ring in **[0,3]**0 by GCATGCgctg .

Likewise:

```
                                   ACc
                                 C       t
[1,3]0:             CCCGGGGAG              caCTTAAGAG c
          cgacGAATTGGGCCCCTC      -        GTGAATTCTC
                                   TCt
                                 A       t
[1,3]1:             CCCGGGGCA              atCTTAAGAG c
          cgacGAATTGGGCCCCGT      -        TAGAATTCTC
```

Note that this strand is obtained by replacing GAGCACctca on the ring in **[1,3]**0 by GCAATCttat . Similarly for the other states:

	T	*GCg*	*c*
[2,3]0:	CCCGGGGAG		tgCTTAAGAG C
	gagtGAATTGGGCCCCTC	–	ACGAATTCTC
	C	*ACc*	*t*
[2,3]1:	CCCGGGGCA		caCTTAAGAG C
	gagtGAATTGGGCCCCGT	–	GTGAATTCTC
	T	*TCt*	*a*
[0,2]0:	CCCGGGGAG		ttCTTAAGAG C
	ataaGAATTGGGCCCCTC	–	AAGAATTCTC
	C	*GCg*	*t*
[0,2]1:	CCCGGGGCA		cgCTTAAGAG C
	ataaGAATTGGGCCCCGT	–	GCGAATTCTC
	T	*TCt*	*a*
[1,2]0:	CCCGGGGAG		ttCTTAAGAG C
	cagcGAATTGGGCCCCTC	–	AAGAATTCTC
	C	*GCg*	*t*
[1,2]1:	CCCGGGGCA		cgCTTAAGAG C
	cagcGAATTGGGCCCCGT	–	GCGAATTCTC

This approach is somewhat different from Smith's implementation of nondeterministic Turing machines since we use linear, double stranded (more stable) instead of circular single strands, we do not have to represent an input or worktape (expect for the last input symbol), only the states, and the efficiency and error rates may be better –see section 3.2. Although, the idea of the approach turns out to be essentially his when restricted to a fsm, we spell out the actual enzymes we are using, as well as how the reactions are to take place.

3.1. Computational Protocol. Under these encodings, the machine is implemented as follows. Initially the molecules representing the initial state are coupled to beads, and packed into a conventional chromatography column (washing to get rid of uncoupled adapters).

1. `Step 1`: Input 0 will be an heterogeneous aliquote mixture of the seven distinct adapters ([**0**]0, [**0,3**]0, [**1,3**]0, [**2,3**]0, etc.) loaded on column with ligase. (This mixture plays the role of "the transition matrix" in [**19**].) Which adapter participates in the reaction (transition) depends on the current state molecule. For example, at the beginning of the reaction, either of the two adapters of type [**0**]0 (leading to [**0,3**] or [**0,2**]) has a `tata` hangover that matches `atat` in the hangover of the current state representation (the start molecule). After ligation, the remaining unreacted input adapters are washed away and collected. Exonuclease is applied to chop any residual input adapters and the unreacted previous state, which has a single stranded overhang. The new states [**0,3**]0's and [**0,2**]0's are protected because of the ring at the end, further reducing the amount of contamination and keeping the transitions fault-tolerant.
2. `Step 2`: Adding SmaI, a restriction enzyme which recognizes the palindrome $\begin{matrix}\texttt{CCCGGG}\\\texttt{GGGCCC}\end{matrix}$ and cuts at the middle, between `G` and `C`, giving rise to the blunt end needed for the coupling to the beads. The resulting adapters are coupled to the beads.

3. `Step 3`: Add EcoRI to cut away the ring at the end, exposing the new state information in the remaining **[0,3]**0's and **[0,2]**0's (i.e., `ttat` and `tatt` , respectively). Note that the `ATC` segment in the start molecule is the same as the `ATC` at the end of the ring, to be hybridized to the `TAG` of the complementary strand when the ring opens by restriction digestion of EcoRI. Thus the state representations in the surviving **[0,3]**0's and **[0,2]**0's are restored as the new current states.

The function of the ring segment is two-fold. First, it prevents the false blunt end reaction during a ligation step; second, it allows coupling of the adapter to beads only at the desired end. The EcoRI restriction site is used for the removal of the end ring sequence to expose the state for subsequent reaction. The first input symbol has been processed and now there are two new current states **[0,3]**0 and **[0,2]**0, from which the next input symbol can be processed in a similar fashion. Steps 1-3 are repeated for as many symbols as necessary.

Under these conditions, the biochemical reactions faithfully reflect the nondeterministic transitions of the automaton. For example, if to the start molecule representing the initial state **[0]** we add adapter 0, plus some required chemicals, molecules representing states **[0,3]** and **[0,2]** will be produced (compare with Fig. 1); likewise, on input adapter 1, molecules representing states **[1,3]** and **[1,2]** will be produced. One can check the other transitions to verify that they do indeed reflect the dynamics of the given fsm (some are shown in Fig. 2).

This method for the computational step has been used in [**8**] to implement deterministic fsm's. The techniques are well tested and generally used as standard protocols in a molecular laboratory. This method can be used in principle for arbitrary fsm's by changing the segments representing the state (larger $n-$mers than 4 for more states). Of course, in practice there are restrictions to how many states can actually be used (although the limitation is only on the n, which still yields an exponential number of state representations).

3.2. Extraction. For the extraction process, we can attempt to couple a fluorescent dye to the DNA segments representing the states and use optical extraction to detect final states at the end of the computation, as was done in [**8**]. The problem in the case of a nondeterministic computation, however, is that there may be many copies of the finite-control in different states, and we want to detect whether any one of them is one of the final states. It would be desirable to *maintain the number of molecules representing any particular state within a certain range*, so that all computation paths are taken into account in the extraction process (avoid false negatives), and yet do not become too numerous that the computation reactions fall out of line with the expected transitions.

The key to the success in the subset construction that determinizes a nondeterministic fsm is that whenever two nondeterministic copies of the machine find themselves in the same state, one can safely discard one of them since all their runs will be identical thereafter. Since the purpose of this paper is to enhance the efficiency and reliability of DNA implementation of nondeterministic computation, it is desirable to have a biochemical procedure that renders the implementation efficient in the tube, in the sense that it will self-regulate to produce approximately equal concentrations of the molecules representing the various states. We propose to use the *methylation* process occurring in living cells and readily implementable in the lab, described as follows.

Any living cell that produces a restriction enzyme must identify and protect its own DNA from restriction. In order to prevent restriction of a cell's own DNA when the cell produces a restriction enzyme, the cell also produces a methylase enzyme (methyltransferase) which methylates, i.e., adds a $-CH_3$ chemical group to certain bases within or near the restriction sites. Nearly all restriction enzymes are thus inhibited by certain methylations of cytosine or adenine bases within their restriction sites. Common methylations include: $N4$–methylcytosine (the nitrogen at position 4 of cytosine is methylated); $C5$–methylcytosine (position 5 carbon of cytosine methylated, etc); hydroxymethylcytosine; and $N6$–methyladenine. (More can be found in [**16**].)

Here are some examples: EcoRI recognizes `G'AATTC` (really $\begin{matrix}\texttt{GAATTC}\\\texttt{CTTAAG}\end{matrix}$ but we will refer to the top single strand for simplicity) and cuts it as described above; however, it is not cut if the upper `C` is methylated to $C5$–methylcytosine. Likewise for HindIII, which recognizes and cuts `A'AGCTT`. KpnI recognizes and cuts `GGTAC'C` but only if `A` is not methylated to $N6$–methyladenine. The reader can peruse [**16**] as a general reference listing of known methylation effects for restriction enzymes.

We can thus accomplish the self-regulating effect described earlier by methylating a fraction of the input molecules being put into the reaction (which fraction will depend on the particulars of the nfsm being implemented). If a state is represented by a palindromic restriction site, state molecules will hybridize with other like molecules representing the same state. The presence of the appropriate restriction enzyme in the reaction will then cut and stop further expansion of this copy of the fsm. However, if methylated bases are found within a fraction of the restriction sites formed, the restriction enzyme will be unable to cut those copies of the fsm. Thus, the constant combined presence of restriction enzymes and methylated input bases will guarantee that the number of molecules in the tube can be maintained within a range for each state, so that they will be fairly represented as the implementation proceeds.

3.3. Advantages and Disadvantages. The most important advantages of this implementation are as follows. They are shared with the implementation inherited by specializing the Turing machine in [**19**] to fms (except when noted):
(a) as the computation proceeds, the size of the molecules representing the finite-state control remains unchanged (with fsm we do not need to worry about tapes, only states);
(b) the reactions proceed in a cycle that can be easily automated;
(c) the final state is easily extracted by standard molecular biotechnology once a minimum concentration of each state is present;
(d) strands of arbitrary length can be processed by feeding them, symbol by symbol or segmented in chunks, with finite resources over long periods of time;
(d) false positives are reduced to negligible levels by using fault-tolerant encodings;
(e) false negatives, mostly due to loss of material during tube transfers and extractions, are now controlled because all reactions occur in the same tube. Hence the implementation is fault-tolerant. (This is an advantage with respect to the implementation in [**19**] which require intertube transfers. No PCR is used in the implementation);
(f) the molecules representing states and transitions can be standardized so as to make them independent of specific states and inputs required by the machine,

For example, the transition diagram of a fsm on binary inputs and 6 states (e.g., Fig. 1) is a subset of the fully connected digraph on the 6 states. Thus a library of molecules which represent all possible input directed transitions on this fully connected digraph could be created once and for all for a given number inputs b and states m. The states and transitions of an arbitrary fsm up to m states can then be obtained by defining the input directed transitions as an appropriate subset of the molecular library. Molecular programming can then be accomplished by selecting and assembling molecular instructions from this library. The molecules can be, in fact, reusable components;

In practice, although the biochemical reactions do not occur at a 100% efficiency, mishaps in the hybridization can be detected with an optical flag during washout to keep the computation on track. For very long strings, the computation would be very time-consuming since the input has to be fed manually and time must be allowed for biochemical reactions to occur (orders of magnitude slower than electronic implementations in the deterministic case, but better in the nondeterministic case – see the analysis in Smith [**19**])). On the other hand, it can be said that, for medical applications for example, an organism may potentially interact much more easily with a biomolecule than with a larger electronic devise inside the body. One might just introduce the molecule representing the initial state of the finite-control, appropriately isolated, into a host, and let it supply its own inputs and support of the environmental conditions for the implementation.

4. Discussion

The fundamental idea in Adleman's approach is to exploit the capability of DNA to implement nondeterministic computations. It is therefore reasonable to expect that exploring implementations of known nondeterministic computational devices will allow us to better gauge the real potential of DNA computing. We have presented a fault-tolerant self-regulating implementation of nondeterministic finite-state machines. Self-regulation is accomplished by constantly exposing methylated and nonmethylated input strands to the restriction enzymes in the tube.

The final test of the quality of this approach ultimately relies on its success in the lab, which is yet to be verified experimentally. Even if this success is limited, it is well known that the number of states required by a nondeterministic machine may be exponentially smaller than the deterministic equivalents in the worst case. In practice, the subset construction does not seem to blow out the number of states exponentially on the average [**12**], although it is probably superpolynomial. Furthermore, nondeterministic fsm can be much more compact and semantically preferable for applications (see, for example, [**11, 18**]).

An important open question concerning any implementation of nondeterminism is whether it can be extended successfully to cases where the relationship is unknown. Perhaps consideration of an intermediate case before looking at a problem such as HAMILTONIAN PATH may yield more decisive results. One important such case is the well-known open problem of the relationship between determinism and nondeterminism for logspace constrained Turing machines. The techniques discussed in this paper suggest a way to obtain experimental evidence about the true answer to this question. Ultimately the answer resides in whether or not the computation runs of a nondeterministic machine are redundant enough that a deterministic device could run them in the same amount of tape allotted (it is known that squaring the amount of tape to $O(\log^2 n)$ suffices for a deterministic equivalent). The experimental performance of DNA implementation with methylated input may at least provide an answer in the range of inputs that would be feasible in practice.

References

[1] L.M. Adleman (1994). Molecular Computation of Solutions to Combinatorial Problems. *Science* **266**, 1021–1024.

[2] E. Baum (1996). Running Dynamic Programming algorithms on a DNA computer. in [**23**], 141-147.

[3] D. Beaver (1996), A Universal Molecular Computer. In: [**15**] 29-36.

[4] D. Boneh, C. Dunworth, R.J. Lipton (1996), Breaking DES Using a Molecular Computer. In: [**15**] 37-65.

[5] C.H. Bennett, The Thermodynamics of computation –a Review. *Int. J. of Theoretical Physics* **21** (1982) 905-940.

[6] R. Deaton, R.C. Murphy, M. Garzon, D.R. Franceschetti, S.E. Stevens, Jr. (1995). Good Encodings for DNA-based Solutions to Combinatorial Problems. In [**23**], 159-171.

[7] R. Deaton, D.R. Franceschetti, M. Garzon, J.A. Rose, R.C. Murphy, S.E. Stevens, Jr. Information Transfer through Hybridization Reactions in DNA based Computing. Proceedings of the Genetic Programming Conference GP-97, Morgan Kaufmann (in press).

[8] J.A. Rose, Y. Gao, M. Garzon, R.C. Murphy, R. Deaton, Franceschetti, S.E. Stevens, Jr. DNA Implementation of Finite-state Machines. In [**13**], 479-486

[9] M. Garzon, P. Neathery, R. Deaton, D.R. Franceschetti, R.C. Murphy, S.E. Stevens, Jr. A New Metric for DNA Computing. In [**13**], 472-478.

[10] F. Guarnieri, M. Fliss, C. Bancroft (1996): Making DNA Add. *Science* **273** 22,-223.

[11] J.E. Hopcroft, J.F. Ullman: Introduction to automata theory, languages and computation. Addison-Wesley, Reading MA, 1979

[12] J.H. Johnson, D. Wood (1997): Instruction Computation in Subset Construction. In [**18**], 1-9.

[13] J.R. Koza, K. Deb, M. Dorigo, D.B. Fogel, M. Garzon, H. Iba, R.L. Riolo, eds. (1997). *Proc. 2nd Annual Genetic Programming Conference*, Morgan Kaufmann.

[14] R. Lipton (1996). Speeding Up Computations via Molecular Biology. In [**15**], 67-74.

[15] R. Lipton, E. Baum, eds. (1996). *DNA Based Computers.* DIMACS Series in Discrete Mathematics and Theoretical Computer Science, vol. **27**. American Mathematical Society, Providence RI.

[16] M. Nelson, E. Raschke, M. McClelland (1993): Effect of site-specific methylation on restriction endonucleases and DNA modification methyltranferases. *Nucleic Acids Research*, **21**:13, 3139.

[17] J.S. Oliver (1996): Computation with DNA-Matrix Multiplication. in [**23**], pp 236-248.

[18] D. Raymond, D. Wood, S. Yu (eds) (1997). *Proc. First Workshop on Implementing Automata*, U. of Western Ontario, London, 1996. Lecture Notes in Computer Science, Springer-Verlag 1260. viii+189 pp.

[19] W. Smith (1996), DNA Computers in vitro and vivo. In: [**15**] 121-185.

[20] L. Streyer (1995). *Biochemistry.* Freeman & Co.

[21] Watson, J. D., Hopkins, N. H., Roberts, J. W., Steitz, J. A., and Weiner, A. M. (1987). *Molecular Biology of the Gene.* The Benjamin/Cummings Publishing Co., Inc, Menlo Park, CA fourth edition.

[22] P. Wilhem, K. Rothemund (1996), A DNA and restriction enzyme implementation of Turing Machines. In: [**15**] 75-119.

[23] E. Baum, D. Boneh, P. Kaplan, R. Lipton, J. Reif, N. Seeman (eds.), *Second Annual Meeting on DNA based computers*, workshop, Princeton University, 1996. To be published in DIMACS series of the American Mathematical Society.

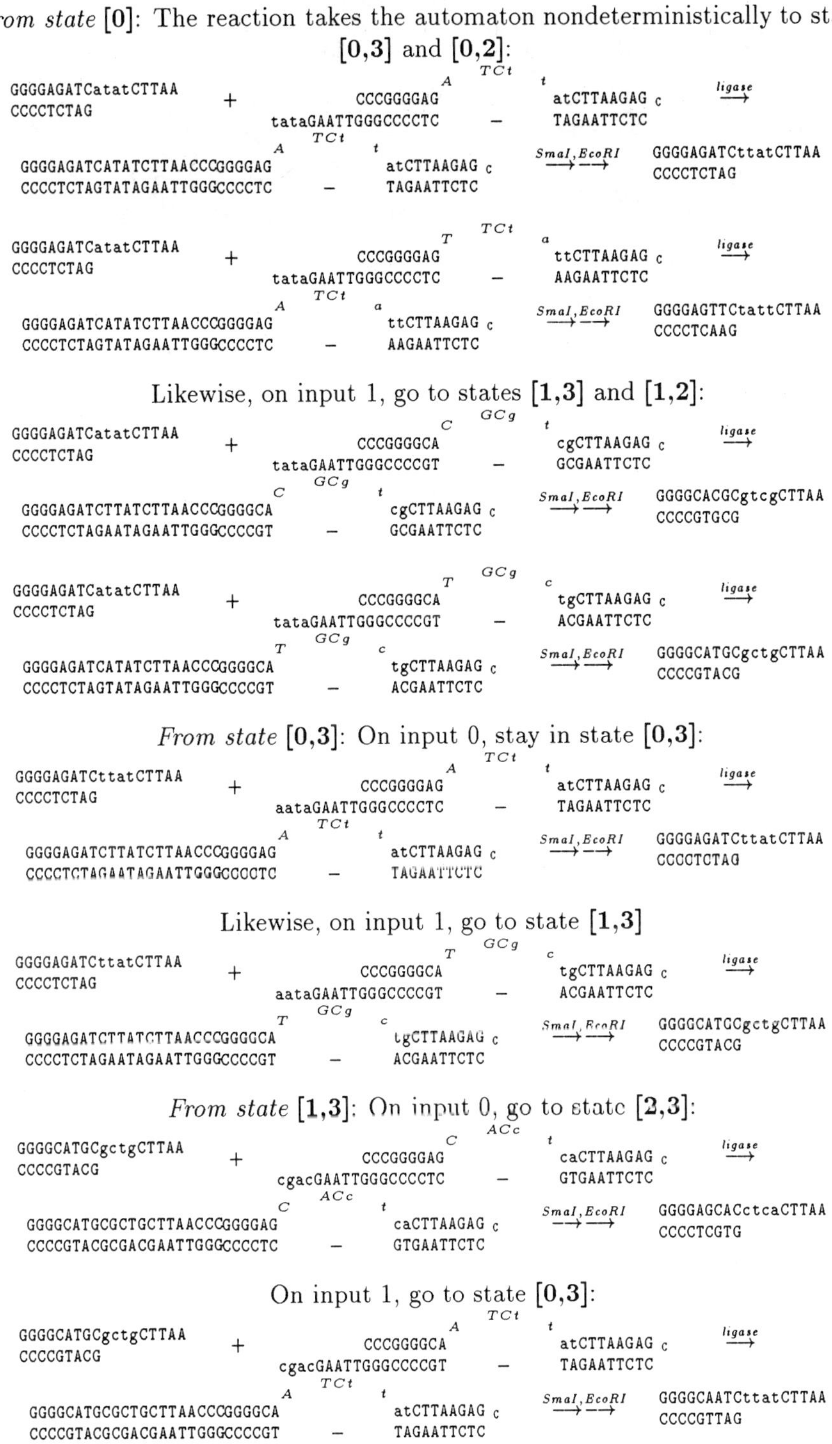

FIGURE 2. All the reactions reflect the transition table of the fsm (only a sample is shown). The protocol is, on input d, to add seven adapters $\mathbf{p}d$, add T4 DNA ligase, cut with *SmaI*, couple with beads, and cut with *EcoRI*.

(Y. GAO) THE UNIVERSITY OF MEMPHIS, MEMPHIS, TN 38152
E-mail address: yuangao@msuvx2.memphis.edu@memphis.edu

(M. GARZON) THE UNIVERSITY OF MEMPHIS, MEMPHIS, TN 38152
E-mail address: mgarzon@memphis.edu

(R. C. MURPHY) THE UNIVERSITY OF MEMPHIS, MEMPHIS, TN 38152

(J. A. ROSE) THE UNIVERSITY OF MEMPHIS, MEMPHIS, TN 38152

(R. DEATON) THE UNIVERSITY OF MEMPHIS, MEMPHIS, TN 38152

(D. R. FRANCESCHETTI) THE UNIVERSITY OF MEMPHIS, MEMPHIS, TN 38152

(S. E. STEVENS, JR.) THE UNIVERSITY OF MEMPHIS, MEMPHIS, TN 38152

DIMACS Series in Discrete Mathematics
and Theoretical Computer Science
Volume **48**, 1999

Arithmetic and Logic Operations with DNA

Vineet Gupta, Srinivasan Parthasarathy, and Mohammed J. Zaki

ABSTRACT. A lot of current research in DNA computing has been directed towards solving difficult combinatorial search problems. However, for DNA computing to be applicable on a wider range of problems, support for basic computational operations such as logic operations like AND, OR and NOT and arithmetic operations like addition and subtraction is necessary. Unlike search problems, which can be solved by generating all possible combinations and extracting the correct output, these operations mandate that only a unique output be generated by specific inputs. The question of suitability of DNA for such simple operations has so far largely been unaddressed. In this paper we describe a novel method for using DNA molecules to solve the basic arithmetic and logic operations. We also show that multiple rounds of operations can be performed in a single test tube, utilizing the output of an operation as an input for the next. Furthermore, the operations can be performed in a linear series or a series-parallel fashion and operators can be mixed to form any operation sequence.

1. Introduction

In recent work, Adleman [**Adl94**] and Lipton [**Lip95**] presented the idea of solving difficult combinatorial search problems using DNA molecules. These studies showed that DNA computing may have an advantage over electronic computers for such problem domains due to the massive parallelism inherent in DNA reactions. However, for DNA computing to be applicable on a wider range of problems, support for simple computational operations is necessary [**Gif94, Poo95, Rub96**]. Boolean operators such as AND, OR and NOT, and arithmetic operators such as addition and subtraction are the fundamental operations of an electronic computer. In contrast to search problems, which can be solved by generating all possible combinations and extracting the correct output, these operations mandate that only a unique output be generated by specific inputs. Although DNA computers might not improve on current silicon technology for these operations, as Adleman [**Adl95**] has pointed out, "they can contribute to our understanding of the nature of computation." There has been some work in simulating boolean circuits and performing additions and matrix multiplications with DNA. Ogihara and Ray [**OR97**] have shown how DNA computers can simulate Boolean circuits with a small overhead. Oliver [**Oli96**] has shown how DNA based methods can be used

1991 *Mathematics Subject Classification.* Primary 68Q10,92C40; Secondary 68Q05,92-08.

to calculate the product of Boolean matrices or matrices containing positive, real numbers. Guarnieri *et al.* [**GFB96**] have proposed a clever way to add two binary numbers. The novelty of their approach is the introduction of a place holder for the carry position while performing additions. [1] As they themselves point out, a limitation of their approach is that the output strand of one operation cannot serve as the input strand for another round of addition. This has been a constraining factor so far in performing a sequence of operations in a single vessel. In this paper we show that this limitation can be overcome (in DNA), allowing a number of basic series and series-parallel bit operation sequences in solution.

The rest of the paper is organized as follows. We present our bit encoding scheme, and our approach to performing a single bit operation in solution in section 2. We extend this framework to handle a *series* or a linear sequence of mixed operations in section 3. We then show how to combine multiple series to perform *series-parallel* operations in section 4. Finally we conclude in section 5, with a discussion on the advantages and limitations of our approach, as well as directions for future work.

2. Bit Encoding and Single Operations

In an electronic computer an operator is succinctly represented by a truth table, a table of all possible combinations of the input bit values and their corresponding output values, as shown in table 1 A. Our approach is to encode these truth tables in DNA using a three-level scheme. An operation is represented in terms of DNA hybridization. For each binary operation, the two bit strings are represented with two different DNA single strands. The first string is called the "input" and the second the "operand" strand. Each bit is represented with a dinucleotide unit, and a bit string with a sequence of dinucleotides. The natural DNA bases Adenine (A), Thymine (T), Uracil (U) and a non-natural base 7-deaza-adenine (P) (shown in figure 2) are used for constructing the dinucleotides. A non-natural base is needed in order to realize all possible input and operand bit values in DNA; a base which can pair with the specificity of A and yet is chemically distinct. 7-deaza-adenine [2] base-pairs selectively with U and T just like A which makes it ideal for our application. The input DNA strand is constructed with dinucleotides 5'-AU representing bit 1, and 5'-UA, the bit 0. The operand strand is constructed with dinucleotides 3'-TA and 3'-PT, representing bit 0, and the dinucleotides 3'-AT and 3'-TP, representing bit 1, where 5' and 3' denote directionality of the DNA strand. The table 1 B shows how these dinucleotides can base-pair to allow for all possible combinations of the input strand and the operand strand. Since, as we will show later in the paper, the operand strands do not have to be sequenced to obtain the final result, but instead carry an output strand, we do not have to chemically distinguish between A and P at any stage. DNA strands carrying TA (similarly AT) might have a different output appended to it than the DNA strand carrying TP (similarly PT) depending upon the result of the calculation. If the input strand is constructed in the 5' to 3' direction the operand strand is constructed in the 3' to 5' direction and vice versa. The input strand is constructed using only A and U

[1] In contrast to boolean operators, ADD can generate two output bits (e.g. $1+1 = 10$). In this case the current result bit position is assigned a value of zero (0) and the one (1) is carried over. The carried over bit is added to the next bit position and the value of that position is adjusted accordingly. The operation is repeated as many times as there is a carry over.

[2] 2-aminopurine and 2-aminoadenine could potentially be used too.

Table 1 A. The truth table for various binary operations

Input 1	Input 2	Output NAND	Output AND	Output XOR	Output ADD
0	0	1	0	0	0
0	1	1	0	1	1
1	0	1	0	1	1
1	1	0	1	0	10

Table 1 B. The truth table encoding using dinucleotide "bits"

Level 1 Input	Level 2 Operand	Pair	Level 3 (Interpretation of the Duplex for the Output) NAND	AND	XOR	ADD
UA = 0	PT = 0	UA PT	1	0	0	0
	AT = 1	UA AT	1	0	1	1
AU = 1	TA = 0	AU TA	1	0	1	1
	TP = 1	AU TP	0	1	0	10

FIGURE 1. Truth Table Encoding

(level 1), while the operand strand with A, T, and P (level 2). This keeps the input distinct from the operand, yet retains the same base-pairing structure and allows all possible combinations of the input and operand bits (see table 1). As a result of an operation the input strand hybridizes with its complementary operand strand to form a double stranded DNA complex. This output is interpreted according to the truth table (level 3) of the operator applied. Table 1 B shows our encoding scheme along with the truth tables for different boolean and arithmetic operations such as NAND, AND, XOR and ADD (addition). The truth table encoding can easily be extended to a number of other operators.

The NAND operator is a *universal* boolean operator, which means that any boolean operation can be represented in terms of a sequence of NAND operators. Our first example shows the execution, in DNA, of the operator NAND on the two binary strings 1001 and 0101, as depicted in table 3. The input bit string 1001, is represented as the DNA sequence 5'-AUUAUAAU. All 16 possible DNA sequences are used to represent the operand 0101, since there are two alternative dinucleotides for each bit in the level 2 encoding. However, only one of the operand strands carries a DNA sequence complementary to the sequence of the input strand. This strand (3'-TAATPTTP) is thus the only one that can hybridize with the input strand to yield a unique output duplex. The output, when decoded using the truth table for NAND, yields the correct answer of 1110. In fact, the same output can be interpreted differently if decoded using a different operator's truth table. For instance, if we were to perform the operation (1001 XOR 0101) instead of (1001 NAND 0101), the input and all of the operand strands would still be the same,

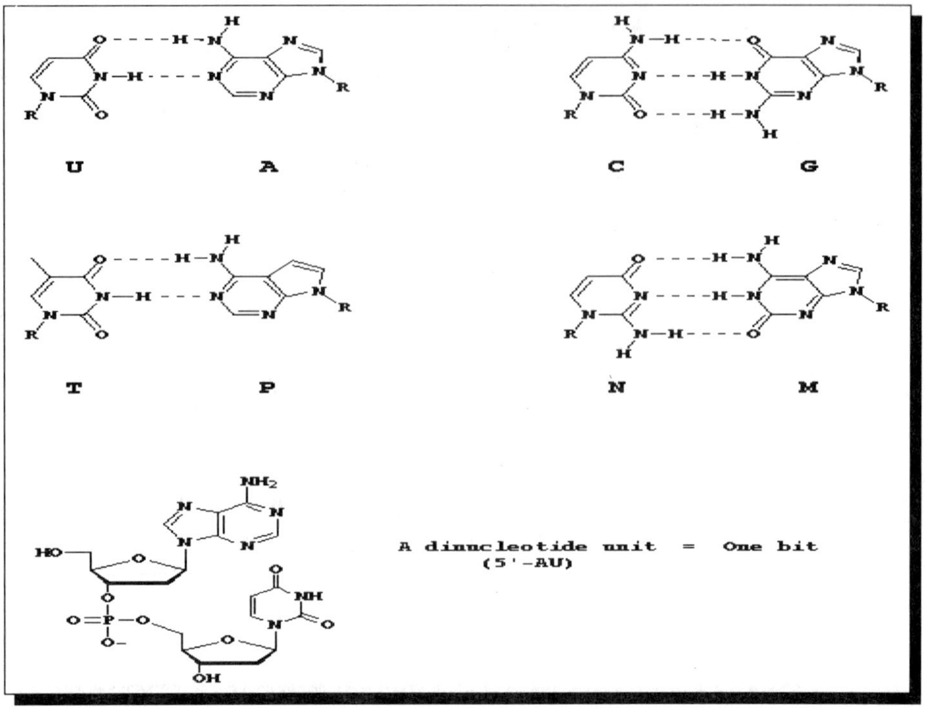

FIGURE 2. Chemical structure of the bases used; Adenine (A), Thyamine (T), Uracil (U), 7-deaza-adenine (P), Guanine (G), Cytidine (C), iso-Guanine (M) and iso-Cytidine (N). Also shown are some example base-pairs, and a dinucleotide "bit" unit.

and would yield the same output duplex. However, interpreting this output duplex according to the truth table for XOR (see table 1 B) would generate a different but correct answer of 1100. This shows that single operations can be performed efficiently with DNA using the proposed approach.

3. Mixed Operation Series

In order to extend this framework to handle a *series* operation, i.e. a linear sequence of operations, we have to address two main issues: 1) how can the output of one operation, a double strand, be used as the input for the next operation, and 2) how to ensure that the operations take place only in the desired order. As shown in the first example from table 3, *each input strand can bind with only one of the operand strands producing a duplex representing the unique output.* As a solution to the first problem we attach, *a priori*, the corresponding output value to each of the operand strands in the form of a single stranded DNA. The DNA sequence of the output is constructed using the level 1 encoding from table 1 B. Thus, each operand strand carries a covalently linked output strand which does not interfere with its operation of hybridizing with the input strand, but instead results in a duplex with a sticky-end upon one such complexation. This sticky-end can then act as the input strand for the next operation. Not only does this allow a series

Table 2 Example of a NAND operation

Input Strand	Operand Strands		Unique Output
1 0 0 1	0 1 0 1		**1 1 1 0**
5'-AUUAUAAU	3'-TATPPTAT	3'-TATPPTTP	5'-AUUAUAAU 3'-TAATPTTP
	3'-TATPTAAT	3'-TATPTATP	
	3'-TAATPTAT	**3'-TAATPTTP**	
	3'-TAATTAAT	3'-TAATTATP	
	3'-PTTPPTAT	3'-PTTPPTTP	
	3'-PTTPTAAT	3'-PTTPTATP	
	3'-PTATPTAT	3'-PTATPTTP	
	3'-PTATTAAT	3'-PTATTATP	

FIGURE 3. Single NAND Operation

operation involving a single operator, but also mixed operators, since the output strand attached to the operand is dependent only on a particular operator.

To solve the second problem, we use a unique DNA sequence tag on each strand. This tag is just a length 4 DNA segment comprised of four bases, the natural bases Guanine(G) and Cytidine (C) along with the non-natural bases iso-Guanine (M) and iso-Cytidine (N), shown in figure 2. This allows for a maximum of 256 unique tags, which gives us the option of performing 256 consecutive operations. This number can be increased by simply using a longer tag sequence. Each input strand is linked with a unique tag, while the operand strand next in order in the sequence, is linked with the complementary tag. The input can thus only hybridize with the next operand. A unique tag for the next step is also linked to the output strand, since it serves as the input for the next operation. The tags also prevent the formation of unproductive duplexes. For example, since each of the strands is made up of an operand and an output, they may become self-complementary. The tags prevent this from happening. Using this approach, we generate unique DNA sequences as inputs and obtain a unique final output resulting from the desired sequence of operations.

In figure 4, we present an example of how three different operators – NAND, XOR and ADD – can be performed in succession on DNA. The example solves the series operation sequence (((1001 NAND 0101) XOR 0001) ADD 0001). The first two bit strings are the same as those used in our first example from table 3. The input strand (5'-AUUAUAAU-GCMN) has the bit string 5'-AUUAUAAU (1001) linked with the tag 5'-GCMN. Figure 4 A shows the 16 possible representations of the operand strand (0101) that could possibly bind with the input (similar to our first example). Only one strand hybridizes with the input, namely the strand 3'-TAATPTTP-CGNM-AUAUAUUA-CCGG, where 3'-TAATPTTP is the bit string 0101, 3'-CGNM is the complement of the tag 5'-GCMN, 3'-AUAUAUUA is the attached output (1110) of the operation and 3'-CCGG is the tag for the next operation. The result of the first operation is thus a duplex with a single stranded

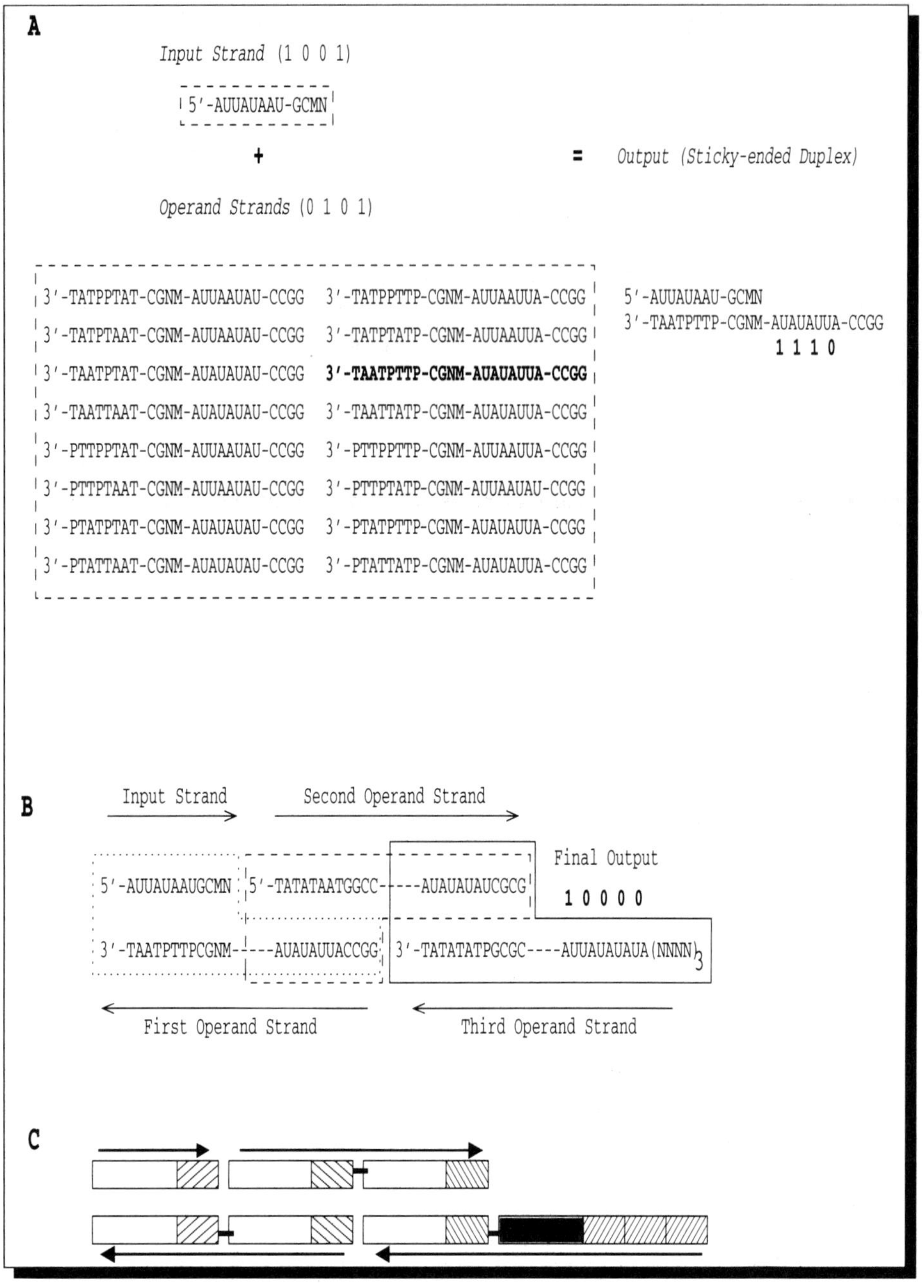

FIGURE 4. Example of a series operation. A) The first operation : (1001 NAND 0101); B) The whole series : (((1001 NAND 0101) XOR 0001) ADD 0001); C) A template of the structure generated (The arrows denote directionality, the unfilled boxes the input and operands, the black box the output, and the pattern-filled boxes the tags).

overhang (3'-AUAUAUUA-CCGG), which is ideally suited to serve as the input strand for the next operation – XOR 0001.

Our second operand strand, representing 0001, also has 16 possible representations (not shown). Only 5'-TATATAATGGCC-AUAUAUAUCGCG (see figure 4 B) hybridizes with the output of the previous operation, producing a duplex with the overhang 5'-AUAUAUAUCGCG as the output (1111) of the current operation. Out of the 16 possible third operand strand (0001) DNA sequences, for ADD 0001, only 3'-TATATATPGCGCAUUAUAUAUA$(NNNN)_3$ hybridizes with the output of the previous round of operations. 5'-$(NNNN)_3$ is a special tag signifying that the sequence of operations has completed and is used for extracting the result. The unique result of our operation 3'-AUUAUAUAUA, when decoded using the truth table, is 10000, which is the correct result of our operation sequence. This examples shows that our scheme generates a unique output from a series operation in solution.

4. Series-Parallel Operations

In order to emulate the more complex boolean circuits as in an electronic computer, support for *series-parallel* operation sequences is desirable. A series-parallel operation sequence merges two or more series operation sequences. In figure 5 we show how the above architecture can be extended to handle a series-parallel operation sequence in a single test tube. The output of the series operation (from the previous example, figure 4 B), a duplex with an overhang is used as both the input and the operand complex, performing the operation I NAND O, where $I = O =$ (((1001 NAND 0101) XOR 0001) ADD 0001). However, instead of the typical operand strand which has an output attached to it, the operand complex for the series-parallel operation has two attached outputs – the output of the series-parallel operation followed by the output of the series operation sequence, shown in figure 5 A). The final step is a copy-operation – AND 11111 (which simply replicates the output). The result of this operation sequence is a three-arm junction [**See82, WYS96**] with the correct output (01111) as the overhang, ideally suited for performing more operations.

5. Discussion and Conclusion

Only a theoretical model is presented in this paper. However, it is based on well-established techniques of biological chemistry. To practically carry out the operations in a test tube, we can attach the input strand 5'-AUUAUAAUGCMN to a magnetic bead, and add all the operand strands for the different operations to be performed, to the solution. The solution can then be heat-denatured, annealed and ligated. [3] All strands with the magnetic bead attached can be filtered off, and the output strand end-sequenced to produce the result.

For an n bit long string ($2n$ length DNA strand), with tag length m, our technique must satisfy some constraints. The length of an input/operand strand is $m + 2n$. Since the mismatch tolerance of a duplex increases with its length,

[3]Ligation can be performed either enzymatically, using DNA ligase enzyme, or chemically using Letsinger's method [**Hea95**]. The DNA strands will have to be properly end-modified for either of the cases. The modifications should not affect their hybridizing properties.

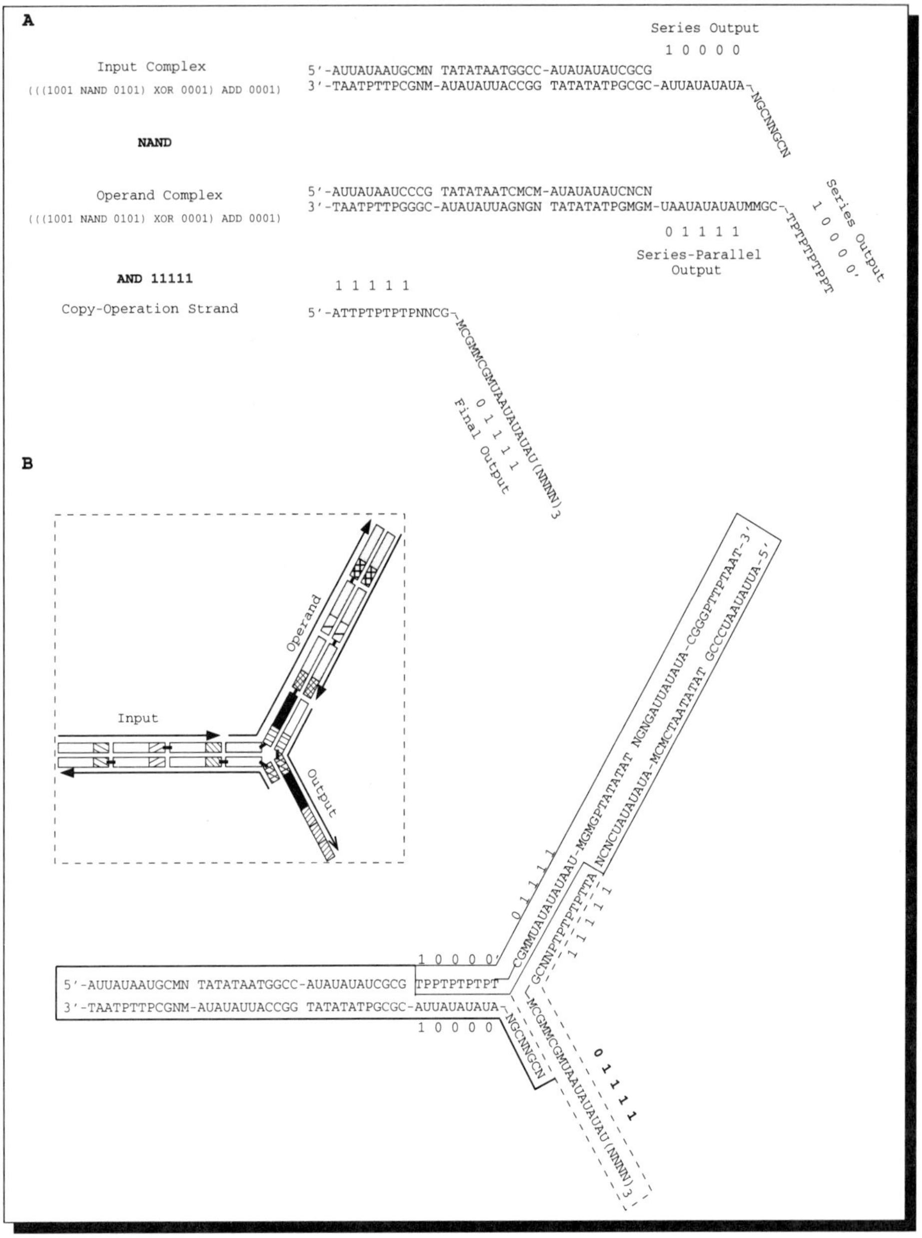

FIGURE 5. Example of a series-parallel combination operation (Input NAND operand); **Input**: (((1001 NAND 0101) XOR 0001) ADD 0001) = 10000 (same as in figure 4); **Operand**: same as **Input**; **Intermediate Result**: 10000 NAND 10000 = 01111; **Copy-operation**: 01111 AND 11111 = 01111 (Final Output). A) The sequence of the input complex, operand complex, and the copy-operation strands; B) The three-arm junction generated as a result of the series-parallel operation. (Inset) A template of the three-arm junction structure.

at present duplexes only about 20 base-pairs long can distinguish single-base mismatches [**Wea79**], introducing the constraint $m + 2n \leq 20$ on our system. Furthermore, for m operations, we need at least $m \cdot 2^n$ distinct DNA strands in the solution. For nanomolar quantities of each, thermodynamical limitations would put the upper bound at approximately 10^{14} different strands. The first constraint, though, clearly subsumes the second. The total length of the output complex, $2n \cdot m$, might also be limited by various factors that affect the reliability, such as the hybridization conditions, nearest-neighbor interactions (sequence context), and the efficiency of both the successive ligations and the result extraction process. We use two-fold (internal and end) mismatch discrimination system to prevent formation of unwanted (mismatched) duplexes/complexes. Each internal mismatch is at least two base long since we use dinucleotides as single "bits". Thus the stability of the perfectly matched duplex is much higher than that of a mismatched duplex. Moreover, ligases are very sensitive to end mis-matches. Only perfectly matched duplexes are ligated by the enzyme. This gives us a very good mismatch protection at the ends of each duplex. To improve the reliability, the end-tag can be used for PCR amplification of the output. The PCR amplified output can be sequenced to also verify the operations performed. Finally, the number of series-parallel operations may be limited due to the steric constraints (branching-out) that three-arm junctions introduce.

There are 4^{2n} different sequences possible for a $2n$ length DNA strand, if random encodings are used. An advantage of using our fixed encoding scheme is that we need to synthesize only 2^n DNA sequences per operation [4], which is a very small subset of 4^{2n}. The increase in demand of DNA strands is also linear in the number of operations rather than exponential ($m \cdot 2^n$ instead of $2^m \cdot 2^n$). Furthermore, since all possible representations of the operands, for each of the operations, are added to the solution in the beginning, all possible outputs are produced in the solution. We can then simply add a given input to the solution, which will bind to its complementary sequence, and then extract the unique output. The same solution can then be reused with different inputs, for sequential extraction of the corresponding outputs. Thus the vessel serves as a black-box (representing some function to be performed), which takes in different input values to produce the corresponding outputs, without having to perform the operation again. Moreover, the fact that all the intermediate results are present in the final output, prevents the loss of any information about the computation, and can be used for implementing reversible logic gates [**BL85**].

There are some other advantages to our approach. The use of dinucleotides as single bit units lets us represent information at a much higher bit density than any previous methodology, and brings it very close to the theoretical limit of a bit per base. [5] Our methodology also makes possible a succinct representation of basic computational operators. The use of different bases for encoding the bits and the operation order permits series as well as series-parallel operation sequences in one test tube, producing a unique result, and simplifies the extraction process. The unique tags help keep the operation sequence in order. They also avoid the *fan-out*

[4] All the strands can be synthesized in a single combinatorial synthesis cycle using techniques of photolithography and nucleic acid chemistry [**Fea91, Pea94**].

[5] Although, in principle mono-nucleotides can be used as single bits, there may be some complications to using such strands for performing different operations. For example, a long sequence of a mono-nucleotide can form unwanted triplexes.

problem where a randomized sequence of molecules can produce an exponentially increasing number of product strands with the progression of the operation sequence. We use 2^n molecules for each step, which interact with 2^n molecules from the previous step to generate the same number of molecules for the next step.

Generation of DNA molecules carrying sticky-ends as the output of each operation, in the reaction vessel, is another nice feature of our scheme. This output is well suited to serve as the input for the next operation which enables us to accomplish the generic input/output semantics of an electronic computer. This is an important first step in solving problems using DNA computing where the DNA molecules have to go through multiple rounds of computation. Moreover, in most of the prior approaches where DNA was used for computing, the encoding was application specific and based on the particular application in mind. One of the most important advantages of our proposed mechanism is that it makes both encoding and computation more general (uniform) and application independent. As our examples show, almost any basic computational operation can be carried out on our "DNA computer". The addition of a memory is an important issue and we plan to incorporate that into our model. See [**RWB+96**] for a solution to this problem. They propose a *sticker* based model which has a random access memory.

In conclusion, we present a new approach for computing with DNA. The ability to perform complex bit operations in solution might help us learn more about the nature of computation and lead to the development of better DNA based computers, capable of solving a wide range of complex problems.

Acknowledgements

We thank R. Jakubiak, Prof. E. T. Kool, H. S. Malik, Prof. B. L. Miller, Prof. M. Ogihara and Prof. A. Ray for their insightful comments and helpful discussions.

References

[Adl94] Leonard M. Adleman, *Molecular computation of solutions to combinatorial problems*, Science **266** (1994), 1021–1024.

[Adl95] Leonard M. Adleman, *On the potential of molecular computing*, Science **268** (1995), no. 5210, 483–484.

[BL85] C. H. Bennet and R. Landauer, *The fundamental physical limits of computation*, Scientific American **253** (1985), 48.

[Fea91] S. P. A. Fodor and et al., *Light-directed, spatially addressable parallel chemical synthesis*, Science **251** (1991), 767.

[GFB96] Frank Guarnieri, Makiko Fliss, and Carter Bancroft, *Making DNA add*, Science **273** (1996), no. 5272, 220–223.

[Gif94] David K. Gifford, *On the path to computation with DNA*, Science **266** (1994), 993–994.

[Hea95] M. K. Herrlein and et al, *A covalent lock for self-assembled oligonucleotide conjugates*, Journal American Chemical Society **117** (1995), 10151.

[Lip95] Richard J. Lipton, *DNA solution of hard computational problems*, Science **268** (1995), 542–545.

[Oli96] John S. Oliver, *Computation with DNA-matrix multiplication*, 2nd Annual Meeting on DNA Based Computers, June 1996.

[OR97] Mitsunori Ogihara and Animesh Ray, *Simulating boolean circuits on a DNA computer*, 1st Annual Conference on Computational Molecular Biology, 1997, pp. 326–331.

[Pea94] A. C. Pease and et al., *Light-generated oligonucleotide arrays for rapid dna sequence analysis*, Proc. National Academy of Science, USA **91** (1994), 5022.

[Poo95] Robert Pool, *A boom in plans for DNA computing*, Science **268** (1995), 498–499.

[Rub96] Harvey Rubin, *Looking for the dna killer app*, Nature Structural Biology **3** (1996), no. 8, 656–658.

[RWB+96] Sam Roweis, Erik Winfree, Richard Burgoyne, Nickolas V. Chelyapov, Myron F. Goodman, Paul W. K. Rothemund, and Leonard M. Adleman, *A sticker based architecture for DNA computation*, 2nd Annual Meeting on DNA Based Computers, May 1996.

[See82] N. C. Seeman, *Nucleic acid junctions and lattices*, Journal Theoretical Biology **99** (1982), 237.

[Wea79] R. B. Wallace and et. al., *Hybridization of synthetic oligodeoxyribonucleotides to ϕ-x 174 dna: The effect of single base pair mismatch*, Nucleic Acid Research **6** (1979), 3543.

[WYS96] Erik Winfree, Xiaoping Yang, and Nadrian C. Seeman, *Universal computation via self-assembly of DNA: Some theory and experiments*, 2nd Annual Meeting on DNA Based Computers, May 1996.

336 Flexner Hall, Rockefeller University, 1230 York Ave., New York, NY 10021
E-mail address: guptav@portugal.rockefeller.edu
URL: http://kool.chem.rochester.edu/myhome.html

Department of Computer Science, University of Rochester, Rochester, NY 14627
E-mail address: srini@cs.rochester.edu
URL: http://www.cs.rochester.edu/u/srini

Department of Computer Science, University of Rochester, Rochester, NY 14627
E-mail address: zaki@cs.rochester.edu
URL: http://www.cs.rochester.edu/u/zaki

DIMACS Series in Discrete Mathematics
and Theoretical Computer Science
Volume **48**, 1999

DNA^{2}DNA Computations: A Potential "Killer App?"

Laura F. Landweber, Richard J. Lipton, and Michael O. Rabin

1. Introduction

Ever since Adleman's seminal paper [**1**] there has been a flood of ideas on how one could use DNA to compute. Lipton first showed that DNA could be used to solve more than just a variation of the famous travelling salesman problem [**13**]. Since then there have been many other papers on using DNA to solve various computational problems [**3, 5, 4, 6, 7, 16**].

At the top level all these papers are similar: they all attempt to use DNA computation to solve some large search problem. Since a liter of water can hold 10^{22} bases of DNA, there is the possibility that it may one day outperform electronic machines.

However, this is currently problematic. There are several reasons for this. First, electronic machines are very fast; moreover, they are getting faster every day. Second, there are many models of how to do DNA computations. Yet, it is unclear if any of these models will be practical. The problem is mainly that DNA technology is not perfect. DNA operations are distinctly not error free.

Finally, there is the lack of a *killer app*. A killer app is an application that fits the DNA model; cannot be solved by the current or even future electronic machines; and is *important*. The latter is critical: to be a killer app the problem must be one for which people are willing to "pay money" for solutions. To date there are no viable candidates for the killer app.

We propose a new way to use DNA computations. This will allow us to use DNA computations to solve important and potentially killer applications. The potential applications include:

1. DNA sequencing;
2. DNA fingerprinting;
3. DNA mutation detection or population screening;
4. Other fundamental operations on DNA.

The key new idea is to use DNA computation to operate on *unknown* pieces of DNA. This is a fundamental change in the way that we use DNA computation. We

---

The first author is a Burroughs Wellcome Fund New Investigator in Molecular Parasitology and is supported in part by NSF MCB-9520253.

R. Lipton was supported in part by NSF CCR-9633103 and AFOSR F49620-97-0190.

1991 *Mathematics Subject Classification.* Primary 68Q05, 68Q42, 92C40, 92D15.

call these DNA^2DNA computations: DNA *to* DNA computations. This idea was first proposed in [**8**] and called "analog" DNA computations there.

The key idea is the following. Suppose that one has a test tube that contains multiple copies of some unknown strand X of DNA. By *unknown* we mean that we do not know the sequence of the strand. Suppose further that we wish to compute some property of X, i.e. for some function $f()$ we wish to obtain the value of $f(X)$. The current way to do this is: (i) sequence the strand X in the laboratory; (ii) then, determine the value of $f(X)$. The difficulty with this method is that it requires an electronic computer to sequence the strand X.

Our new idea is to avoid the expensive step of sequencing the strand X. In particular, we plan to operate as follows: We will add to the test tube certain *known* strands of DNA and use these to perform a DNA computation on X. The result of this computation will be the answer $f(X)$.

The advantage of this method is that it avoids the sequencing step. Our hope is that this direct method of computing with unknown strands of DNA could be the key to finding "killer apps."

There is one huge advantage to our approach: since the problems we are solving are not digital, there is no way that electronic machines can compete. It's not that DNA based computation is faster, but that there is no way for electronic computers to approach the problems at all. One way to say this dramatically is that there is no place on a PC to "pour" in the test tube of unknown DNA. Without input, the problem cannot be solved at all on an electronic computer.

Our method is based on a new transformation that allows us to "encode" an unknown piece of DNA. All of the DNA computations to date use special redundant codes. It is critical that the DNA be redundantly encoded. Without such a coding the computations cannot be performed. Indeed the main contributions of [**1, 13**] were the construction of methods for creating and managing such codes.

Of course naturally occurring DNA is *not* coded in this redundant manner. This is a major roadblock: without redundant codes the methods of DNA computation do not apply. However, we propose a method for the transformation of a strand of DNA of unknown sequence. This transformation causes the DNA to be *re-coded* into any redundant code that we choose.

There are many advantages to this re-coding. Mainly, one could now extend all of the "tricks" of DNA computation to problems that involve unknown DNA. Since the DNA is coded the way that we choose we can operate on it much more freely. For example, one important application of this method is the following: (Note, the *exact* theorem statements are in section 4.)

THEOREM 1.1. *Suppose that X and Y are unknown strands in distinct test tubes. Then, it is possible to check whether $X = Y$ in $O(\log n \log\log n)$ bio-steps where both strands are at most length n.*

Note, we mean that we test *exactly* whether or not X and Y are equal: the method will discover if they differ in even *one* base. Further, this is only a simple example of a more general type of theorem:

THEOREM 1.2. *Suppose that $X^{(1)}, ..., X^{(k)}$ are unknown strands of length at most n that are in distinct test tubes. Then, in $O(\log n \log\log n)$ bio-steps we can compute the value of $F(X^{(1)}, ..., X^{(k)})$ where $F()$ is an NC^1 function.*

It is important to point out that our results avoid one of the key difficulties that face "classic" DNA computations. By "classic" we mean DNA computations that attempt to do purely digital problems. The advantage is that our results are much more error tolerant. The reason is that in classic DNA computations there is often "one" strand that the experimenter seeks to find. In our new type of DNA computations, there are many many copies. Thus, small error rates or partial rates of completion for some of the operations should not be a problem.

We prove these results by combining our re-coding methods with a generalization of the pretty simulation method of Ogihara and Ray [**14**]. Other methods could be used but their method is perfect for our needs. (See [**11**] for related ideas.) Note, in [**2**] there is a criticism of [**14**] for using an unrealistic model. We feel that this criticism is interesting but misses the essential point. They feel that the cost of the pour operation is not correctly included in [**14**]. The answer seems to be two-fold: First, even if the methods are linear in "pour" it's so fast that essentially the time is still logarithmic. Second, one can imagine using robots so that the pours can actually be done all at once.

2. Model

In this section we introduce our model of DNA computations. It is related to, but fundamentally different from, the models used in papers on classic DNA computations [**3, 5, 4, 6, 7, 16**]. The key point is that in DNA^{2}DNA computations the operations need not work perfectly. For example, we will only require assumptions about how selective DNA is when single strands anneal/ligate together. This is a major advantage of DNA^{2}DNA computations. Of course the hope is that this weakening in the required models will make DNA^{2}DNA computations really work in the laboratory. (Note, we are just beginning experiments at Princeton University that we hope will show that this is correct.)

All our computations are described in terms of operations that are performed on *test tubes.* The state of a test tube is, thus, a critical concept. At any time a test tube will contain a multi-set of different pieces of DNA. Some pieces will be single strands, some double strands and others more complex structures. Clearly, in order to describe mathematically such a state, we need to supply the following information:

1. The types of pieces of DNA that are in the test tube;
2. The total number of pieces that are in the test tube;
3. The number of pieces of each type that are in the test tube.

We will use string terminology to describe single strands of DNA. More precisely, we will identify strings S over the alphabet $\{A, T, C, G\}$ with the single strand of DNA of the form: $5' - S_1, ..., S_n - 3'$. Also by the *Watson-Crick* complement of S we will mean the string that is the reverse of S with each element changed into its complement, i.e. "A" with "T" and "C" with "G".

Suppose that a test tube T only contains single strands of DNA: note, this is an important special case. Clearly, its mathematical definition requires that we supply the following:

1. A collection of strings that correspond to the single strands in T, *i.e.* $S^{(1)}$, ..., $S^{(k)}$;
2. A integer M that is the total number of strands in T;

3. A collection of frequencies $q_1, ..., q_k$ so that the ith strand $S^{(i)}$ occurs $q_i M$ times where $q_1 + ... + q_k = 1$.

One of the key insights about DNA^{2}DNA computation is that we can simplify this definition: *we do not need to supply M*. That is we need not worry about the exact number of strands that are in the test tube. We need only to keep track of the frequency of each strand.

This is an important point about the difference between some classic DNA computations and DNA^{2}DNA computations. In some classic computations the number of types of strands k is the same order of magnitude as the total number of strands M. This is because in classic computations each strand is performing a separate computation: we need to have both k and M as large as possible.

On the other hand, in DNA^{2}DNA computations k will often be relatively small. For example, $k = 1,000$ and $M = 10^{15}$ are quite reasonable parameters. Since M/k is so large we can essentially ignore the exact value of M. Of course it is critical for all DNA computations that there be enough material available to make the operations feasible. Note, if in some situation M becomes too small, then a standard "trick" is to use PCR to increase the number of total strands and thus restore M to a large enough value.

In summary, for the rest of the paper we will only supply the frequencies of each piece not the total number of pieces in describing a test tube. A common situation is the following: Say that a test tube T contains $S^{(1)}$, ..., $S^{(k)}$ in *equal amounts* provided T contains approximately the same number of copies of each the given single strands of DNA.

Now let us turn to consider the class of operations that we require: (Each is a *bio-step* in our computations.)

1. **Cut.** This operation cuts or cleaves double strands of DNA at a certain pattern. This is done by using a restriction enzyme.
2. **Gel Separate.** This operation uses denaturing polyacrylamide gel electrophoresis to separate DNA molecules by length.
3. **Anneal.** This operation allows single strands to form double strands based on Watson-Crick pairing, i.e. "A" with "T" and "C" with "G."

As stated earlier we do *not* assume that each operation works perfectly. Let us now discuss the exact error model that we assume. Let $\tau > 0$ be a fixed small constant: we expect that it may be smaller than 10^{-2}. We will use τ to bound the error rate of all the operations that we perform. Note, we really have a collection of τ's: one for each operation. However, to avoid statements that are overly complex we will consider all the error rates together. Of course, one can in principle unravel this and get the exact dependence on each error rate, if one needs finer resolution.

Now let us discuss the error rates of each type of operation. A cut can fail in two basic ways. First, a pattern that should be cut may not be cut. Second, some place that does not match the pattern may be incorrectly cut. We assume that at least $1/2$ of the correct sites are cut; we assume that at most τ of the incorrect ones are cut. Note, the activity of most restriction enzymes is usually stated in terms of how long they take to cut $1/2$ of the population. One can increase this amount by either adding more enzyme or increasing the time of incubation.

Next let us discuss the separation of DNA by length. As in other papers we will arrange things so that no separation is ever required to separate strands that are too close in length. Further, we will arrange it so that the lengths are quite short.

Operation	.	Correct	Incorrect
cut	.	1/2	τ
gel separate	.	1/2	τ
anneal	far-apart	1/2	0
anneal	near	p_1	p_2

TABLE 1. A Table of Potential Errors

Gel methods work best for very short lengths. For lengths below several hundred one can distinguish i from $i+1$. We will assume that at least 1/2 of the strands of the given length are correctly extracted; we also assume that at most τ strands of the wrong length are also extracted. Note, this means that we do *not* assume that strands are not lost in performing the gel. As long as 1/2 of the correct strands are not lost the operation fits our model.

Finally, we must discuss the error model used for annealing. This is the most complex. There are two cases. The first case is the "far-apart" case. In this case the single strands are either the exact Watson-Crick complements or are such that they agree in at most 1/4 of their positions. Furthermore we assume that the length of the match is above a fixed threshold. In this case we assume that at least 1/2 of the correct pairs form; we assume that effectively *none* of the incorrect ones form. Note, that we are implicitly assuming that there are ample DNA strands for these reactions to actually take place. However, we have already stated that there will always be "enough" material.

The second case is the "near" case. In this case, the correct and incorrect strands agree in more than 1/4 of their positions. Now we can no longer assume that incorrect pairs will not form. For example, if two strands α and β are Watson-Crick complements except for one position, then they will likely bind each other. This is even more likely if the one place at which they differ is at the end. In this case they will bind almost as well as a perfectly matched pair. Thus, in this case we cannot assume that the rates of formation are vastly different for the correct and the incorrect case.

In this case we make the following weak assumption. We assume only that the rate or probability of forming a perfect match is strictly greater than that of a partial match. This is themodynamically reasonable: More matches will be better. We make no assumption about the exact difference. However, it is important to make a small assumption that the gap is at least $\delta > 0$ for a fixed small value of δ.

In summary, the error model is listed in Table 1. In each case, we claim p_1 is strictly larger than p_2 by an amount that is at least δ. This is somewhat analogous to a selection coefficient. The big surprise, perhaps, is that we can assume so little about annealing accuracy in the near case. It is unclear that such a weak assumption is enough to produce any results. However, it turns out that it is enough: the process is the subject of the next section.

3. Re-Coding of Unknown DNA

In this section we will show how to re-code an unknown strand of DNA X by one that is coded as we wish. Suppose that we have a test tube that contains X.

We must show how to create a new test tube that contains a *re-coded* version of the strand X. We will do this in two stages. Since our operations are only approximate we will *not* be able to do this exactly. Rather, we will be able to construct a test tube that "approximates" the desired one.

DEFINITION 3.1. *Let test tube T contain the single strands $S^{(1)}, ..., S^{(k)}$ with frequencies $q_1, ..., q_k$ and let T' contain $S^{(1)}, ..., S^{(k)}$ with frequencies $q'_1, ..., q'_k$. Then, say that T ϵ-approximates T' provided for all $i = 1, ..., k$, $|q_i - q'_i| \leq \epsilon$.*

Also k is the *support* of the test tube, *i.e.* is the number of distinct strands that occur with non-zero frequency. This is often called the "complexity" of the test tube; however, to avoid confusion with computer science theory terminology we will use support.

Note the following simple fact: if T_1 ϵ_1-approximates T_2 and T_2 ϵ_2-approximates T_3, then, T_1 ϵ-approximates T_3 where $\epsilon = \epsilon_1 + \epsilon_2$. Next a string definition:

DEFINITION 3.2. *A string α is* in *the string S provided α occurs as a consecutive substring of S,* i.e. *that for some i, $\alpha = S_i, ..., S_{i+l-1}$ where l is the length of the string α. A string α is* in *the strand X provided α is in X.*

There are two "tricks" that allow us to improve the quality of our basic operations. The first is that we can repeat a length separation multiple times. Clearly, less "correct" DNA is selected but also less "incorrect" DNA survives. For example, the following is useful:

LEMMA 3.1. *Suppose that test tube T contains equal amounts of $S^{(1)}, ..., S^{(k)}$ where each string is a different length. Then, for any $\epsilon > 0$ and each i we can in $O(\log(1/\epsilon))$ bio-steps construct a test tube that is an ϵ-approximation to the test tube that only contains $S^{(i)}$.*

PROOF. Let $(p_1, ..., p_k)$ be the frequencies of the k strings. By assumption each p_i is equal to $1/k$. Now after l gel separations the frequencies are $(q_1, ..., q_k)$. Without loss of generality let the first string be the one that we wish to select. Then, after the l steps, $q_1 \geq \frac{1}{2^l k}$ and $q_i \leq \frac{\tau^l}{k}$. However, these frequencies need to sum to one. In order to do this we re-normalize, i.e. we multiply by $1/t$ where $t = q_1 + \cdots + q_k$. Let the resulting frequencies be $(q_1^*, ..., q_k^*)$. Then, for each $i = 1, ..., k$, $q_i^* = q_i/t$.

Now t satisfies $\frac{1}{2^l k} \leq t \leq \frac{1}{k}$ for τ small. We require that the error is at most ϵ. Therefore we require that for $i = 2, .., k$, $q_i^* \leq \epsilon$ or that $\tau^l \leq tk\epsilon$. A simple calculation shows that this is true for $l = O(\log(1/\epsilon))$ since t is at least $\frac{1}{2^l k}$. □

We do not know in general how to produce a similar lemma for cuts. Repeating a cut, for example, will cause more correct material to be cut but will also cut more incorrect material. However, there is a very important case where we can essentially do this. Suppose that we have a test tube T and we plan to first apply a cut step and then a separation step. If all the pieces from the cut have the same length then we can apply Lemma 3.1 after the cut. The effect of this will be that incorrectly cut material will be filtered out. In a sense we have made the cut appear to have an error rate of ϵ rather than τ.

THEOREM 3.2. *Suppose that a test tube contains the unknown strand X. Also let $l = O(\log n)$ and let $f()$ be a function that is defined on length l strings. Then, for any $\epsilon > 0$ in $O(\log n + \log(1/\epsilon))$ bio-steps we can create a test tube T' that is* *an ϵ-approximation to the test tube that contains in equal amounts the strands that correspond to $f(\alpha)$ where α ranges over the length l consecutive substrings of X.*

Thus, we can go from a test tube that contains one string to another that contains re-coded versions of all the consecutive pieces of X. This is an important first step. For example, if we construct an (l, j)-*encoding function* $e(\alpha)$ that maps strings of length l into strings of length j, we could use this function to uniquely assign a different string to each string, for $j \geq l$. Here, j could be much larger than l, which would make our encoding functions highly redundant.

A simple example is shown Table 2. In this case, the value of l is 1 and the value of j is 3. Now suppose that $e()$ is some (l, j)-encoding function and that X

α	$e(\alpha)$
A	ATG
T	GCC
G	TTA
C	TGA

TABLE 2. An Encoding Function

is some string that is longer than l. Then, it is natural to extend the definition of the encoding function so that $e(X)$ denotes the following *set*: if α is a string that is of length l and α occurs somewhere in X as a consecutive substring, then $e(\alpha)$ is in the set. Thus, if X is of the form

$$\cdots \alpha \cdots$$

where $\cdots$ stands for an arbitrary string, then $e(\alpha)$ is in the set. In this example, $e(AT)$ is equal to the set $\{ATG,\ GCC\}$.

PROOF. Recall that T is a test tube that contains many copies of the strand X. Our plan is to add to this test tube additional pieces of DNA. These will be from a set we call the *probe* set. In particular, assume that we have already created the following probe set in another test tube. The probe set will contain for each string α of length l, the strand that corresponds to the Watson-Crick complement of $\alpha f(\alpha)$. We then will anchor the strands of the test tube T to a surface of another test tube T'. Then, add the probe set and allow them to anneal. Now, wash off the excess. Next elute the bound probes from the solid support. Then, allow them to re-attach. Repeat these steps, i.e. perform a molecular selection procedure and call this collection of probes T'.

The result is that T' will contain those probes that survived the repeated washing steps. We claim that these will be almost exclusively the correct ones, i.e. a probe $\alpha f(\alpha)$ that survives will have with high probability α in X.

Let us calculate the survival probability in the correct and the incorrect case. In the correct case the probability that a probe survives is p_1^m where m is the number of iterated cycles of selection by binding; in the incorrect case it is p_2^m. Here $p_1 > p_2 + \delta$.

Note, this assumes that only the α part of the probes are available for binding. We can arrange this in a number of ways. The simplest way is to use bases that are "blocked" so that they cannot bind at all.

Since p_2 is bounded below p_1 for $m = O(\log n + \log(1/\epsilon))$ we will expect that the fraction of incorrect probes that survive is at most ϵ.

Finally, we can arrange the probes so that we can cut away and release the $f(\alpha)$ part. Then, provided we have arranged that the length of the $f(\alpha)$ is much larger than l, a separation yields the desired test tube of DNA. In order to make the error rate low we use Lemma 3.1 to repeat the selection and separation steps. $\square$

We plan now to use Theorem 3.2 to allow us to re-code the whole of the unknown X. Note, however, that already the transformation is quite useful. For example, it could potentially be used to increase the power of "DNA chips." These chips attempt to sequence unknown DNA via hybridization.

Our plan is to use Theorem 3.2 to build a special test tube of DNA. It will contain pieces for each l consecutive substring of X. Moreover, these pieces will be able to anneal together to form the encoding of X. The key is that this method will only work on "reasonable" X's. The problem is that Theorem 3.2 only allows us to work with short parts of X, so that X must have the property that it is determined by its short pieces. If X is not, then this approach cannot succeed.

DEFINITION 3.3. *Say that a string X is l-*determined *provided that X is uniquely reconstructible from its l long subsequences.*

Note, if X is random then certainly for l approximately $2 \log n$ all the l pieces are likely to be unique; in this case X is trivially l-determined. However, subsequences can be repeated and X can still be l-determined. This notion is already in use in DNA chips [**17**]. The method of sequencing via hybridization can only work for sequences that are l-determined for a small value of l. Also, we do not require that it is easy to find X from its pieces, only that it is possible.

Let us fix two functions $h_L(\beta, i)$ and $h_R(\beta, i)$ where β is a string of length $l-1$ and i is from 1 to n. These functions are the "re-coding" or "hashing" functions that we will use. Thus, one can think of them as assigning a hash to a length $l-1$ string and an index i. Further, for a string X let $X^{(i)}$ denote the ith substring of X of length $l-1$.

In the following proof it is helpful to have a notation for handling DNA and its complements. We, therefore, we use the notation we introduced in [**6**]. If $\alpha_1, ..., \alpha_m$ is a string over the alphabet $\{A, T, C, G\}$, then $\uparrow \alpha$ denotes the single strand of DNA of the form $5' - \alpha_1, ..., \alpha_m - 3'$. We use $\downarrow \alpha$ to denote the single stand of DNA of the form $5' - \beta_m, ..., \beta_1 - 3'$ where β_i is the Watson-Crick complement of α_i. Finally, we use $\updownarrow \alpha$ to denote the double strand of DNA. Note, that $\uparrow \alpha + \downarrow \alpha$ can ligate together to form $\updownarrow \alpha$. The power of the notation is that one is freed from thinking directly about Watson-Crick relationships; instead one "names" then indirectly. For example, the following is a typical use of the notation: $\uparrow \alpha\beta + \downarrow \beta\delta$ can form $\uparrow \alpha \updownarrow \beta \downarrow \delta$.

THEOREM 3.3. *Suppose that T is a test tube with unknown DNA X that is $l-1$-determined for $l = O(\log n)$ where n is its length. Suppose also that $h_L()$ and $h_R()$ are hash functions as above. Then, for any $\epsilon > 0$ in $O(\log n + \log(1/\epsilon))$ bio-steps we can form the test tube T' that is an ϵ-approximation to the test tube that contains $\updownarrow h_L(X^{(1)}, 1)h_R(X^{(1)}, 1)$, $h_L(X^{(2)}, 2)h_R(X^{(2)}, 2)$, ...*

PROOF. Construct a test tube via Theorem 3.2 that contains the following pieces of DNA:

1. For each i and each j, $\uparrow h_L(X^{(j)}, i)h_R(X^{(j)}, i)$;
2. For each i and each j, $\downarrow h_R(X^{(j)}, i)h_L(X^{(j+1)}, i+1)$.

Then, we let the pieces ligate together. If the hash functions are "random" enough we expect the ligation to form one long path. We expect that

$$\uparrow h_L(X^{(1)}, 1)h_R(X^{(1)}, 1) \text{ and } \downarrow h_R(X^{(1)}, 1)h_L(X^{(2)}, 2)$$

will ligate together to form

$$\uparrow h_L(X^{(1)}, 1) \updownarrow h_R(X^{(1)}, 1) \downarrow h_L(X^{(2)}, 2).$$

Then, $\uparrow h_L(X^{(2)}, 2)h_R(X^{(2)}, 2)$ will add on to form

$$\uparrow h_L(X^{(1)}, 1) \updownarrow h_R(X^{(1)}, 1)h_L(X^{(2)}, 2) \uparrow h_R(X^{(2)}, 2)$$

and so on.

If Theorem 3.2 were perfect, then because we are in the far-apart case only the correct strands would form. The key is that as long as X is $l-1$-determined, then no other strand should form that is of the correct length. Then, we could finish up the proof by using Lemma 3.1 to perform a length separation.

However, Theorem 3.2 only creates an approximation to the test tube that contains the pieces according to (1) and (2). Thus, we need to take into account the fact that ϵ of the test tube is incorrect.

Consider how the pieces anneal and ligate together to form the one correct long strand of the correct length. If there are errors sometimes incorrect pieces will assemble. Call these "mis-steps." The point is that these occur but the frequency is at most ϵ. Let m be the support of the test tube and N the total number of strands. (Recall that m is the number of different types of strands that occur in the test tube.) Thus, the expected number of ways to assemble 2 correct pieces together with a mis-step is at most $N\binom{m}{2}\epsilon$. In general with l mis-steps it is $N\binom{m}{l}\epsilon^{l-1}$. Thus, the total fraction of wrong strands is at most

$$\frac{1}{N}\sum_{k\geq 2} N\binom{m}{k}\epsilon^{k-1}.$$

An easy calculation, then shows that the fraction of incorrect strands allowing mis-steps is at most $O(m\epsilon)$. Thus, for ϵ small enough this will complete the proof. □

Essentially, the proof of this theorem uses the same method Adleman used in his original paper [**1**]. The main difference is that we do not place the pieces explicitly into the test tube. Rather they are generated by the action of the molecular selection step of Theorem 3.2. Another key difference is that we allow errors.

The reason that Theorem 3.3 is useful is that it allows us to essentially recode the DNA in any way that we like. It constructs

$$\updownarrow h_L(X^{(1)}, 1)h_R(X^{(1)}, 1), h_L(X^{(2)}, 2)h_R(X^{(2)}, 2), \ldots$$

Clearly, this allows us to recode each separate base with a coding function that depends on the base *and* on the base's position. Thus, an "A" in position 1 is coded differently from an "A" in any other position. Of course this is exactly what we need to make DNA computations work.

4. Applications

Once DNA is re-coded the full power of DNA computation can be used to solve many interesting problems. In particular, we can now show that any reasonable computation can be efficiently performed on unknown DNA.

One of our key ideas is a general method for converting a DNA computation that assumes no error into one that uses our error model. We cannot do this in general, but we are able to do this for an important class of computations. The class is large enough to allow us, for example, to use the result of [**14**]. It seems like an important open problem to find methods that do this in general.

Suppose that $T_1, ..., T_m$ are a series of test tubes. Suppose further that each test tube T_{i+1} is obtained from T_i by one of the following operations:

1. a gel separation;
2. a ligation step followed by a gel separation. The ligation is in the far apart case;
3. a cut followed by a gel separation.

Assume that *no* errors occur: equivalently assume that $\tau = 0$. Call such a DNA computation *special*. Then, the following is the key fact:

LEMMA 4.1. *Let $T_1, ..., T_m$ be a series of test tubes that forms a special DNA computation. Then, from T_1 we can construct a test tube T' that is an ϵ-approximate to T_m in $O(m \log m + m \log 1/\epsilon)$ steps.*

Note, the construct is carried out in our model where errors occur, i.e. where τ is non-zero.

PROOF. Fix $\epsilon_0 > 0$. We construct a test tube T'_i that is an ϵ_i-approximation of T_i for each $i = 1, ..., m$ where ϵ_i is equal to $(i-1)\epsilon_0$. Clearly, this is trivial for $i = 1$.

Now let $i > 1$. Then, T_i is an ϵ_i-approximation of T_i. Suppose that T_{i+1} follows from T_i by a gel separation; the other cases are similar. Then, we can construct a test tube T'_{i+1} from T'_i that is an ϵ_0-approximation of T'_{i+1} in at most $O(\log(1/\epsilon_0))$ steps. This implies by induction that T'_{i+1} is as required.

Now the final error is $(m-1)\epsilon_0$. In order to make this small enough we need that $\epsilon_0 = \epsilon/m$. Thus, the construction requires at most $O(m \log m + m \log 1/\epsilon)$ DNA steps. □

THEOREM 4.2. *Suppose that $X^{(1)}, ..., X^{(k)}$ are unknown strands of length at most n that are in distinct test tubes. Moreover, assume that they are $O(\log n)$-determined. Then, in $O(\log n \log \log n + \log l)$ bio-steps we can compute the value of $F(X^{(1)}, ..., X^{(k)})$ where $F()$ is an NC^1 function with error probability less than 2^{-l}.*

PROOF. Use Theorem 3.3 to construct a test tube that contains the information required by [**14**] to compute F. Think of the input as kn bits and select a coding method as required by [**14**]. Then, by Theorem 3.3 we can replace each of the test tubes by an approximation to one that contains $X^{(j)}$ as one long strand. Then, we cut this into n pieces: one for each of its bits. We pour equal amounts of these k test tubes together. Then, we perform the operations as in [**14**] to compute the value of F.

The key point is that we can use Lemma 4.1 to argue that this can all be done in the presence of errors. This follows since the simulation method of [**14**] is *special* in our sense. Since the computation takes $O(\log n)$ steps the bound on the number of DNA steps follows. □

Note, we needed a new operation here: the ability to take test tubes $T_1, ..., T_k$ and create a new test tube T that contains equal amounts of material from each test tube. We claim that this is a reasonable operation that can be done again with error rate at most τ.

COROLLARY 4.3. *Suppose that X and Y are unknown strands both at most n long in distinct test tubes. Moreover assume that they are $O(\log n)$-determined. Then, it is possible to check whether or not $X = Y$ in $O(\log n \log\log n + \log l)$ bio-steps with error probability less than 2^{-l}.*

5. Conclusions

Before discussing whether or not these results are practical, there is a generalization that should be mentioned. The main one is the case of *partially unknown* DNA. In many interesting situations the DNA in a test tube is not unknown. Rather we know that it is equal to a known X_0 except for perhaps a few bases. This occurs in the case of mutation detection, for example. In this case the same theorems of section 4 can apply. However, now the probe set can be dramatically reduced in size.

There are a number of issues that must be solved before we can claim that these methods are practical. We view them as the starting point of a new direction for DNA computation. We believe that they should be viewed as an "existence" proof. That is our results are not immediately demonstrable in the laboratory. However, the idea of re-coding unknown DNA and then directly computing with it, DNA^{2}DNA computations, has potential importance.

Some of the practical issues are the following:

1. How should we build the probe sets?
2. Can we weaken the assumption on annealing in the far-apart case?
3. Can we weaken the assumption that the DNA is l-determined?

Clearly, one cannot use DNA synthesizers directly to build large probe sets. At least two interesting methods seem possible. For one, we may be able to use the same technology that is used to create DNA chips. The micro-robotic methods used to create these chips might be useful for generating our probe sets. For another, we may be able to exploit the structure of the probe sets. The probe sets we need are very regular sets. Indeed the following seems to be an important open problem: *Given a set of strings what is the cost in bio-steps to create a test tube that contains only those strings?*

Second, an important question is how far can we weaken the assumption of how far-apart strands anneal and ligate? One of the most important questions for much of DNA computation is to better model annealing and ligation. Obviously, the more realistic we are in modeling how strands mis-pair, the more practical our results will be. In particular, can we prove Theorem 3.3 in the case where incorrect annealing/ligations occur but with a low probability?

Finally, there is the problem of the assumption we made that the DNA is l-determined. As stated earlier, random or even approximately random strings are $l = O(\log n)$-determined. However, there are two problems with this. First, the size of l may be logarithmic but too large for practice. Second, real DNA is not random. Can the methods of Theorem 3.3 be improved to handle real DNA? We are currently investigating these questions.

6. Acknowledgements

We thank Chris Dunworth for a number of comments about this paper. We also like to thank Dan Boneh for his help in creating the original idea of "re-coding" DNA.

References

[1] L. Adleman. Molecular Computation of Solutions to Combinatorial Problems. *Science*, 266:1021–1024, 1994.

[2] M. Amos, A. Gibbons, P. Dunne. The Complexity and Viability of DNA Computations, *CTAG-97001*.

[3] S. Roweis, E. Winfree, R. Burgoyne, N. Chelyapov, M. Goodman, P. Rothemund, L. Adleman, A Sticker Based Architecture for DNA Computation. *DNA Computers: Proceedings of the Second DIMACS Workshop*, L. Landweber and E. Baum *editors*, DIMACS Series in Discrete Mathematics and Theoretical Computer Science, American Math Society, 1998.

[4] E. Bach, A. Condon, E. Glaser and C. Tanguay, DNA Models and Algorithms for NP-Complete Problems. *Proceedings of the 11th Annual IEEE Conference on Computational Complexity*, 1996.

[5] D. Beaver, Molecular Computing. *Penn State University Tech Report CSE-95-001.*

[6] D. Boneh, C. Dunworth, R. Lipton, Breaking DES Using a Molecular Computer, *DNA Computers: Proceedings of a DIMACS Workshop*, R. Lipton and E. Baum *editors*, DIMACS Series in Discrete Mathematics and Theoretical Computer Science, American Math Society, 37-65, 1996.

[7] D. Boneh, C. Dunworth, R. Lipton, J. Sgall, On the Computational Power of DNA, *CS-TR-499-95.*

[8] D. Boneh, L. Landweber, R. Lipton. Analog DNA Computations, unpublished manuscript, July 1996.

[9] D. Boneh and R. Lipton, Making DNA Computers Error Resistant. *DNA Computers: Proceedings of the Second DIMACS Workshop*, L. Landweber and E. Baum *editors*, DIMACS Series in Discrete Mathematics and Theoretical Computer Science, American Math Society, 1998.

[10] R. Dramanac, DNA sequence determination by hybridization: a strategy for efficient large-scale sequencing, *Science*, 263:596-596, 1994.

[11] F. Guarnieri and C. Bancroft, Making DNA Add. *Science*, 273:220-223, 1996.

[12] R. Karp, C. Kenyon, and O. Waarts, Error-Resilient DNA computation. *Proceedings of the 7th ACM-SIAM Symposium on Discrete Algorithms*, pages 458-467. ACM Press/SIAM, 1996.

[13] R. Lipton. DNA Solution of Hard Computational Problems. *Science*, 268:542-545, 1995.

[14] M. Ogihara and A. Ray. Simulating Boolean Circuits on a DNA Computer. *University of Rochester Technical Report 631*, August 1996.

[15] C. Papadimitrou, private communication, 1994.

[16] J. Reif, Parallel Molecular Computation: Models and Simulations. *Seventh Annual ACM Symposium on Parallel Algorithms and Architectures*, 1995.

[17] M. Waterman, *Introduction to Computational Biology,* Chapman Hall, 1995.

(L. Landweber) Department of Ecology and Evolutionary Biology, Princeton University, Princeton, NJ 08544

E-mail address: `lfl@princeton.edu`

(R. Lipton) Department of Computer Science, Princeton University, Princeton, NJ 08544, and, Bellcore Research

E-mail address: `rjl@princeton.edu`

(M. Rabin) Harvard University, Division of Applied Sciences, Cambridge, MA 02138, and Institute of Computer Science, Hebrew University, Jerusalem, Israel

DIMACS Series in Discrete Mathematics
and Theoretical Computer Science
Volume **48**, 1999

Strategies for DNA Computing

Andrew D. Ellington, Michael P. Roberston, Ken D. James, J. Colin Cox

ABSTRACT. Based on the seminal insights of Len Adleman, it has been recognized that Watson-Crick base-pairing or hybridization can perhaps be used as a means to implement and evaluate algorithms. A variety of schemes have been proposed, but these schemes have for the most part revolved around the problem of how to encode information in nucleic acids and how to use hybridization to evaluate the encoded information. We have carried out experiments to evaluate the fidelity of hybridization logic, and the likely error rates of DNA computations. Based on this analysis, it seems likely that DNA may be limited as a media for the solution of computationally complex problems. We therefore suggest a new direction in nucleic acid computing, a method by which the nucleic acids themselves play a far more active role than merely serving as templates for hybridization. In this paradigm for computing, ribozymes and other functional nucleic acids serve functions equivalent to elements of hardware such as insruction sets or transistors.

1. Introduction

Following the demonstration by Len Adleman[1] that a directed Hamiltonian Path Problem (HPP) could be encoded in DNA and evaluated, the use of nucleic acids for computational purposes has been the focus of extensive speculation. Proponents hypothesize that since nucleic acid hybridization is akin to massively parallel computation that DNA computers may have an inherent, although currently unrealized, superiority over silicon-based computers. Thus, computationally hard problems (like the nondeterministic polynomial time completeness (NP complete) set of problems) may be solvable with a linear increase in time for additional variables in an equation or nodes in graph, rather than the polynomial or exponential increase in time that constrains silicon-based computers[6]. The lure of bypassing mathematical constraints has led DNA to be touted as an appropriate computational media for problems ranging from the simplistic addition of two binary integers[7] to the development of a working Turing machine[4,5,14,16]. The reality is that to date no large, computationally complex problem has yet been encoded in DNA and solved by molecular biology methods[13]. The failure of DNA to rival silicon may in part be the result of the nascent and evolving nature of interactions between the molecular biology and computer science communities[2], but is more likely a reflection of the inherent intractability of DNA as a computational media.

All DNA computations so far envisioned rely on what might be called 'hybridization logic,' the ability to encode and solve computational problems by forming correct base-pairings and avoiding incorrect base-pairings. However, the potential fidelity of hybridization logic has until recently been unclear. We have developed a

1991 *Mathematics Subject Classification.* Primary 92B08, 92B99.

search algorithm in which oligonucleotide solutions are extracted from pools of random sequences and have used this assay to assess the potential fidelities of DNA computations under a variety of conditions. While under some conditions the fidelity of extraction and template copying were relatively high, under most conditions numerous mismatches and mutations were incorporated into the copied DNA strand. Taken together, these results suggest that DNA computing based on hybridization logic may be limited to situations in which low copying fidelity is required or to situations in which there are specific advantages to having the computation device itself be constructed of DNA.

In contrast, it may be possible to develop computational devices that are not inherently based on hybridization logic. Functional nucleic acids, aptamers and ribozymes, can be selected from random sequence populations. These functional nucleic acids may serve the same purpose as instruction sets in traditional central processing units or as transistors in circuit boards, in that they can carry out specific, designed tasks. As an example, we have evolved a ribozyme that can perform a ligation reaction in response to a specific oligonucleotide effector. This ribozyme in effect represents an informational transducer which can 'read' an 'input' sequence and transmogrify it into a different 'output' sequence.

2. The fidelity of template-directed oligonucleotide hybridization: an evaluation of 'hybridization logic.'

2.1 A search algorithm encoded in DNA. The basic molecular biology operations performed during the evaluation of HPPs or other NP complete problems encoded in DNA have been the template-directed juxtaposition and ligation of oligonucleotides, as shown in Figure 1a, followed by amplification and sieving by hybridization. We have mimicked this DNA computation by encoding a search algorithm in DNA. The evaluation of the encoded search requires the hybridization and ligation of oligonucleotides drawn from a pool of sequences, as shown in Figure 1b, followed by amplification and sequencing. In this case, however, one of the hybridization partners (the template or query sequence) is a single sequence and the other partner (the substrate) is a multitude of different sequences (substrate; up to 4^{12} = 16,777,216 sequences). The solution of the search algorithm requires the evaluation of more than 32 million possible ligation events (ca. 16 million sequences x 2 ligations / sequence). Therefore, we believe the procedures required for the evaluation of the search algorithm approximate the procedures that would be required for the evaluation of a large, computationally complex problem. The fidelity of evaluation of the search algorithm should provide some estimate of the fidelity of evaluation of other DNA computations.

Operationally, the query sequence was a dodecamer string embedded within a gapped DNA duplex (Figure 1c). The query sequence contained the four canonical bases in equal measure and thus allowed the assessment of all possible base-pairings. A random sequence DNA pool that contained all possible dodecamers served as a potential solution set. The random sequence pool was mixed with the gapped duplex and oligonucleotide substrates that hybridized across from the gap were fixed in place by ligation. While we had previously carried out similar studies using chemical ligation (9), most DNA computations that have been proposed or implemented rely on enzymatic ligation. Therefore, *E. coli* DNA ligase was employed as a coupling reagent. A mesophilic ligase was chosen rather than a thermophilic ligase because the melting temperatures of even perfectly paired dodecamer substrates were much lower than the temperature optima of commercially available thermostable ligases, such as the DNA ligase from *Thermus aquaticus*. In addition, the *E. coli* ligase was chosen over more popular molecular biology enzymes, such as T4 DNA ligase, because it had been shown to inefficiently catalyze blunt-end ligation events (New England Biolabs, Beverly, MA) that could have given rise to artefactual products.

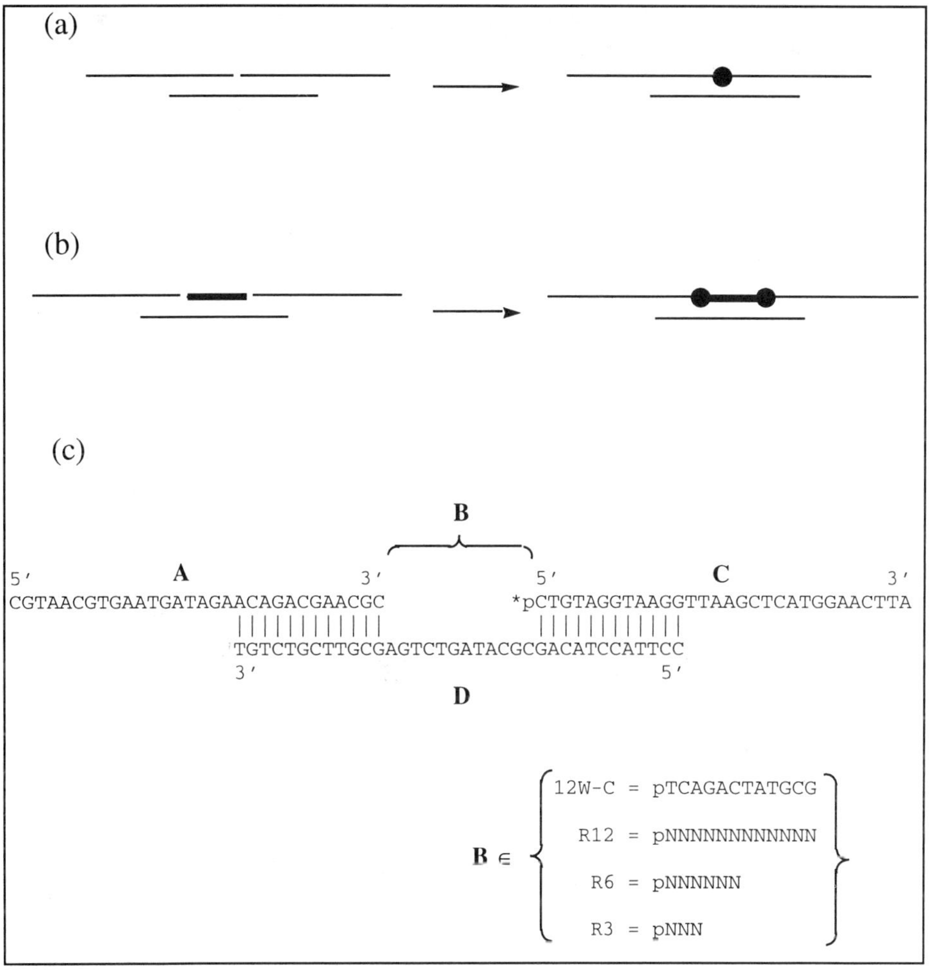

FIGURE 1. Schemes for DNA computation. (a) The basic molecular biology operations required for the solution of DNA computations such as the Hamiltonian Path (1), Maximal Clique, and addition problems. Two oligonucleotides are juxtaposed by a template and joined to one another by ligation. (b) The basic molecular biology operations required for the solution of the search algorithm described in this paper. Three oligonucleotide substrates are juxtaposed by a template and joined to one another by ligation. The terminal oligonucleotide substrates are constant sequences, while the internal oligonucleotide substrate (bold line) is a pool of dodecamers. (c) The sequences of the oligonucleotides used for the evaluation of the search algorithm. The precise Watson-Crick complement of the single-stranded region in the gapped duplex is 12W-C. Each position in the random sequence pools R12, R6, and R3 is shown as 'N,' which represents an approximately equimolar mixture of the four canonical deoxyribotides. The 5' ends of oligonucleotides C, 12W-C, R12, R6, and R3 are phosphorylated (p). In several experiments, the 5' end of C was radiolabeled (*p).

Sequence solutions to the search algorithm were preferentially amplified using the polymerase chain reaction. Several steps were taken to avoid amplification artefacts and cross-contamination of amplification reactions. First, primer binding sites were presented within single-stranded overhangs to guard against the inadvertent amplification of template strands. Second, only those oligonucleotides that were programmed to ligate were phosphorylated. Third, full-length ligation products were gel-isolated prior to amplification and sequencing. Fourth, control ligation products (e.g., Figure 2) were never present on the same gel as experimental ligation products. Finally, portions of gels adjacent to lanes containing ligation products were also excised and putative ligation products were isolated. When amplification reactions were seeded with these extracts no amplification products were observed.

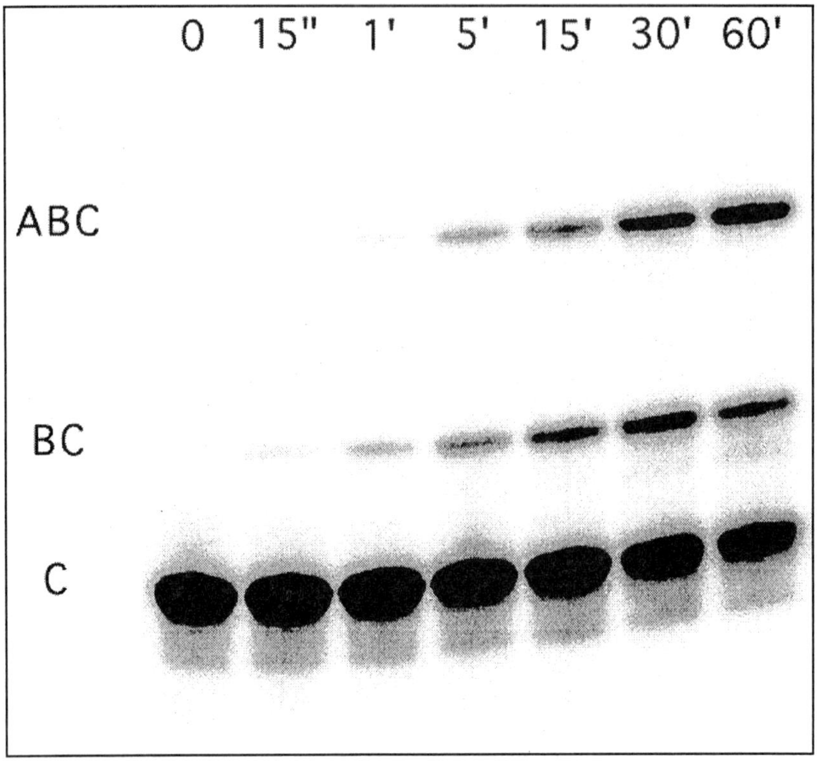

FIGURE 2. Time course of enzymatic ligation. Oligonucleotides A, 12W-C, and *pC (Figure 1) were mixed with template D at 25° C, a ligation reaction was initiated, and reaction products were separated on a denaturing polyacrylamide gel as a function of time. The intermediate ligation product between pB and C (BC) appeared first, followed by the full-length ligation product between A, pB, and *pC (ABC). Because of the position of the radiolabel, the intermediate ligation product between A and B could not be observed, but is likely present..

In order to determine whether our search algorithm could be successfully evaluated, a constant sequence dodecamer that corresponded to the correct sequence solution was first mixed with the gapped duplex containing the query sequence, a ligation reaction was initiated, and the products were analyzed by gel electrophoresis. Only the correct full-length product and an intermediate leading to the correct full-length product were observed (Figure 2). The formation of full-length product was monitored as a function of time. The ligation product could be observed as early as 1', but continued to accumulate for over an hour. Based on these experiments, sequence solutions to the search algorithm were extracted at two time points: one minute and one hour. By examining both time points we could potentially determine whether there was a trade-off between the efficiency of evaluation of the search algorithm (the amount of product produced) and the accuracy of evaluation of the search algorithm (the proportion of fully complementary products obtained). Since large, computationally complex problems encoded in DNA may require the efficient joining of hundreds to thousands of different oligonucleotides, relatively low reaction temperatures (16° C and 25° C) were chosen to maximize the efficiency of ligation [9]. The choice of relatively low temperatures was also prompted in part by a desire to compare results with the dodecamer pool to results with hexamer and trimer pools (see below); the shorter pools would not have efficiently hybridized to the gapped duplex template at temperatures significantly above 25° C [8]. The fact that the *E. coli* ligase was active at reduced temperatures again made this enzyme an ideal choice for these experiments. Finally, the search algorithm was performed at the two slightly different temperatures to determine whether its evaluation was robust to reaction conditions or was easily perturbed by subtle alterations in the molecular biology 'hardware.'

2.2. Solutions to the search algorithm. The sequences of four sets of amplified ligation products, corresponding to ligation at 16° C for one minute or for one hour were acquired; a sample of the sequence data from ligation reactions at 16°C is shown in Figure 3a. As is obvious from even a cursory examination of the sequence data, the fidelity of evaluation of the search algorithm under these conditions was relatively low. These results are further categorized in terms of potential base-pairings in Figure 3b. The most frequent mismatches that were observed were G:T wobble pairings, but almost all other mismatches were observed as well. On average, each dodecamer sequence solution to the search algorithm contained 3.3 mistakes (mismatches, deletions, or insertions; Table I). While erroneous answers represent a significant fraction of the set of sequence solutions, it should be recalled that from a computational point of view that the correct 'answer' to the 'query' would have been unknown prior to the experiment and that averaging the flawed answers would yield a completely correct consensus sequence solution to the search algorithm.

12 Min. 16•

	A	B	C
W-C	ACAGACGAACGC	TCAGACTATGCG	CTGTAGGTAAGG
A01	ACAGACGAACGC	TTGGGCTATGCG	CTGTAGGTAAGG
A02	ACAGACGAACGC	TCAGTATACGCG	CTGTAGGTAAGG
A03	ACAGACGAACGC	TTGGACTGTGCG	CTGTAGGTAAGG
A04	ACAGACGAACGC	TCATACGATGCG	CTGTAGGTAAGG
A05	ACAGACGAACGC	TCAGGCTATGCA	CTGTAGGTAAGG
A06	ACAGACGAACGC	ACGGAATATACG	CTGTAGGTAAGG
A07	ACAGACGAACGC	TTAGAATATGCG	CTGTAGGTAAGG
A08	ACAGACGAACGC	TCGGTCTCTGCG	CTGTAGGTAAGG
A09	ACAGACGAACGC	TTGGTCTATACG	CTGTAGGTAAGG
A10	ACAGACGAACGC	CCGGAAGATGCA	CTGTAGGTAAGG
A11	ACAGACGAACGC	TCGGAATGCGCG	CTGTAGGTAAGG
A12	ACAGACGAACGC	TTGGATTATGCG	CTGTAGGTAAGG
A13	ACAGACGAACGC	TTGGAATATGCG	CTGTAGGTAAGG
A14	ACAGACGAACGC	TTAGCATATGCG	CTGTAGGTAAGG
A15	ACAGACGAACGC	TTGGAATATGCG	CTGTAGGTAAGG
A16	ACAGACGAACGC	TCGGAATGCGCG	CTGTAGGTAAGG
A17	ACAGACGAACGC	TCAGATTACGCG	CTGTAGGTAAGG

R12 Hr. 16•

	A	B	C
W-C	ACAGACGAACGC	TCAGACTATGCG	CTGTAGGTAAGG
B01	ACAGACGAACGC	TCTTAGTAGGCG	CTGTAGGTAAGG
B02	ACAGACGAACGC	TGTGAGTACGCG	CTGTAGGTAAGG
B03	ACAGACGAACGC	TCATTCCAGGCG	CTGTAGGTAAGG
B04	ACAGACGAACGC	CCTGAATGTGTG	CTGTAGGTAAGG
B05	ACAGACGAACGC	ACGGAGAATGCG	CTGTAGGTAAGG
B06	ACAGACGAACGC	TTCGAATATGTG	CTGTAGGTAAGG
B07	ACAGACGAACGC	TCATGATATGCG	CTGTAGGTAAGG
B08	ACAGACGAACGC	TGTGAGAATGCG	CTGTAGGTAAGG
B09	ACAGACGAACGC	CCCGCCTTTGCG	CTGTAGGTAAGG
B10	ACAGACGAACGC	TGAGACAATGTG	CTGTAGGTAAGG
B11	ACAGACGAACGC	TTTG---ACTCG	*T*TGTAGGTAAGG
B12	ACAGACGAACGC	TTAGGCTAAGCG	CTGTAGGTAAGG
B13	ACAGACGAACGC	TCGGA--A-GCG	*T*TGTAGGTAAGG
B14	ACAGACGAACGC	TGGGCCTATGCG	CTGTAGGTAAGG
B15	ACAGACGAACGC	ATAGAGTTTGCG	CTGTAGGTAAGG
B16	ACAGACGAACGC	TCAGTTTCTGCG	CTGTAGGTAAGG
B17	ACAGACGAACGC	TTAGAT-GTGTG	CTGTAGGTAAGG
B18	ACAGACGAACGC	TTTGTCTAAGCG	CTGTAGGTAAGG
B19	ACAGACGAACGC	TCTGGGAATGCG	CTGTAGGTAAGG
B20	ACAGACGAACGC	TCGGGCTAACCG	CTGTAGGTAAGG
B21	ACAGACGAACGC	TTGTACTACGCG	CTGTAGGTAAGG
B22	ACAGACGAACGC	TTGTAGGATGCG	CTGTAGGTAAGG

FIGURE 3a. Sequences of PCR products derived from enzymatic ligations at two different time points using random dodecamer substrates (R12) at 16° C. Constant sequences are shown in grey; randomized regions are in black. While many of the residues could potentially form canonical base pairs with the template, those that are likely to form mismatches are in **bold** and deletions are indicated by dashes (-). The fully complementary strand (W-C) is presented at the head of each list.

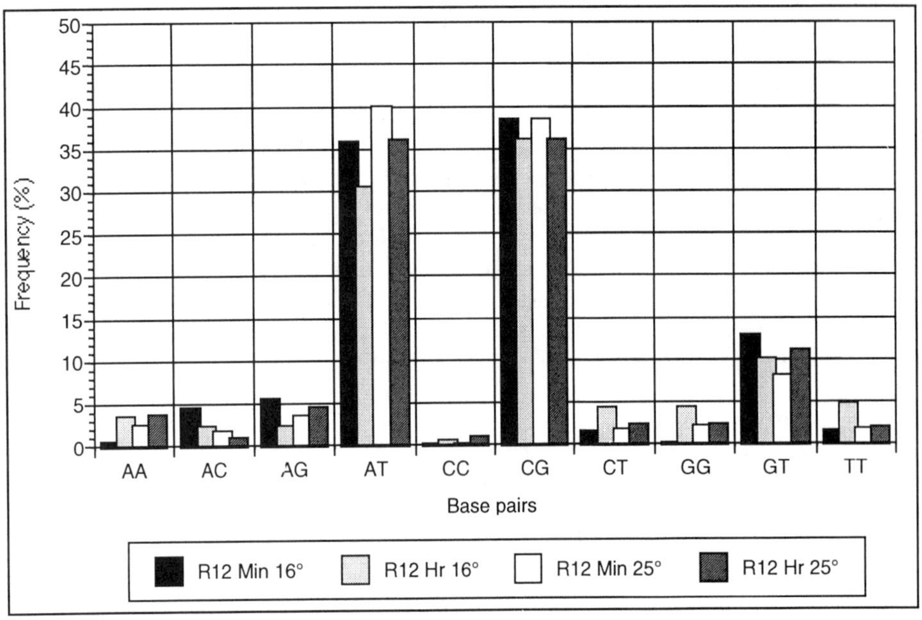

FIGURE 3b. The relative frequency of predicted base pairings in sequence solutions. The data from Figure 3a is tabulated in terms of predicted base-pairings. Because the template region contained an equal number of each of the four bases, perfect fidelity would have resulted in 50% of the pairs being A:T and 50% being G:C. (c) The fidelity of pairing as a function of position. The data from Figure 3a is tabulated as a function of position along the dodecamer template. The residues listed along the bottom of the graph are the correct pairing partners for the template, from 5' to 3'. The lines chart the frequencies of these residues under the different reaction conditions employed

	16°		25°	
	Min.	Hr.	Min.	Hr.
R12	3.1	4	2.6	3.4
R6	0.6	1.6	0.1	0.8
R3	1.8	4.7	0	10.4

TABLE 1. Fidelity of sequence solutions as a function of time and temperature. The data shown in Figure 3a, and the corresponding data used to derive Figures 4 and 5, were used to derive relative fidelities. Each mismatch, deletion, or insertion relative to the wild-type sequence solution was counted as a single error. The number of errors was then divided by the number of sequence solutions to determine the error rate. For example, the search algorithm was evaluated with the dodecamer pool at 16° for one minute with an error rate of 3.1 mismatches, deletions, or insertions per sequence solution

The trends within the fidelity data (Table I) can be readily rationalized. There is less fidelity at 16° C than at 25° C. The free energy of hybridization would have been greater at 16° C than at 25° C, and thus mismatches would have been better tolerated at the lower temperature. There is less fidelity after ligation for one hour than after ligation for one minute. The correct products would have formed and been fixed in place most quickly, while incorrect products would have accumulated with time. These trends can also be observed for individual DNA base-pairs (data not shown). Taken together, these results indicated that the evaluation of the search algorithm and other DNA computations can potentially be optimized by adjusting the thermodynamic and kinetic parameters of the DNA computer itself.

2.3. Optimization of the search algorithm with shorter pools. Based on our initial results, we hypothesized that further changes in temperature or time parameters could yield somewhat more accurate evaluations of the search algorithm, but that truly robust changes could best be obtained by changing the DNA components. The search algorithm was therefore performed again using shorter (hexamer and trimer) random sequence pools. Shorter pools would likely yield higher fidelity sequence solutions for three reasons: first, the melting temperatures of each oligonucleotide species would be correspondingly lower, and thus individual mismatches would be expected to have a proportionately larger effect on the ability of a substrate to hybridize to the query sequence at a given temperature. Second, the number of members of the random sequence pools was much smaller (4,096 hexamers or 64 trimers). While the overall

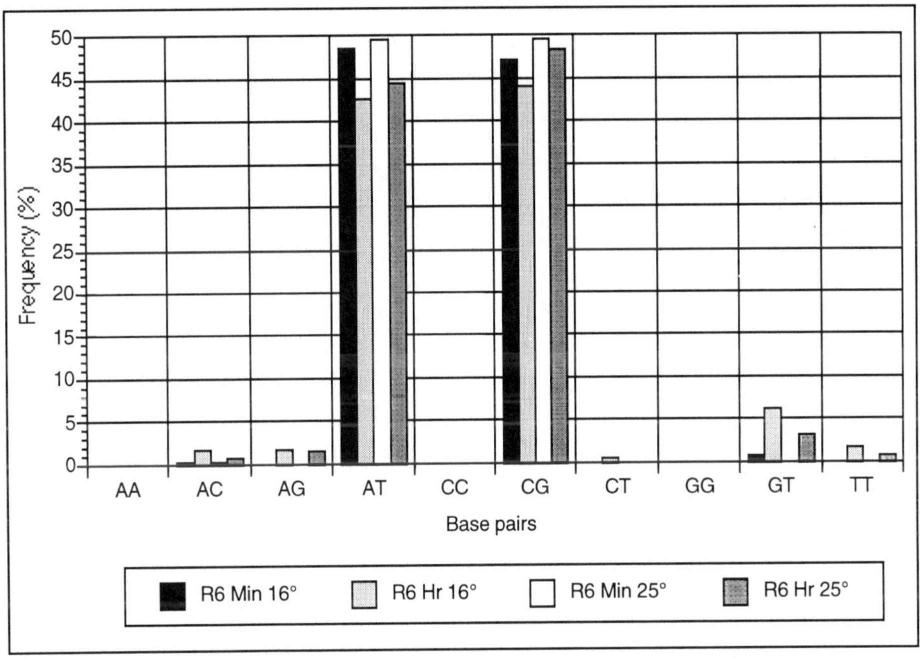

FIGURE 4. Fidelity of hexamer sequence solutions. Data from the evaluation of the search algorithm with hexamer pools is presented as in Figure 3b.

probability of finding correct sequence solutions would have been the same irrespective of whether hexamer or dodecamer pools were used (i.e., 4096 x 4096 = 16,777,216), the fidelity of each individual ligation event may have been higher because fewer 'incorrect' substrate:template pairings would have been assessed each time by the ligase. Third, the distance between ligation junctions would be smaller for shorter oligonucleotide pools. Given that the enzyme ligase appeared to be enforcing complementarity at and near ligation junctions, the shorter distance between ligation junctions might give the enzyme a greater chance to proofread each hybridization event before locking it in place.

The hypothesis that the extraction of correct sequence solutions from the potential solution set would be greatly enhanced by using shorter oligonucleotide pools as substrates was borne out in part by the experimental data. From 18 to 24 clones from each experiment were sequenced, and the relative fidelities of sequence solutions selected from the hexamer and trimer pools were determined, as shown in Figures 4 and 5, respectively. A numerical summary of the fidelity data is again provided in Table I. The observed fidelity of the sequence solutions drawn from the hexamer pool has improved greatly from those drawn from the dodecamer pool; there are now on average only 0.78 mistakes / sequence solution. The trends within the hexamer data were similar to those seen for the dodecamer pool: higher temperatures and shorter ligation times promoted fidelity. In contrast, the overall fidelity of the sequence solutions drawn from the trimer pool (4.2 mistakes / sequence solution) was now the same or worse than those drawn from the dodecamer pool. However, this average masks the fact that when the search algorithm was carried out at 25° C for 1 minute with the trimer pool it was evaluated perfectly, at least for the 18 independent clones that were examined. This result is especially impressive given that 90 individual ligation events must have occurred without mistake to generate the ensemble of clones. As before, the error rate with the trimer pool was higher at lower temperatures and longer times, but the increase in the relative number of errors was much larger than had previously been observed with the dodecamer or hexamer pools.

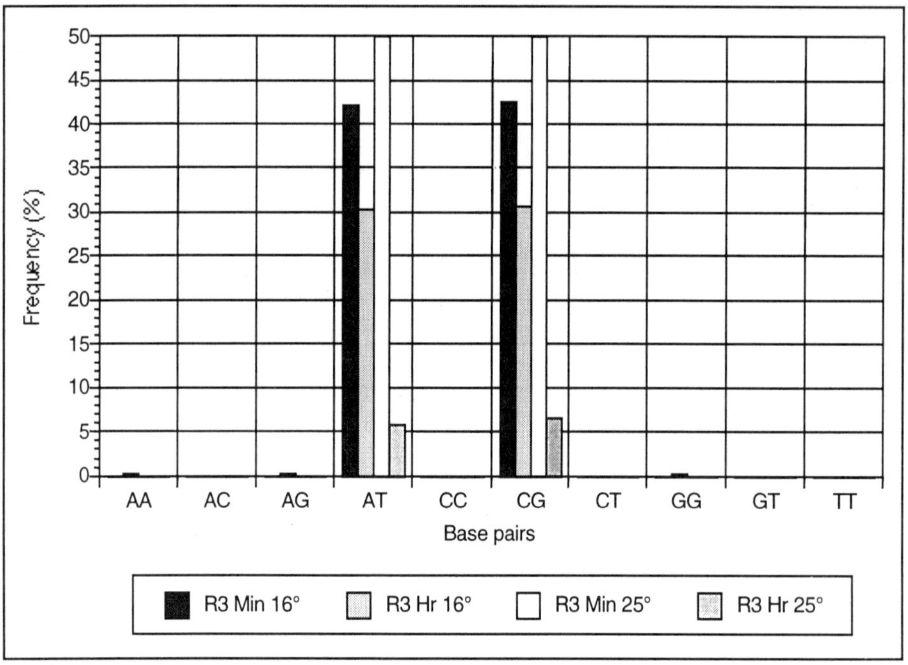

FIGURE 5. Fidelity of trimer sequence solutions. Data from the evaluation of the search algorithm with trimer pools is presented as in Figure 3b. Because of the large number of deletion variants that were present amongst the sequence solutions the values represented in the bars will not sum to 100%. This was also true for Figure 3b and Figure 4, but was not readily apparent because deletion variants were a much smaller proportion of the acquired sequence solutions..

2.4 Implications for DNA computing. The results of the evaluation of our search algorithm have both favorable and unfavorable implications for DNA computations. In the most favorable light, the number of mistakes yielded by a DNA computation would be the number of wrong answers generated per operation performed, or in this case the number of wrong oligonucleotides ligated in place per oligonucleotide considered. When the search algorithm is evaluated with a dodecamer pool there are on the order of three mistakes per sequence solution (Table I). Since there are 5940 three mutant variants ({12! / [9! 3!]} x 3^3) of the correct sequence solution the error rate for the search algorithm carried out with a dodecamer pool could be calculated as 5940 three mutant variants / 16,777,216 oligonucleotides considered or approximately one wrong oligonucleotide per 2,824 oligonucleotides considered. This number compares favorably with some estimates of the frequency of mistakes generated by Intel's formerly flawed Pentium chip (up to 1 mistake / 10^3 divisions), although not with Intel's own estimates of the frequency of mistakes (1 mistake / 9×10^9 divisions) [11,12]. Alternatively, the number of mistakes yielded by a DNA computation can be estimated as the number of wrong answers produced per correct answer, rather than per operation performed. In this case, the error rate for DNA computation with a dodecamer pool would be on the order of 5,940 wrong oligonucleotides ligated in place per every correct sequence solution. While we have identified conditions for the evaluation of the search algorithm where few or no (within the limits of our sequencing data) incorrect answers were returned, the fact that the fidelities of copying were extremely sensitive to ligation conditions and substrate lengths does not bode well for the possibility that a computationally complex calculation that involves multiple DNA strands will eventually be faithfully carried out.

3. Ribozymes and functional RNAs as alternatives to hybridization logic.

For several years our lab has contributed to the field of nucleic acid selection techniques and molecular evolution (see Osborne *et al.*, [10], for a review of the field). We have recently carried out a selection for a ribozyme ligase, a RNA enzyme that can catalyze the joining of RNA molecules (just at T4 DNA ligase, a protein enzyme, catalyzes the joining of DNA molecules). We started with a RNA population that contained 90 random sequence positions flanked by constant sequence regions. The population spanned over 10^{15} different variants. We selected for those members of the population that could covalently append a particular constant sequence region to themselves (Figure 6). Those variants that could catalyze this reaction were selectively amplified by a combination of reverse transcription, the polymerase chain reaction (PCR), and *in vitro* transcription. After five cycles of selection and amplification, we identified a ribozyme ligase that was approximately 300,000-fold faster than the background reaction. Interestingly, this ribozyme ligase was dependent on a DNA oligonucleotide present in the reaction mixture for its activity. Additional experiments have suggested that the ribozyme ligase is an allosteric enzyme: that binding of the DNA oligonucleotide changes the conformation of the ribozyme ligase from inactive to active (Figure 7).

The relevance of these results for nucleic acid computation is immediate. The ribozyme serves as an 'operator' that recognizes one sequence (the DNA effector) and promotes the formation of another sequence (the ligated product). In other words, any one ribozyme can 'read' one sequence and 'write' another. A series of such ribozyme operators could perhaps be used as the equivalent of an instruction set to effect re-write rules, and to implement transformational grammars (Figure 8a). Irrespective of the immediate utility of the allosteric ribozyme ligase that has been produced, it is important to recognize that selection techniques can be used to define such operators, and that collaboration between computer science and molecular biology communities will be essential to identify what types of operations should be generated and engineered.

From another vantage, the ribozyme ligase can be viewed as a new piece of molecular 'hardware,' similar to a transistor (Figure 8b). Since the ribozyme ligase is

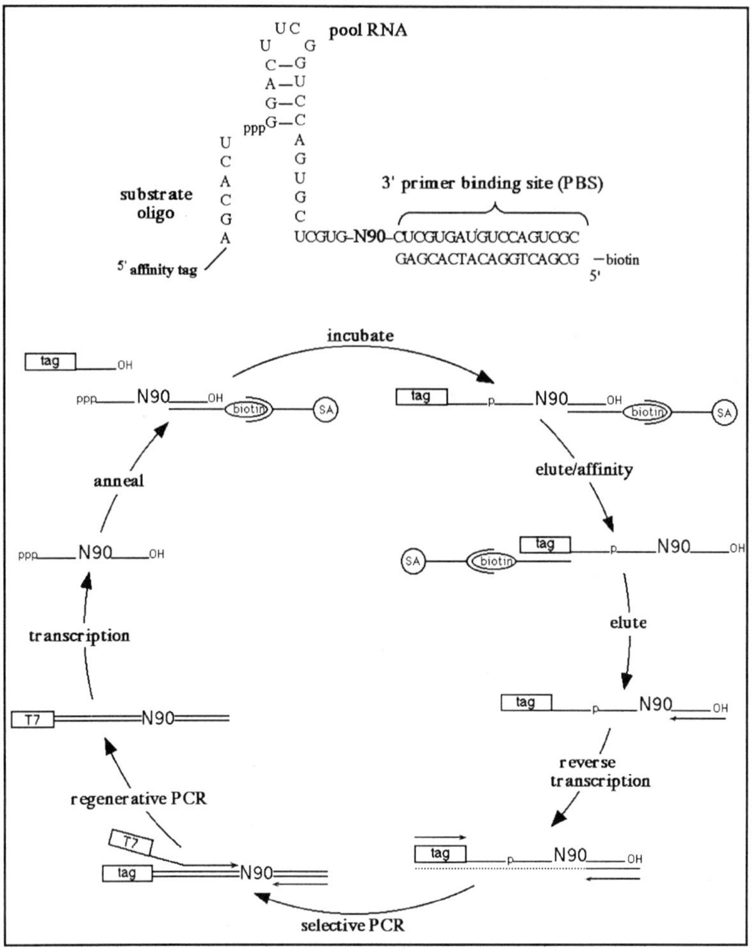

FIGURE 6. Scheme for the selection of ribozyme ligases. This scheme is similar to one employed by Bartel and Szostak [3]. A shorter random sequence pool (90 versus 220 randomized residues) and a different ligation junction (see stem-loop at top of figure) were used in our selection.

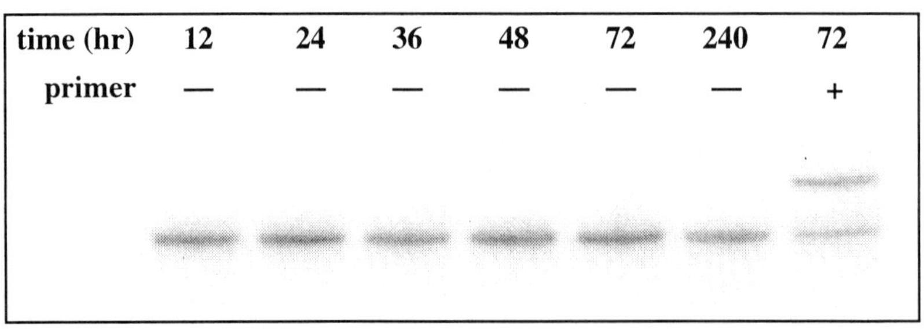

Figure 7. Allosteric ribozyme ligase. In the absence of a cDNA primer that was present during the ligation reaction and during the selection, no ligation product is observed over 240 hours. However, in the presence of the oligonucleotide effector a ligation product is readily observed, even at time points much shorter than 72 hours

catalytic, recognition of the DNA effector can potentially lead to the generation of many copies of a new (re-written) sequence string. The new string can in turn serve to activate another ribozyme ligase, which can further generate additional copies of a re-written sequence string, and so forth. Such amplification cascades could be implemented in series or in parallel, and could be further modulated by 'inhibitor' oligonucleotides.

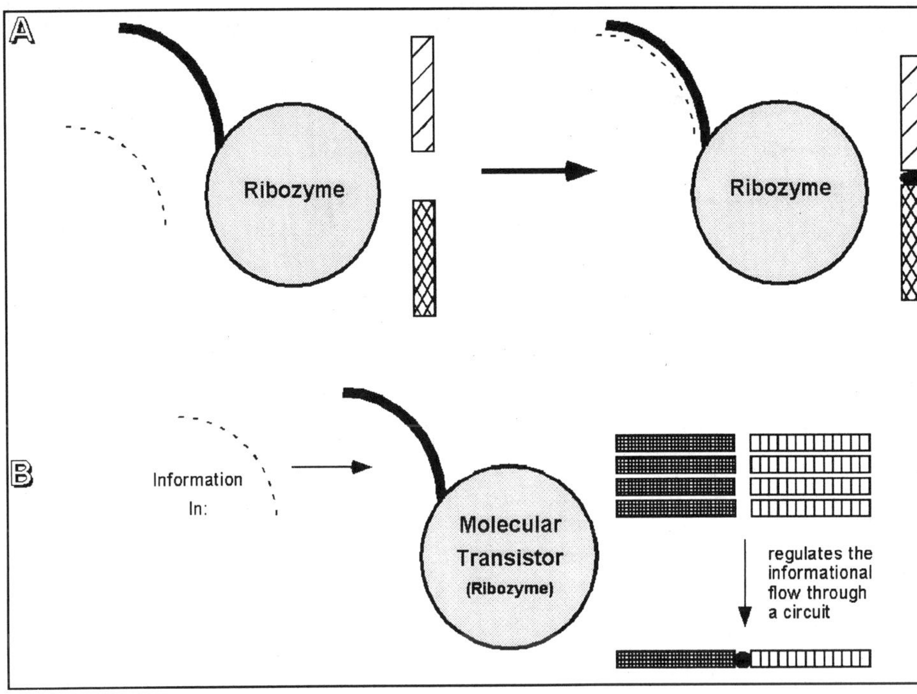

FIGURE 8. Computational applications of an allosteric ribozyme ligase (a) The allosteric ribozyme ligase binds an effector oligonucleotide (dashed line) and may generate new sequence information by covalently joining two oligonucleotide substrates (hashed and cross-hatched). (b) The allosteric ribozyme ligase may also produce a gain in information by sensing an effector oligonucleotide and catalytically producing multiple new ligated oligonucleotides. The joining of multiple allosteric ligases in computational circuits (e.g., 'and' and 'or' gates) is an obvious extension of these schemes.

Finally, the active participation of the nucleic acid structures (such as the folded, catalytically active ligase) in computations, as opposed to just nucleic acid sequences, may allow the 'inputs' and 'outputs' of nucleic acid computations to be extended far beyond sequence strings. Our and other labs have generated a wide variety of folded nucleic acid shapes that can specifically recognize a wide variety of molecules, from small organic drugs to proteins [10]. Ribozymes, such as the one described above, can potentially be selected or engineered to recognize such non-nucleic acid molecules as well (see for example, Tang and Breaker, [15]). In this instance, it may be possible to engineer ribozymes that sense metabolically relevant molecules, such as insulin or glucagon, and 'compute' an output based on the levels that are sensed.

4. Conclusion

DNA computations based on hybridization logic may be inherently far more error prone than comparable calculations in silicon, and thus may enjoy limited utility for the solution of computationally complex problems. However, other principles from mathematics and computer science may be adapted to the engineering of functional nucleic acids. These 'hard-wired' nucleic acid instruction sets, operators, or transistors may be particularly useful as biologically compatible components of therapeutic and diagnostic devices.

References

1. L. M. Adleman, *Molecular Computation of Solutions to Combinatorial Problems*, Science **266** (1994), 1021-1024.
2. L. M. Adleman, *On the Potential of Molecular Computing*, Science **268** (1995), 483-484.
3. D. P. Bartel and J. W. Szostak, *Isolational of New Ribozymes from a Large Pool of Random Sequences*, Science **261** (1993), 1411-1418.
4. D. Beaver, available at http://www.cse.psu.edu/~beaver/publications/tm.ps.z
5. C. H. Bennett, Int. J. Theor. Physics, **21** (1982), 905-940.
6. D. K. Gifford, *On the Path to Computation with DNA*, Science **266** (1994), 993-994.
7. F. Guarnieri, M. Fliss, and C. Bancroft, *Making DNA Add*, Science **273** (1996), 220-223.
8. K D. James and A. D. Ellington, *Surprising fidelity of template-directed chemical ligation of oligonucleotides*, Chem Biol **4** (1997), 595-605.
9. T. Maniatis, E. F. Fritsch, and J. Sambrook, *Molecular Cloning: A Laboratory Manual*, Cold Springs Harbor University Press, Cold Springs Harbor, N.Y., 1989.
10. S. E. Osborne, I. Matsumura, and A. D. Ellington, *Aptamers as therapeutic and diagnostic reagents: problems and prospects*, Curr Opin Chem Biol **1** (1997), 5-9.
11. V.R. Pratt, available at http://boole.stanford.edu/pub/PENTIUM/bugs.
12. V.R. Pratt, available at http://boole.stanford.edu/pub/PENTIUM/individual.bugs/bug21.
13. D. E. Rosen, S. McGrew, and A. D. Ellington, *Molecular Computing: Does DNA Compute?*, Current Biology **6** (1996), 254-257.
14. P. W. K. Rothemund, available at http://www.ugcs.caltech.edu/~pwkr/oett.html
15. J. Tang and R. R. Breaker, *Rational design of allosteric ribozymes*, Chem Biol **4** (1997), 453-549.
16. E. Winfree, available at http://dope.caltech.edu/winfree/papers/ligation.ps

DIMACS Series in Discrete Mathematics
and Theoretical Computer Science
Volume **48**, 1999

A Surface-Based DNA Algorithm for Minimal Set Cover

Tony L Eng and Benjamin M Serridge

ABSTRACT. We present an algorithm that solves the Minimal Set Cover problem within the framework of a surface-based model of computation. Affixing DNA strands to a solid surface reduces the possibility of error resulting from the loss of DNA strands in solution. An algorithm solving the Minimal Set Cover problem has previously been proposed using the sticker model of DNA computation [**Roweis**]. We show that, by taking advantage of the flexibility afforded by the surface-based model, we can design an encoding that eliminates the need for the separation steps required by the sticker model, thereby reducing both the likelihood of error and the number of "biological manipulation steps" required by the algorithm. To our knowledge, the algorithm we present is the first published DNA computing algorithm to make use of surface-based techniques.

1. Introduction

A practical problem faced by solution-based models of DNA computation is the loss of DNA strands during computation. Especially problematic are separation steps, such as those required by the sticker model [**Roweis**], where repeated wash cycles are prone to lose DNA. The loss of DNA strands that encode solution witnesses can lead to erroneous or partial answers. In response to these issues, a model has been proposed [**Cai**] that performs computation on DNA strands that are attached to a surface.

In this paper, we propose a surface-based algorithm that solves the Minimal Set Cover problem. In doing so, we demonstrate the feasibility of the surface-based model and highlight its performance advantages (for this problem) over the sticker model. Section 2 provides background regarding the Minimal Set Cover problem and the model of surface-based DNA computation proposed in [**Cai**]. Section 3 describes the algorithm itself, including a description of the encoding and the steps necessary for creating the DNA strands representing the possible solutions. Finally, Section 4 provides some discussion of the tradeoffs involved, followed by some conclusions in Section 5.

2. Background

2.1. The Minimal Set Cover Problem. Informally, the Minimal Set Cover problem is as follows: Given an initial set of B bags, where each bag contains a set

1991 *Mathematics Subject Classification.* Primary 92B05, 68Q05; Secondary 03D15.

The author was supported in part by NIH/NHCRI Training Grant in Genomic Science #HG00039 and DARPA Grant #N00014-95-1-1246.

of objects of different types (objects come in A types), find the smallest collection (subset) of these bags whose contents include at least one example of every type of object. See [**Garey**] for a more formal description of the problem.

2.2. Surface-Based DNA Computation. In the surface-based computing model [**Cai**], DNA strands, rather than floating free in solution, are attached to a surface such as silicon or glass. The goal of surface-based computing is to simplify the handling of the DNA strands and to remedy the problem of losing strands, at the expense of the ability to physically separate the strands from each other.

The DNA are initially single-stranded, with the 3' end attached to the surface and the 5' end free in solution. A strand is considered *marked* if the free 5' end is double-stranded.

2.2.1. *Operations.* We extend the operations available within the surface-based framework [**Cai**] with a detach operation. We review all operations briefly here; note, however, that the operations described below represent a slightly simplified version (in particular, with respect to append-marked and append-unmarked).

1. **mark(x):** Mark all DNA strands that contain the sequence denoted by x.
2. **unmark():** Unmark any marked strands (previously unmarked strands are unaffected).
3. **erase():** Erase any bases that have been appended since computation began, resetting the strands to their initial sequence.
4. **append-marked(y):** Append the double-stranded DNA sequence represented by y to all marked strands.
5. **append-unmarked(x,y):** Append the double-stranded DNA sequence represented by y to all unmarked strands that contain sequence x at their free end.
6. **destroy-unmarked():** Eliminate all strands that are not currently marked.
7. **detach():** Detach all DNA strands from the surface.

The biological implementations for these operations are found in [**Cai**], except for detach, which is new. Detaching all strands from the surface can be accomplished by designing the strands to include a special sequence near the base which can later be used as a restriction enzyme site for severing the strands.

3. Solving Minimal Set Cover with the Surface-Based Model

3.1. Algorithm. An approach for solving the Minimal Set Cover problem is described below.

The following four stages constitute the algorithm:

1. **Initialization:** Construct input strands representing all possible collections of bags.
2. **Recording:** For each strand, record all objects contained in the collection of bags encoded by the strand.
3. **Filtering:** Terminate strands that do not represent collections containing at least one of every object.
4. **Isolation:** Isolate the strand (if any) representing the collection with the smallest number of bags.

The initialization step generates all possible collections of bags. The objects that are represented in each collection are then examined, and only those strands

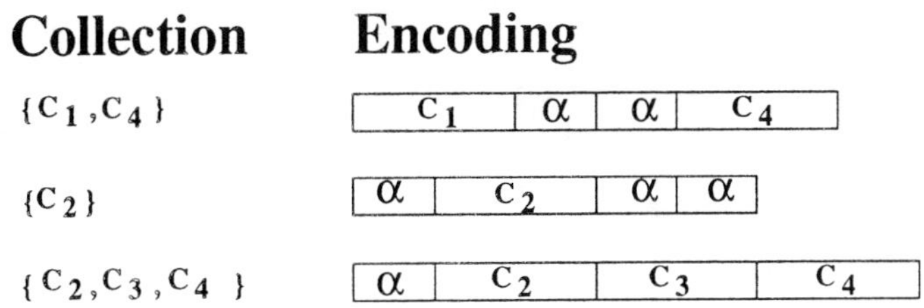

FIGURE 1. Examples of Encodings for a Collection of Bags

that represent collections in which all objects are present (at least once) are retained. Of these, the desired strand is that which encodes the collection with the fewest bags. Note that at this level the algorithm is essentially identical to the one proposed in [**Roweis**]. The improvements over the sticker model derive from an increased flexibility in the design of the encoding. Whereas in the sticker model the encoding of a subset of bags is of a fixed length, the surface-based model permits a design whereby DNA strands encoding smaller subsets are *physically* shorter, a property often exploited for solving optimization problems using solution-based techniques [**Boneh**]. Gel electrophoresis takes advantage of this property to perform the isolation step, thus bypassing the counting steps required by the sticker-based model.

3.2. Encoding. As in the sticker model [**Roweis**], the encoding divides each DNA strand into two parts, one encoding the collection of bags and the other encoding the objects contained within that collection. The following sections describe each part separately.

3.2.1. *Encoding a Collection of Bags.* Assume a problem size of B bags and A object types. Design for each of the B bags in the original set a sequence C_i to represent the presence of that bag in a collection. In addition, choose a single sequence α to encode the absence of a bag. A subset of bags is represented by the concatenation of B DNA sequences, each one either a C_i or the α sequence, corresponding to the presence or absence, respectively, of that particular bag. See Figure 1 for some example collection encodings.

As suggested by the figures, the α sequence should be designed to be shorter than the C_i sequences, in order that the length of a given DNA strand be proportional to the number of bags in the collection represented by that strand. This property facilitates detection by enabling the use of gel electrophoresis to separate DNA strands based on their length, which is particularly useful for the Minimal Set Cover problem because the desired solution is the one represented by the shortest strand. In this manner we are able to avoid explicitly counting the number of bags in each collection.

3.2.2. *Encoding the Objects in a Collection.* Each of the A object types is represented by a unique sequence of DNA bases. The contents of a given bag are represented by the concatenation (in any order) of the sequences representing the objects contained in that bag. These sequences are constructed in a "pre-computation" step before the algorithm begins: for each bag C_i, construct a test-tube TT_i containing double stranded DNA strands encoding the objects contained in C_i.

Step 1: Initialization
create initial set as described above
Step 2: Recording
for each bag C_i,
mark(C_i)
append-marked(TT_i)
unmark()
Step 3: Filtering
for each object a_j,
mark(a_j)
destroy-unmarked()
Step 4: Isolation
erase()
detach()
sort-based-on-size()

FIGURE 2. Algorithm for Minimal Set Cover Problem

3.3. Surface-Based Algorithm. The surface-based algorithm for computing Minimal Set Cover is detailed in Figure 2.

3.3.1. *Initialization.* To construct strands representing collections of bags, the sequences corresponding to each bag are concatenated together one by one, building upward from the surface. To facilitate this procedure, each bag sequence is constructed as a double-stranded DNA molecule with two overhangs (see Figure 3), one to anneal with the overhang from the previous bag sequence, and one to anneal with the overhang of the next bag sequence. We call the overhang sequences x and y, with $\overline{x}$ and $\overline{y}$ representing their respective complementary sequences. To avoid the possibility of bag sequences annealing to themselves in any given step, the following strategy is employed: All bags C_i with odd i use the overhang sequence $\overline{x}$ to anneal to the previous bag and the sequence y to anneal to the next one. Similarly, all C_i with even i use the overhang sequence $\overline{y}$ to anneal to the previous bag and the sequence x to anneal to the next one. Thus, no bag sequence can anneal to itself.

Similarly, there are two types of α's as well; one with $\overline{x}$ and y overhangs, and another with $\overline{y}$ and x overhangs.

We attempt to generate all 2^B possible choices of collections. Each collection is determined by a series of "coin flips" which decide, for each bag, whether or not to include that bag in the collection. The procedure is a B step process, one step for each bag. Specifically, for each bag C_i, an equal proportion of C_i (the sequence for a given bag i) and α (the sequence for the absence of a bag) are mixed and poured onto the surface with conditions conducive to annealing. The C_i sequences and the α sequences will attach to the dangling base sequences by hybridization,

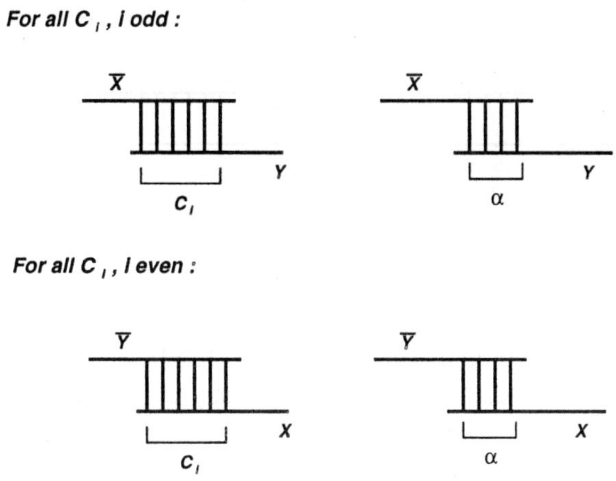

FIGURE 3. Bag Encoding

dividing themselves equally among the bases. Note that because the overhangs are distinct (x incompatible with $\overline{y}$ and vice-versa), multiple copies of C_i and α are unable to join together. At the end of this step, strands will have been extended by either exactly one α sequence, (signifying that the set does not contain this bag) or exactly one C_i sequence (signifying that it does).

The surface is then washed, and the next bag C_{i+1} can be appended. This process continues until all B bags have been added. At the end, each strand will represent a collection where the probability that each bag is present in the collection is approximately one-half.

3.3.2. *Recording.* This step records all the objects present in a collection of bags. For each bag C_i, mark all the strands that contain bag C_i and append the contents of TT_i to these marked strands. Recall that TT_i contained the double-stranded DNA sequence corresponding to the contents of C_i. These strands are then unmarked, and this process is repeated for the next bag. This will add to each strand a DNA sequence representing the union of the objects in all the bags in the collection encoded by that strand. For a collection, objects that occur in more than one bag will be represented more than once.

3.3.3. *Filtering.* The goal of this step is to remove all strands that do not encode a collection of bags that collectively contain all A objects. With this in mind, for each object, mark all strands that contain the sequence representing this object, and then destroy all unmarked strands. At the end only those strands that included (at least) one copy of every object will remain.

3.3.4. *Isolation.* When this stage is reached, all surviving strands encode collections that contain all the objects. Here, we are able to take advantage of our choice of length encoding because the answer we seek, if one exists, is encoded by the shortest strand(s).

First, erase all appended information by cleaving the DNA strands at a restriction enzyme site separating the initial encoding of bags from the encoding of the objects contained therein, thereby returning the strands to the state they were in after the Initialization step.

Next, detach all strands from the surface and sort them based on length using gel electrophoresis.

4. Discussion

4.1. Analysis of the Algorithm. The total number of biological manipulation steps for this algorithm is $O(A+B)$.

In the first step, a process linear in the number of bags is used to construct the initial strands. Next the objects of each bag are noted and recorded on each strand, taking $O(B)$ biological operations in all. The third step requires $O(A)$ operations to test the presence of each object in turn. The final step is $O(1)$, and thus the algorithm as a whole runs in $O(A+B)$ operations.

The "pre-computation," in which the strands encoding the contents of each bag are created, requires $O(AB)$ bio-steps (worst case), although in practice the tubes can be created in parallel, reducing the actual time spent to $O(B)$.

Note that after the Recording step, the DNA strands are extended by $O(AB)$ bases (the case in which every object appears in every bag).

4.2. Problem Encoding. The choice of encoding enabled us to avoid counting the number of bags in each collection. By choosing different lengths for the sequences representing the presence or absence of a bag, the encoding guarantees that collections containing fewer bags will be represented by shorter strands of DNA. This strategy is not possible with the sticker model, since the presence or absence of a bag in a collection is determined not by the length of the sequence corresponding to it, but by its strandedness (single or double).

One concern is that strand length may grow during computation, up to a length of order AB. Long strands of DNA are problematic for many reasons, including the high potential for unwanted secondary structures (self-annealing), annealing with neighboring strands, and shearing. Furthermore, basic operations such as mark, which run the length of a given strand (via polymerase extension), will take longer to run and have higher error probabilities when applied to long strands. These limitations in practice will limit the maximum problem size that can be solved.

Finally, the fact that the choice of encoding was the key to solving the problem efficiently suggests that the surface-based model may not be appropriate for every problem. Specifically, problems in which the validity of a solution is not proportional to its size may be better suited to alternative models of DNA computation. In many cases, a combination of the two techniques may be the best strategy.

4.3. Detection of Strands. The possibility remains that only a single strand of DNA is a witness for the correct solution. In the process of detaching the strands from the surface and running them on a gel, this strand may be lost or even fragmented. To avoid this situation, one can amplify any strands that remain on the surface after the filtering steps. Using the attached single strands as a template, they can be polymerized in their entirety by the exponential PCR amplification process or the linear amplification process consisting of polymerize-melt-wash cycles.

5. Conclusion

We have proposed an algorithm for solving the Minimal Set Cover problem using a surface-based model of DNA computation. We employed a problem encoding that enabled us to use gel electrophoresis to select out the correct answer, replacing the separation steps required by the sticker model algorithm. In doing so, we hope to have reduced the potential for error due to the loss of DNA strands resulting from the repeated application of separation operations, and have demonstrated that surface-based algorithms may be a possible alternative paradigm for DNA computing.

6. Acknowledgments

The authors would like to express their appreciation to Professor David Gifford for the "Biological Computation" class that prompted the research and for his revision of an earlier draft of the paper; to Jyotsna Iyer for her research into the chemistry of surface-attachment; and to Professor Michael Sipser for his consultation on matters of theoretical computer science.

References

[Adleman] Adleman, L. M. (1994). "Molecular Computation of Solutions to Combinatorial Problems", Science, 266, 1021-1024.

[Boneh] Boneh, D., Dunworth, C., Lipton, R., Sgall, J., "On the Computational Power of DNA", manuscript, *http://www.cs.princeton.edu/ dabo/papers/biocircuit.ps.Z*

[Cai] Cai,W., Condon, A., Corn, R., Glaser, E., Fei, Z., Frutos, T., Gul, Z., Lagally, M., Liu, Q., Smith, L., Thiel, A., "The Power of Surface-Based DNA Computation", extended abstract, Recomb, 1997.

[Roweis] Roweis, S., Winfree, E., Burgoyne, R., Chelyapov, N., Goodman, M., Rothemund, P., Adleman, L., "A Sticker Based Model for DNA Computation", May 1996.

[Garey] Michael R. Garey and David S. Johnson. Computers and Intractability: A Guide to the Theory of NP-Completeness. W. H. Freeman and Co., 1979

Laboratory for Computer Science, Massachusetts Institute of Technology, Cambridge, Massachusetts 02139

Current address: 545 Technology Square, NE43-338, Cambridge, MA 02139

E-mail address: `tlong@theory.lcs.mit.edu`

Laboratory for Computer Science, Massachusetts Institute of Technology, Cambridge, Massachusetts 02139

Current address: 545 Technology Square, NE43-338, Cambridge, MA 02139

E-mail address: `serridge@alum.mit.edu`

DIMACS Series in Discrete Mathematics
and Theoretical Computer Science
Volume **48**, 1999

Solid Phase DNA Solution to the Hamiltonian Path Problem

Nobuhiko Morimoto, Masanori Arita, and Akira Suyama†

Abstract. A solid-phase method for solving the Hamiltonian path problem (HPP) is described. The method employs only fast and simple DNA operations amenable to full automation. Single-stranded DNA molecules representing paths with no city visited twice are synthesized city-by-city from the start city on the surface of a solid support. The solution can thus be found in the computation time proportional to the number of cities. As well as the stepwise path synthesis, a pruning technique developed for the removal of looping paths helps the reduction of DNA molecules necessary for the computation; thus definitely increasing the size of problems solvable on a DNA-based computer. Experiments using Adleman's seven-city instance of the HPP showed that the path extension cycle was very accurate and took only about 45 min. Our solid-phase method has originally been developed for solving the HPP, but it could also be applied to other problems requiring a massive parallelism in computation.

1. Introduction

DNA-based computation utilizes the massive parallelism of DNA molecular reactions to solve the NP-complete problem in the computation time proportional to the size of a problem. In 1994, Adleman first demonstrated DNA molecules can be used to find the solution to a tiny instance of the directed Hamiltonian path problem (HPP) [**1**]. He generates all possible paths in a graph with DNA molecules at once. Then he finds the solution out of them by simultaneously subjecting all DNA molecules to a sequence of selection methods commonly used in molecular biology experiments. After that, Lipton and other researchers proposed DNA-based algorithms for solving other combinatorial NP-complete problems [**2**] [**3**]. These studies have blazed a trail in the field of DNA-based computation.

The massive parallelism of DNA-based computation proposes a new architecture of high-speed parallel computers. There are about 6×10^{15} DNA molecules in a DNA oligonucleotide solution with a volume ($100\,\mu$l) and a concentration (0.1 mM) usually used in biological experiments. So, if letting every DNA molecule represent

1991 *Mathematics Subject Classification.* Primary 68-06; Secondary 68M99, 68R99.

The second author was supported by Research Fellowships of the Japan Society for the Promotion of Science for Young Scientists. This work was supported in part by the Japan Society for the Promotion of Science "Research for the Future" Program (JSPS-RFTF96100101).

†To whom correspondence should be addressed.

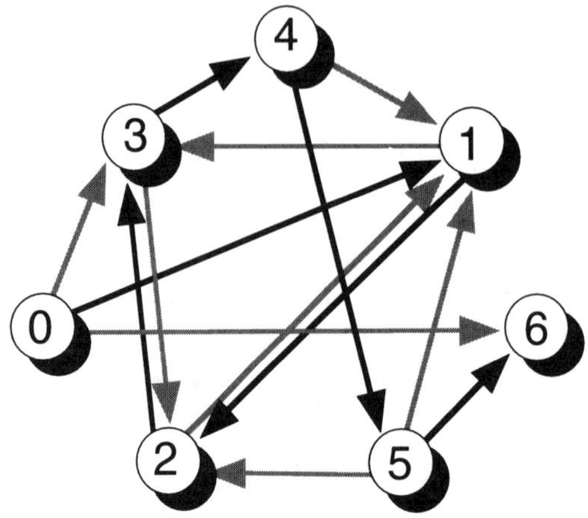

FIGURE 1. Adleman's graph for the HPP

a data or instruction stream, 6×10^{15} operations could be performed simultaneously. However, no DNA-based computations so far reported are actually fast. For example, Adleman spent a week finding a solution to a seven-city instance of the HPP. This is because the computation is performed by means of slow and laborious biological methods. Furthermore, the parallelism in a DNA solution is indeed huge, but it is not sufficient to represent all possible data streams for a large instance of the NP-complete problem. Therefore, to build up feasible DNA-based computers, both the speedup of each biological method and the great reduction in the number of data and instruction streams simultaneously represented with DNA molecules are needed

In the present study we describe a new experimental method for solving the HPP, which is much faster and requires much smaller number of DNA molecules than Adleman's method. Our method is based on a solid phase chemistry. DNA strands representing paths with no city visited twice are synthesized city-by-city from the start city on the surface of a solid support. An n-city instance of the HPP can be solved in $n-1$ extension cycles, which take about $45 \times (n-1)$ min. As well as the stepwise path synthesis, a pruning technique developed for the removal of looping paths helps the reduction of DNA molecules representing possible paths; thus definitely increasing the size of problems solvable on a DNA-based computer.

2. Materials and Methods

2.1. DNA oligomers. The same seven-city instance of the HPP as that given by Adleman (Fig. 1) was used in the present study. However, the set of base sequences that represent vertices (cities) and edges in the graph was different from that used by Adleman. The new set of base sequences that was suitable for the present method was designed and prepared.

Single-stranded DNA oligomer representing the start city was 40 bases long. It was composed of two parts: the $5'$ end part of 25 bases long called a root sequence (RS) and the $3'$ end part of 15 bases long specifying the start city, i.e. city 0 (v_0). Thus the start city oligomer is designated by RS-v_0. The $5'$ end of RS-v_0 was biotinylated to be attached onto streptavidin-coated magnetic beads through the biotin-streptavidin interaction. An RS part has the recognition site of a restriction enzyme *Eco*RI which allows RMVDC (remove duplicate cities) operation to get rid of looping path strands from the beads.

Single-stranded DNA oligomers representing other cities than the start city were also a 40-base oligomer composed of two parts but with a slightly different structure. The $5'$ end part was 15 bases long and used to specify city i (v_i) in the graph. The $3'$ end part was a common sequence of 25 bases long called a tail sequence (TS), which was used to block coupling of more than one city to the end while executing EXT (extend) operation. These city oligomers are thus designated by v_i-TS ($i = 1, 2, \cdots, 6$). The $5'$ end of v_i-TS was phosphorylated for a ligation reaction in EXT operation. There is the recognition site of a restriction enzyme *Bam*HI in a TS part so that it can be removed by DEBLK (deblock) operation.

Base sequences specifying cities and those of RS and TS were designed to share no common subsequences of more than 5 bases long in order to prohibit inaccurate path extension and unfavorable association of single-stranded DNA oligomers in a reaction mixture. In addition, they were designed to have no stable secondary structure to ensure efficient annealing of complementary DNA oligomers. Their GC-content was not greatly biased, ranging from 0.33 to 0.67, so that all complementary DNA oligomers are subjected to annealing conditions with almost the same stringency.

The base sequences of the city oligomers used in the present study are as follows:

RS-v_0: `5'-b-aaacgtctctccacgagcgcGAATT-Cggcctacttaagag-3'`
v_1-TS: `5'-p-ttgttacgcagcccG-GATCCataggagactaagagagacg-3'`
v_2-TS: `5'-p-attgccgaggtactG-GATCCataggagactaagagagacg-3'`
v_3-TS: `5'-p-taccgaaagtatgaG-GATCCataggagactaagagagacg-3'`
v_4-TS: `5'-p-tcgtgtgccgtgcaG-GATCCataggagactaagagagacg-3'`
v_5-TS: `5'-p-tgtgaggttcgagaG-GATCCataggagactaagagagacg-3'`
v_6-TS: `5'-p-ttgagcttgaaactG-GATCCataggagactaagagagacg-3'`,

where "b" and "p" designate a biotin and a phosphate group, respectively. The recognition sites of *Bam*HI and *Eco*RI are capitalized. The sites cleaved with these restriction enzymes are designated by "-".

A base sequence of single-stranded DNA oligomer representing an edge was the reverse complement of the two base sequences of connected cities lining up side by side. For example, the base sequence of edge oligomer connecting the start city (city 0) and city 1 is the reverse complement of the base sequence of v_0-v_1, and is designated by $rev(v_0\text{-}v_1)$. The base sequence of $rev(v_0\text{-}v_1)$ was thus `5'-cgggctgcgtaacaactcttaagtaggccg-3'`.

All DNA oligomers, synthesized on an automated DNA synthesizer and purified with an OPC cartridge, were obtained from Sawaday Technology, Inc. (Japan).

2.2. Experimental procedures for executing basic operations. The solid phase Hamiltonian path synthesis method was composed of seven basic operations: ASC, EXT, CAP, DEBLK, RMVDC, AMP and GPCR operations. ASC operation performs the attachment of the start city oligomer to a solid support. EXT operation extends path strands on a solid support by a city. CAP operation covers the $3'$ end of path strands on a solid support with an unreactive group to prohibit further extension of the paths. DEBLK operation removes the unreactive group to allow further extension of path strands. RMVDC operation removes path strands with a looping path, i.e. a path with cities visited twice, from a solid support. AMP operation increases the number of path strands on a solid support. GPCR operation is a graduated PCR to display the order of cities on a path. All operations except GPCR were implemented by only simple and fast methods, which are amenable to full automation.

2.2.1. *ASC operation.* Single-stranded start city oligomer RS-v_0, which was biotinylated at the 5′ end, was immobilized on streptavidin-coated magnetic beads. Twenty microliters (containing 20 μg) of Dynabeads M-280 Streptavidin (DYNAL A. S., Norway) were washed twice with 100 μl of binding and washing buffer (B&W buffer: TE buffer (10 m*M* Tris-HCl (pH 7.6), 0.1 m*M* EDTA) containing 1 *M* NaCl) using a Dynal MPC, and resuspended in 50 μl of B&W buffer. Ten pmol of biotinylated RS-v_0 in an equal volume of B&W buffer was added and incubated at 25℃ for 15 min while gently shaking the tube. The beads were then washed once with 100 μl of B&W buffer and twice with 100 μl of TE buffer at room temperature.

2.2.2. *EXT operation.* Beads were mixed with 1 pmol each of city oligomers v_i-TS ($i = 1, 2, \cdots, 6$) and edge oligomers $rev(v_i\text{-}v_j)$ (i, j; all pairs of connected cities) in 50 μl of *Taq* ligation buffer (20 m*M* Tris-HCl (pH 7.6), 25 m*M* CH_3COOK, 10 m*M* $(CH_3COO)_2Mg$, 10 m*M* DTT, 1 m*M* NAD, 0.1% Triton X-100). To avoid a misligation reaction, the reaction mixture was heated up to 94℃ before an addition of 20 units of *Taq* DNA ligase (New England Biolabs, USA). The beads were then incubated at 65℃ for 10 min while gently shaking the tube. After the ligation reaction, the beads were washed twice with 100 μl of TE buffer at 94℃ to remove excess city and edge oligomers.

2.2.3. *CAP operation.* Beads were incubated at 37℃ for 10 min in 50 μl of cacodylate buffer (120 m*M* sodium cacodylate (pH 7.2), 1.2 m*M* $CoCl_2$, 0.6 m*M* 2-mercaptoethanol) containing 10 nmol of dideoxyribosyl thymidine triphosphate (ddTTP) and 20 units of terminal deoxynucleotidyl transferase (TdT) (TOYOBO, Japan). After the capping reaction, the beads were washed twice with 100 μl of TE buffer at room temperature.

2.2.4. *DEBLK operation.* Beads were mixed with 5 pmol each of $rev(v_i\text{-}v_i\text{-}$TS) oligomers ($i = 1, 2, \cdots, 5$) in 50 μl of H buffer (50 m*M* Tris-HCl (pH 7.5), 100 m*M* NaCl, 10 m*M* $MgCl_2$, 1 m*M* DTT). The reaction mixture was heated to 94℃ for 1 min and then cooled down to 37℃ in 5 min to facilitate annealing of the $rev(v_i\text{-}v_i\text{-}$TS) oligomers to path strands on the beads. After adding 200 units of *Bam*HI, the beads were incubated at 37℃ for 5 min. The beads were then washed twice with 100 μl of TE buffer at 94℃, making the path strands single-stranded.

2.2.5. *RMVDC operation.* Beads were suspended in 50 μl of Ex *Taq* buffer (TAKARA, Japan) containing 4 μl each of the 0.2 m*M* dNTPs and 5 pmol each of $rev(v_i\text{-}v_i\text{-}$TS) oligomers ($i = 1, 2, \cdots, 5$), which anneal to path strands on the beads. The reaction mixture was heated up to 94℃ and 1.25 units of Ex *Taq* DNA polymerase (TAKARA, Japan) was added to initiate the polymerization. The beads were then incubated at 80℃ for 1 min. After stopping the polymerization reaction with the addition of 2 μl of 500 m*M* EDTA, the beads were washed once with 100 μl of TE buffer at room temperature. The beads were then resuspended in 50 μl of H buffer and treated for 5 min at 37℃ with 200 units of *Eco*RI to remove looping path strands from the beads. After an *Eco*RI digestion, the beads were washed twice with 100 μl of TE buffer at 94℃, making the path strands single-stranded.

2.2.6. *AMP operation.* Beads were subjected to a PCR in 50 μl of Ex *Taq* buffer containing 4 μl each of the 0.2 m*M* dNTPs and 10 pmol each of RS and *rev*(TS) primers. PCR amplification was performed on UNOII Thermocycler (Biometra,

Germany) using the temperature profile of the initial denaturation at 94℃ for 1 min followed by 20 cycles of the denaturation at 94℃ for 20 sec, the annealing at 40℃ for 1 min, and the extension at 72℃ for 1 min. The hot start technique was used, so 1.25 units of Ex *Taq* DNA polymerase was added just after the initial denaturation step.

After the first PCR amplification, 1 μl of the supernatant was subjected to the second PCR in the same way as the first PCR but the RS primer biotinylated at the 5′ end was used. After ethanol precipitation and removal of excess primers by centrifugal filtration, the PCR product was suspended in 50 μl of B&W buffer and then attached to 20 μg of streptavidin-coated magnetic beads through biotin at the 5′ end of the RS-primer strand. The double- stranded path strands immobilized on the beads were denatured in 100 μl of TE buffer for 2 min at 94℃. Dissociated strands were washed away twice with 100 μl of TE buffer at 94℃.

2.2.7. *GPCR operation.* The length of a path and the order of cities on a path were analyzed using a PCR with appropriate primer pairs. To determine the length of a path, RS and *rev*(TS) primers were used. To determine the position of city i, RS (or RS-v_0) and $rev(v_i)$ (or $rev(v_i$-TS)) primers were used. For the analysis, the beads were suspended in 50 μl of Ex *Taq* buffer containing 4 μl each of the 0.2 mM dNTPs and 10 pmol each of the primers. PCR amplification was performed using the temperature profile of the initial denaturation at 94℃ for 1 min followed by 15-25 cycles of the denaturation at 94℃ for 20 sec, the annealing at 40℃ for 1 min, and the extension at 72℃ for 1 min. When RS-v_0 and $rev(v_i$-TS) primers were used, the annealing step was omitted. The hot start technique was used, so 1.25 units of Ex *Taq* DNA polymerase was added just after the initial denaturation step.

PCR products in the supernatant were resolved on a P/ACE5510 capillary electrophoresis system (Beckman, USA) using LIFluor dsDNA1000 Kit. The laser-induced fluorescence detection equipped with a P/ACE5510 system was used for the high sensitivity detection of PCR products. The intensity of fluorescence signal was calibrated using a double-stranded DNA solution, the concentration of which was determined by a UV absorbance at 260 nm. The peak area of 700 RFU·ETU (RFU: relative fluorescence unit; ETU: elution time unit) on an electropherogram corresponded to 20 fmol/ μl double-stranded DNA solution of 40 bp. The elution time of PCR products on an electropherogram was calibrated using the standard molecular weight marker supplied with LIFluor dsDNA1000 Kit.

3. Results

3.1. Solid phase Hamiltonian path synthesis. Sold phase DNA synthesis is an automated technique for the synthesis of polynucleotides, in which the chain grows while it is attached to a solid support. Our method for solving the HPP is very similar to the solid phase DNA synthesis (Fig. 2). Single-stranded DNA molecules representing paths without duplicate cities are synthesized city-by-city from the start city to the end city on the surface of a solid support. Excess DNA oligomers as well as enzymes and buffers are easily washed away without losses of path strands while the synthesis cycle is repeated. In the solid phase DNA synthesis, all nucleotides in a reaction mixture are coupled to the end of growing chains. In contrast, only those cities connected to the city at the end of growing paths are coupled in the solid phase path synthesis. This is a major difference

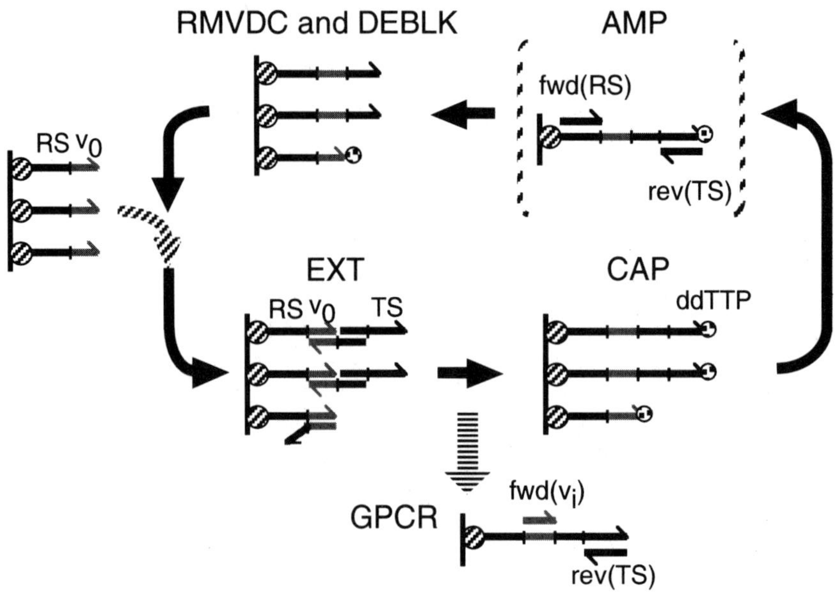

FIGURE 2. Solid phase Hamiltonian path sysnthesis

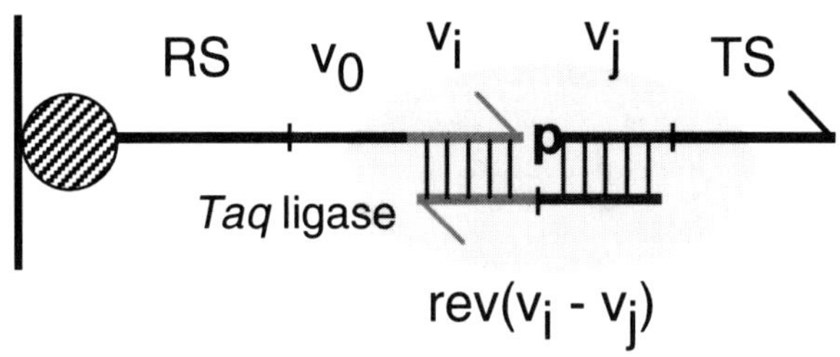

FIGURE 3. EXT operation

between the standard solid phase DNA synthesis and the solid phase Hamiltonian path synthesis.

As shown in Fig. 2, the process of the solid phase Hamiltonian path synthesis is initiated by attaching the $5'$ end of the start city oligomer RS-v_0 to the surface of a solid support. The start city oligomer is then coupled to connected city oligomers using EXT operation. *Taq* DNA ligase couples only connected cities to the $3'$ end of path strands because an edge oligomer $rev(v_i\text{-}v_j)$ stacks only a city oligomer v_j-TS at the $3'$ end of path strands terminating with v_i (Fig. 3). After the coupling reaction, the $3'$ end of path strands is capped with an unreactive group (ddTTP) using CAP operation in order to prohibit further path extension from the end.

When the length of paths increases to more than two cities, looping paths with duplicate cities begin to be synthesized. To remove the looping paths, RMVDC operation is performed after CAP operation. Single-stranded DNA oligomer $rev(v_i\text{-}v_i\text{-TS})$ is used to remove the duplication of city i. This oligomer has a tandem repeat of the complementary base sequence to city i. When a path has the duplication, the path strand and a tandem-city oligomer $rev(v_i\text{-}v_i\text{-TS})$ form a stable hybrid

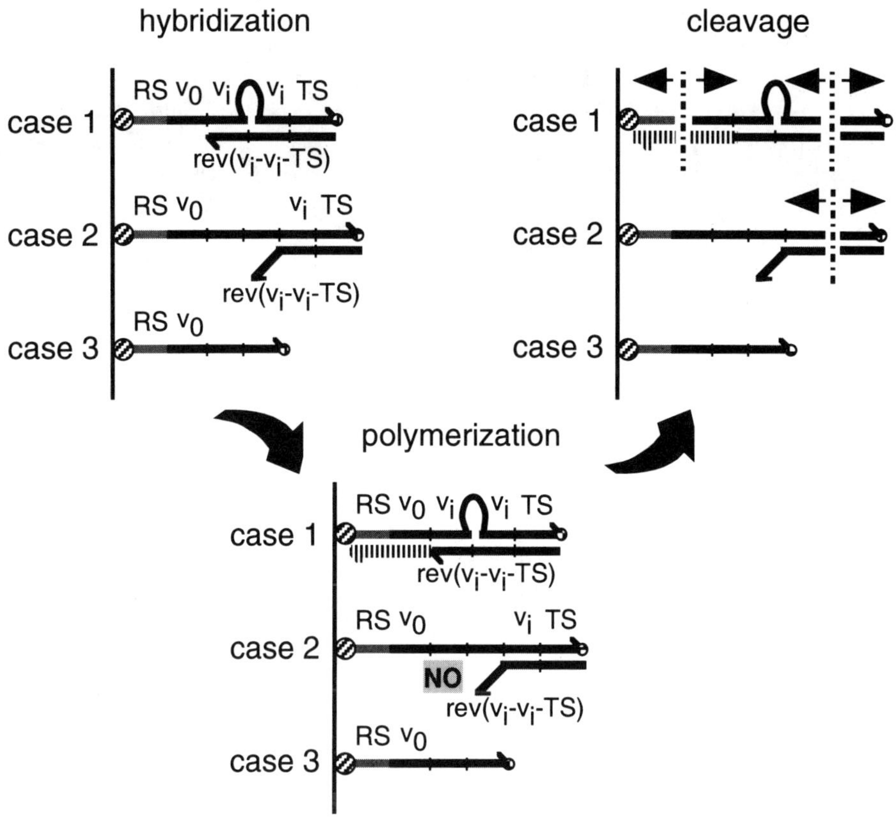

FIGURE 4. RMVDC and DEBLK operation

with a bulge loop (case 1 in Fig. 4). In the hybrid the 3′ end of $rev(v_i\text{-}v_i\text{-}\mathrm{TS})$ is base-paired to the path strand. So extension of $rev(v_i\text{-}v_i\text{-}\mathrm{TS})$ is initiated by *Taq* DNA polymerase, making an RS part double-stranded. The RS part at the root of a path strand has the recognition sequence of an *Eco*RI restriction enzyme. A path strand with the duplication is therefore removed from a solid support by *Eco*RI digestion. In contrast, when a path has no duplication, the path strand and $rev(v_i\text{-}v_i\text{-}\mathrm{TS})$ oligomer form a stable hybrid without bulge loop (case 2 in Fig. 4). In the hybrid the 3′ end of $rev(v_i\text{-}v_i\text{-}\mathrm{TS})$ is not base-paired to the path strand. So no extension of $rev(v_i\text{-}v_i\text{-}\mathrm{TS})$ is initiated by *Taq* DNA polymerase, leaving the RS part single-stranded. A path without the duplication therefore remains on a solid support after executing RMVDC operation.

DEBLK operation follows RMVDC operation in order to advance to the next extension cycle. Annealing of $rev(v_i\text{-}v_i\text{-}\mathrm{TS})$ oligomer to a path strand makes a TS part double-stranded. The TS part at the 3′ end of a path strand has the recognition sequence of a *Bam*HI restriction enzyme. The 3′ end of a path strand is thus activated to the next path extension by cutting the TS part off with *Bam*HI (case 2 in Fig. 4). In contrast, a path strand failed to be extended by EXT operation has no TS part, thus staying inactive to the next path extension (case 3 in Fig. 4). Restriction enzymes *Eco*RI and *Bam*HI used in RMVDC and DEBLK operations, respectively, have the activity in the same buffer (H buffer). RMVDC and DEBLK operations can thus be executed at once as shown in Fig. 4. The simultaneous execution helps the reduction of the computation time.

The pruning of looping paths with RMVDC operation as well as the blocking of failed paths with CAP operation gradually reduce the number of path strands on a solid support. AMP operation should be carried out when the number of path strands remaining on a solid support is suspected to be too small to cover the combinatorial variety of intermediate paths. AMP operation increases the number of path strands on a solid support

EXT, CAP, (AMP), RMVDC and DEBLK operations are repeated until the length of paths extends to the number of cities (vertices) in the graph. Finally, the order of cities on the path is displayed by GPCR operation (graduated PCR) to get a solution to an instance of the HPP. Therefore, the HPP can be solved in the computation time proportional to the number of cities.

3.2. Accuracy and efficiency of EXT operation. EXT operation extends the length of paths by a city. Only connected cities are coupled to the $3'$ end of path strands on a solid support because an edge oligomer $rev(v_i\text{-}v_j)$ stacks only a city oligomer v_j-TS at the $3'$ end of path strands terminating with city i. The accuracy of this coupling reaction performed with a *Taq* DNA ligase determines the accuracy of EXT operation. For correctly solving the HPP using the present method, the high accuracy of EXT operation is essential. Thus we examined how accurate EXT operation was.

To examine the accuracy of EXT operation, we analyzed the order of cities on path strands using GPCR operation just after the first EXT operation. In Adleman's graph (Fig. 1) the start city (city 0) is connected to city 1, 3 and 6. So only such path strands as RS-v_0-v_1-TS, RS-v_0-v_3-TS and RS-v_0-v_6-TS are generated after the first EXT operation if a ligation reaction in EXT operation correctly couples the connected cities. As shown in Fig. 5, only those correct path strands were generated. GPCR operations using RS-v_0 and $rev(v_i\text{-TS})$ ($i = 1$, 3 or 6) primers gave the product of 80 base pairs eluted around an elution time of 14 min. In contrast, other GPCR operations using RS-v_0 and $rev(v_j\text{ -TS})$ ($j = 2$, 4 or 5) primers did not produce any significantly amplified DNA fragments. Therefore, EXT operation was shown to be accurate enough for DNA-based computation of the HPP using the present method when the ligation reaction was carried out at 65℃. At a lower temperature, the condition of the hybridization between edge and city oligomers is less stringent. So the accuracy of EXT operation might be reduced at a lower temperature. However, an EXT operation performed at 60℃ still showed the accurate path extension (data not shown).

In addition to the accuracy of EXT operation, its efficiency is important for the performance of DNA-based computation using the present method. The efficiency of EXT operation was measured by the peak area of a PCR product obtained by GPCR operation performed after the EXT operation. The peak area is likely to reflect the amount of a path strand generated by EXT operation because the number of PCR cycles used in the GPCR operation was as small as 15. Figure 5 shows that the peak area of a PCR product RS-v_0-v_1-TS was larger than that of the other two PCR products, RS-v_0-v_3-TS and RS-v_0-v_6-TS. Therefore, a city oligomer v_1-TS was more efficiently coupled to the $3'$ end of RS-v_0 on beads than the other two city oligomers, v_3-TS and v_6-TS. The GC- content of the base sequence of city 1

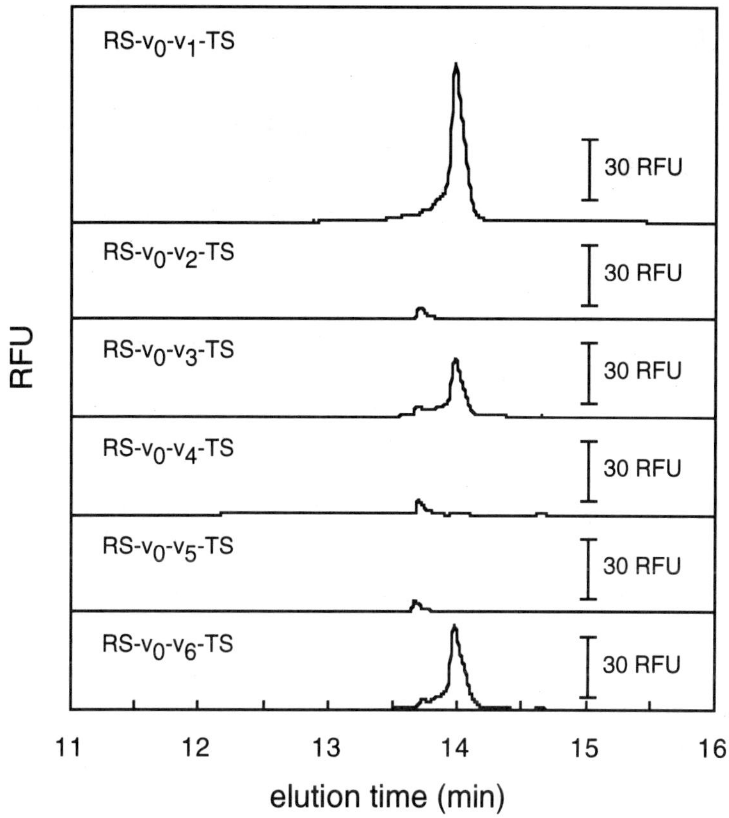

FIGURE 5. Electropherograms of graduated PCR products obtained by GPCR operation performed after the first EXT operation. The path sequence of a PCR product is designated at the upper left of each electropherogram.

is 0.6, which is higher than that of city 3 (0.4) and city 6 (0.4). An edge oligomer $rev(v_0\text{-}v_1)$ thus is able to form a more stable hybrid duplex with a city oligomer v_1-TS than with the other two city oligomers. This may cause the difference in the coupling efficiency.

The fact that the efficiency of EXT operation depends on the GC-content of a city oligomer means that the path search in a graph does not proceed uniformly; some paths are more preferentially generated than others. This biased search reduces the performance of DNA-based computation of the HPP using the present method. The design of city oligomers with more uniform GC-content may solve this problem.

3.3. Efficiency of CAP operation. For correctly solving the HPP using the present method, the efficiency of CAP operation is important in addition to the accuracy of EXT operation. CAP operation covers the $3'$ end of path strand on a solid support with an unreactive group to prohibit further extension of the path strand. The unreactive group is removed by DEBLK operation for a path strand with a TS part, while it is not removed for a path strand without a TS part. A

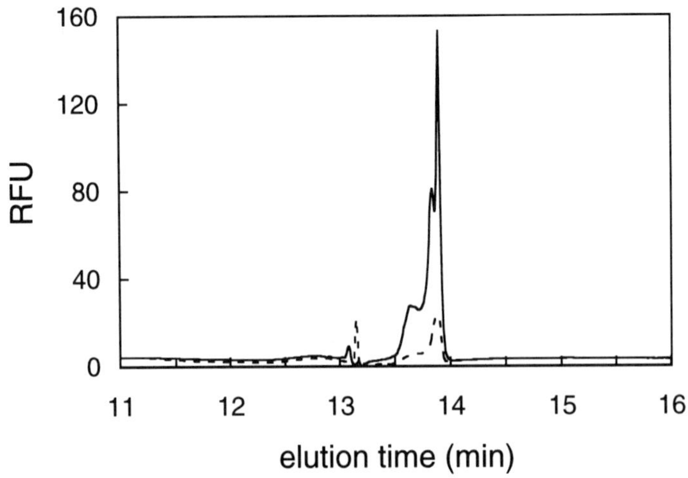

FIGURE 6. Electropherogram of graduated PCR products obtained by GPCR operation that follows EXT operation. Beads were subjected to CAP operation (broken line) or not (solid line) in advance of the EXT operation.

path strand that has failed to be extended by the previous EXT operation has no TS part. CAP operation thus prohibits the failed path strand from extending in the next EXT operation. The extension of the failed path strand causes errors in the solution of the HPP. Therefore, a high efficiency of CAP operation is essential for the solution of the HPP using the present method.

The efficiency of CAP operation was examined by treating beads with CAP operation in advance of EXT operation. The beads had the start city oligomer RS-v_0 on the surface. If the efficiency of CAP operation is high enough to cover completely the $3'$ end of a RS-v_0 strand with an unreactive group, no extension of the strand occurs in the following EXT operation. Figure 6 shows that the amount of an 80-bp PCR product was greatly reduced when the beads were treated with CAP operation in advance of EXT operation. The 80-bp PCR product corresponds to a path strand RS-v_0-v_1-TS, which is generated by extension of a path strand RS-v_0. The number of PCR cycles in the GPCR operation was as small as 15 to ensure the quantitative PCR. Therefore, a large reduction in the amount of the 80-bp PCR product means that the efficiency of CAP operation was sufficiently high to suppress the path strand extension although it was unlikely to be 100%.

3.4. Efficiency of DEBLK operation. An advance to the next extension cycle requires that DEBLK operation activates the $3'$ end of path strands by removing an unreactive group at the end. Without the activation, no EXT operation extends path strands. Generation of long path strands thus requires a high efficiency of DEBLK operation.

The efficiency of DEBLK operation was examined by repeating the solid phase path synthesis cycle. After the first EXT operation, one-fiftieth of beads were analyzed by GPCR operation to display the sequences of path strands on the beads. As shown in Fig. 7a, a GPCR operation using the primer pair of RS-v_0 and rev(v_1-TS) gave a product of 80 base pairs eluted at 14.0 min, which corresponds to a path

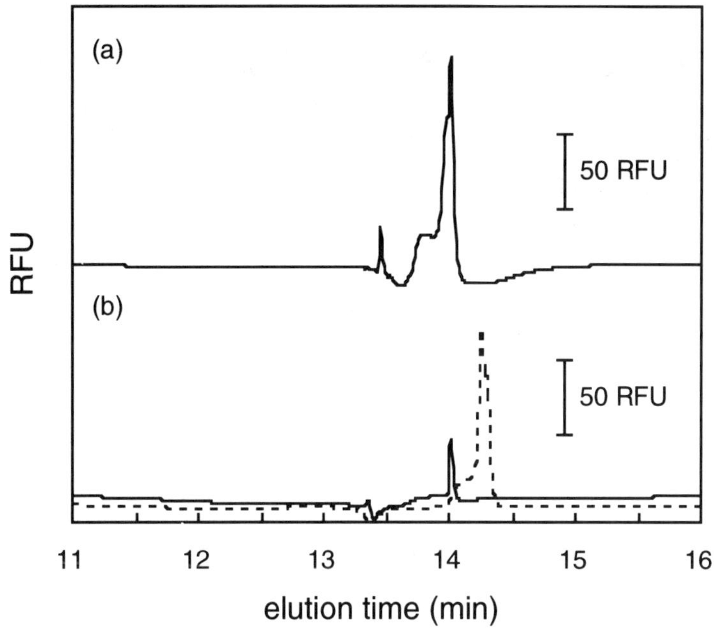

FIGURE 7. Electropherograms of graduated PCR products after the first cycle (a) and the second cycle (b) of the path synthesis. Graduated PCRs were performed using a primer pair of RS-v_0 and $rev(v_1\text{-TS})$ (solid line) or that of RS-v_0 and $rev(v_2\text{-TS})$ (broken line).

strand RS-v_0-v_1-TS. The rest of the beads were then subjected to CAP, DEBLK and EXT operations. After the second EXT operation, a GPCR operation using the primer pair of RS-v_0 and $rev(v_2\text{-TS})$ gave a product of 95 base pairs eluted at 14.3 min, which corresponds to a path strand RS-v_0-v_1-v_2-TS (Fig. 7b). The occurrence of the 95-bp strand proves that DEBLK operation activated the 3′ end of path strands to some extent. On the other hand, a GPCR operation using RS-v_0 and $rev(v_1\text{-TS})$ primers gave a product of 85 base pairs, which corresponds to a path strand RS-v_0-v_1-TS (Fig. 7b). This fact means that some of path strands remained unactivated after DEBLK operation. The peak size of the 95-bp product is about four times as large as that of the 85-bp product. The number of PCR cycles used in the GPCR operation was as small as 15 to ensure the quantitative graduated PCR. Therefore, about 80% of the path strand on the beads were deblocked by DEBLK operation.

3.5. Free energy consideration on RMVDC operation. RMVDC operation utilizes the structure of a hybrid formed between a path strand and a tandem-city oligomer $rev(v_i\text{-}v_i\text{-TS})$. Elimination of path strands with duplicate cities by RMVDC operation requires the stable formation of the hybrid with a bulge loop (Fig. 4). Before starting extensive experiments on RMVDC operation, we theoretically examined the stability of a hybrid formed between a path strand and the tandem-city oligomer.

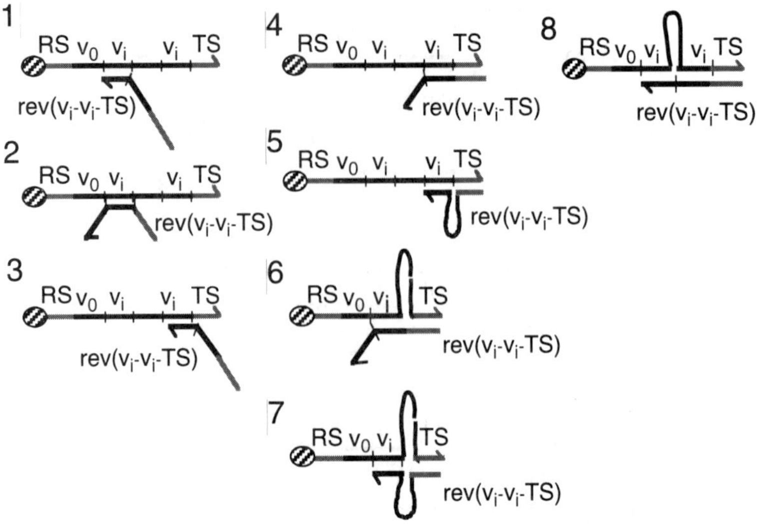

FIGURE 8. Possible hybrid structure formed between a path strand and a tandem-city oligomer $rev(v_i\text{-}v_i\text{-TS})$.

Figure 8 shows the structure of a possible hybrid formed between a path strand and a $rev(v_i\text{-}v_i\text{-TS})$ oligomer. In hybrids 1, 3, 5, 7 and 8, *Taq* DNA polymerase can initiate extension of the $rev(v_i\text{-}v_i\text{-TS})$ oligomer because the $3'$ end part of the oligomer is base-paired to the path strand. Among these hybrids, only hybrid 8 involves the duplicate parts on the path strand. Therefore, hybrid 8 must be the most stable among them to remove correctly a looping path using RMVDC operation in the way shown in Fig. 4.

The stability of a hybrid depends on the free energy of the hybrid. The lower the free energy is, the higher the stability is. The free energy of a hybrid is the sum of the free energy of each part composing the hybrid. In general, single-stranded regions are regarded as the standard state of the free energy calculation; thus a zero free energy is assigned to them. Double-helical regions have a negative free energy; these regions stabilize the hybrid. Loops such as bulge, internal and bifurcation loops have a positive free energy; thus the loops destabilize the hybrid. The free energy of a double-helical region is approximately proportional to the length of the region because the free energy is the sum of the free energy of each base-pair stack making the region up. The free energy of a loop increases monotonously with its size; thus a larger loop destabilizes the hybrid more extensively. From this rough consideration, the following order of the stability of the possible hybrids was derived. Hybrid 8 is most stable; hybrid 4 is less stable than it. Hybrids 5, 6 and 7 follow hybrid 4, leaving hybrids 1, 2 and 3 least stable.

To be more precise, hybrid 8 is most stable only when the length of v_i is large enough to compensate the free energy increase caused by the bulge loop formation. Otherwise, the most stable is not hybrid 8 but hybrid 4. Let the length of v_i be n bases. Then the free energy of the double-helical region of a v_i part is given by $-2 \times (n-1)$ kcal/mol, assuming that the base sequence of v_i has the uniform and average stability of base-stacking [**4**]. On the other hand, the free energy of a bulge loop of m bases long is given by $+1.5 \ln m$ for large m as shown in Fig. 9

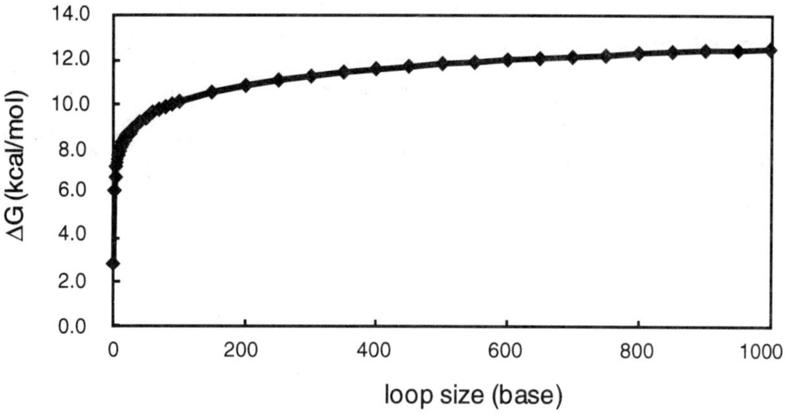

FIGURE 9. Free energy of a bulge loop formation at 25 ℃ in $[\mathrm{Na}^+] = 1\,M$, pH 7.

[**5**]. Therefore, for a bulge loop of up to 1000 bases long, the v_i of 15 bases long is sufficient to make hybrid 8 most stable.

4. Discussion

The present method for solving the HPP is based on a solid phase chemistry and thus is similar to the surface-based approach to a memory for Boolean circuits [**6**]. The method has three advantages over Adleman's method [**1**], which has pioneered the field of DNA-based computation. First, the method is actually much faster than Adleman's method. Only simple and fast DNA operations were employed to solve the problem. Secondly, it can be performed on a fully automated machine. The full automation is essential not only for the speedup of computation but also for the error-free computation. Thirdly, the number of DNA molecules necessary for the computation is extensively reduced. Adleman's method generates all possible paths in a graph at once. The generated paths thus contain many straight and looping paths of any length, which do not always start with the start city nor end with the end city. In contrast, the present method stepwise generates only no-looping paths that start exactly with the start city. The extensive reduction allows the present method to be applied to a larger instance of the HPP.

The present method has been developed for solving the HPP. However, it could be applied to other computational problems. The method is composed of seven basic operations to manipulate DNA molecules on the surface of a solid support. These basic operations have some generality. Thus they can be used for the implementations of other DNA-based algorithms. For example, the satisfaction problem (SAT) could be solved efficiently by using some of these basic operations.

Experiments to examine the feasibility of the present method using Adleman's seven-city instance of the HPP is still going on. Especially, the full results of the experiments on RMVDC operation were not mentioned in the present study because many details on the experimental procedure of RMVDC operation still remain to be improved. These improvements are fixed soon and then we will be ready for starting the experiment to solve a larger instance of the HPP. The solid phase method thus would go a step toward the realization of feasible DNA-based computers.

References

[1] L. M. Adleman, *Molecular Computation of Solutions to Combinatorial Problems*, Science **266** (1994), 1021-1024.

[2] R. J. Lipton, *DNA Solutions of Hard Computational Problems*, Science **268** (1995), 542-544.

[3] R. J. Lipton, E. B. Baum (eds.) *DNA based computers: Proc. of a DIMACS workshop, April 4, 1995*, DIMACS Series in Discrete Mathematics and Theoretical Computer Science vol. 27, Amer. Math. Soc., Providence, RI, 1996.

[4] K. J. Breslauer, R. Frank, H. Blöcker, L. A. Marky, *Predicting DNA duplex stability from the base sequence*, Proc. Natl. Acad. Sci. USA **83** (1986), 3746-3750.

[5] C. DeLisi, D. M. Crothers, *Prediction of RNA Secondary Structure*, Proc. Natl. Acad. Sci. USA **68** (1971), 2682-2685.

[6] W. Cai, A. E. Condon, R. M. Corn, E. Glaser, Z. Fei, T. Frutos, Z. Guo, M. G. Lagally, Q. Liu, L. M. Smith, A. Thiel, *The Power of Surface-based DNA Computation*, Proc. RECOMB'97 (1997), 67-74.

Institute of Physics, Graduate School of Arts and Sciences, University of Tokyo, 3-8-1 Komaba, Meguro-ku, Tokyo 153-8902, Japan

E-mail address: morimo@genta.c.u-tokyo.ac.jp

Department of Information Science, Graduate School of Science, University of Tokyo, 7-3-1 Hongo, Bunkyo-ku, Tokyo 113-0033, Japan

E-mail address: arita@is.s.u-tokyo.ac.jp

Institute of Physics, Graduate School of Arts and Sciences, University of Tokyo, 3-8-1 Komaba, Meguro-ku, Tokyo 153-8902, Japan

E-mail address: suyama@dna.c.u-tokyo.ac.jp

DIMACS Series in Discrete Mathematics
and Theoretical Computer Science
Volume **48**, 1999

On Molecular Approximation Algorithms for NP Optimization Problem*

Bin Fu   Richard Beigel

Abstract

We develop a general technique for constructing molecular-based approximation algorithms for NP optimization problems. Our algorithms exhibit a useful volume–accuracy tradeoff. In particular we solve the Covering problem of Hochbaum and Maass using polynomial time and $O\left(\ell^2(\log \ell)n^2\binom{n(n-1)}{\ell^2/2}\right)$ volume with error ratio $(1+\frac{1}{\ell})^2$. We also present the first candidate for a problem that can be solved more efficiently with the Amplify operation than without.

1. Introduction

Molecular computers were introduced by Adleman [1, 6], but so far the field lacks a "killer application." It is well known that a DNA computer can solve SAT in linear time [6], but using an exponential number of DNA strands. The number of strands used by an algorithm is called the "volume." Although recent papers [3, 2, 7] solve NP problems using smaller exponential volume, we believe that it is essential to find applications of DNA computers that use subexponential volume.

In this paper we apply DNA computers to the approximate solution of NP-optimization problem. We show how DNA algorithms using subexponential volume and polynomial time can achieve much better accuracy than classical polynomial time approximation algorithms. We illustrate our method on Hochbaum and Maass's classical approximation algorithm [5] for the Covering problem. Our molecular algorithm runs in polynomial time and uses $O\left(\ell^2(\log \ell)n^2\binom{n(n-1)}{\ell^2/2}\right)$ volume to achieve error ratio $(1+\frac{1}{\ell})^2$. This algorithm does not use controversial operations like Intersection or even Amplify.

We also improve their algorithm in the classical computation model and implement it in a more complicated molecular computation model using the Amplify operation. Its volume is decreased to $O\left(c^{\ell^2/2}\ell^2(\log \ell)n^2\binom{n(n-1)}{\ell^2/2}\right)$,

*Research supported in part by the National Science Foundation under grants CCR-8958528 and CCR-9415410. AMS classification number is 68Q25.

where $c = 0.823$. This is the first example we know in which the Amplify operation seems to enable DNA computers to solve a problem more efficiently.

We also study molecular approximation algorithms for dense instances of many NP-hard problems. Our results suggest a volume–accuracy tradeoff in DNA-based approximation algorithms for NP optimization problems. We think that approximation algorithms may be the "killer application" that will motivate the construction of DNA computers.

2. Hochbaum and Maass's Approximation Algorithm and Molecular Implementation

In this section we give a description of Hochbaum and Maass's approximation algorithm for the Opt-covering problem [5] and its implementation via molecular computation.

Definition 1.

- Covering *problem:* Given a set N of points in the plane, a positive integer m, and a real number D, is it possible to cover all points of N with m disks of diameter D.
- Opt-covering *problem:* Given a set N of points on the plane and a real number D, find the minimum number of disks of diameter D that cover all points in N.
- An approximation A for Opt-covering has *error ratio* $1+\epsilon$ if

$$1 \leq \left| \frac{A(N,D)}{\mathrm{OPT}(N,D)} \right| \leq 1+\epsilon$$

for all N, where $\mathrm{OPT}(N,D)$ is the least number of disks covering all points in N.

Throughout this paper, the size of a problem instance is the number of points $|N|$, which we denote by n.

2.1. Hochbaum and Maass's Approximation Algorithm

Definition 2. Let H be $aD \times bD$ rectangle with $b \leq a$, l be an integer and N be a set of points on plane.

- Partition H into $D \times bD$ strips: $I_1, I_2, \ldots, I_a$. For each $1 \leq i \leq \ell$, define

$$t_i = \begin{cases} 1 + \left\lfloor \frac{a-i}{\ell} \right\rfloor & \text{if } \left\lfloor \frac{a-i}{\ell} \right\rfloor = \frac{a-i}{\ell} \\ 2 + \left\lfloor \frac{a-i}{\ell} \right\rfloor & \text{otherwise} \end{cases}$$

$$H_{i,k} = \begin{cases} \bigcup_{j=1}^{i}\{I_j\} & \text{if } k=1 \\ \bigcup_{j=(k-2)\ell+i+1}^{(k-1)\ell+i}\{I_j\}, & \text{if } 2 \le k < t_i \\ \bigcup_{j=(t_i-2)\ell+i+1}^{a}\{I_j\} & \text{if } k = t_i \end{cases}$$

Each $H_{i,j}$ is called an *ℓ-strip* of H.

- For $i = 1, \ldots, \ell$, $P_i(H) = \{H_{i,1}, \ldots, H_{i,t_i}\}$ is an *ℓ-partition* of H. Clearly each $P_i(H)$ is a partition of H, induced by grouping ℓ small $D \times bD$-strips into a big $lD \times bD$ strip (boundary strips may be smaller).
- An *isolated* point is a point in N that has distance more than D with any other point in N.
- $\mathrm{OPT}(H)$ is the least number of disks to cover all points of N in H.
- Fix Q as a optimal cover all points of N in H, $\mathrm{OPT}_i(H)$ is the number of disks in Q touching two adjacent l-strips in $P_i(H)$.

Assume A is a approximation of Opt-covering with error ratio r_A. For each $P_i(H)$, let

$$h_i = \sum_{E \in P_i(H)} A(E) \le r_A \cdot \sum \mathrm{OPT}(E) = r_A \cdot (\mathrm{OPT}(H) + \mathrm{OPT}_i(H)).$$

Since no disk in Q can contribute to both $\mathrm{OPT}_i(H)$ and $\mathrm{OPT}_j(H)$ with $i \ne j$,

$$\sum_{1 \le i \le l} \mathrm{OPT}_i(H) \le \mathrm{OPT}(H).$$

So, $h_1 + \cdots + h_l \le r_A \cdot (l \cdot \mathrm{OPT}(H) + \sum_{i=1}^{l} \mathrm{OPT}_i(H)) \le r_A \cdot (l+1) \cdot \mathrm{OPT}(H)$.

$$\text{Hence, } \min_{1 \le i \le l} h_i \le r_A \cdot (1 + \frac{1}{l}) \cdot \mathrm{OPT}(H).$$

We have a new approximation for $\mathrm{OPT}(H)$, with error ratio $r_A \cdot (1 + \frac{1}{l})$, from an approximation to $\mathrm{OPT}(E)$ for $E \in P_i(H)$ with error ratio r_A. Suppose all points of N are in area I, a $uD \times vD$ retangle. The strategy is used in vertical and horizatal direction. I has partitions $P_1(I), \ldots, P_l(I)$ and each element $E \in P_i(I)$ is a $l'D \times vD$ retangle ($l' \le l$). For each $E \in P_i(I)$, E has partitions $P_1(E), \ldots, P_l(E)$. Each element $F \in P_i(E)$ is a $l_1 D \times l_2 D$ retangle ($l_1, l_2 \le l$). Use brute-force method to compute $\mathrm{OPT}(F)$ (each isolated point is covered with a disk and all of other disks have tow points on it boundary). Since $(l/\sqrt{2})^2 = l^2/2$ disks can cover a whole $lD \times lD$

retangle, the number of cases for brute-force method is bounded by $\binom{n(n-1)}{l^2/2}$. The algorithm is formally described as follows whose error ratio is $(1+1/l)^2$.

Algorithm

Input N;

begin

find a smallest $uD \times vD$ retangle I to include all points in N;

for each $1 \le i \le l, E \in P_i(I), 1 \le j \le l, F \in P_j(E)$ and $F \cap N \ne \emptyset$

do $a_F = \text{OPT}(F)$ (via brute force method);

for each $1 \le i \le l, E \in P_i(I)$, and $1 \le j \le l$ do

$b_{E,j} = \sum_{F \in P_j(E) \text{ and } F \cap N \neq \emptyset} a_F$;

for each $1 \le i \le l, E \in P_i(I)$ and $E \cap N \ne \emptyset$ do $b_E = \min_{1 \le j \le l} b_{E,j}$;

for each $1 \le i \le l$ do $c_{I,i} = \sum_{E \in P_i(I) \text{ and } E \cap N \neq \emptyset} b_E$;

$c_I = \min_{1 \le i \le l} c_{I,i}$;

output(c_I);

end;

End of Algorithm

2.2. Molecular Algorithm

Definition 3.

- An NPinit($s(n)$) machine is a nondeterministic Turing machine that nondeterministically chooses a number between 1 and $s(n)$ and then proceeds deterministically.
- NPinit($s(n)$) is the class of languages accepted by NPinit($s(n)$) machines.

Theorem 4. Covering $\in$ NPinit$\left(\binom{n(n-1)}{m}\right)$, *where* $n = |N|$.

Proof: Let $\langle N, m\rangle$ be the input for Covering. Let M be the set of all isolated points in N. Let $N' = N - M$. For each isolated point, use one disk to cover it. Let $m' = m - |M|$ and $n' = |N'|$. In order to cover the points in N' we only consider to put each disk with two points at it boundary. Consider all of the $\binom{n'(n'-1)}{m'}$ arrangements, $\langle N, m\rangle \in$ Covering iff one of them can cover all points in N'.

It is easy to see there is a map g between the $\binom{n'(n'-1)}{m'}$ arrangements and $\{1, \ldots, \binom{n'(n'-1)}{m'}\}$ such that g is 1-1 and polynomial time computable. Nondeterministically choose an integer $i \le \binom{n(n-1)}{m}$, if $i > \binom{n'(n'-1)}{m'}$, then return *reject*; otherwise, return *accept* iff the arrangement $g(i)$ can cover all points in N'. ∎

Proposition 5.

i. Opt-covering *can be computed in polynomial time via* $|N|$ *parallel queries to* Covering *and for each query* $\langle N, m\rangle$, $m \le n$.

ii. Opt-covering *can be computed in polynomial time via* $\log_2 |N|$ *sequential queries to* Covering *and for each query* $\langle N, m\rangle$, $m \le n$.

Proof:

i. Opt-covering(N) is equal to the smallest $i \le |N|$ such that $(N, i) \in$ Covering.

ii. Use binary search to find the smallest i above via query Covering. ∎

Theorem 6.

i. There is an approximation algorithm with error ratio $(1+\frac{1}{\ell})^2$ *for* Opt-covering *that runs in* $n^{O(1)}$ *time and makes* $\frac{\ell^2}{2}(\ell \cdot n)^2$ *parallel queries to* Covering *where for each query* $\langle M, m\rangle$, $|M| \le n$ *and* $m \le \frac{\ell^2}{2}$.

ii. There is an approximation algorithm with error ratio $(1+\frac{1}{\ell})^2$ *for* Opt-covering *that runs in* $n^{O(1)}$ *time and makes* $(\log \frac{\ell^2}{2})(\ell \cdot n)^2$ *adaptive queries to* Covering *where for each query* $\langle M, m\rangle$, $|M| \le n$ *and* $m \le \frac{\ell^2}{2}$

Proof: By Hochbaum and Maass's algorithm and Proposition 5. ∎

We are going to transform the approximation algorithm into a molecular algorithm. Some molecular operations are defined as follows:

Separate(T_1, c, i, T_2, T_3)
  $T_2 :=$ the set of all strings in T_1 whose ith character is c;
  $T_3 :=$ the set of all strings in T_1 whose ith character is not c;
  $T_1 := \emptyset$.

Append(T, c)
  $T := \{xc : x \in T\}$.

Merge(T_1, T_2, T_3)
  $T_3 := T_1 \cup T_2$;
  $T_1 := \emptyset$;
  $T_2 := \emptyset$.

Amplify(T_1, T_2, T_3)
  $T_2 := T_1$;
  $T_3 := T_1$;
  $T_1 := \emptyset$.

Test(T)
  If T is empty then "true" else "false".

Init(T, m)
$T := \{i : 1 \leq i \in m\}$.

Definition 7. In a MOL$(s(n))$ computation on input x, one test tube T_0 is initialized to hold encodings of the numbers $1, \ldots, s(|x|)$ (it can be done via Init$(T_0, s(|x|))$). A sequence of molecular operations $o_1, \ldots, o_{t(|x|)} = f(x)$ is then performed, where f is a conventional polynomial-time computable function (that is, the program is uniform in a weak but appropriate sense). The computation accepts if T_0 is nonempty after the last operation is performed. MOL$(s(n))$ is the class of languages accepted by such a computation where the running time $t(n)$ is polynomial bounded and the operations permitted are Append, Separate and Merge.

Theorem 8 (Beigel and Fu [3]). NPinit$(s(n)) \subseteq$ MOL$(s(n))$.

Theorem 9. Covering $\in$ MOL$(\binom{n(n-1)}{m})$.

Proof: Follows from Theorem 8 and Theorem 4. ∎

Definition 10. A MOL$'(s(n))$ algorithm is an algorithm with Init, Test, Separate, Append and Merge that runs polynomial steps and uses no more than $s(n)$ DNA strands.

Theorem 11. *For each* $\ell > 0$, *there is a* MOL$'(O(\ell^2(\log \ell)n^2\binom{n(n-1)}{\ell^2/2}))$ *algorithm of error ratio* $(1 + \frac{1}{\ell})^2$ *for* Opt-covering.

Proof: By Theorem 9 and Theorem 6 and Hochbaum and Maass's algorithm. ∎

3. Improving Hochbaum and Maass's Algorithm

In order to decrease the volume needed for molecular computation, we improve Hochbaum and Maass's algorithm in the classical computation model. The most expensive part of their algorithm is at the brute-force method in finding the optimal solution for Opt-covering. Since Opt-covering can be computed via querying Covering, we focus on improving the algorithm for Covering.

Definition 12. NPpath$(s(n))$ is the class of languages accepted by polynomial time nondeterministic Turing machine with no more than $s(n)$ paths.

Theorem 13. Covering $\in$ NPpath $\left(c^m\binom{n(n-1)}{m}\right)$, *where* $c = 0.823$ *and* $n = |N|$ *for input* $\langle N, m\rangle$.

Proof: We use a new nondeterministic algorithm to determine Covering. A *position* is either an isolated point of N or else a pair of non isolated points of N with one more bit that determines the direction to put a disk with the two points at its boundary. Clearly the number of positions of N is bounded by $n(n-1)$. If N can be covered by the m disks, there is a position where a disk can cover at least $\frac{|N|}{m}$ points of N. The size of the problem can be quickly decreased after a disk is put on the position.

```
function Covering(N, m)
begin
    unmark all of the positions;
    for i = 1 to n(n − 1) do
    begin
        if N = ∅ then return(accept);
        if m = 0 then return(reject);
        find an unmarked position p where a disk can cover at least |N|/m
            points;
        mark the position p;
        nondeterministically choose a_i ∈ {0, 1};
        if a_i = 1 then
        begin
            put a disk at position p;
            let M be the set of all points covered by the disk above;
            N = N − M;
            m = m − 1;
            mark all positions that are adjacent to any point in M;
        end;
    end;
end;
```

The algorithm is a nondeterministic. Every computation path corresponds to a string $a_1 \dots a_{n(n-1)}$ (append some 0s if it stops before $i = n(n-1)$) with $\sum_{i=1}^{n(n-1)} a_i \le m$. Let $\alpha = 0.198, \beta = 1.0 - \alpha$.

Lemma 14. *If* $2m \le k$, *there are at most* $0.5c^m\binom{k}{m}$ *strings in the set*

$$\{a_1 \cdots a_k : (a_1 + \cdots + a_{k/2} \le \alpha m) \wedge (a_1 + \cdots + a_k \le m)\}.$$

Proof: The number of strings in

$$\{a_1 \cdots a_k : (a_1 + \cdots + a_{k/2} \le \alpha m) \wedge (a_1 + \cdots + a_k \le m)\}$$

is at most

$$\begin{aligned} m\sum_{i=0}^{\alpha m}\binom{k/2}{i}\binom{k/2}{m-i} &\le m(\alpha m+1)\binom{k/2}{\alpha m}\binom{k/2}{\beta m} \\ &= \frac{1}{(2-o(1))^m(\alpha^\alpha\beta^\beta)^m}\binom{k}{m} \\ &\le 0.5c^m\binom{k}{m}. \end{aligned}$$

∎

Lemma 15. *There are at most $0.5c^m\binom{n(n-1)}{m}$ paths $a_1\cdots a_{n(n-1)}$ with the property $a_1+\cdots+a_{n(n-1)/2}\ge \alpha m$.*

Proof: For every path $a_1\cdots a_{n(n-1)}$ with the property $a_1+\cdots+a_{n(n-1)/2}\ge \alpha m$, it has used at least αm disks before $i=n(n-1)/2$ in the algorithm. After $i=n(n-1)/2$, there are at most

$$(1-\frac{1}{m})(1-\frac{1}{m-1})\cdots(1-\frac{1}{m-\alpha m+1})n=\beta n$$

points left in N and the number of unmarked positions is $\le(\beta n)^2$. So, the total number of positions chosen in the path is at most $\frac{n(n-1)}{2}+(\beta n)^2\le(\frac{1}{2}+\beta^2+\frac{2}{n})n(n-1)$. The number of paths of this kind is at most

$$m\binom{(\frac{1}{2}+\beta^2+\frac{2}{n})n(n-1)}{m}\le m(\frac{1}{2}+\beta^2+\frac{2}{n})^m\binom{n(n-1)}{m}\le 0.5c^m\binom{n(n-1)}{m}.$$

∎

For $m>|N|$, it is trivial. We can assume $m\le|N|$. Combining the two lemmas above we get the theorem.

∎

Definition 16.

- In the MOL-A$(s(n))$ computation, T_0 is initialized to contain a single copy of the empty string, the Separate, Merge, Append, Amplify operations are permitted, and the total number of strings in all tubes must be at most $s(n)$ at all times and the number of operations is bounded by polynomial of n. The computation is accepted if T_0 is not empty in the final step.
- MOL-A$(s(n))$ is the class of all languages accepted by a MOL-A$(s(n))$ computation.

- A MOL-A$'(s(n))$ algorithm is an algorithm with Separate, Merge, Append, Amplify, Init(1) and Test operations.

Theorem 17 (Fu and Beigel [4]). NPpath$(s(n))$ = MOL-A$(s(n))$.

Theorem 18. Covering $\in$ MOL-A$'\left(c^m\binom{n(n-1)}{m}\right)$, *where* $c = 0.823$.

Proof: By Theorem 17 and Theorem 13. ∎

Theorem 19. *For each* $\ell > 0$, *there is a* MOL-A$'\left(O(\ell^2(\log \ell)n^2 c^{\ell^2/2}\binom{n(n-1)}{\ell^2/2}))\right)$ *algorithm of error ratio* $(1+\frac{1}{\ell})^2$ *for* Opt-covering, *where* $c = 0.823$.

Proof: By Theorem 18 and Theorem 6 and Hochbaum and Maass's algorithm. ∎

References

[1] L. Adleman. Molecular computation of solutions to combinatorial problems. *Science*, 266:1021–1024, Nov. 1994.

[2] E. Bach, A. Condon, E. Glaser, and C. Tanguay. DNA models and algorithms for NP-complete problems. In *Proceedings of the 11th Annual Conference on Structure in Complexity Theory*, pages 290–299, 1996.

[3] R. Beigel and B. Fu. Molecular computing, bounded nondeterminism, and efficient recursion. In *In Prodeedings of 24th International Colloquium on Automata, and Programming (July 7 - 11, 1997, Bologna, Italy), Lecture Notes of Computer Science 1256*, pages 816–826, 1997.

[4] B. Fu and R. Beigel. A comparison of resource-bounded molecular computation models. In *Proceedings of the 5th Israeli Symposium on Theory of Computing and Systems*, pages 24–35, 1997.

[5] D. S. Hochbaum and W. Maass. Approximation schemes for covering and packing problems in image processing and VLSI. *J. ACM*, 32(1):130–136, 1985.

[6] R. Lipton. Using DNA to solve NP-complete problems. *Science*, 268:542–545, Apr. 1995.

[7] M. Ogihara. Breadth first search 3SAT algorithms for DNA computers. Technical Report 629, University of Rochester, July 1996.

Bin Fu
Department of Computer Science,
Yale University, New Haven, CT 06520; and

Department of Computer Science,
University of Maryland, College Park, Maryland 20740.

Richard Beigel
Department of Computer Science,
Yale University, New Haven, CT, 06520; and

Department of Electricial Engineering and Computer Science,
Lehigh University, Bethlehem, PA 18015; and

The Human–Computer Interaction Lab,
University of Maryland at College Park,
College Park, Maryland, 20742.

DIMACS Series in Discrete Mathematics
and Theoretical Computer Science
Volume **48**, 1999

Local Parallel Biomolecular Computation

John H. Reif

Abstract. *Biomolecular Computation(BMC)* is computation at the molecular scale, using biotechnology engineering techniques. Most proposed methods for BMC used *distributed (molecular) parallelism (DP)*; where operations are executed in parallel on large numbers of distinct molecules. BMC done exclusively by DP requires that the computation execute sequentially within any given molecule (though done in parallel for multiple molecules). In contrast, *local parallelism (LP)* allows operations to be executed in parallel on each given molecule. Winfree, et al [**W96, WYS96**]) proposed an innovative method for LP-BMC, that of computation by *unmediated self-assembly* of $2D$ arrays of DNA molecules, applying known domino tiling techniques (see Buchi [**B62**], Berger [**B66**], and [**R71, LP81**]) in combination with recombinant DNA nano-fabrication techniques of Seeman et al [**SZC94**]. We develop improved techniques to more fully exploit the potential power of LP-BMC. To increase the likelihood of success of assembly, we propose instead a refined *step-wise assembly* method. We also introduce the *assembly frame,* a rigid nano-structure which binds the input DNA strands in place and constrains the shape of the assembly. Our main results are LP-BMC algorithms for some fundamental problems. For these problems we decrease the assembly size to linear in the input size and and significantly decrease the number of time steps. We give LP-BMC algorithms with linear assembly size and logarithmic time, for the parallel prefix computation problems, which include integer addition, subtraction, multiplication by a constant number, finite state automata simulation, and fingerprinting (hashing) a string. We also give LP-BMC methods for perfect shuffle and pair-wise exchange using a linear size assembly and constant time. This provides logarithmic time LP-BMC algorithms for the large class of normal parallel algorithms [**S71, U84, L92**] on shuffle-exchange networks, e.g. DFT, bitonic merge, fixed permutation of data, as well as evaluation of a bounded degree Boolean circuit in time bounded by a logarithmic times the circuit depth. Our LP-BMC methods require much smaller volumes than previous DP-BMC algorithms [**R95, OR97**] for these problems. All our LP-BMC assembly techniques can be combined with DP-BMC parallelism to simultaneously solve multiple problems with distinct inputs (e.g. do parallel arithmetic on multiple inputs, or determine satisfying inputs of a circuit), so they are an enhancement of the power of DP-BMC.

Supported by NSF/DARPA CCR-9725021, CCR-96-33567, NSF IRI-9619647, ARO contract DAAH-04-96-1-0448, ARPA/SISTO contracts N00014-91-J-1985, and N00014-92-C-0182 under subcontract KI-92-01-0182. Postscript versions of this paper and its figures are at http://www.cs.duke.edu/~reif/paper/Assembly.ps and http://www.cs.duke.edu/~reif/paper/Assembly.fig.ps.

1991 *Mathematics Subject Classification.* Primary 92–08, 68Q10, 68Q22, 94C05, 94C10.

1. Introduction

Biomolecular Computation. We will use the term *nano-computation* to denote computation executed at the molecule scale. See Reif [**R97**] for an extensive survey of BMC. To be effective, nano-computing will depend on a technology that allows for manipulation of objects at a molecular scale. One promising such technology is that of microbiology, specifically by use of well established recombinant DNA [**WHR87, WHR87, S94, SM97**] and RNA operations (appropriately modified to insure the required reliability). A key operation is *hybridization* of single-stranded DNA which are Watson-Crick complementary and oriented in opposite directions, into double-stranded DNA in a helical structure. We will use the term *Biomolecular computation (BMC)* to denote nano-computation using biotechnology. The first implementation of BMC by Adleman [**A94**] solved a small Hamiltonian path problem by use of recombinant DNA technologies.

Certain fundamental principles for doing BMC may be (to at least a degree) independent of the underlying technology. This motivates us to develop parallel algorithms for BMC in the context of well defined models, that represent the key aspects and technology constraints of BMC. It is still not clear what the fundamental techniques for doing BMC may be. In particular, the literature in BMC utilizes only a few distinct classes of methods.

Distributed Parallel Biomolecular Computation. In the following, let a *Distributed Parallel (DP) operation* be a single operation done in parallel on multiple distinct molecules (for example the use of restriction enzymes to do editing of these molecules, each at the corresponding same labeled site). Up to now, generally operations in BMC are executed within individual molecules in a sequential fashion, and the potential advantage is due to the parallel execution on multiple molecules, which we call *Distributed Parallel BMC (DP-BMC)*. DP-BMC provides massive parallelism due to the vast numbers of molecules in a test tube.

BMCs [**SM97**] have utilized certain *base methods* which include:

- generation of a large number of sequences by random assembly; see Adleman [**A95**] and Lipton [**L95**] (interestingly, this is the only method now in use in BMC that could be viewed as a combined DP-BMC and LP-BMC method) and
- DP-BMC methods for string editing operations, executed sequentially within each molecule.

Many known proposed methods for BMC then use a combination of these base methods and DP-BMC. They can be categorized as follows:

- simulation of sequential computations (e.g. execution of bit serial arithmetic or more generally universal Turing Machines)
- solution of NP search problems, by the use of DP-BMC processing.

Note that the papers [**Be94, A95, L95, BDL95, SM97, BCGT96, ARRW96**] propose to solve NP search problems in essentially the same manner, as follows:

1. a *generation phase* for assembling all possible solutions,
2. *verification* of the validity of each possible solution; this is done in parallel by DP-BMC for each possible solution, but is done sequentially for each solution.
3. *detection* of a valid solution.

In many cases, the potential speed-up benefits from DP-BMC are somewhat limited because each circuit or formula defining the search problem is locally evaluated sequentially, and so such DP-BMC methods for BMC require many steps to do this evaluation locally. This presents a possible disadvantage, since the step duration in BMC is much longer than the step duration for conventional VLSI, but this disadvantage may be overcome by the use of massive parallelism due to DP-BMC.

For example, consider the recent stickers project of Adleman et al [**ARRW96**] (see also an earlier proposal of [**BDL95**]) for breaking the DES standard cryptosystem via BMC. Their proposed method for breaking DES makes use of DP-BMC to evaluate in parallel the DES circuit for all possible key values. Their method uses at least as many operations as required by the sequential evaluation of the DES circuit, which consists of a few thousand Boolean gates. In contrast, a conventional VLSI circuit for DES takes time which is the depth of the DES circuit, which can be done in conventional VLSI at a very high step rate, in just a few dozen parallel steps. [**ARRW96**] still argue, quite convincingly, that the massive parallelism due to DP-BMC may allow them a sufficient advantage over conventional computers to allow them to break DES. Nevertheless, this provides a motive for us to attempt to also exploit DP-BMC where possible to improve the advantages of BMC, even in the case where the more obvious major advantage is due to DP.

Circuits can be evaluated in parallel by use of DP-BMC algorithms* [**R95, OR97**] requiring a large volume; in contrast our LP-BMC methods require a considerably smaller volume.

Local Parallel Biomolecular Computation. Let a *Local Parallel (LP) operation* be a parallel operation done on a given molecule (for example the editing of a single molecule at multiple sites). For large molecules, such as long DNA or RNA strands, LP-BMC is the natural analog to the parallelism as provided by conventional VLSI for parallel circuit evaluation.

Another motivation for developing LP-BMC methods is to allow us to efficiently work with large biological DNA strands. In contrast, most methods proposed for BMC deal with relatively short DNA and RNA strands, due to the sequential processing within individual strands. However, many biological applications, for example DNA sequencing, involve large DNA strands. LP-BMC methods may allow for the efficient processing for such large DNA and RNA strands.

LP-BMC via Unmediated Self-assembly. Domino tiling techniques were developed by Buchi [**B62**], Berger [**B66**], and others [**R71, LP81**] in the early 1960s, and allow simulation of a 1 tape Turing Machine with $S \geq n$ space (or a $1D$ parallel array of S processors) running in time $T \geq n$ by construction of an assembly of an $S/2 \times T$ array of $A = ST/2 \geq n^2/2$ tiles. Winfree [**W96**] (also Winfree, et al [**WYS96**]), proposed to execute these domino tiling techniques in LP-BMC

*Reif [**R95**] gave first known polylog DP-BMC algorithm for the parallel evaluation Boolean circuits. If a circuit is constructable in s space with unbounded fan-out, and depth d, then Reif's algorithm had time cost $O(d + s \log s)$ for circuit evaluation in a PAM model for DP-BMC where long complementary matching is allowed (with a factor s further cost for a more restricted RDNA model where only constant length matches are allowed). Recently Ogihara and Ray [**OR97**] gave the similar bounds for circuit evaluation assuming implicitly a similar model for DP-BMC where long complementary matching is allowed (but do not consider more refined models such as the RDNA where only constant length matches are allowed). Both these results required large amount of communication between distinct molecules and thus require large volume growing at least as 2^{2s}.

by assembling $2D$ arrays of DNA molecules, using the recombinant DNA nanofabrication techniques of Seeman et al [**SZC94**]. Winfree proposed the computation be done by *unmediated self-assembly,* that is the assembly of tiles proceeds without mediation (i.e., external control). While the idea of computation via unmediated self-assembly is very appealing, it is not clear what the likelihood is for a successful assembly. This is a critical issue, for if the likelihood of successful of assembly is very low, possible remedies by use of multiple parallel assembly attempts or repetitions would require either unfeasible large volumes or take a correspondingly very large time. We consider a simple model for random assembly, which assumes assembly growth by tile placement at random untiled locations, using only tiles which have maximal matching with previously placed tiles. We show that for certain settings of the parameters in our random assembly model, computations via unmediated self-assembly can fail with extremely high likelihood $1 - 2^{-cA} \geq 1 - 2^{-cn^2/2}$. Since that random assembly model does not include the possibility of disassembly, our analysis is not conclusive, but does illustrate the possible pitfalls of unmediated self-assembly. We also consider a more sophisticated model that includes disassembly as well as assembly. Although recent computer simulations [**W97**] indicate that careful choice of settings of the parameters may allow unmediated self-assembly to succeed in such a model, unmediated self-assembly remains unproved and experimentally untested.

The Goals of this Paper. The central task which we investigate in this paper, is to *develop and refine methods for LP-BMC* and to combine these methods with the already known DP-BMC techniques (e.g., as provided naturally by recombinant DNA techniques). Our approach is to utilize certain parallel algorithm design techniques, including parallel prefix computation and parallel shuffle-exchange operations. With the use of some form of global parallel communication between distributed molecules or an associative match operation on sets of molecules, these parallel algorithm design techniques can be applied to BMC (Reif [**R95**]). However, we require in this paper that DP-BMC not utilize global parallel communication between distributed molecules, and in this case application of these parallel algorithm design techniques to LP-BMC is not obvious or straightforward.

It should be emphasized that although our molecular assembly techniques have some experimental basis (due the extensive DNA-nano-fabrication work of Seeman [**SZC94, WYS96**], *who has constructed many of the basic components (tiles) of our proposed assemblies, this paper is theoretical in nature and should be evaluated from that perspective.*

Our Improvements to LP-BMC:

(1) Step-Wise Assembly. We propose instead a refined assembly method, which we call *step-wise assembly.* This provides control of the assembly of an $S/2 \times T$ array in distinct T steps, to increase the likelihood of success. It does not require large volumes or convergence time, even in the worst case assembly model that we assume for our own assembly algorithms. This simulation is also extended to simulate a $2D$ processor array.

(2) Framing. We also introduce a new method for constraining the assembly, which we call an *assembly frame,* which is a rigid nano-structure whose interior boundary binds the the input DNA strand and constrains the shape of the initial assembly to a given polygonal shape, thus providing for inputs and constraining the geometry of the subsequent tiling assembly.

(3) Compact Assemblies with Small Depth. Using these assembly techniques, we propose LP-BMC methods that decrease the size of the assembly to linear in the input size and have small assembly depth (time steps for assembly). These techniques for LP-BMC can be used for many applications: basic arithmetic and register level operations, data rearrangements and permutations, as well as more sophisticated operations done on large sequences of data. We use LP-BMC to solve the following problems with linear assembly size:

- logarithmic time *prefix computation:* given a composition operation that is associative and a sequence of n inputs, compute the composition of all prefixes of the inputs (e.g., prefix sums of n numbers; n-bit arithmetic such as addition, and subtraction, multiplication by a constant number; simulation of a finite state automata on an input of length n; fingerprint a length n string),
- constant time *perfect shuffle and pair-wise exchange permutations*, which allows constant time emulation of the shuffle-exchange network,
- logarithmic time execution of a large class of *normal parallel algorithms* [**S71, U84, L92**] on shuffle-exchange networks (e.g., DFT and bitonic merge of n-vectors),
- logarithmic time *general permutations*, by use of known Beneš network techniques [**B65**],
- *evaluation of a bounded degree Boolean circuit* of size n in time bounded by the circuit depth times $O(\log n)$ (without a large amount of communication between distinct molecules, in contrast with known DP-BMC algorithms [**R95, OR97**]).

Using LP-BMC to Enhance DP-BMC. LP parallelism complements the more apparent DP parallelism available in BMC. We feel that BMC needs to develop methodologies that combine the use of LP-BMC and DP-BMC, just as conventional electronic computational systems combine the use of VLSI and more distributed modes of computing. Let a *LPDP* operation be a LP operation done on multiple distinct molecules, that is via a combination of LP and DP. Let *LPDP-BMC* be the resulting mode of BMC computation combining LP-BMC and DP-BMC . With the use of surface attachments, all the LP-BMC techniques we propose can be executed in parallel by separate molecules in a distributed fashion. Thus they can simultaneously solve multiple problems with distinct inputs. For example, our LP-BMC methods for parallel prefix, addition, and circuit evaluation can be done simultaneously in parallel for all possible inputs using LPDP-BMC, so they can execute parallel arithmetic on multiple inputs, and also determine satisfying inputs of a circuit. Thus our LP-BMC assembly techniques are an enhancement of the power of DP-BMC.

1.1. Organization of this Paper. An introduction to BMC is given in Section 1. Section 2 provides useful DNA notation (subsection 2.1), preliminary descriptions of previous work on DNA nano-fabrication techniques (subsection 2.2) and known Turing Machine simulations using tiling (subsection 2.3). Section 3 describes Winfree's proposed use of unmediated BMC using DNA self-assembly. In the Appendix (Section 10) we consider some random assembly models and give bounds on the success likelihood of unmediated self-assembly. Section 4 presents our refined method for BMC via step-wise assembly. We discuss emulation of a

$1D$ array using step-wise $2D$ assembly (subsection 4.1), methods for DNA nano-construction of the tiling assembly (subsection 4.2) (including nano-construction of of tiles, encoding methods readout methods after assembly, and rigid framing of the initial assembly), 3-way and 4-way pairing of DNA strands, and extensions to emulations of $2D$ processor arrays (subsection 4.3). The further Sections provide our main results: the construction of *very compact, linear size assemblies, with progressively small depth*. Section 5 proposes a step-wise local assembly method for prefix computation with linear size and depth and Section 6 describes application of these techniques to integer addition. Section 7 describes step-wise local assembly with linear size and logarithmic depth for parallel prefix computation and integer addition. Section 8 describes methods for emulating shuffle-exchange operations with linear size and constant depth using unmediated assembly, including pair-wise exchange (subsection 8.1), and perfect shuffle operations (subsection 8.2). This allows us to efficiently emulate shuffle-exchange networks (subsection 8.3), execute normal parallel algorithms (subsection 8.4), and execute general permutations and parallel evaluation of circuits (subsection 8.5). Section 9 gives conclusions and acknowledgments.

2. Preliminaries

2.1. DNA notation. We denote single-stranded DNA by *ssDNA* and *dsDNA* to denote double-stranded DNA. We denote DNA double crossover molecules as *DX*. We denote the orientation of ssDNA by $3' - 5'$ or $5' - 3'$.

2.2. DNA Nano-Fabrication Techniques. Feynman [**F 61**] proposed nano-fabrication of structures of molecular size. Nanotechnology, without use of DNA, is discussed in the texts [**CL92, M93**].

Nano-fabrication of structures in DNA was pioneered by Seeman (e.g., see [**SZC94**]) in the 1990s (also see [**CRFCC+96, MLMS96**] for other work in DNA nanotechnology). These ingenious constructions used such basic constructive components as:

- *DNA junctions:* i.e., immobile and partially mobile DNA n-armed branched junctions [**SCK89**], antijunctions and mesojunctions [**DZS92**],
- *DNA knots:* i.e., ssDNA knots [**MDS91, DS92**] and Borromean rings[**MS97**],
- *DNA crossover molecules:* i.e., DX molecules of Fu and Seeman[**FS93**].

Many, but not all, of Seeman's constructions used:

- DX molecules for rigidity (or dsDNA for partial rigidity), and
- *solid-support (e.g., immobilized via surface attachments* [**S88**]*)*, to avoid interaction between constructed molecules [**ZS92**] (in contrast, some of his early work [**SC91**] did the construction in solution, with considerable more difficulty).

Using these techniques, Seeman and his students such as Chen and Wang nano-fabricated in DNA (see [**ZS92, ZS94, SWLQ+96, SQLYL+96, SZDC95, SZC94, SC91**] and [**SZDWM+94, SQLYL+96, SWLQ+96**])

- $2D$ *polygons,* including interlinked squares,
- $3D$ *polyhedra,* including a cube and a truncated octahedron.

The ultimate goal in Seeman's work seems to have been to construct regular arrays to immobilize proteins in regular arrays so as to be able to do crystallography analysis of proteins. However, his work may well be of central importance to the progress of the emerging field of BMC.

2.3. Known Tiling Results. Consider the square regions in the plane defined by horizontal and vertical lines which are integer units apart from the horizontal and vertical axis respectively.

A class of *domino tiling problems* were defined by Wang [**W61**] as follows: We are given a finite set of tiles of unit size square tiles each with top and bottom sides labeled with symbols over a finite alphabet. These labels will be called *pads.* We also specify the initial placement of a specified subset of these tiles, and the borders of the region where tiles must be placed defining the *extent of tiling.* The problem is to place the tiles, chosen with replacement, in all these square regions within the specified borders, so that each pair of vertical abutting tiles have identical symbols on their contacting sides. Let the *size* of the tiling assembly be the number of tiles placed.

Wang [**W61**](and similarly Buchi [**B62**]) proved that given a finite set of tile types, the domino tiling problem is undecidable if the extent of tiling is the positive quadrant of the plane and a single initial tile is required to be at a fixed location. Berger [**B66**] gave a proof that removed the latter condition, thus proving the undecidability of the tiling problem with no initial placement of tiles. Later simplified proofs of Robinson [**R71**], and the text book of Lewis and Papadimitriou [**LP81**], (pages 296–300) provide a direct simulation of a single tape deterministic Turing Machine to prove this undecidablility result.

Garey, Johnson, and Papadimitriou [**GJP77**] (see citation of [**GJ79**], page 257) proved that the domino tiling problem is NP-complete if the extent of tiling is a rectangle of polynomial size. They used a direct reduction from the directed Hamiltonian path problem (known to be NP-complete). Later, the textbook of Lewis and Papadimitriou [**LP81**](pages 345–348) gave a direct proof of this NP-completeness result, providing a simulation of a single tape nondeterministic Turing Machine running in time $T \geq n$ and space $S \leq n$ by assembly of an $S/2 \times T$ array of tiles. More recently, Grunbaum, Branko, and Shepard [**GBS87**] surveyed these and related results on the complexity of tiling, including these direct simulations of Turing Machines.

3. BMC using Unmediated Self-Assembly

Winfree [**W96**] defined a *cellular automaton* as a Turing Machine with a single tape of S cells such that disjoint pairs of adjacent cells are updated in parallel by a transition function depending on those cells and the state of the machine. The disjoint pairs of cells which are updated are at consecutive positions that are either odd, even or even, odd (these two cases alternate each step). He observed that the tiling constructions cited above ([**LP81**](pages 345–348), [**GBS87**]) also give the direct simulation of a cellular automaton running in time $T \geq n$ and space S by assembly of an $S/2 \times T$ array of tiles.

Winfree [**W96**] then proposed a very intriguing idea: to do these tiling constructions by application of the DNA nano-fabrication techniques of Seeman et al [**SZC94**], which may be used for the construction of small DNA molecules that can function as square tiles with pads on the sides. The pads are ssDNA (they are also denoted *sticky ends* in some literature). If two pads are Watson-Crick complementary and $3' - 5'$ and $5' - 3'$ oriented in opposite directions, we consider them to be *matching*. At the appropriate conditions (determined by temperature and salinity, etc.), they may hybridization into doubly stranded DNA. The assembly of the tiles is due to this hybridization of pairs of matching pads on the sides of the tiles. We will call this idea *BMC via unmediated DNA self-assembly*, since the computations advance with no intervention by any controllers. The advantages of the unmediated DNA assembly idea of Winfree, et al, are potentially very significant for BMC: the computations advance with no intervention by any controllers, and require no thermal cycling.

Winfree, et al [**WYS96**] then provided further elaboration of this idea, to solve a variety of computational problems using unmediated DNA self-assembly. For example, they propose the use of these unmediated DNA assembly techniques to directly solve the NP-complete directed Hamiltonian path problem, using a construction similar to that of Garey, Johnson, and Papadimitriou's [**GJP77**] (see also [**GJ79**], page 257) NP-completeness proof for tiling of polynomial size extent. Winfree, et al [**WYS96**], also provided a very valuable experimental test validating the preferential pairing of matching DNA tiles over partially non-matching DNA tiles, but did not at that time experimentally demonstrate a DNA self-assembly for a (non-trivial) computation.

Simulation of a $1D$ Cellular Automaton. We now consider Winfree's [**W96**] direct simulation of a $1D$ cellular automaton running in time $T \geq n$ and space $S \geq n$ by assembly of an $S/2 \times T$ array. Each tile π has two distinguished pads in_1, in_2 ordered left to right on its bottom side, and two distinguished pads out_1, out_2 ordered left to right on its top side (see Figure 1a). The pads are defined from the transitions of the cellular automaton. We adapt the convention of considering the bottom of the tile to be associated with the current step and the top of the tile associated with subsequent step, via its pads in these positions. In Winfree's [**W96**] intended assembly, for each $t, 0 \leq t < T$ and $i, 1 \leq i \leq S$ where $(i = t) \bmod 2$ there will be placed a tile $\pi(i,t)$ representing the transition from time $t-1$ to time t at the two consecutive tape cells at positions $i, i+1$. The bottom of the tile is associated with time $t-1$ and the top of the tile is associated with time t, via its pads in these positions. The construction was intended to result in an assembly so tile $\pi(i,t)$ is placed to abut and match its pads in_1, in_2 at the out_2, out_1 respective pads of the neighboring tiles $\pi(i-1,t-1), \pi(i+1,t-1)$, which abut just below $\pi(i,t)$. Also, tile $\pi(i,t)$ is placed to abut and match its pads out_1, out_2 at the in_2, in_1 respective pads of the neighboring tiles $\pi(i-1,t+1), \pi(i+1,t+1)$ which abut just atop $\pi(i,t)$. The size (number of tiles placed) of the assembly is $A = ST/2$.

Is Unmediated DNA Self-Assembly Feasible for Computation? We have noted that the idea of unmediated DNA assembly has theoretical and mathematical foundations provided by the known tiling complexity results which were quite well established in the 1960s and 1970s. However, the idea of unmediated DNA assembly can not necessarily rely only on these, for it needs some sort of analysis or experimental validation of the random assembly process itself.

Possible Blockages during Unmediated Self-Assembly. A *blockage* is an untiled location where no tile can be placed there without a *pad mismatch* (a conflict between a pad of the tile and the pad of an abutting tile at that location). We assume there are no partial or total mismatches allowed between the pads of abutting placed tiles. In the assembly process, a blockage will eventually result in a situation where further tiling is not possible (without some disassembly) due to inconsistent pads on nearby tiles.

Homogeneous Assemblies without Blockages During Unmediated Self-Assembly. We will consider some cases where blockages may not occur. An assembly will be called a *homogeneous* if the tile of the assembly have identical pads, and will otherwise be called *heterogeneous*. Unmediated DNA self-assembly may often be used successfully for homogeneous assembly, since subassemblies can compose without blockages due to pad mismatches (however, they may not be able to compose due to geometry mismatches). Such a special case is similar to the homogeneous crystallization process found in physics. This is an important application in its own right from the perspective of biochemists and for the general goal of nano-fabrication of large structures. In the case of a homogeneous crystal assembly with 1, or a small number of tiles, Seeman [**S85**] contended that there is an entropic driving force operative to solve small inconsistencies between the ideal contacts and ideal periodicity. An assembly frame, as proposed in this paper, may be used to seed an initial homogeneous crystal assembly, and to constrain subsequent unmediated homogeneous assembly.

Assumptions that Rule out Assembly Blockages During Unmediated Self-Assembly. The papers of Winfree [**W96**] and Winfree, et al [**WYS96**] do not consider the possibility of a blockage, as defined above. It is implicitly assumed that blockages do not occur in the assembly process. Instead, only assemblies without blockages are considered as end products of the assembly process. Let a tile be *a k-pad-matching tile* for a given location adjacent to the assembly if it may be placed into the assembly in this location with some k pads which correctly match other pads in abutting previously placed tiles of the assembly. Winfree [**W96**] assumes the assembly process is restricted to add into the assembly only maximal matching tiles which are k-pad-matching, for $k > 1$, and never to add 1-pad-matching tiles. The unmediated DNA assembly of Winfree [**W96**] succeeds without blockage if the assembly process is so restricted.

Assembly Blockages That Might Occur During Unmediated Heterogeneous Self-Assembly. Let a tile τ be *a m-BP-matching tile* for a given location adjacent to the assembly if τ may be placed into the assembly in this location with a total of m base pairs in pads which correctly match, via Watson-Crick complementation, abutting previously placed tiles of the assembly. Let a tile be *a maximal matching tile* in that given location, if it is a m-BP-matching tile in that given location, and there is no other tile which is a m'-BP-matching tile in that given location, for any $m' > m$. Note that the definition of a maximal matching tile in that given location depends on the current state of the assembly, and a maximal matching tile which is a m-BP-matching tile in the current assembly may be a m'-BP-matching tile in a later state of the assembly, for some $m' \geq m$. When we use unmediated DNA assembly techniques for general computation simulations, the resulting tiles will not generally have identical pads, so the assemblies are heterogeneous. Thus a maximal matching tile τ may not be unique for a given location L; for example if an adjoining location L' (where the tile has a pad which is expected to be matched) is currently untiled, there may be another tile τ' that is also currently a maximal matching tile in that given location L (but placing τ or τ' at L may be incorrect, since they might later have a pad mismatch in any tiling of that adjoining location L'). We shall show that such situations may lead to blockages, which may potentially be endemic to unmediated local assemblies doing computation simulations.

The above example indicates † that at a given state of the assembly, there can be many locations where a tile may be placed so that all maximal matching tiles in that location are only 1-pad-matching tiles, and these can quickly become 2-pad-matching tiles at a subsequent stage in the assembly.
Blockage Probabilities in Random Assembly Models. In Appendix Section 10 we consider various random assembly models for the kinetics of assembly via hybridization at a fixed temperature. For a simple random assembly model, we shall demonstrate in Appendix Section 10 for certain setting of assembly parameters, 1-pad-matching tiles might be initially placed at such locations, and thus blockages in assembly can occur. For a more sophisticated random assembly-disassembly model also considered in Section 10, the issue is unresolved, though there are some recent computer simulations of Winfree [**W97**] which seem to avoid blockages in a assembly-disassembly model for careful choice of settings of the parameters. However, at this time, there is no theoretical proof of the success (without blockage) of unmediated self-assembly and this method remains experimentally untested.

4. BMC via Step-wise assembly

We propose to use instead an improved assembly method for computation which we call *step-wise assembly* that involves control of the assembly in distinct steps. Our step-wise assembly method refines the unmediated self-assembly method of Winfree, et al [**W96, WYS96**] to increase the likelihood that the assembly always succeeds (and does not require such large volumes or convergence time per step). For example, we modify the assembly of Winfree, et al [**W96**], as detailed also in Section 3, to control the assembly of the $S/2 \times T$ tiling array so it is done in breadth first manner from the initial tile placements, in T distinct steps. Let the *depth* of a tile in the intended tiling be the number of matches between tiles in paths of this breadth first assembly tree. Let the *assembly depth* d be the maximum depth of the intended assembly. As in the previous section 3, we idealize the DNA molecules to be assembled as polyhedral *tiles* with pads on the sides that can be matched by Watson-Crick complementation, and the assembly of the tiles is due to hybridization of matching pads on the sides of the tiles.

We specialize the tiles used for each time step t to include *time stamps* on their pads; that is we augment the tile pads with characters distinctly encoding the assembly depth (complemented if need be to insure consistent Watson-Crick complementary matching). The input pads of each such tile are edited to contain a DNA string encoding the step number t (complemented to insure Watson-Crick complementary matching) and the output pads of the tile are edited to contain also a unique DNA string encoding $t + 1$ (non-complemented to insure Watson-Crick complementary matching).

† Note that only 1-pad-matching tiles compete to be placed at such locations. Also note that the placement of such a 1-pad-matching tile τ (see Figure 1b) in the assembly may be shortly later reinforced by subsequent placement of tiles at adjoining locations that increase the number of pad matches of τ above 1 in the resulting assembly (see Figure 1c). Thus, a 1-pad-matching tile can quickly become 2-pad-matching tile at a subsequent stage in the assembly. This can easily happen in unmediated DNA assembly even if the temperature has been set to favor 2-pad-matching over 1-pad-matching. Thus, it does not seem reasonable to us to arbitrarily restrict the assembly process so as to adds tiles into the assembly only at locations where there are only k-pad-matching tiles, for $k > 1$.

The number of tile types is increased by a multiplicative factor of d, and each resulting tile has an associated depth. We presume that the initial tiles are initially placed as specified, without any initial blockage. To insure distinct assemblies to not interfere with each other, we assume each initial assembly is solid-supported using standard surface biotechnology (e.g., immobilized via surface attachments [**S88**]).

We will consider the entire assembly to be done with a maximal matching tile in each given location, and assume that there are no partial or total mismatches allowed between the pads of abutting placed tiles.

For our assembly algorithms, for simplicity we assume a *worst case assembly* model: we assume that the further assembly is done by choosing a worst case unit region which adjoins at least one region so far tiled and where further tiling is possible (the region must be so far untiled, and not a blockage), and then choose (the worst case) maximal matching tile to place there, where that tile is chosen among all tiles where there are no pad mismatches among abutting tiles. We repeat this worst case choice of regions to tile, until every such region that can be tiled, is tiled in this worst case manner. (Note that our assumption that there are no pad mismatches is intended to approximate the actual case where the likelihood of pad mismatch is nonzero, but very low, by appropriate choice of key parameters such as temperature and match lengths.)

Our idea is to provide control for the assembly in discrete steps $t = 1, 2, \ldots$. We add all the tiles (providing multiple copies of each type of tile) of depth t on step t and after sufficient time for the completion of this step's further DNA assembly, wash away the unattached tiles of step t. We will insure no blockage, as previously described, can ever occur, in our worst case assembly model. As in the previous assembly methods, we require no thermal cycling.

4.1. Emulation of a $1D$ Array Using Step-wise $2D$ Assembly. To illustrate the idea of step-wise assembly, we now refine Winfree's [**W96**] direct simulation of a cellular automaton running in time $T \geq n$ and space $S \geq n$ by assembly of an $S/2 \times T$ array of $A = ST$ tiles. We again assume the reader is familiar with his construction.

We augment the pads of each of the tiles defined in Winfree's construction by including *time stamps*: that is for each $t, 0 \leq t \leq T-1$, and for each tile, the bottom pads in_1, in_2 are edited to contain a unique DNA time stamp encoding of t (complemented to insure Watson-Crick complementary matching) and the top pads out_1, out_2 are edited to contain also a unique DNA time stamp encoding $t+1$ (non-complemented to insure Watson-Crick complementary matching). Thus, in our intended assembly, the neighboring tiles $\pi(i-1, t-1), \pi(i+1, t-1)$ have both of their upper side pads out_1, out_2 composed with time stamp encoding t, and the neighboring tiles $\pi(i-1, t), \pi(i+1, t)$ have both of their bottom side pads in_1, in_2 composed with time stamp encoding t. The assembly is controlled as follows: the tiles representing the initial configuration of the machine at time $t = 0$ can be assume to initially placed in correct location, as $\pi(1, 0), \ldots, \pi(S, 0)$. For each $t = 1, \ldots, T$ we must add to the assembly only the step t tiles which are time-stamped with $t-1$ in pads in_1, in_2 and t in pads out_1, out_2. These time-stamped tiles can and must be unique placed as $\{\pi(i,t)|, 1 \leq i \leq S$ where $(i = t) \bmod 2\}$, thus allowing the assembly to advance from time $t-1$ to time t on all the tape positions $1 \leq i \leq S$ where $(i = t) \bmod 2\}$. The resulting step t addition to the assembly (see Figure 1d) is in the shape of a $2D$ rectangular tiling of length S and unit height. The time-stamped paddings of each of these tiles do not interfere with any other tiles at this time, so there can be no blockages: the assembly correctly places all tiles associated with each time step $t > 0$. Thus, the resulting tiling assembly is unique.

Note that we can somewhat simplify time stamping by using time stamps modulo 2. This requires us to use solid-support methodology [**ZS92**] for the assemblies and to wash away all the tiles of previous steps $t' < t$ before proceeding to step t.

4.2. Nano-Fabrication of the Tiling Assembly. We now breifly discuss how the tiling assembly is to be done via DNA nano-fabrication.

Nano-Fabrication of Tiles. The tiles need be made quite rigid, in a shape that conforms with the intended multi-tile assembly. The tiles can be contructed by the nano-fabrication techniques of Seeman, et al. as surveyed in [**SZC94, SWLQ+96, SQLYL+96**] (see also subsection 2.2). Interaction between tiles during their construction can be avoided using solid-support methodology described in [**ZS92**]. Winfree, et al [**W96, WYS96**] made use of DX molecules [**FS93**] for nano-fabrication of rigid tiles, and it may also be possible to make use of DNA poly-crossover molecules. The square DNA tiles considered later in the paper are similar to 'Form V' DNA, which is half-right and half-left handed, and results when two complementary (large) circles are mixed together in the absence of topoisomerases. It should be cautioned that there may be considerable care must be made to insure the geometry is correct so that (i) and the tiles is rigid as intended, (ii) the shape that conforms with the intended multi-tile assembly, (iii) the pads of matching tiles are correctly aligned, and (iv) the helical twist of the doubly stranded DNA is respected. Since the nano-fabrication techniques of Seeman, et al. are at this time more an art than a science (see [**SQLYL+96**]), considerable experimentation is required.

Encoding Methods for Pads. Throughout this paper, we will assume distinct encoding functions such as E from tuples of small integers to short distinct ssDNA words. We assume the ssDNA words are not degenerate (I.e., with repeated subsequences or reverse subsequences to avoid mismatching), and are chosen using the DNA word design techniques as described in [**A94, L94, BL95, B96, GFBCL+96**] and [**M96, KCL96, GFBCL+96, DMGFS96, JK97a, DMRGF+97**].

These known ssDNA word design techniques have the dual goals of (i) minimization secondary structure (e.g., unwanted folding) and (ii) maximization of binding specificity and discrimination efficiency. We let $\bar{E}(-)$ denote the Watson-Crick complementation of these encodings.

Readout Methods After Assembly. The final configuration of the simulation machine is given by the linear sequence of pads of the bottom sides of the *final tiles:* $\{\pi(i,T)|, 1 \leq i \leq S$ where $(i = T) \bmod 2\}$ of step T. This output might be constructed as a ssDNA in $O(1)$ BMC steps by a variety of well known techniques:

1. By the addition of short segments of ssDNA, each containing the composition of a pair of ssDNA which are Watson-Crick complements of a consecutive pair of pads on the top side of the final tiles, and then ligating these short segments together to form the required output ssDNA.
2. By adding dangling ssDNA segments attached to the top ends of the final tiles, ligating consecutive pairs of these together to form a single ssDNA,

and disconnecting (by appropriate use of restriction enzymes) the side edges of the final tiles.

It should be again cautioned again that in these constructions, considerable care must be made (and experimentation), to insuring the geometry is correct so that the readout ssDNA is correctly aligned with the appropriate pads of the tiles, and the helical twist of the dsDNA is respected.

Rigid Framing of the Initial Assembly. The assembly frame is one of the key innovations of this paper and is also employed in many of our other assembly constructions given in this paper. An assembly frame is a rigid DNA nano-structure. The purpose of the assembly frame is (i) to constrain the placement of the input ssDNA on the boundaries of the tiling assembly, and (ii) to initiate and further constrain the geometry of the subsequent assembly. The input ssDNA strand binds to the assembly frame at prescribed places along the strand. conforming to the boundaries of the assembly frame. The initial tiling assembly also binds to these input ssDNA strands (at other prescribed places). In this way the assembly frame constrains the initial tiling assembly, and thus constrains the geometry of the subsequent tiling assembly.

For example, in our simulation of a $1D$ automata, to insure the initial tiles are placed as specified, as a $S/2 \times 1$ rectangle, we can employ a rigid DNA structure, which we will call an *assembly frame.* In this case, this can be easily done in this case by adding additional pads on the left and right sides of each of the initial tiles for first time step $t = 1$. We then define these pads so that the left pad of each tile is the same as the right pad of the tile intended on its right, and so that the right pad of each tile is the same as the left pad of the tile intended on its left.

In general, a frame may constrain the assembly to a prescribed polygonal path, e.g., a straight line, or constrain the assembly within a prescribed polygon, e.g., a rectangle. (See Figures 3a and 7a, giving a ssDNA encoding the input n-vector. See Figures 3b and 7b, giving a possible DNA nano-structure for the rectangular frame.) The construction of an assembly frame in this case is less straightforward, but never-the-less can be done by known recombinant DNA techniques and/or DNA nano-fabrication techniques of Seeman et al [**SZC94**], as discussed in subsection 2.2:

- We can make use of dsDNA, where possible, which is somewhat more rigid than ssDNA for moderate lengths.
- We can make use of the DNA junctions developed by Seeman, et al [**SCK89, DZS92**] and the DX molecules of Fu and Seeman[**FS93**] which are quite rigid, to form a DNA superstructure to hold the DNA strand in the required initial position.
- We can insert into the input, which we generally assume is ssDNA, some unique pads at regular distances, and then can build a DNA superstructure (built of DX molecules) containing a polygon of the required initial shape with matching pads (provided by Watson-Crick complementation) to hold the input ssDNA in place.
- Each initial assembly frame can be solid-supported using standard surface biotechnology, to insure distinct assemblies to not interfere with each other.

3-way and 4-way Pairing of DNA Strands. The construction of an assembly frame can be facilitated in certain cases by 3-way pairing of DNA strands. There are number of known methodologies for 3-way pairing of DNA strands.

- The DNA triple helix can be constructed by the Dervan system [**DD92**].
- Another option is the use of a motif known as DX+J [**LYQS96**], which is almost as stiff as DX, and provides the option of having DNA come out the side of a tile.
- Seeman [**S97**] further suggests to ligate from the perpendicular arm a second 3-arm junction, which would provide a location for a complementary ssDNA to bind, with 4-ary recognition.

All of these allow for 3-way pairing of DNA strands, which is obtained by restricting the encoding of DNA strands to two bases, rather than the usual 4 bases, thus resulting in a binary (rather than the usual 4-ary) encoding in DNA. The input ssDNA and/or the DNA superstructure can be constructed using this encoding.

4.3. Extensions to Emulation of $2D$ Processor Arrays. Our step-wise assembly techniques can also be extended to $3D$ DNA assembly to simulate a $2D$ processor array where each processor is an FSA with the same finite state control. Let a $2D$ *cellular automaton* be a Turing Machine with a single $2D$ square tape of shape $S' \times S'$ such that disjoint pairs of adjacent cells are updated in parallel by a transition function depending on those cells and the state of the machine. We describe a T step $3D$ DNA array assembly of total size $ST/4$, for simulating T steps of a processor array with $S = (S')^2$ cells, refining a similar (but unmediated) $3D$ assembly method briefly discussed in [**W96**] and also [**WYS96**]. In the intended assembly, for each $t, 1 \leq t \leq T$ and $i, 1 \leq i, j \leq S'$ where $(i = t) \bmod 2$ and $(j = t) \bmod 2$ there will be a cubic tile $\pi(i,j,t)$ representing the transition from time $t-1$ to time t at the four adjacent tape cells at positions $(i,j), (i+1,j), (i+1,j+1), (i,j+1)$ (we will adopt this rotational order for positioning of pads as well). We again adapt the convention of considering the bottom of the tile to be associated with time $t-1$ and the top of the tile associated with time t, via its pads in these positions. This requires the construction of $3D$ DNA cubes using the nano-technology developed by Seeman et al [**SQLYL+96**]. Each cubic tile $\pi(i,j,t)$ has (i) four distinguished pads in_1, in_2, in_3, in_4 rotationally positioned at the four corners on its bottom side augmented with time stamps encoding $t-1$ (complemented to insure Watson-Crick complementary matching) and also (ii) four distinguished pads out_1, out_2, out_3, out_4 similarly rotationally positioned at the four corners on its bottom side augmented with time stamps encoding t (non-complemented to insure Watson-Crick complementary matching). This results in an assembly that has tile $\pi(i,j,t)$ abut and match its pads in_1, in_2, in_3, in_4 at the out_3, out_4, out_1, out_2 respective pads of the neighboring tile $\pi(i-1,j-1,t-1), \pi(i+1,j-1,t-1)$, $\pi(i+1,j+1,t-1), \pi(i-1,j+1,t-1)$, which abut just below $\pi(i,j,t)$. Also, the assembly that has tile $\pi(i,j,t)$ abut and match pads $out_1, out_2, out_3, out_4$ at the in_3, in_4, in_1, in_2 respective pads of the at neighboring tiles $\pi(i-1,j-1,t+1), \pi(i+1,j-1,t+1), \pi(i+1,j+1,t+1), \pi(i-1,j+1,t+1)$, which abut just atop $\pi(i,j,t)$. The size of the assembly is $A = ST/4$.

It should be cautioned that such $3D$ assemblies may entail considerably more difficulties than $2D$ assemblies, in particular (i) the cubic tiles need to be designed to be rigid (a multiple DX is suggested by [**W96**], but may not be rigid), and (ii) the geometry of the assembly must be carefully designed to facilitate diffusion and access of polyhedral tiles into the untiled locations adjacent to the $3D$ assembly (this may be improved my our use of step-wise assembly).

5. A Linear Size and Depth Prefix Computation Assembly

One drawback of the step-wise assembly method, if applied using the standard regular square tiling simulations of T step computations, is that it results in a large assembly of $A = ST$ tiles. In the rest of this paper we give our main results, which are tilings that are much more compact, and have considerably smaller total size which is linear in the input. We apply these tilings to solve some restricted, but fundamental, problems that arise in parallel computation.

Monoid Sum and Prefix Computation. Let $(D, \cdot)$ be a *monoid*, that is D is a set with an identity element λ over the operation $\cdot$ and the operation $\cdot$ is associative. Suppose we are given as input an n-vector $(a_1, \ldots, a_n)$ with elements from some domain D. The *monoid sum problem* is given $(a_1, \ldots, a_n)$, to compute $a_1 \cdot \cdots \cdot a_n$. The *prefix computation problem* is given $(a_1, \ldots, a_n)$, to compute each prefix: $b_i = a_1 \cdot \cdots \cdot a_i$, for $i = 1, \ldots, n$. The monoid sum and prefix computation problems occur in many applications, for example:

1. *Prefix sums:* in this case the domain D are assumed to be small b-bit numbers, the operation $\cdot$ represents integer addition, and n is the number of integers to be summed.
2. *Integer arithmetic*, including addition, subtraction, and multiplication times a constant. In the case of n bit binary addition, the domain is $D = \{$carry, no-carry, carry-propagate $\}$ and the operation $\cdot$ represents carry-sums from a bit position to the next higher bit position. Once the propagated carries are determined at each bit position by the prefix computation, the output bits giving the bits of the addition can be immediately determined.
3. *Finite State Automaton (FSA) simulation:* in this case the domain D is the set of states of the FSA paired with the input symbols, the operation $\cdot$ represents the finite state transitions, and n is the length of the input string. Once the resulting states at each time step are determined by the prefix computation, the output of the FSA can be immediately determined.
4. *fingerprinting strings:* given a length n string, the operation $\cdot$ represents is defined to be an associative hash function, and the monoid sum gives a hash encoding of the input string.

All these problems have obvious sequential algorithms with linear work for many conventional models of computation such as circuits and random access machines.

Linear Assembly for Prefix Computation. To solve the prefix computation problem $(a_1, \ldots, a_n)$ by LP-BMC, we propose a linear size assembly. We assume distinct encoding functions E, E' from pairs in $D \times \{1, \ldots, n\}$ to short distinct ssDNA words, as briefly discussed in subsection 4.2. We let $\bar{E}(-), \bar{E}'(-)$ denote the Watson-Crick complementation of these encodings. We assume the input ssDNA encodes the vector $(a_1, \ldots, a_n)$ as a sequence of unit length ssDNA words of the form $E(a_1, 1), \ldots, E(a_n, n)$. This input ssDNA is positioned into a straight line of length n (see Figure 2a) by the use of rigid framing techniques described in subsection 4.2, e.g., the use of a DNA DX superstructure with pads.

Each tile in this case is a unit square as given in Figure 2b. Each tile will have pads, as defined below.‡ For each $a, b \in D$, $a' = b \cdot a \in D$ and $t \in \{1, \ldots, n\}$ (where if $t = 1$ then $b = \lambda$), there is a *step t* tile $\tau(a, b, t)$ with:

- the *input* pad $\bar{E}(a, t)$ on the bottom side (with the intent to match with the tth input),
- the *output* pad $\bar{E}''(a', t)$ on the top side (with the intent to match with the tth output),
- the *carry-in* pad $\bar{E}'(b)$, on the left side (with the intent, for $t > 1$, to match with the $(t-1)$th prefix given by the carry-out pad on the left side of the $t-1$ tile), and
- the *carry-out* pad $E'(a')$ on the right side (with the intent to match with the tth prefix given by the carry-in pad on the left side of the $t+1$ tile).

We use step-wise assembly, adding all step t tiles at time $t = 1, \ldots, n$. Suppose the prefix computation problem $(a_1, \ldots, a_n)$ has output $(b_1, \ldots, b_n)$, where $b_i = a_1 \cdot \cdots \cdot a_i$. For each $t = 1, \ldots, n$ the tile $\tau(a_t, b_t, t)$ is the unique tile that can be placed on the tth step, and that must be placed so its input pad matches the tth element of the input, and for $t > 2$, this tile also abuts the previous tile $\tau(a_{t-1}, b_{t-1}, t-1)$ matching the respective carry-in and carry-out pads. Thus, the tiling is unique and there can be no blockages.

The resulting $2D$ assembly (see Figure 2c) is in the shape of rectangular tiling of length n and unit height, containing tiles $\tau(a_1, b_1, 1), \tau(a_2, b_2, 1), \ldots, \tau(a_{n-1}, b_n, n)$ in this order, with the encoded input matched to the input pads on the bottom sides of these tiles and the output pads on the top sides of the tiles giving the encoded output sequence $(\bar{E}(b_1, 1), \ldots, \bar{E}(b_n, n))$.

6. Application to Bit-Serial Integer Addition

Lipton et al. [**BDLS95**] developed circuit evaluation algorithms using BMC which allow for binary addition at the cost of a number of steps linear in the size of the circuit. Bancroft and Guarnieri [**GFB96**] demonstrated the first experimental execution of a DNA-based arithmetic calculation on a single pair of bits. Their methods, if extended to n-bit arithmetic, require at least n thermal cycles and a considerable number of recombinant DNA operations per bit.

Here, we give a step-wise assembly method requiring no repeated thermal cycling. The *integer addition problem* can be considered to have as input two length n vectors $(a_1, a_2, \ldots, a_{n-1}, a_n)$ and $(a'_1, a_2, \ldots, a'_{n-1}, a'_n)$ vectors of bits representing the binary representation of two numbers; the output is a length $n+1$ vector $(s_1, s_2, \ldots, s_n, s_{n+1})$ of bits representing the sum of these two numbers.

We apply our proposed step-wise method for prefix computation to synthesize a step-wise method for n-bit binary addition. Our method will take n steps and will construct an $O(n)$ size assembly (the number of tiles in the assembly can be further reduced to $O(n/\log n)$ by a somewhat more complex construction where we use a k-arry encoding on the input numbers, for $k = \log n$, and define tiles to do the appropriate k-arry arithmetic). We assume the same distinct encoding functions E, E', E'' and assume the input is a ssDNA encoding the vectors $(a_1, a_2, \ldots, a_{n-1}, a_n)$ and $(a'_1, a_2, \ldots, a'_{n-1}, a'_n)$ as a sequence of unit length ssDNA words of the form $E(a_1, 1), \ldots E(a_n, n), E(a'_1, n+1), \ldots, E(a'_n, 2n)$.

‡To reduce notation in this and all further Figures of this paper, we simply label each tile pad within the Figures without use of the encoding notation E, and drop complementation notation.

It will be useful to define the $n+1$ bits of the input $a_{n+1} = 0, a'_{n+1} = 0$. For technical reasons, we need to reverse the order of the encoding of the bits of the second input number. Again by known recombinant DNA techniques (as described in our subsection 8 on assemblies for shuffle permutations), we can reverse the latter part of the input ssDNA containing the latter n elements $E(a'_1, n+1)\ldots E(a'_n, 2n)$, yielding a ssDNA with a sequence of unit length ssDNA words of the form $E(a_1, 1)\ldots E(a_n, n)E(0, n+1)\beta$, $\bar{E}(0, 2n+1)^R \bar{E}(a'_n, 2n)^R, \ldots \bar{E}(a'_1, n+1)^R$. Also, a unit length ssDNA dummy (distinct from all the other strings used in DNA encodings) can easily be inserted into this ssDNA just after the nth element. This ssDNA is positioned into the shape of a rectangle (see Figure 3a) of length $n+1$ and unit height by the use of rigid framing techniques described in subsection 4.2 (see Figure 3b).

Each tile in this case is a unit square with a unit length internal middle segment as given in Figure 3c. For each $a, a', b \in \{0,1\}$ and $i \in \{1, \ldots, n+1\}$ (where if $i = 1$ then $b = 0$), there is a tile $\tau^*(a, a', b, i)$ with:

- a unit square with
  - *input*$_1$ pad $\bar{E}(a, i)$ on the bottom side, (with the intent of giving the ith bit of the first input number),
  - *input*$_2$ pad $E^R(a', n+i)$ on the top side, (with the intent of giving the ith bit of the second input number),
  - *carry-in* pad $\bar{E}'(b, i-1)$ on the left side, (with the intent of giving the Watson-Crick complement of the $(i-1)$th carry bit b), and
  - *carry-out* pad $E'(b', i)$ on the right side, for $b' = (a \vee a') \wedge (a \vee b) \wedge (b \vee a')$ (with the intent of giving the ith carry bit b'),
- an internal middle segment, running between the left and right sides, with an *output* pad $\bar{E}''(s, i)$, where $s = a \oplus a' \oplus b$ (with the intent of giving the ith bit of the intended output).

Thus, $\tau^*(a, a', b, i)$ determines the carry-sum at the ith position from both the inputs at the ith position and the carry from the $i-1$ bit position, and then propagates the carry-sum at the ith position to the $i-1$ bit position.

We will use a step-wise mediated assembly, adding all tiles of the form $\tau^*(a, a', b, i)$ on distinct steps $i = 1, \ldots, n+1$. For each $i \in \{1, \ldots, n\}$, the tile $\tau^*(a_i, a'_i, i)$ is the unique tile that can be placed so

- the bottom side abuts (input$_1$ pad matches) the ith element $E(a_i, i)$ of the input giving the ith bit of the first input number,
- the top side abuts (input$_2$ pad matches) the element $\bar{E}(a'_i, n+i)$ giving the ith bit of the second input number,
- the left side abuts (the carry-in pad matches)
  $E'(b_{i-1}, i-1)$, giving the $(i-1)$th carry bit b_{i-1} (which is 0 if $i = 1$),
- the right side abuts (the carry-out pad matches) $E'(b_i, i)$, providing the ith carry bit $b_i = (a_i \vee a'_i) \wedge (a_i \vee b_i) \wedge (b_i \vee a'_i)$, and
- the middle segment has output pad $\bar{E}''(s_i, i)$, where
  $s_i = a_i \oplus a'_i \oplus b_i$, giving the ith bit of the intended output.

The location of this tile does not interfere with the placement of any other tiles. The resulting $2D$ assembly (see Figure 3d) is unique and contains tiles $\tau^*(a_1, a'_1, 1)$, $\tau^*(a_2, a'_2, 2), \ldots, \tau^*(a_n, a'_n, n)$, $\tau^*(a_{n+1}, a'_{n+1}, n+1)$ in this order, with the inputs matched to the top and bottom side input pads input$_1$, input$_2$ of these square tiles, with the carry pads of consecutive tiles matched, and with the output pads of the middle segment of the tiles giving the required encoded output sequence $\bar{E}''(s_1, 1), \bar{E}''(s_2, 2), \ldots, \bar{E}''(s_{n-1}, n-1), \bar{E}''(s_n, n)$, where $s_i = a_i \oplus a'_i \oplus b_i$ is the ith bit of the sum of the input numbers. Again, this output can be constructed as a ssDNA in $O(1)$ BMC steps by a variety of known techniques, as described at the end of subsection 8.2. (Alternatively, the output can again be facilitated by redefining each tile to be a $3D$ polyhedron resembling a pup tent as given in Figure 3e, with the same square base as the square tiles described above, and with sloping rectangular sides meeting at a common segment with an output pad.)

7. Logarithmic Depth Assembly for Prefix Computation

A drawback of our proposed step-wise assembly method given in Section 5, if applied using the standard regular square tiling simulations of T step computations, is that it involves T assembly steps, which is the depth of such a regular tiling.

We now consider tilings that are not regular, and thus can have considerably smaller depth. We apply these tiling to some restricted, but fundamental, problems that arise in parallel computation. The prefix computation problem has known optimal parallel algorithm (for circuits and parallel random access machines), with $O(\log n)$ time and n work due to Ladner and Fischer [**LF80**].

Fortune and Wyllie [**FW 78**] developed a technique known as *parallel pointer jumping* which contracts consecutive pairs of an input list $\mathcal{L}^{(0)} = (a_1, \ldots, a_n)$, into single elements. Let us assume n is a power of 2. Let $(D, \cdot)$ be the monoid associated with the problem. For example the sublist (a_i, a_{i+1}), for an odd i, is contracted by pointer jumping into $(a_i \cdot a_{i+1})$. When executed in parallel on every consecutive odd-even pair of an even length list $(a_1, a_2, \ldots, a_{n-1}, a_n)$, parallel pointer jumping contracts this list into the list $\mathcal{L}^{(1)} = (a_1 \cdot a_2, a_3 \cdot a_4, \ldots, a_{n-1} \cdot a_n)$ of length $n/2$. The result of t stages of parallel pointer jumping is the list $\mathcal{L}^{(t)} =$
$(a_1^{(t)}, a_2^{(t)}, \ldots, a_{n^{(t)}}^{(t)})$, where $n^{(t)} = n/2^{t-1}$, and for each odd $i \in \{1, \ldots, n^{(t)}\}$, we have $a_i^{(t)} \cdot a_{i+1}^{(t)} = a_{(i+1)/2, t+1}$. Repeating this parallel pointer jumping for $T = \log n$ stages contracts a length $n = 2^T$ list to a single element.

To solve the monoid sum problem $(a_1, \ldots, a_n)$, (computing output $a_1 \cdot \cdots \cdot a_n$) where n is a power of 2, we propose a linear size assembly by LP-BMC in $\log n$ steps using self-assembling tiles to emulate parallel pointer jumping.

We slightly redefine the distinct encoding functions E, E' from pairs in $D \times \{1, \ldots, n\} \times \{0, \ldots, \log n\}$ to short distinct binary ssDNA words. We assume the input ssDNA encodes the vector $\mathcal{L}^{(0)} = (a_1, \ldots, a_n)$ as a sequence of unit length ssDNA words of the form $E(a_1, 1, t) \ldots E(a_n, n, t)$. This ssDNA (Figure 4a) initially positioned as a straight line (again, our assembly may be facilitated by the use of rigid framing techniques described in subsection 4.2).

The tiles are rectangular in shape of size $2^t \times 1$, for stages $t = 1, \ldots, \log n$, as given in Figure 4b. The width of the rectangular tiles of step t is 2^t, and the height is 1.

For each $a, a' \in D$ and $t \in \{1, \ldots, \log n\}$, and odd $i \in \{1, \ldots, n^{(t)}\}$ there is a *step* t tile $\tau(a, a', i, t)$ with:

- on the lower side of the tile, they have two consecutive pads: the $input_{odd}$ pad $\bar{E}(a, i, t-1)$ followed by the $input_{even}$ pad $\bar{E}(a, i+1, t-1)$ (with the intent to match these pads with values a, a' at consecutive positions $i, i+1$ of $\mathcal{L}^{(t-1)}$ computed in a previous stage $t-1$), and

- *ouput* pad $E(a'', (i+1)/2, t)$ on the upper side of the rectanglar tile, where $a'' = a \cdot a' \in D$ (with the intent to match this pad with the composition of the values at positions $i, i+1$ and provide this resulting value a'' to the $(i+1)/2$th position of $\mathcal{L}^{(t)}$ on the subsequent step).

We have already noted that rigid square tiles have been nano-fabricated in DNA, and by composing these these, we can construct rigid rectangular square tiles.

We use step-wise assembly, adding all step t tiles at times $t = 1, \ldots, \log n$. The monoid sum problem $\mathcal{L}^{(0)} = (a_1, \ldots, a_n)$ has intended output $b_n = a_1 \cdot \cdots \cdot a_n$. Let $a_{i,j}$ denote $a_i \cdot a_{i+1} \cdots \cdot a_j$. The resulting assembly (see Figures 4c,d,e) is a polyhedral tiling of size $O(n)$ and depth $O(\log n)$.

An inductive argument shows that on the tth step of assembly, the unmatched bases of the rectanglar tiles involved in the assembly at step t form a unique (note that the chain is uniquely determined since the pairing due to pointer jumping is between only odd and even consecutive elements) length $n^{(t)} \leq n/2^t$ chain of output pads encoding the exactly the vector $\mathcal{L}^{(t)} = (a_1^{(t)}, a_2^{(t)}, \ldots, a_{n^{(t)}}^{(t)})$ derived from from the input vector

$\mathcal{L}^{(0)} = (a_1, a_2, \ldots, a_{n-1}, a_n)$ after t stages of parallel pointer jumping. This clearly holds on the 1st stage, and remains true after each stage.

The step t rectanglar tiles will form a sequence:
$\tau(a_1^{(t-1)}, a_2^{(t-1)}, 1, t), \tau(a_3^{(t-1)}, a_4^{(t-1)}, 3, t), \ldots,$
$\tau(a_{n^{(t-1)}}^{(t-1)}, a_{n^{(t-1)}}^{(t-1)}, n^{(t)}, t)$. The ith rectangle
$\tau(a_i^{(t-1)}, a_{i+1}^{(t-1)}, i, t)$, of this sequence has

- input$_{odd}$ pad $\bar{E}(a_i^{(t-1)}, i, t-1)$ which uniquely matches the ith element of $\mathcal{L}^{(t-1)}$,
- input$_{even}$ pad $\bar{E}(a_{i+1}^{(t-1)}, i+1, t-1)$ which uniquely matches the $(i+1)$th element of $\mathcal{L}^{(t-1)}$, and
- output pad $E(a_{i,(i+1)/2}^{(t)}, (i+1)/2, t)$, where $a_i^{(t-1)} \cdot a_{i+1}^{(t-1)} = a_{(i+1)/2}^{(t)}$ which provides the $i(i+1)/2$th element of $\mathcal{L}^{(t)}$.

The location of each of these tiles do not interfere with the placement of any other tiles at this step t, so the step-wise assembly insures no blockages.

On each stage, the number of elements of the vector $\mathcal{L}^{(t)}$ decrease by a factor of 1/2, so $n^{(t)} \leq n^{(t-1)}/2$. Since $n^{(\log n)} = 1$, on the final stage $T = \log n$ there is only a single step T tile which is placed in the assembly and its output pad at its upper side is unmatched, with encoded output $E(a_1 \cdot a_2 \cdots \cdot a_n, 1)$ giving the solution $b_n = a_1 \cdot a_2 \cdots \cdot a_n$ of the monoid sum problem. We now extend this $O(\log n)$ assembly depth solution of the monoid sum problem to solve prefix computation in $O(\log n)$ assembly depth. (Note that this assembly can be simplified somewhat by letting the index i of each tile $\tau(a_1, a_2, i, t)$ be taken $\bmod 2$, if the input and output values do not require indexing.)

3D Tiling for Prefix Computation. To solve the prefix computation problem $(a_1, \ldots, a_n)$ in $O(\log n)$ steps we propose an additional linear size assembly by LP-BMC. We will use self-assembling $3D$ polyhedra tiles to do reverse propagation of intermediate values computed by the rectanglular tiles in the earlier monoid sum assembly process. Recall the prefix computation problem $(a_1, \ldots, a_n)$ has intended output $(b_1, \ldots, b_n)$, where $b_i = a_1 \cdot \cdots \cdot a_i$. Also let $b_0 = \lambda$.

Each $3D$ polyhedron forming a tile in this case is given in Figure 5a, consisting of two horizontal rectangles (each with the same shape of the previously defined $2^t \times 1$ rectangular tiles for parallel monoid sum: with a base of length 2^t and sides of length 1) connected by unit length line segments between each of their corresponding vertices.

To allow matching of the polyhedra tiles defined here with the sides of the rectangular tiles for parallel monoid sum, we can simply install additional pads on the sides of those original rectangular tiles. Alternatively, we can assume a known methodology for 3-way pairing of DNA strands, as discussed in subsection 4.2, obtained by restricting the encoding of DNA strands to two bases. We let $\bar{E}(-), \widehat{E}(-)$ denote the (possibly 3-way) Watson-Crick complementation of these encodings.

In this phase of the assembly, it is convenient to have the step numbers decrease from $\log n$ to 1. For each $a, a', b \in D$, and $t \in \{1, \ldots, \log n\}$ (where if $t = \log n$ then $b = \lambda$), and odd $i \in \{1, \ldots, n^{(t)}\}$ there is a *step* t polyhedron tile $\tau^+(a, a', b, i, t)$ with:

- on the lower rectangle (which is intended to match with an already placed rectanglar tile $\tau(a, a', i, t)$) there are:
  - *monoid sum input*$_{odd}$, *input*$_{even}$ pads $\widehat{E}(a, i, t-1)$, $\widehat{E}(a, i+1, t-1)$ on the lower side of the lower rectangle, (with the intent to match onto the previously placed rectanglar tile $\tau(a, a', i, t)$ with values a, a' at consecutive locations $i, i+1$ of $\mathcal{L}^{(t-1)}$ computed in a previous stage $t-1$), and
  - *monoid sum ouput* pad $E(a'', (i+1)/2, t)$ on the upper side of the lower rectangle, where $a'' = a \cdot a' \in D$ (with the intent to match with the composition of the values at locations $i, i+1$ given by the upper side pad of the previously placed rectanglar tile $\tau(a, a', i, t)$).
- on the upper rectangle (which is intended to propagate the partially computed prefix sums backward) there are:
  - *propagate-in* pad $\bar{E}'(b, (i+1)/2, t)$ on the upper side of the upper rectangle,
  - on the lower side of the upper rectangle, two consecutive pads: *propagate-out*$_{odd}$ pad $\widehat{E}'(b, i, t-1)$ (with the intent to match with b and propagate it as the prefix-sum to the ith tile of step $t-1$), followed by *propagate-out*$_{even}$ pad $\widehat{E}'(b \cdot a, i+1, t-1)$ (with the intent to match with $b \cdot a$ and propagate it as the prefix-sum to the to the $(i+1)$th tile of step $t-1$).

After completion of the step-wise assembly of the rectangular tiles for parallel monoid sum, as previously described above, we next use a step-wise assembly, adding all step t polyhedron tiles in reverse order $t = \log n, \log n - 1, \ldots, 1$. The prefix computation problem $(a_1, \ldots, a_n)$ has intended output $a_1 \cdot \cdots \cdot a_n$. The resulting $3D$ assembly (see Figures 5b,c,d) is a polyhedral tiling of size $O(n)$ and depth $O(\log n)$.

To aid us in the analysis of our assembly, observe that if the result of t stages of parallel pointer jumping is the vector $\mathcal{L}^{(t)} = (a_1^{(t)}, a_2^{(t)}, \ldots, a_{n^{(t)}}^{(t)})$, then the solution of the prefix problem for $\mathcal{L}^{(t)}$ is given by $(b_1^{(t)}, b_2^{(t)}, \ldots, b_{n^{(t)}}^{(t)})$, with $b_0^{(t)} = \lambda$, where for each odd $i \in \{1, \ldots, n^{(t)}\}$, and $t = 2, \ldots, \log n$ we have $b_i^{(t-1)} = b_{(i-1)/2}^{(t)}$ and

$b_{i+1}^{(t-1)} = a_i^{(t-1)} \cdot b_{(i-1)/2}^{(t)}$. Hence each $b_i = b_{i,1}$ can be computed by these recurrence equations; in fact we will determine them by matching pads of the assembly of tiles.

The step t polyhedral tiles will form a sequence:
$\tau^+(a_1^{(t-1)}, a_2^{(t-1)}, b_0^{(t)}, 1, t), \tau^+(a_3^{(t-1)}, a_4^{(t-1)}, b_1^{(t)}, 1, t), \ldots,$
$\tau^+(a_{n^{(t-1)}}^{(t-1)}, a_{n^{(t-1)}, b^{(t)}_{n^{(t)}-1}}^{(t-1)}, n^{(t)}, t)$. By induction we have that the ith tile
$\tau^+(a_i^{(t-1)}, a_{i+1}^{(t-1)}, b_{(i-1)/2}^{(t)}, i, t)$ has

- a lower rectangle with:
  - monoid sum input$_{odd}$, input$_{even}$ pads $\widehat{E}(a_i^{(t-1)}, i, t-1)$,$\widehat{E}(a_{i+1}^{(t-1)}, i+1, t-1)$ which match the (odd, even) i,$(i+1)$th elements of $\mathcal{L}^{(t-1)}$, and
  - monoid sum ouput pad $E(a_{i,(i+1)/2}^{(t)}, (i+1)/2, t)$, since $a_i^{(t-1)} \cdot a_{i+1}^{(t-1)} = a_{(i+1)/2}^{(t)}$ which provides the $(i+1)/2$th element of $\mathcal{L}^{(t)}$.
- and an upper rectangle with:
  - propagate-in pad $\bar{E}'(b_{(i-1)/2}^{(t)}, (i+1)/2, t)$ on the upper side, and
  - on the lower side: propagate-out$_{odd}$ pad $\widehat{E}'(b_{(i-1)/2}^{(t)}, i, t-1)$ $= \widehat{E}'(b_{i-1}^{(t-1)}, i, t-1)$, since $b_{(i-1)/2}^{(t)} = b_{i-1}^{(t-1)}$, followed by propagate-out$_{even}$ pad $\widehat{E}'(b_{(i-1)/2}^{(t-1)} \cdot a_i^{(t-1)}, i, t-1) = \widehat{E}'(b_i^{(t-1)}, i, t-1)$ on the left side, on the top right side, since $b_{(i-1)/2}^{(t-1)} \cdot a_i^{(t-1)} = b_i^{(t-1)}$.

This clearly holds on the 1st stage, and remains true after each stage.

The output to the prefix computation problem is provided by the upper rectangle of the stage 1 polyhedral tiles, that give, for odd i, on their (left side) propagate-out$_{odd}$ pad $\widehat{E}'(b_{i-1}^{(t-1)}, i, t-1)$ encoding value $b_{i-1} = b_{i-1,1}$ and on their (left side) propagate-out$_{even}$ pad $\widehat{E}'(b_i^{(t-1)}, i, t-1)$ encoding value $b_i = b_{i,1}$. So, the polyhedral assembly provides $(b_1, \ldots, b_{n-1})$. Recall that we already have the monoid sum value b_n given by the base pad of the last (step $\log n$) previously placed rectanglular tiles. Thus, the assemblies together provide the entire solution $(b_1, \ldots, b_n)$ of the prefix computation problem. Again, this output can be constructed as a ssDNA in $O(1)$ BMC steps by a variety of known techniques given in subsection 4.2.

8. Emulating Shuffle-Exchange Networks

8.1. Pair-wise Exchange using Unmediated Assembly. Given an even length n vector $(a_1, a_2, \ldots, a_{n-1}, a_n)$ of elements from any domain D, the *pair-wise exchange* problem is to form a vector $(a_2, a_1, \ldots, a_n, a_{n-1})$, that is exchange every odd-even pair.

We can use nearly the same assembly construction as used for sequential prefix in Section 5 to solve this problem, but in this case we do not require control of the assembly.

We assume distinct DNA encoding functions E, E' and again assume the input ssDNA encodes the vector $(a_1, \ldots, a_n)$ as a sequence of unit length ssDNA words of the form $E(a_1, 1), \ldots, E(a_n, n)$. This input ssDNA is positioned into a straight line of length n (see again Figure 2a), by the use of rigid framing techniques described described in subsection 4.2. Each tile is a 2×2 square as given in Figure 6a. For each $a, a' \in D$ and $i \in \{1, \ldots, n\}$ where i is odd, there is a tile $\tau'(a, a', i)$ with:

- the *input* pad $\bar{E}(a,i), \bar{E}(a',i+1)$ on the bottom side (with the intent to match the i,$(i+1)$th pair of inputs),
- *output* pads $E'(a',i), E'(a,i+1)$ on the top side (with the intent to output the $i,(i+1)$th pair of outputs in reverse order).

We use unmediated assembly, adding all tiles at the same time. For each $i \in \{1,\ldots,n\}$ where i is odd, the tile $\tau'(a_i,a_{i+1},i)$ is the unique tile that can be placed so its bottom side abuts and its pad matches the $i,(i+1)$th pair of elements $E(a_i,i), E(a_{i+1},i+1)$ of the input. Its location and pad positions do not interfere with any other tiles at this time, so there can be no blockages, and the tiling is unique.

The resulting $2D$ assembly is unique and is in the shape of a rectangular tiling (see Figure 6b) of length n and height 2 containing tiles

$$\tau'(a_1,a_2,1), \tau'(a_3,a_4,3), \ldots, \tau'(a_{n-1},a_n,n-1)$$

in this order, with the encoded input matched to the input pads on the bottom sides of these tiles and the output pads on the bottom sides of the tiles giving the required encoded output sequence
$\bar{E}(a_2,1), \bar{E}(a_1,2), \bar{E}(a_4,3), \bar{E}(a_3,4),$
$\ldots, \bar{E}(a_n,n-1), \bar{E}(a_{n-1},n)$. Again, this output can be constructed as a ssDNA in $O(1)$ BMC steps by a variety of well known techniques; see subsection 4.2. (Note that this tiling assembly construction can be simplified somewhat by taking the time stamps $\bmod 2$, but the integer time stamps will in general be needed if we apply repeated pair-wise exchange and perfect shuffle operations.)

8.2. Perfect Shuffle Operations. Given an even length n vector $(a_1,a_2, \ldots,a_{n-1},a_n)$ of elements from any domain D, the *perfect shuffle problem* is to form a vector $(a_1,a_{n/2+1}, a_2, a_{n/2+2}, \ldots,a_{n/2},a_n)$, i.e., form a vector that shuffles the first $n/2$ elements with the last $n/2$ elements.

To solve the Perfect Shuffle problem we propose a linear size assembly, which will be unmediated and will not require multiple steps in the assembly.

We assume distinct DNA encoding functions E, E' and again assume the input ssDNA encodes the vector $(a_1,\ldots,a_n)$ as a sequence of ssDNA words of the form

$$E(a_1,1),\ldots E(a_n,n).$$

Without loss of generality, let position 1 be the $5'$ end of the input ssDNA, position n be the $3'$ end.

For technical reasons, we need to reverse the order of the encoding of the latter half of the input. By known recombinant DNA techniques we can reverse the latter part of the ssDNA containing the latter $n/2$ elements $E(a_{n/2+1},n/2+1)\ldots E(a_n,n)$, by using the Watson-Crick complement, yielding a ssDNA with a sequence of unit length ssDNA words of the form

$$E(a_1,1)\ldots E(a_{n/2},n/2)\bar{E}(a_n,n)^R \ldots \bar{E}(a_{n/2+1},n/2+1)^R,$$

where the superscript R denotes string reversal [§].

[§] We can do the reversal (with refinements suggested by [**W97**]) as follows:

- Ligate a hairpin region H to the 3' end.
- Use polymerase to extend the hairpin, and denatured, this yields $E(a_1,1),\ldots E(a_n,n)$ H $\bar{E}(a_n,n)^R \ldots$ $\bar{E}(a_1,1)^R$.

By known recombinant DNA techniques (via circularization and appropriate use of restriction enzymes), ssDNA is then edited so as to increase its length to $n+2$, by insertion of a dummy (distinct from all the other strings used in DNA encodings) ssDNA just after the $(n/2)$th element. The resulting ssDNA can be positioned in $2D$ into the shape of a rectangle (the dummy ssDNA was inserted so it forms the right side of this rectangle) with a missing left side (see Figure 7a) of width $n/2$ and height 2 by the use of rigid framing techniques described in subsection 4.2 (see Figure 7b).

Each tiles in this case is a rectangle of width 1 and height 2 with an internal middle segment. The internal middle segment consists of two consecutive straight segments each of length 1 forming a (raised upsidedown) V, and bridging between the left and right sides of the tile, as given in Figure 7c. For each $a, a' \in D$ and $i \in \{1, \ldots, n/2\}$, there is a tile $\tau''(a, a', i)$ with:

- a 1×2 rectangle with
  - the *input*$_{low}$ pad $\bar{E}(a, i)$ on the bottom side (with the intent to match the ith input a_i) and
  - the *input*$_{high}$ pad $E(a', n/2+i)^R$ on the top side (with the intent to match the $(n/2+i)$th input $a_{n/2+i}$), and
- an internal middle section, bridging between the left and right sides, with an *output* pad
  $\bar{E}'(a,i)\bar{E}'(a,n/2+i)$ (with the intent to provide the outputs).

We will use unmediated assembly, adding all tiles at the same time. For each $i \in \{1, \ldots, n/2\}$, the tile $\tau''(a_i, a_{n/2+i}, i)$ is the unique tile that can be placed so the bottom side abuts (and its input$_{low}$ pad matches) the ith element $E(a_i, i)$ of the input and the top side abuts (and its input$_{high}$ pad matches) the $(n/2+i)$th element $E(a_{n/2+i}, n/2+i)$ of the input. The location of this tile does not interfere with the placement of any other tiles. The resulting $2D$ assembly (see Figure 7d) is unique and is a rectangle of length n and height 2, containing tiles $\tau''(a_1, a_{n/2}, 1), \tau''(a_2, a_{n/2+1}, 2), \ldots,$

$\tau''(a_{n/2}, a_n, n/2)$ in this order, with the inputs matched to pads on the bottom and top sides of these tiles and the output pads of the middle segments of the tiles giving the required encoded output sequence $\bar{E}'(a_1, 1)$,

$\bar{E}'(a_{n/2}, 2)$,$\bar{E}'(a_2, 3)$, $\bar{E}'(a_{n/2+1}, 4)$,$\ldots$, $\bar{E}'(a_{n/2}, n-1)$, $\bar{E}'(a_n, n)$. This output can be constructed as a ssDNA in $O(1)$ BMC steps by a variety of known techniques; see subsection 4.2. For example, we may insert some matching ssDNA to the ends of the middle segments, ligate the consecutive middle segments to form a single ssDNA, then cut out, by restriction enzymes, all but this middle segment ssDNA containing the output pads. (Note that again this tiling assembly construction can be simplified somewhat by taking the time stamps $\bmod 2$, but the integer

---

- Cut between $\bar{E}(a_{n/2+1}, n/2+1)^R$ $\bar{E}(a_{n/2}, n/2)^R$, using an excess of complementary oligos to form the dsDNA target site for the restriction enzyme, yeilding $E(a_1, 1), \ldots E(a_n, n)$ H $\bar{E}(a_n, n)^R \ldots$ $\bar{E}(a_{n/2+1}, n/2+1)^R$.
- Circularize and then cut between $E(a_{n/2}, n/2)$ $E(a_{n/2+1}, n/2+1)$ and between H $\bar{E}(a_n, n)^R$, so the longer DNA is $\bar{E}(a_n, n)^R \ldots$ $\bar{E}(a_{n/2+1}, n/2+1)^R$ $E(a_1, 1), \ldots E(a_{n/2}, n/2)$.
- Circularizing again, and cutting between $E(a_{n/2+1}, n/2+1)^R$ $E(a_1, 1)$, we get $E(a_1, 1) \ldots E(a_{n/2}, n/2)$ $\bar{E}(a_n, n)^R \ldots$ $\bar{E}(a_{n/2+1}, n/2+1)^R$, as required.

time stamps will in general be needed if we repeatedly apply pair-wise exchange operations in combination with these perfect shuffle operations.) To further facilitate readout, each tile can be alternatively defined to be $3D$ polyhedron in the shape given in Figure 7e, resembling pup tent, with the same rectangular base as the 1×2 tiles described above, and a top segment with an output pad consisting of two consecutive straight segments each of length 1 forming a V, and bridging between the left and right sides of the pup tent. These $3D$ polyhedron tiles might be fabricated the use of three-axis multiple DX molecules i.e., a DX with one additional double helix axis stacked on top.

8.3. Shuffle-Exchange Networks. A *shuffle-exchange network* (Stone [**S71**]) is a graph with node set $\{1, \ldots, n\}$ and with a set of edges that effect both

1. a perfect shuffle permutation on the nodes, and
2. a pair-wise exchange operation on the nodes.

See Ullman [**U84**] and Leighton [**L92**] for discussion of the the use of shuffle-exchange network for emulations of other networks such as the butterfly and CCC networks and see Schawbe [**S90**] for a proof of the computational equivalence of hypercube-derived networks, such as the butterfly network.

In subsections 8.1, 8.2 we have given constant time unmediated LP-BMC algorithms for executing perfect shuffle and pair-wise exchange on length n vectors encoded into DNA. By simple edits (insertion of short dummy sequences to and reversal of subsequences) to the DNA as described in these subsections, the output encoding (say from a shuffle-exchange step) is made compatible with the input encoding required for the following step (say a pairwise exchange). By the definition of a shuffle-exchange network, it follows that these LP-BMC algorithms can be used to emulate a shuffle-exchange network, in constant time. Furthermore, we can combine each shuffle-exchange with a computational step on consecutive odd-even pairs of data, using one step of the step-wise tiling assembly technique described in Section 4.

8.4. Normal Parallel Algorithms. *Normal parallel algorithms* are a well-known class of parallel algorithms; for details see Ullman [**U84**] and Leighton [**L92**] (note they are also sometimes called *ascend-descend* algorithms). A normal parallel algorithms can be executed (see details in [**S71, U84, L92**]) on an n-vector of data using n processors executing in parallel in $O(\log n)$ stages. Each stage consists of the following parallel operations:

- a shuffle-exchange permutation and
- parallel execution of a computational step (costing each processor $O(1)$ work) on each of the $n/2$ consecutive odd-even pairs of data.

We can do each of these stages using LP-BMC within constant time, as follows:

- The techniques of Subsection 8 executes the shuffle-exchange permutation by LP-BMC in constant time.
- The parallel execution of an $O(1)$ work computational step on $n/2$ consecutive odd-even pairs of data, can be done by LP-BMC in constant time by emulation of one time step of an n processor $1D$ array, using a step-wise tiling assembly as given in Subsection 4.1. In this assembly, which will have constant depth, we use again square tiles, each with two input pads which match to consecutive odd-even pairs of data, and two output pads providing

the these computed values to the next stage (we can apply the constant time readout methods of 4.2, to provide these computed values to the next stage).

Well known example applications of normal parallel algorithms are

- the $O(\log n)$ time, n processor bitonic merge algorithm of Batcher [**B68**], which merges two length n vectors using a normal parallel algorithm and
- the $O(\log^2 n)$ time, n processor [**B68**] bitonic sort algorithm, which sorts a vector of length n using $O(\log n)$ normal algorithm passes.

Assuming that the elements to be operated on are integers of b bits, we can use 2^{2b} tiles to effect a key comparison in one step of tiling. Thus, the bitonic merge and sort algorithms can be executed by our LP-BMC methods with these same time bounds and using $O(n)$ size assemblies.

8.5. Executing General Permutations and Parallel Evaluation of Circuits. A known parallel algorithm, known as the Beneš network [**B65**], allows the execution of an (arbitrary) fixed permutation of n data elements by use of a normal parallel algorithm (Waksman [**W68**]). This requires $O(\log n)$ LP-BMC steps by our previous results of Subsection 8.4 for normal parallel algorithms.

This also implies LP-BMC can do parallel evaluation of a fixed, bounded degree Boolean circuit of size n and of depth d. The LP-BMC assembly will have size $O(n)$ and depth $O(d \log n)$. To do this, we evaluate the circuit in stages $t = 1, \ldots, d$ where on stage t we:

- apply, in $O(\log n)$ LP-BMC steps, the Beneš technique (precomputed for the permutations of each level of the fixed circuit) to move the data required for inputs to each node of the circuit of level t, and
- use the techniques of subsection 4.1 to do by LP-BMC in constant time, the parallel Boolean operations required to evaluate each level t node of the circuit (again, this is done by emulation of an n processor $1D$ array, using a constant number of steps of our step-wise $2D$ assembly given in Subsection 4.1), and
- after assembly, we apply the constant time methods of subsection 4.2, to provide these node evaluation values to the next level of the circuit.

9. Conclusion

We summarize the advantages of our step-wise techniques for LP-BMC:

- it requires no thermal cycling,
- results in no blockages, and
- it provides greater flexibility and control over the chemical conditions under which assembly occurs, as compaired to unmediated self-assembly, and
- whereas the kinetic models of Winfree [**W97**] suggest that unmediated self-assembly should proceed as slowly as possible, in step-wise self assembly, we may be able to speed up considerable by washing over high concentrations of tiles.

There are a number of insights provided by this paper that are new: (1) different geometrical sizes and shapes of tiles can have a very significant effect upon computational speed, (2) DNA folding over on itself using rigid frames can allow communication between memory bits in long strands of DNA, thus allowing fast circuit simulation, and (3) stepwise methods in which the assembly is partially deconstructed after completion (e.g. the rigid frame and tiles are removed, keeping only the output strand) allow for considerably greater flexibility in the computation.

Our main constructive results provided assemblies which are compact ($O(n)$ size) with small $O(\log n)$ depth for a number of fundamental problems, including prefix computation, integer addition, list merging, fixed permutations, and many other normal parallel algorithms. This small depth and size may decrease probability of errors in the assembly process.

Acknowledgments. N. C. Seeman gave us many very valuable and insightful suggestions on our assemblies and nano-fabrication techniques in general, particularly with respect to the geometry and DNA nano-construction of our tiles and assembly frame, and also informed us of the useful biochemistry, e.g. methods for 3-way pairing of DNA strands. E. Winfree gave helpful suggestions on many aspects of this paper, for which we are very grateful, including our lower bound results for unmediated assembly, the use of Boltzman energy functions, time stamping modulo 2, methods for fabrication of $3D$ polyhedron tiles via multiple DX molecules, a refinement of our reversal method for segments of DNA strands, and a further enumerating the potential advantages of our step-wise techniques. Also, T. LaBean and A. Gehani provided useful suggestions for improvements to this paper. K. Redding provided substantial aid in preparation of the paper. Thanks to my father A. Reif, a biochemist who in my early years provided access to his lab, thus providing me with an introduction to basic biological lab techniques.

References

[A94] L. Adleman, *Molecular Computation of Solution to Combinatorial Problems*, Science, **266**, 1021–1024, (1994).

[A95] L. Adleman, *On Constructing a Molecular Computer*, U.S.C. Dept of CS Draft (1995). Available via anonymous ftp from *ftp.usc.edu/pub/csinfo /papers /adleman /molecular_computer.ps*

[ARRW96] L. M. Adleman, P. W.K. Rothemund, S. Roweis, E. Winfree, *On Applying Molecular Computation To The Data Encryption Standard*, 2nd Annual Meeting on DNA Based Computers, Princeton University, June 1996.

[AJPWL+96] A. P. Alivisatos, K. P. Johnsson, X. Peng, T. E. Wilson, C. J. Loweth, M. P. Bruchez Jr., P. G. Schultz, *Organization of 'nanocrystal molecules' using DNA*, Nature, bf 382, 609–611, August 1996.

[A96] J.-T. Amenyo, *Mesoscopic computer engineering: Automating DNA-based molecular computing via traditional practices of parallel computer architecture design*, Proceedings of the Second Annual Meeting on DNA Based Computers, June 1996.

[AGH96] M. Amos, A. Gibbons, and D. Hodgson, *Error-resistant Implementation of DNA Computations*, 2nd Annual Meeting on DNA Based Computers, Princeton University, June 1996.

[AH97] M. Arita and M. Hagiya, *Joining and Rotating Data with Molecules*, ICEC'97 Special Session on DNA Based Computation, Indiana, April 1997.

[BCGT96] E. Bach, A. Condon, E. Glaser, and C. Tanguay, *Improved Models and Algorithms for DNA Computation*, Proc. 11th Annual IEEE Conference on Computational Complexity, Submitted (by invitation) to: J. Computer and System Sciences, May 1996.

[B68] K. Batcher *Sorting Networks and their applications*, Spring Joint Computer Conference, **32**, 307–314, AFIPS Press, Montvale, N. J. (1968).

[B95] E. B. Baum, *How to build an associative memory vastly larger than the brain*, Science, April 28, 1995.

[B96] E. B. Baum, *DNA Sequences Useful for Computation*, 2nd Annual Meeting on DNA Based Computers, Princeton University, June 1996.

[Be94] D. Beaver, *Factoring: The DNA Solution*, Advances in Cryptology, Asia Crypt94 Proceedings, Lecure Notes in Computer Science, (1994), Springer Verlag,

[BeA95] D. Beaver, *A Universal Molecular Computer*, revised as *Molecular Computing*, Penn State University Technical Memo CSE-95-001, Pond Lab, (1995).

[BeB95] D. Beaver, *Computing with DNA*, J. Comp. Biol., **2**, 1–7, (1995).

[B65] V. Beneš, *Mathematical Theory of Connecting Networks and Telephone Traffic*, Academic Press, New York, NY (1965).

[B66] R. Berger *The Undecidability of the Domino Problem*, Memoirs of the American Mathematical Society, **66**, (1966).

[BDL95] D. Boneh, C. Dunworth, R. Lipton, *Breaking DES Using a Molecular Computer*, Princeton CS Tech-Report number CS-TR-489-95, (1995).

[BDLS95] D. Boneh, C. Dunworth, R. Lipton, J. Sgall, *On the Computational Power of DNA*, Princeton CS Tech-Report number CS-TR-499-95, (1995). Also published in Discrete Applied Math, (Dec 96)

[BL95] D. Boneh, R. Lipton, *Making DNA Computers Error Resistant*, Princeton CS Tech-Report CS-TR-491-95, Also in 2nd Annual Meeting on DNA Based Computers, Princeton University, June 1996.

[BKP90] C. Brooks, M. Karplus, M. Pettitt, *Proteins, A Theoretical Perspective of Dynamics, Structure & Thermodynamics*, John Wiley & Sons,

[B62] J.R Buchi, *Turing Machines and the Entscheidungsproblem*, Mathematiche Annalen, **148**, 201–213, (1962).

[CCCFF+97] W. Cai, A. Condon, R. M. Corn, Z. Fei, T. Frutos, E. Glaser, Z. Guo, M. G. Lagally, , Q. Liu, L. M. Smith, and A. Thiel, *The Power of Surface-Based Computation*, Proc. First International Conference on Computational Molecular Biology (RECOMB97), January 1997.

[CRFCC+96] W. Cai, E. Rudkevich, Z. Fei, A. Condon, R. Corn, L.M. Smith, M. G. Lagally, *Influence of Surface Morphology in Surface-Based DNA Computing*, Submitted to the 43rd AVS National Symposium, Abstract No. BI+MM-MoM10, (1996).

[CCTKS88] J.-H. Chen, M. E. A. Churchill, T. D. Tullius, N. R. Kallenbach, N. C. Seeman, *Construction and Analysis of Monomobile DNA Junctions*, Biochemistry, **27**, (1988)

[CRWEC95] J. Chen, C. A. Rauch, J. H. White, P. T. Englund, N. R. Cozzarelli, em The Topology of the Kinetoplast DNA Network, rm Cell, **80**, 61–69, January 1995

[CL92] B. Crandall and J. Lewis (eds.), *Nanotechnology*, MIT Press, (1992).

[DMGFS96] R. Deaton, R.C. Murphy, M. Garzon, D.R. Franceschetti, and S.E. Stevens, Jr., *Good encodings for DNA-based solutions to combinatorial problems*, Proceedings of the Second Annual Meeting on DNA Based Computers, June 1996.

[DMRGF+97] R. Deaton, R. C., Murphy, J. A. Rose, M. Garzon, D. R. Franceschetti, and S. E. Stevens, Jr., *A DNA Based Implementation of an Evolutionary Search for Good Encodings for DNA Computation*, ICEC'97 Special Session on DNA Based Computation, Indiana, April 1997.

[DHK96] A. L. Delcher, L. Hood, R. M. Karp, *Report on the DNA/Biomolecular Computing Workshop*, June 1996.

[DD92] M. D. Distefano, P. B. Dervan *Ligand promoted dimerization of oligonucleotides binding cooperatively to DNA*, J.Am.Chem.Soc., **114**, 11006-11007, (1992).

[DDSPL+ 93] , Drmanac, R, Drmanac, S., Strezoska, Z., Paunesku, T., Labat, I., Zeremski, M., Snoddy, J., Funkhouser, W. K., Koop, B., Hood, L., Crkenjakov, R., *DNA Sequence Determination by Hybridization: A Strategy for Efficient Large-Scale Sequencing*, Science, **260**, 1649–1652, (1993).

[DS92] S.M. Du and N.C. Seeman, *The Synthesis of a DNA Knot Containing both Positive and Negative Nodes*, J. Am. Chem. Soc., **114**, 9652–9655, (1992).

[DZS92] S.M. Du, S. Zhang and N.C. Seeman, *DNA Junctions, Antijunctions and Mesojunctions*, Biochem., **31**, 10955–10963, (1992).

[EC] Eisenberg and Crothers, *Physical Chemistry with applications to the life sciences.*

[ER82] M. Engler and C. Richardson, *The Enzyme*, (P. Boyer, ed.) Academic Press, 3–29, (1982).

[F 61] R. Feynman, Minaturization (D. Gilbert, ed.), Reinhold, 282–296, (1961).

[FW 78] S. Fortune and J. Wyllie, *Parallelism in random access machines*, Proc. 10th Annual ACM S.T.O.C., San Diego, CA, 114–118, (1978).

[FS93] T.-J. Fu and N.C. Seeman, *DNA Double Crossover Structures*, Biochemistry, **32**, 3211–3220, (1993).

[GJ79] M. R. Garey and D. S. Johnson *Computers and Intractability: A Guide to the Theory of NP-Completeness*, W.H Freeman and Company, page 257, (1979).

[GJP77] M. R. Garey, D. S. Johnson, and C. H. Papadimitriou, unpublished manuscript, (1977).

[GFBCL+96] J. M. Gray, T. G. Frutos, A. M. Berman, A. E. Condon, M. G. Lagally, L. M. Smith, R. M. Corn, *Reducing Errors in DNA Computing by Appropriate Word Design*, University of Wisconsin, Department of Chemistry, October 9, 1996.

[GBS87] S. Grunbaum, Branko, and G.C. Shepard *Tilings and Patterns*, H Freeman and Company, **Chapter 11**, (1987).

[GB96] F. Guarnieri, and C. Bancroft, *Use of a Horizontal Chain Reaction for DNA-Based Addition*, Proceedings of the Second Annual Meeting on DNA Based Computers., June 10-12, 1996, American Mathematical Society, Providence, RI (in press), (1996).

[GFB96] Guarnieri, F., Fliss, M., and C. Bancroft, *Making DNA add*, Add. Science, **273**, 220–223, (1996).

[J92] J. JáJá, *An Introduction to Parallel Algorithms*, Addison Wesley, (1992).

[LF80] R.E. Ladner, and M.J. Fischer, *Parallel Prefix Computation*, JACM, **27(4)**:831–838, (1980).

[L92] F. T. Leighton, *Introduction to Parallel Algorithms and Architectures*, Morgan Kaufmann Press, San Mateo, CA, Chapter 3, (1992).

[LP81] H. R. Lewis and C.H. Papadimitriou *Elements of the Theory of Computation*, Prentice-Hall, pages 296–300 and 345–348 (1981).

[JK96] N. Jonoska, S. A. Karl, *A Molecular Computation of the Road Coloring Problem*, 2nd Annual Meeting on DNA Based Computers, Princeton University, June 1996.

[JK97a] N. Jonoska, S. A. Karl, *Ligation Experiments in Computing with DNA*, ICEC'97 Special Session on DNA Based Computation, Indiana, April 1997.

[JK97b] N. Jonoska and S. A. Karl *Creating 3-Dimensional Graph Structures with DNA*, 3nd Annual Meeting on DNA Based Computers, University of Pens., (June 1997).

[LL97] L. F. Landweber and R. Lipton *DNA 2 DNA Computations: A Potential 'Killer App'?*, 3nd Annual Meeting on DNA Based Computers, University of Pens., (June 1997).

[KCL96] P. Kaplan, G. Cecchi, and A. Libchaber, *DNA based molecular computation: Template-template interactions in PCR*, The Second Annual Workshop on DNA Based Computers, American Mathematical Society, To appear, (1996).

[KMRS96a] S. A. Kurtz, S. R. Mahaney, J. S. Royer, J. Simon, *Active Transport in Biological Computing*, 2nd Annual Meeting on DNA Based Computers, Princeton University, June 1996.

[KMRS96b] S. A. Kurtz, S. R. Mahaney, J. S. Royer, J. Simon, *Biological Computing*, to appear, (1996).

[LYQS96] X.J. Li, X.P. Yang, J. Qi, and N.C. Seeman, *Antiparallel DNA Double Crossover Molecules as Components for Nanoconstruction*, J. Am. Chem. Soc., **118**, 6131–6140, (1996).

[L94] R. Lipton, *Speeding Up Computations via Molecular Biology*, Princeton University Draft, (1994).

[L95] R. J. Lipton, *DNA Solution of Hard Computational Problems*, Science, **268**, 542–545, (1995).

[L96] , R. J. Lipton. *DNA Computations Can Have Global Memory*, unpublished manuscript, April 1996.

[LBL 96] R. J. Lipton, D. Boneh, and L. Landweber, *Analog DNA Based Computation*, in preparation. (1996).

[LGCCL+96] Q. Liu, Z. Guo, A.E. Condon, R.M. Corn, M.G. Lagally,, and L.M. Smith, *A Surface-Based Approach to DNA Computation,*, Proc. Second Annual Princeton Meeting on DNA-Based Computing, June 1996.

[MS97] C. Mao and N.C. Seeman, *Construction of Borromean Rings from DNA*, Nature, **386**(6621), 137–138, (March,1997).

[MH87] J. McCammon, S. Harvey, *Dynamics of Proteins and Nucleic Acids*, Cambridge University Press, (1987).

[M93] R. Merkle, Nanotechnology **4** 21, (1993).

[MR85] G. Miller, J.H. Reif, *Parallel Tree Contraction and its Application*, 26th Annual IEEE Symposium on Foundations of Computer Science, Portland, OR, 478–489, October 1985. Also published as *Parallel Tree Contraction Part I: Fundamentals*, Advances in Computing Research, **5**, 47–72, (1989), and *Parallel Tree Contraction Part II: Further Applications*, SIAM Journal on Computing, **20:6**, 1128–1147, (1991).

[M96] K. U. Mir, *A Restricted Genetic Alphabet for DNA Computing*, 2nd Annual Meeting on DNA Based Computers, Princeton University, June 1996.

[MLMS96] C. A. Mirkin, R. L. Letsinger, R. C. Mucic, J. J. Storhoff, *A DNA-based Method for Rationally Assembling Nanoparticles Into Macroscopic Materials*, Nature, **382**, 607–611, August 1996.

[MDS91] J.E. Mueller, S. M. Du and N.C. Seeman, *The Design and Synthesis of a Knot from Single-Stranded DNA*, J. Am. Chem. Soc., **113**, 6306–6308, (1991).

[MDFS97] R. C. Murphy, R. Deaton, D. R. Franceschetti, S. E. Stevens, *A New Algorithm for DNA Based Computation*, ICEC'97 Special Session on DNA Based Computation, Indiana, April 1997.

[OR97] M. Ogihara, A. Ray, *Simulating Boolean circuits on a DNA computer*, 1st Annual International Conference On Computational Molecular Biology (RECOMB97), Santa Fe, New Mexico, January 1997.

[OP94] R. Old and S. Primrose, *Principles of Gene Manipulation, An Introduction to Genetic Engineering*, Blackwell Scientific Publications, Fifth Edition, (1994).

[PP97] G. E. Plum and D. S. Pilch, *Nucleic Acid Hybridization: Triplex Stability and Energetics.*, Annu. Rev. Biophys. Biomol. Struct. , **24:**, 319–350, (February 1997).

[Po97] R. Pool, *Dr. Tinkertoy*, Discover, **18:2**, 50–57, (February 1997).

[QYS96] J. Qi, X. J. Li, X. P. Yang, and N. C. Seeman, *Ligation of triangles built from bulged 3-arm DNA branched junctions*, J. Am. Chem. Soc., **v118:26**, 6121–6130, (July, 1996).

[R93] J. Reif (ed.), *Synthesis of Parallel Algorithms*, Morgan Kaufmann, (1993).

[R95] J. Reif, *Parallel Molecular Computation: Models and Simulations*, Seventh Annual ACM Symposium on Parallel Algorithms and Architectures (SPAA95), ACM, Santa Barbara, 213–223, June 1995. Also accepted and to appear in *Algorithmica*, special issue on Computational Biology, 1998. Postscript versions of this paper and its figures are at http://www.cs.duke.edu/~reif/paper/paper.html.

[R97] J. Reif, *Paradigms for Biomolecular Computation*, First International Conference on Unconventional Models of Computation, Auckland, New Zealand, January 1998. Postscript versions of this paper is at *http : //www.cs.duke.edu/r̃eif/paper/paradigm.ps.*

[R 78] R. J. Roberts, *Restriction and modification enzymes and their recognition sequences*, Gene, **4**, 183–194, (1978).

[R71] R. M. Robinson *Undecidablility and Nonperiodicity for Tilings of the Plane*, Inventiones Mathematicae, **12**, 177–209, (1971).

[Ro95] P.W.K. Rothemund, *A DNA and restriction enzyme implementation of Turing Machines*, manuscript available at http://www.ugcs.caltech.edu/ pwkr/oett.html, (1995).

[RW95] D. Rooβ and K.W. Wagner, *On the power of Bio-Computers*, unpublished manuscript.

[RWBCG+96] S. Roweis, E. Winfree, R. Burgoyne, N. V. Chelyapov, M. F. Goodman, P. W.K. Rothemund, L. M. Adleman, *A Sticker Based Architecture for DNA Computation*, 2nd Annual Meeting on DNA Based Computers, Princeton University, June1996, Also as Laboratory for Molecular Science, Univ. of So. California technical report *A Sticker Based Model for DNA Computation*, May 1996.

[RS97] G. Rozenberg, A. Salomaa, ICEC'97 Special Session on DNA Based Computation, Indiana, April 1997.

[R96] H. Rubin, *Looking for the DNA killer app.*, Nature, **3**, 656–658, (1996).

[SFM89] J. Sambrook, E. Fritsch, T. Maniatis, *Molecular Cloning*, Cold Spring Harbor Lab, NY, (1989).

[S90] E. Schawbe *On the computational equivalence of hypercube-drived networks*, Proceeding of the 2nd Annual Symposium on Parallel Algorithms and Architectures, 388–397 (1990).

[S82] N.C. Seeman, *Nucleic Acid Junctions and Lattices*, J. Theor. Biol., **99**, 237-247, (1982).

[S97] N.C. Seeman, *personal communication to J.H. Reif*, (1997).

[S85] N.C. Seeman, *Macromolecular Design, Nucleic Acid Junctions and Crystal Formation*, Journal of Biomolecular Structure and Dynamics, **3**, 1-34, (1985).

[SC91] N. C. Seeman and J. Chen, *Synthesis from DNA of a molecule with the connectivity of a cube*, Nature, **350**, 631–633, (1991).

[SCDMZ+93] N. C. Seeman, J. Chen, S. M. Du, John E. Mueller, Yuwen Zhang, Tsu-Ju Fu, Yinli Wang, Hui Wang, Siwei Zhang, *Synthetic DNA knots and catenanes*, New Jour. of Chemistry, **17**, 739–755, (1993).

[SCK89] N. C. Seeman, J.-H. Chen, N. R. Kallenbach, *Gel electrophoretic analysis of DNA branched junctions*, Electrophoresis, **10**, 345–354, (1989).

[SQLYL+96] N. C. Seeman, J. Qi, X. Li, X. Yang, N. B. Leontis, B. Liu, Y. Zhang, S. M. Du, and J. Chen, *The control of DNA structure: From topological modules to geometrical modules*, Modular Chemistry, J. Michl, ed., Kluwer, To appear, (1996).

[SWLQ+96] N. C. Seeman, H. Wang, B. Liu, J. Qi, X. Li, X. Yang, F. Liu, W. Sun, Z. Shen, R. Sha, C. Mao, Y. Wang, S. Zhang, T.-J. Fu, S. Du, J. E. Mueller, Y. Zhang, and J. Chen, *The Perils of Polynucleotides: The Experimental Gap Between the Design and Assembly of Unusual DNA Structures*, The Second Annual Workshop on DNA Based Computers, American Mathematical Society, June 1996.

[SZC94] N. C. Seeman, Y. Zhang, and J. Chen, *DNA nanoconstructions*, J. Vac. Sci. Technol., **12:4**, 1895–1905, (1994).

[SZDC95] N. C. Seeman, Y. Zhang, S. M. Du, and J. Chen, *Construction of DNA polyhedra and knots through symmetry minimization*, Supramolecular Stereochemistry, J. S. Siegel, ed., 27–32, (1995).

[SZDWM+94] N. C. Seeman, Y. Zhang, S. Du, H. Wang, J. E. Mueller, and J. Chen, *The control of DNA structure and topology: An overview*, Mat. Res. Soc. Symp. Proc., **356**, 57–66, (1994).

[SM97] J. Setubal and J. Meidanis *Introduction to Computational Molecular Biology*, PWS Pub. Co., Chapt 9, (1997).

[S94] R. Sinden, *DNA Structure and Function*, Academic Press, (1994).

[S88] L. M. Smith, *Automated Synthesis and Sequence Analysis of Biological Macromolecules*, Anal. Chem., **60**, 381A–390A, (1988).

[SS95] W. Smith, A. Schweitzer, *DNA Computers in Vitro and Vivo*, NEC Research Inst. Tech Report 95-057-3-0058-3. (1995).

[S71] H.S. Stone *Parallel Processing with the perfect shuffle*, IEEE Trans. on Computers, **C-20**:2, 153-161 (1971).

[U84] J. Ullman, *Computational Aspects of VLSI*, Computer Science Press, (1984), Chapter 6.

[W68] A. Waksman *A Permutation Network*, JACM, **15(1)**,159-163 (1968).

[W61] H. Wang, *Proving Theorems by Pattern Recognition*, Bell System Technical Journal, **40**, 1–141, (1961).

[WDS96] H. Wang, R.J. Di Gate, and N.C. Seeman, *An RNA Topoisomerase*, Proc. Nat. Acad. Sci. (USA), **93**, 9477–9482, (1996).

[WGWZ92] J. Watson, M. Gilman, J. Witkowski, M. Zoller, *Recombinant DNA (2nd ed.)*, Scientific American Books, W.H. Freeman and Co., (1992).

[WHR87] J. Watson, N. Hoplins, J. Roberts, et. al., *Molecular Biology of the Gene*, Benjamin/Cummings, Menlo Park, CA, (1987).

[W95] E. Winfree, *Complexity of Restricted and Unrestricted Models of Molecular Computation*, California Institute of Technology technical report May, 1995, Also Princeton DIMACS Technical Report workshop on DNA-based computers, To appear, April 4, 1995.

[W96] E. Winfree, *On the computational power of DNA annealing and ligation*, DNA based computers, Lipton, R.J. and Baum, E.B. eds., Am. Math. Soc., Providence, RI, (1996).

[W97] E. Winfree, communication to J. H. Reif (June, 1997).

[WYS96] E. Winfree, X. Yang, N. C. Seeman, *Universal Computation via Self-assembly of DNA: Some Theory and Experiments*, 2nd Annual Meeting on DNA Based Computers, Princeton University, June 1996.

[ZS92] Y. Zhang and N.C. Seeman, *A Solid-Support Methodology for the Construction of Geometrical Objects from DNA*, J. Am. Chem. Soc., **114**, 2656–2663, (1992).

[ZS94] Y. Zhang and N.C. Seeman, *The Construction of a DNA Truncated Octahedron*, J. Am. Chem. Soc., **116**, 1661–1669, (1994).

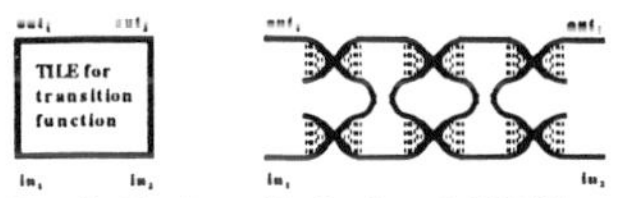

Figure 1a: The tile transition function and a DNA DX nanostructure for the tile.

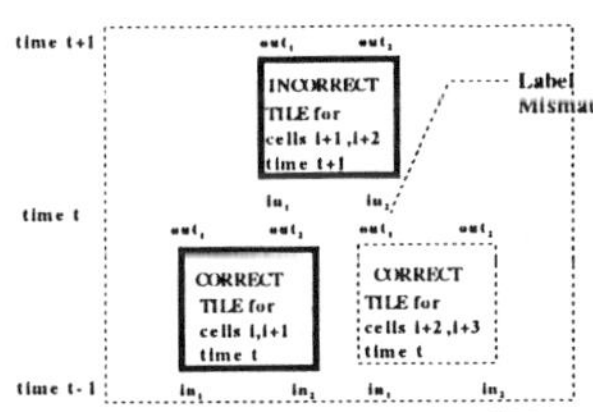

Figure 1b: Unmediated self-assembly correctly places a tile at (i,t) but incorrectly places a tile at (i+1,t+1). Later the tile at (i+2,t) can not be placed, so assembly is blocked.

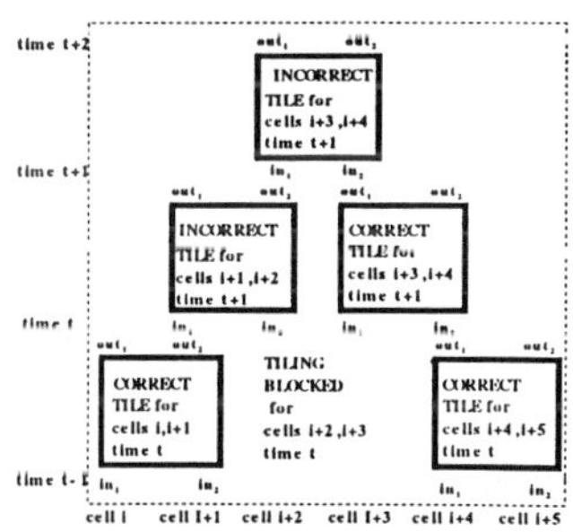

Figure 1c: Unmediated self-assembly correctly places a tile at (i,t), (i+4,t) but incorrectly places a tile at (i+1,t+1). Later tiles placed at (i+3,t+1), (i+2,t+2) reinforce the incorrect tiling, so the tile at (i+2,t) can not be placed, and the assembly is permanently blocked.

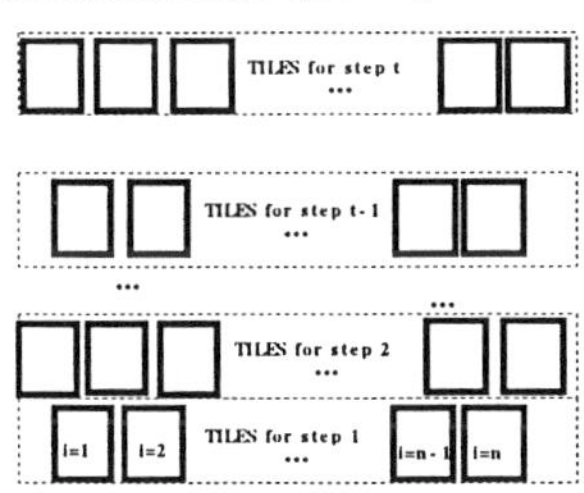

Figure 1d: Step-wise assembly for step t is a rectangular tiling.

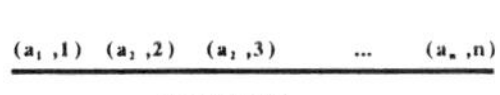

Figure 2a: ssDNA encoding input n-vector fixed in a straight line.

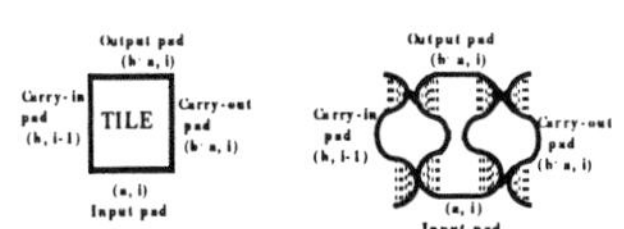

Figure 2b: Square tile for sequential prefix computation and a DNA DX nano-structure for the tile.

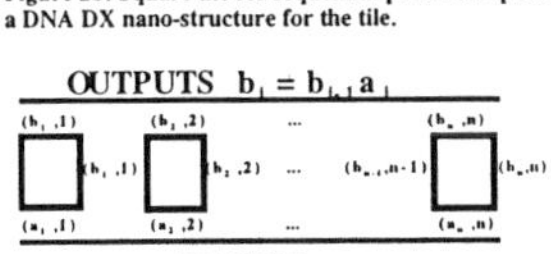

Figure 2c: 2D assembly for sequential prefix computation.

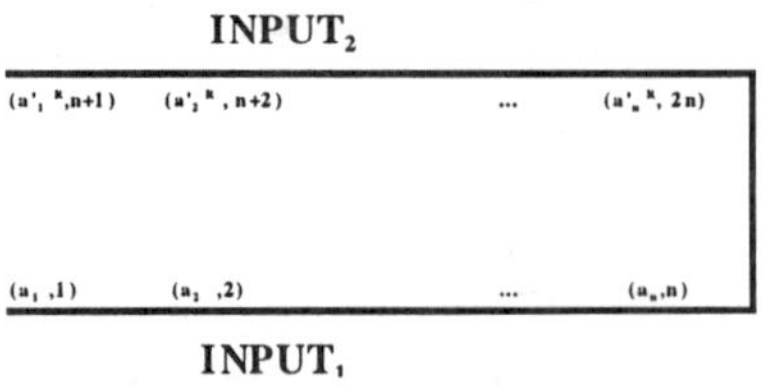

Figure 3a: Input ssDNA encoding two binary numbers.

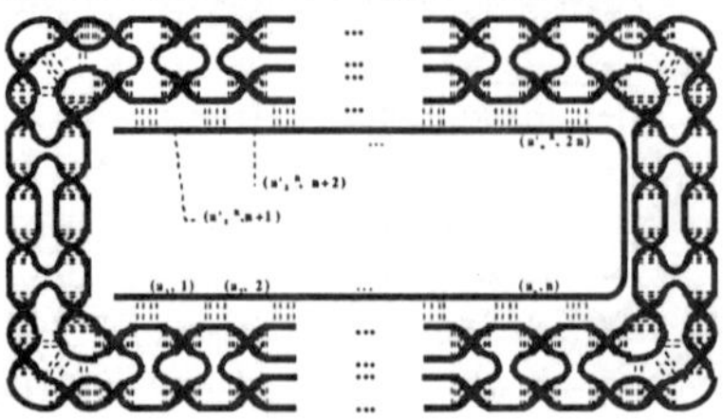

Figure 3b: A DNA DX nano-structure for the rectangular frame.

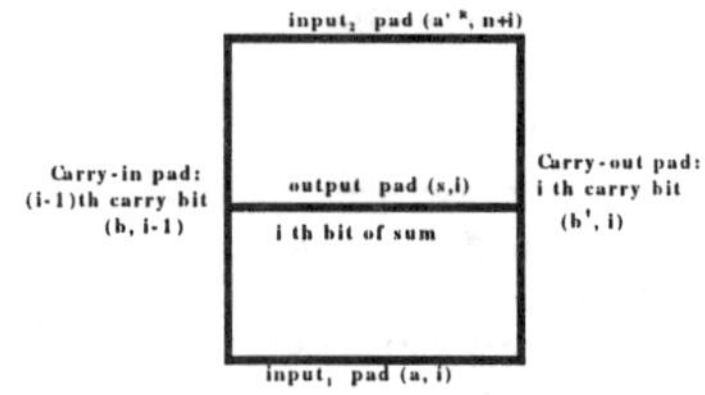

Figure 3c: Square tile with middle segment, used for bit-serial addition.

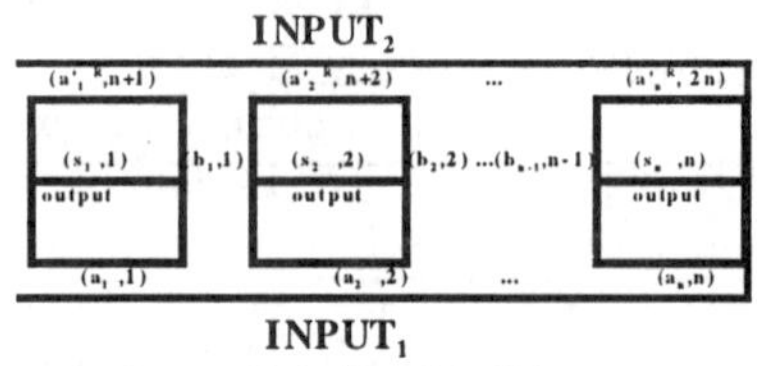

Figure 3d: 2D assembly for bit-serial addition.

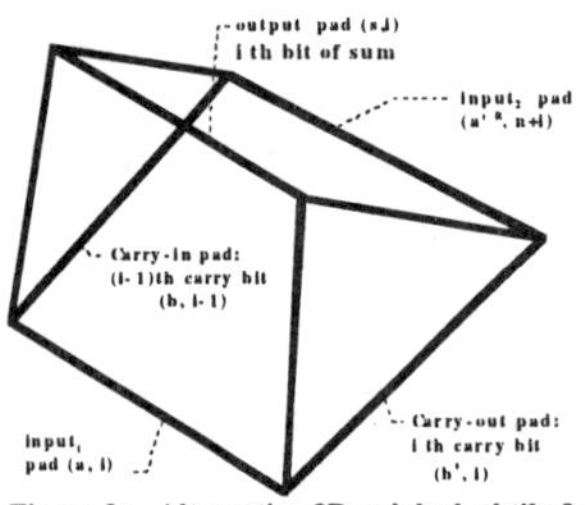

Figure 3e: Alternative 3D polyhedral tile for bit-serial addition.

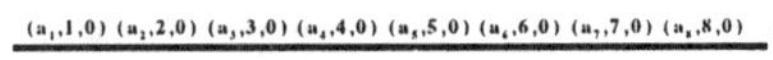

Figure 4a: Input ssDNA encoding an n-vector.

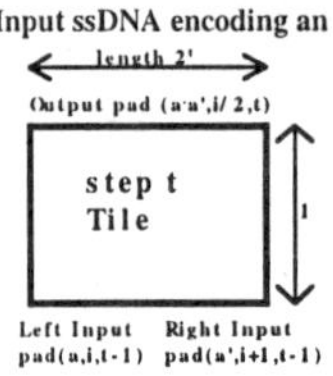

Figure 4b: A Rectangular tile for parallel monoid sum computation.

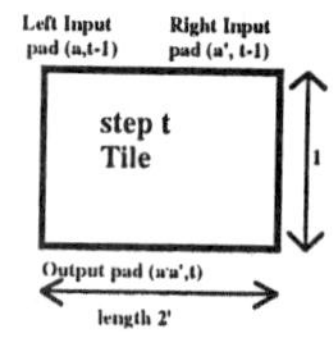

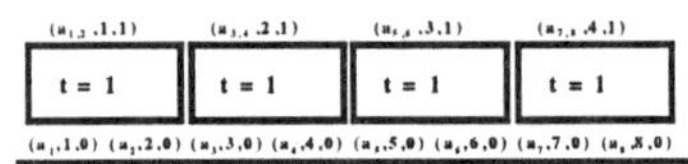

Figure 4c: Step t = 1 of step-wise assembly for parallel monoid sum computes $a_{i,i+1} = a_i \cdot a_{i+1}$ for odd I.

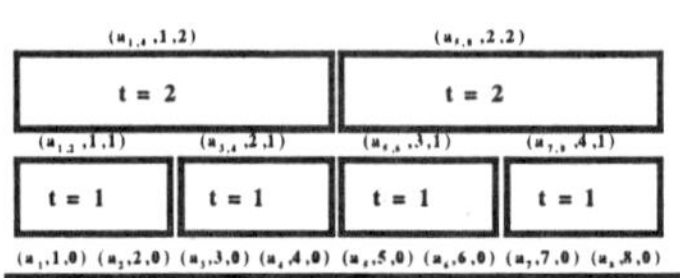

Figure 4d: Step t = 2 of step-wise assembly for parallel monoid sum computes $b_4 = a_{1,4} = a_1 \cdot a_2 \cdot a_3 \cdot a_4$ and $a_{5,8} = a_5 \cdot a_6 \cdot a_7 \cdot a_8$.

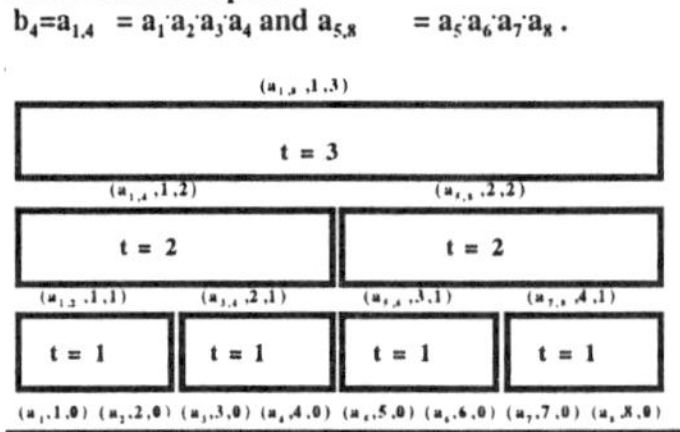

Figure 4e: Final step t =3 of assembly for parallel monoid sum computes $b_8 = a_{1,8} = a_1 \cdot a_2 \cdots a_8$.

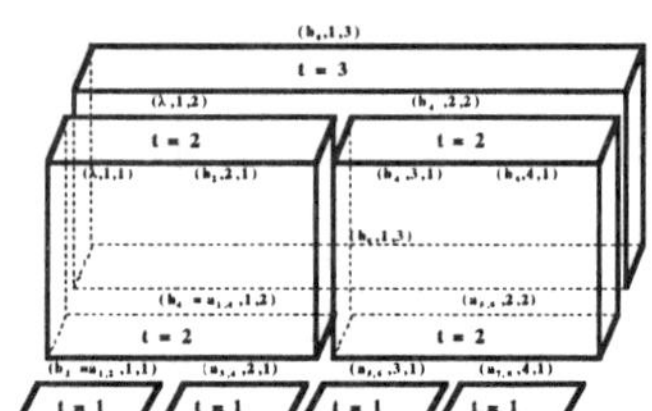

Figure 5c : Stage t=2 of 3D parallel prefix assembly: computation of b_2, b_4, and b_6.

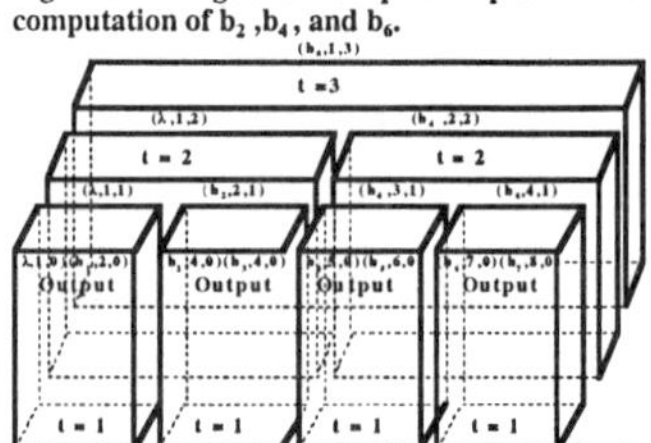

Figure 5d: Final stage t=1 of 3D parallel prefix assembly: computation of b_1,...,b_7.

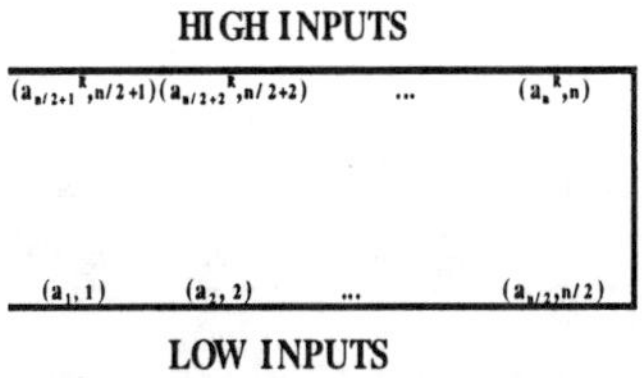

Figure 7a: ssDNA encoding input n-vector.

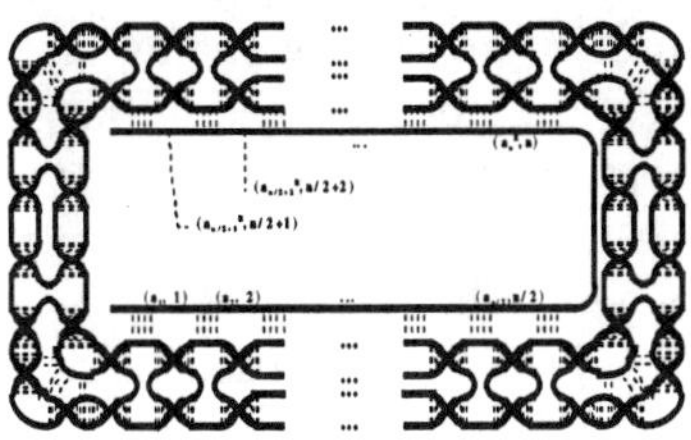

Figure 7b: A DNA DX nano-structure for the rectangular frame.

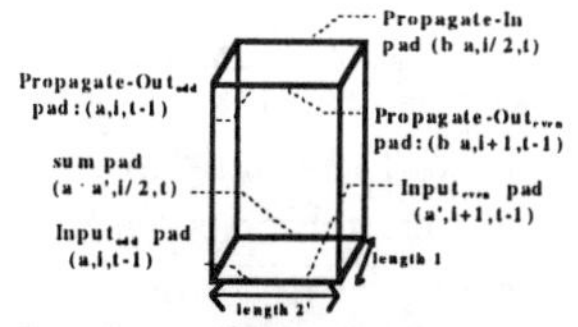

Figure 5a: A 3D polyhedron tile for parallel prefix computation.

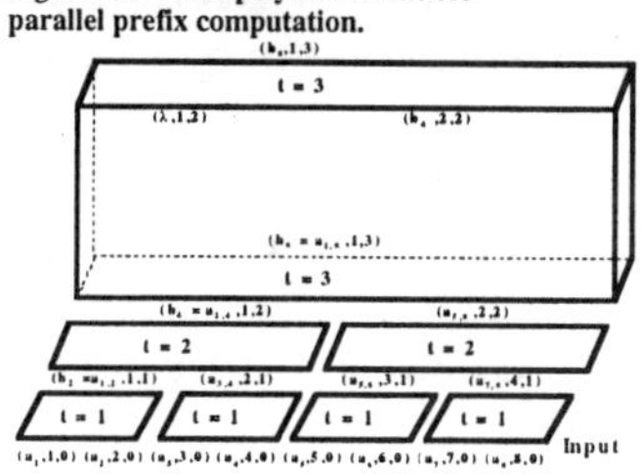

Figure 5b: Stage t=3 of 3D parallel prefix assembly: computation of b_4.

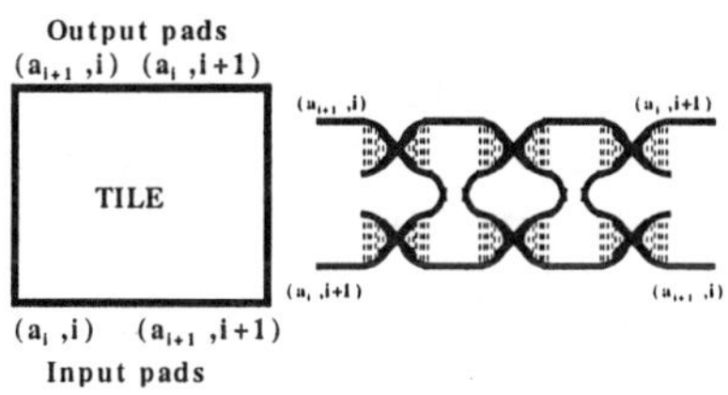

Figure 6a: A square tile for pair-wise exchange and a DNA DX nano-structure for the tile.

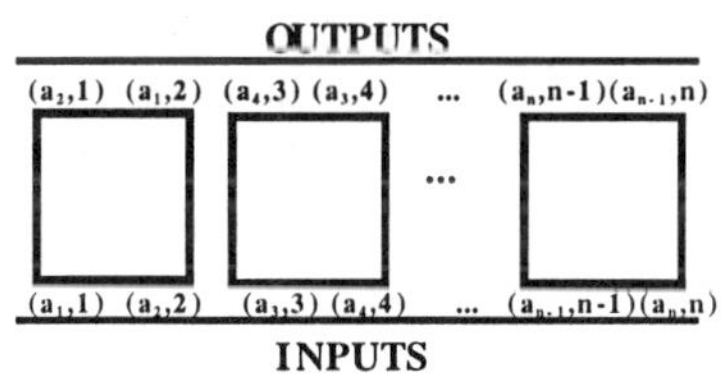

Figure 6b: The 2D assembly for pair-wise exchange.

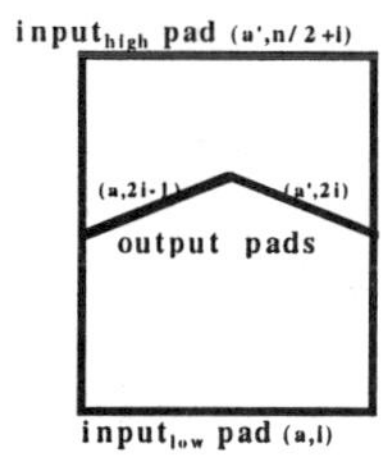

Figure 7c: A square tile for perfect shuffle, with middle segment used for output.

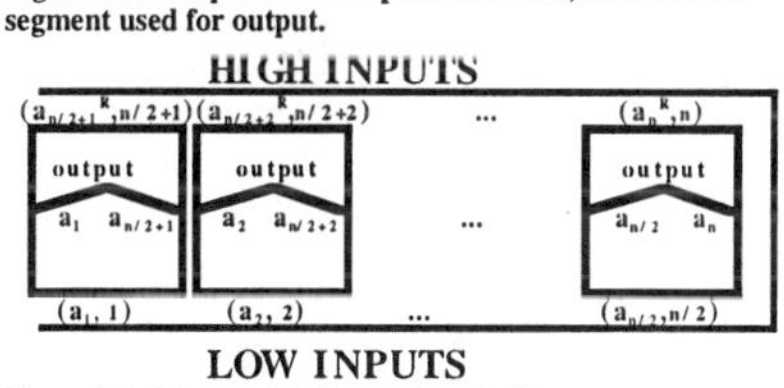

Figure 7d: 2D assembly for perfect shuffle. The middle segments give output.

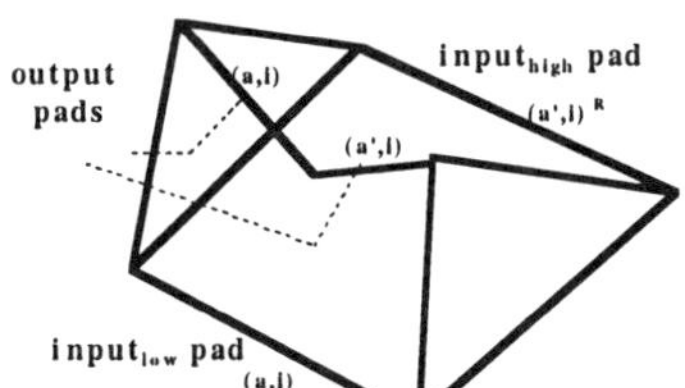

Figure 7e: Alternative 3D polyhedral tile for perfect shuffle.

10. Appendix: The Success Probability of Unmediated Local Assembly

We consider here random assembly models for the kinetics of assembly via hybridization at a fixed temperature. For certain setting of assembly parameters, We shall demonstrate that blockages in assembly can occur in these random assembly models, for certain setting of parameters.

A Random Assembly Model. We now consider a model for random assembly, where the assembly grows by tile placement at random untiled locations adjacent to the assembly, and the tiles are choosen by a random process depending on the number of matches with previously placed tiles. We presume that the initial tiles are initially placed as specified, without any initial blockage. As a *random assembly* model, we assume that the further assembly is done by choosing a random untiled location L at a unit region which adjoins at least one location so far tiled. We then choose a tile τ to place in that chosen location L, where that τ is chosen randomly among all tiles with a relative probability p_m, which we call the *preferential matching* parameter. Here m is the total number of base pairs in pads of τ that correctly match the pads of tiles abutting this location L. To approximately model the kinetics of assembly via hybridization, we define the preferential matching parameter as p_m using the Boltzmann distribution from statistical mechanics [**EC**]; this is an increasing exponential function of m of form $e^{c_0 m - c_{init}}$, where $c_0 > 0$ is the free energy of a single base-pair interaction and $c_{init} > 0$ is the initiation energy, which depend upon the concentration of units, temperature, and other physical parameters. Thus the preferential matching parameter for a tile τ is $p_m(\tau) = e^{c_0 m - c_{init}}$, for a constant $c_0 > 0$. To obtain the probability of choosing a tile τ from the relative probability $p_m(\tau)$, we normalize by dividing by the sum over the preferential matching parameters of all tiles that can be chosen (This random choice is done among all tiles, including an empty tile with $p_m = 1.$). The assembly progresses until the tiling is completed, or there is no position where further tiling is possible.

Let us assume for simplicity that the number of base pairs per pad of each tile is ℓ. If a tile τ has k matching pads at a given site, then $m = k\ell$, so the preferential matching parameter is $p_m(\tau) = e^{c_0 k\ell - c_{init}}$.

(1) When $c_0 k\ell - c_{init}$ is very small, and negative, with high probability the empty tile is chosen; thus rather than placing an incorrect tile, no mistake is made and the same site might be chosen again later, when a correct k-pad-match might be placed (this is the case considered by Winfree [**W97**]).

(2) When $c_0 k\ell - c_{init}$ is moderately small, but positive, (e.g., with constant tile pad lengths), we get a significant likelihood of placement of tiles with incorrect matches and placement of tiles which are not maximal matching.

(3) However, for moderately large values of $c_0 k\ell - c_{init}$, which we would recommend, then the assembly is highly preferential to correct k pad matches, and so pad mismatches among abutting tiles are very unlikely, and hence we are very likely to place a maximal matching tile in a given location. In the case of long sticky ends and appropriate temperature, this leads to a model having correct maximal k pad matches with high probability¶ For example, if say $c_0 k\ell - c_{init} = c' \log n$, for

¶Nevertheless, at constant temperature there will still be some finite likelihood of error due to mismatches. Winfree [**W97**] has suggested the following scenario where maximal matching might be a good model for step-wise assembly: At each step, add the tiles for that step and cool down slowly. (a) At a temperature where k pad matches is a favorable reaction but $k' < k$ pad matches are not favorable, sites where the maximal match has k pads will be filled selectively with the correct tile. (b) At a lower temperature, sites where the maximal match has $k - 1$ pads can

a sufficiently large constant $c' > 0$, then maximal matching tiles are placed with likelihood $\geq 1 - n^{-\Omega(1)}$.

A Simplified Random Assembly Model. We now consider a simplified model for random assembly. As in the previous described model for random assembly, the assembly growths by tile placement at random untiled locations adjacent to the assembly. However, we assume we place at those locations only tiles randomly chosen among all maximal matching tiles in that location. (Note that by the above discussion, for moderate values of $c_0 k\ell - c_{init} = c' \log n$, this simplified model which places maximal matching tiles, only differs with very low likelihood $n^{-\Omega(1)}$, from the previous described model for random assembly with a preferential matching parameter.) Unfortunately, even in this favorable case, where we place in randomly chosen locations only maximal matching tiles, we will show the unmediated assembly of Winfree's [**W96**] can nevertheless fail with high likelihood.

The Success Probability of Unmediated Local Assembly. We now provide a probabilistic analysis of Winfree's [**W96**] direct simulation of a Universal TM (or cellular automata) running in time $T \geq n$ and space $S \geq n$ by assembly of an $S/2 \times T$ array described in detail at the end of Section 3 (also given in detail in [**W96**]). We assume the simple random assembly model described above, where we randomly choose an untiled location (abutting a previously tiled location) and choose a tile to place in that location, randomly among all maximal matching tiles in that location. We will prove that during an attempt to complete an unmediated DNA assembly of size $A = ST/2$, the probability β of failure by blockage can be $\approx 1 - 2^{-cA^2}$, where $c = O(1/\sigma^2)$ and σ is the maximum number of maximal matching tile types for any given location (in the construction of [**W96**], σ is a constant depending on the number of tape symbols and states of the simulated TM). Hence, for $T \geq n$ and $S \geq n$, the probability of failure can be $\beta \geq 1 - 2^{-cn^2/2}$, which very quickly approaches 1 for even small n.

We first will (informally) describe what potentially may go wrong with this unmediated DNA assembly method. We assume the reader is familiar with this tiling construction. For general computation simulations such as these, the resulting tiles do not necessarily have identical pads. Recall a *blockage* is an untiled location where no tile can be placed there without a pad mismatch. Now since the assembly is unmediated, it can proceed both forward and backward in time. Then in general, the reverse time simulation can not be uniquely determined by the current state of the assembly. In particular, a partial assembly may choose a reverse simulation that is not consistent with the rest of the values of the tape cells. This can cause a blockage that halts the assembly locally. This blockage can occur even if the TM as a whole is reversible, since the entire tape contents are not determined by a partial assembly. In the simulation of a Universal TM, or any other universal machine, similar situations can be made to occur.

---

be filled selectively, etc. Note that at this lower temperature, an incorrect partial match of $k-1$ pads could be not be placed into a site with a match of k pads, since that site is already filled. (c) And so on, as the temperature is lowered, maximal matches with fewer and fewer pads may be placed.

However, it should be noted that in this scenario an incorrect tile with $k+1$ pad matches and 1 pad mismatch may be preferred to a correct tile with the intended k pad match and 0 pad mismatches. Thus the assumption that partial mismatches are disallowed is confirmed by this scenario, only if the assembly has a geometry that rules out the possibility of incorrect tile assembly with $k+1$ pad matches and 1 pad mismatch.

This argument can be formalized as follows. Let $\pi(i,t)$ denote a tile placed in location (i,t). Ideally, we would like the assembly process to position tiles to simulate the machine's computation, as follows. For each $t, 1 \leq t \leq T$ and $i, 1 \leq i \leq S$ where $(i = t) \bmod 2$ there is a tile $\pi(i,t)$ placed in location (i,t) representing the transition from time $t-1$ to time t at the two consecutive tape cells at positions $i, i+1$. Note that Winfree's **[W96]** construction has tile $\pi(i,t)$ abut and match pads at neighboring tiles $\pi(i-1,t-1), \pi(i+1,t-1), \pi(i-1,t+1), \pi(i+1,t+1)$, if these neighboring tiles have been placed down. The tiles $\pi(1,0), \ldots, \pi(S,0)$ representing the initial configuration of the TM at time 0 can be assumed to be initially placed in the correct location. However, the assembly may not necessarily correctly position tiles associated with subsequent times > 0. Now suppose that the simulation assembly process has progressed with a partial subassembly at time step $t-1$, but where there remains at least a 3×3 area of the plane to tile associated with tape positions $i, i+1, i+2$ and time steps $t, t+1, t+2$. Further suppose that the simulation assembly process proceeds forward via placing a single tile $\pi(i,t)$ (bridging time $t-1$ to time t at two consecutive tape cells $i, i+1$), and then the random assembly also proceeds via placing a tile $\pi'(i+1,t+1)$ (bridging time t to time $t+1$ at the two consecutive tape cells $i+1, i+2$). Now further suppose the tile $\pi(i+2,t)$ has not yet been positioned by the random assembly, so the value of the tape cell $i+2$ at time t is not determined by immediate context. Thus the tile $\pi'(i+1,t+1)$ abuts only one tile $\pi(i,t)$. Since the random assembly has not yet placed tile $\pi(i+2,t)$, $\pi'(i+1,t+1)$ does not abuts that tile. The key problem is that since the tile $\pi'(i+1,t+1)$ abuts only one tile $\pi(i,t)$ (rather than two tiles) of previous time step t, the tile $\pi'(i+1,t+1)$ may have been incorrectly placed in the assembly, so that if we later attempt (see Figure 1b) to correctly place tile $\pi(i+2,t)$, its pad in_2 abutting (incorrectly placed) tile $\pi'(i+1,t+1)$ may not match, and the assembly is blocked (In contrast, a correctly placed tile $\pi(i+1,t+1)$ would have matched at its pads in_1, in_2 where it abuts both tile $\pi(i,t)$ and (a later placed) $\pi(i+2,t)$. However the random assembly instead incorrectly placed a tile $\pi'(i+1,t+1)$ which matches only at its pad in_1 abutting $\pi(i,t)$ but may not later match at its pad in_2 abutting where $\pi(i+2,t)$ would later be placed.)

The tile pads depend on the transition function of the simulated TM or cellular automaton, and this transition function can be appropriately chosen by us to insure a possible mismatch. Thus, given any partial assembly where there remains a 3×3 area of the plane to tile, then the tiling can in the worst case be so blocked by the further assembly of just 2 additional tiles into this area. The incorrectly placed tile $\pi'(i+1,t+1)$ matches at only one pad, but is a maximal matching tile in that location, since no other tile matches at that location within the current assembly with more than one pad. The other tile $\pi(i,t)$ also is a maximal matching tile in its location. Thus both these tiles are chosen with probability $\geq 1/\sigma$, where σ is the maximum number of distinct tile types that can can be maximal matching tiles in any given location. Hence the probability both of these tiles are chosen, causing a blockage at any point in the assembly process, is at least β_0, where $\beta_0 = \Omega(1/\sigma^2)$. Thus, the probability of no failure by blockage during the $(S-3)(T-3)/2$ assembly steps, where there remains a 3×3 area of the plane to tile, can be $\leq (1-\beta_0)^{(S-3)(T-3)/2} \approx e^{-\beta_0(A-3(S+T)/2)} \leq e^{-cA}$, where the size of the assembly is $A = ST/2$ and $c = \beta_0 + 3$. Hence, since $T \geq n$ and $S \geq n$, the total probability of failure by blockage during an attempt to complete an assembly of size A is $\beta \geq 1 - e^{-cA}\, ge 1 - e^{-cn^2/2}$, which very quickly approaches 1 for even small cn^2.

Further Discussion of Feasibility of Unmediated Local Assembly for Computation. A possible objection to our analysis of unmediated local assembly is that not every Turing Machine would result in such blockage problems; and indeed we have chosen an especially bad one. But we know of no non-trivial computation (other than permutations of the input) via unmediated DNA assembly that does not exhibit this blockage problem, for worst case inputs. Moreover, we found no effective way to insure against these blockage problems a priori, and it may be (we conjecture this) impossible to do so in the case of a Universal Turing Machine, since it simulates all others. This blockage problem might be remedied by the use of DP-BMC to allow for many parallel assemblies. In particular, if we use $1/(1-\beta) \approx 2^{cA}$ parallel assemblies, then the probability of success increases to $1-\beta^{1/(1-\beta)} \approx 1-1/e$, which is constant, but this requires extremely large volumes growing as $1/(1-\beta) \geq 2^{cn^2/2}$, which would overwhelm even DP-BMC for very small n. Alternatively, we might repeat the assembly process until success, but this requires a very large expected number $1/(1-\beta) \geq 2^{cn^2/2}$ repeats.

A Random Assembly-Disassembly Model. A quite valid objection to our analysis is that our random assembly model is too simple to fully model the kinetics of assembly via hybridization at a fixed temperature. Indeed our random assembly model does not allow for later preferential disassembly of partially matching tiles (that may have been incorrectly placed) in favor of better matching tiles (that may now be correctly placed). We now consider a more sophisticated *random assembly-disassembly model* that allows both random preferential pairing of matched tiles and also a probability of preferential disassembly of a partially matched tiles disassociating, where for a randomly chosen region adjacent to the assembly, (i) if the region is untiled, then we choose a tile τ to place in that chosen location, where that tile is chosen randomly among all tiles with relative probability again given by the preferential matching parameter $p_m(\tau)$, and (ii) if the region is tiled, then the probability of disassembly of the tile τ is a function $q_m(\tau)$, where in both cases m is the total number m of base pairs in its matching pads (correctly matching pads of abutting tiles), that the tile has in this placement. To approximately model the kinetics of assembly via hybridization, we can let p_m be defined by the Boltzmann distribution as an increasing exponential function $e^{c_0 m - c'_{init}}$ of m for constants $c_0, c_{init} > 0$ and we let q_m be a decreasing exponential function $e^{-c_0 m + c'_{init}}$ of m for constants $c'_0, c'_{init} > 0$, where the constants depend upon the concentration of units and temperature. In practice, the on-rate for matches and mismatches is expected to be (roughly) identical if the DNA are present in equal concentrations, so there is little or no preference for association, and only preference for dissociation. This can be modeled by setting the constant c_0 very small; and might be considered 0 in practice. Thus p_m is very nearly constant and nearly independent of m.

Such a random assembly-disassembly model is much harder to analyze rigorously. It remains an open problem to provide rigorous upper and lower bounds for such a random assembly-disassembly model. We know of no non-trivial examples of computations (other than the data permutation operations proposed in this paper) via unmediated DNA assembly where there are provable success bounds in any reasonable probabilistic assembly-disassembly model.

Potentially the situation may be worse, rather than improved. Consider the situation of a large assembly composed of (i) a correct initial assembly, composed at a number of locations with (ii) a partial assembly that is incorrect; in that case the random disassembly process might occur preferentially at the mismatched boundaries of these two subassemblies, but would not be able to preferentially distinguish locally which is the correct subassembly and would appear to randomly disassemble both subassemblies. Thus the probability of a blockage at any step of the assembly may remain constant, and rather than halt the assembly, the number of blockages may increase as the assembly proceeds, with high likelihood. Since the probability of any pair of matched tiles disassociating is not dependent on the existence of blockages, we cannot derive a better upper bound than $2^{c'A} \geq 2^{c'ST}$ time for convergence to the required complete $S/2 \times T$ tiling, for a constant $c' > 0$.

In a random assembly-disassembly model, the success likelihood may be a very complicated function of chosen parameters such as temperature and match lengths, ruling out a rigorous analysis. It is conceivable that on certain computations, with carefully chosen parameters, the assembly can have reasonable likelihood of success. The kinetics of assembly via hybridization will vary if the temperature is gradually dropped, and thus we may obtain a transition through such a setting of parameters where the assembly has reasonable likelihood of success. While the lack of a proof for the success of unmediated DNA assembly is unsettling, it may happen that analytic proofs are not essential to resolve the issue. The issue might be resolved (without rigorous analysis) by computer simulations and experimental evidence, (beyond testing preferential matching of pairs of tiles) on a moderate scale, of the random assembly-disassembly process. Recently Winfree [**W97**] has done some promising computer simulations of a related random assembly-disassembly model, indicating that careful choice of settings of the parameters may allow unmediated self-assembly sometimes to succeed. However, unmediated self-assembly remains experimentally untested and we now await experimental evidence of this.

Department of Computer Science, LSRC Building, Duke University, Box 90129, Durham, NC 27708-0129

E-mail address: reif@cs.duke.edu

DIMACS Series in Discrete Mathematics
and Theoretical Computer Science
Volume **48**, 1999

DNA-Based Parallel Computation by "Counting"

Mitsunori Ogihara and Animesh Ray

ABSTRACT. The potential of DNA as a truly parallel computing device is enormous. Solution-phase DNA chemistry, though not unlimited, provides the only currently-available experimental system. Its practical feasibility, however, is controversial. We have sought to extend the feasibility and generality of DNA computing by a novel application of the theory of *counting*. The biochemically equivalent operation for DNA *counting* is well known. We propose a DNA algorithm that employs this new operation. We also present an implementation of this algorithm by a novel DNA-chemical method. Preliminary computer simulations suggest that the algorithm can significantly reduce the DNA space complexity (i.e., the maximum number of DNA molecules that must be present in the test tube during computation) for solving 3SAT to $O(2^{0.4n})$. If the observation is correct, our algorithm can solve 3SAT instances of size up to or exceeding 120 variables.

Introduction

Two major issues in DNA computing.

Adleman [**A**], subsequently Lipton [**L**], demonstrated the possibility of solving difficult problems such as NP-complete problems, typified by Hamiltonian Path and 3SAT, with the help of parallel processing by liquid phase DNA chemistry. Basic operations of DNA computing are biochemical reactions performed on DNA strands. Adleman and Lipton suggested seven basic operations, namely, *synthesis*, *amplification*, *merger*, *appending*, *extraction*, *detection*, and *polymerizations*. Adleman's method and Lipton's method, though different in detail, share the same philosophy: create in a test tube the whole solution space by letting DNA strands represent potential solutions and find real solutions by pattern matching of DNA strands. The advantage of their methods mainly exists in that all the potential solutions can be processed simultaneously because of massive parallelism of chemical reactions. The drawbacks are: (i) the uncertainty of accuracy of the chemical operations involved and (ii) the exponential growth in the maximum number of DNA molecules that must be present in the test tube during computation. First, chemical

1991 *Mathematics Subject Classification.* Primary 68Q10, 68Q22; Secondary 68Q25.

Key words and phrases. NP-Complete Problems, 3SAT, DNA Computing, Parallel Computation, Counting.

Supported in part by NSF CAREER Award CCR-9701911 and by DARPA/NSF-9725021.

Supported in part by National Science Foundation Grant MCB-9630402.

reactions are error-prone. They are limited by diffusion kinetics and mass action laws. Furthermore, recognition of base complementarity, though highly accurate in biochemical contexts, is affected by temperature, ionic strength of the solvent and the base sequence of the interacting strands [[]pages 1109–1264]CS/ These factors have not yet been rigorously tested to define the acceptable bounds of fidelity for molecular computation of large instances.

Second, and more importantly, these DNA algorithms suffer from the "exponential curse" [**H**]. The Avogadro's number gives the absolute upper bound on the number of DNA strands one can place in a given volume of liquid. (In practice, the bound is likely to be much smaller—about 10^{18} per liter.) In Adleman's algorithm, the complexity of oligonucleotide paths approaches n^n while in Lipton's algorithm, the complexity is 2^n. So, both Adleman's method and Lipton's method can handle inputs of size up to 60. However, such small instances are in most cases manageable by electronic computers. Efforts have been made to reduce the amount of volume for DNA computers to solve NP-complete problems [**BCGT,O,BF**]. In this paper, we focus our attention to 3SAT. The current fastest sequential algorithm for 3SAT runs in 1.5^n steps [**K**]. If the algorithm is implemented on a computer that executes a million instructions in a second, the formulas of 60 variables will require approximately 10 hours; so there is no need to use DNA computers. If we double the size and set $n = 120$, then not only the sequential algorithm but also the DNA algorithms will blow up. Indeed, for problems with $n = 120$, Adleman's method needs to generate a complexity involving 2.5×10^{21} grams of nucleotide units and Lipton's method requires 2×10^{19} grams of nucleotide units. These are far beyond the limits of chemical technology.

Search space reduction by quantitation.

The present paper proposes a method for coping with the second problem. Our approach is to add one more operation to the computational basis of DNA, namely, "quantitation" of single DNA strands in a test tube. In computational complexity theory, the operation naturally corresponds to "counting"—counting the number of solutions of NP problems[**V1,V2**]. namely, the class #P. There is a vast literature on this topic and one of the major achievements in the past 5 to 10 years has been the realization and application of the enormous power of "counting" in computation [**T,TO**]. For example, it is known that every language in the polynomial hierarchy can be decided by a single "counting" operation [**T**].

The use of quantitation.

It seems very likely that space complexity of NP problems on DNA computers is significantly reduced when endowed with this new operation. Recently, Ogihara has proposed a breadth first search heuristic for 3SAT [**O**]. The algorithm operates with partial assignments to an input formula and gradually extends them until all partial assignments become full. By preliminary computer experiments, we observe that the ratio of the logarithm of the maximum number of partial assignments in the test tube over the number of variables always stays below 0.45. This implies that the space complexity of 3SAT can be reduced from 2^n to $2^{0.45n}$, thereby allowing us to solve formulas of more than 120 variables. We also observed that the ratio for unsatisfiable formulas is always less than 0.40. Based on this observation, we propose to further reduce the space complexity of 3SAT on DNA computers to $2^{0.40n}$ by means of quantitation. At any time of computation, if we discover that the number of single DNA strands in the test tube exceeds $2^{0.40n}$, then we stop the computation and assert that the input formula is satisfiable.

The Algorithm

A well-known practical algorithm for SAT is the Davis-Putnam procedure [**DP**]. This procedure calls itself after rewriting the input formula according to a number of rules for generating a smaller formula with the same truth value.

(I) if the input formula has no clauses, then it is satisfiable,

(II) if it has a clause with no literals, it is unsatisfiable,

(III) if it has a clause with exactly one literal, then make the literal true and rewrite the formula accordingly, and

(IV) if some variable appears only positively or negatively, then pick one such variable and assign a value to it to make the literal true, and rewrite the formula accordingly.

If none of the rules apply, it picks up an arbitrary variable as a *branching point* to create two formulas by assigning 0 and 1 to the variable. It then calls itself on the two formulas. If one of the calls returns with the positive answer, then the input is satisfiable; otherwise, it is unsatisfiable.

Although the procedure is very simple, its behavior is yet to be analyzed. What has been proven so far is only the average case time complexity analysis of its simplified versions [**GPB,PB,P,FP,DPP,CF,I**]: they use fewer rules and have fixed branching points, that is, branching at the ith level of the search tree is performed with respect to only x_i.

Our algorithm is different from these variants: it is a *breadth first search* variant of those simplified versions and it deals only with 3CNF formulas. The algorithm runs with predetermined branching points; i.e., prior to the actual run, the variables are reordered and the ith branching point is set to the ith variable in the new ordering. The reordering algorithm is simple: For each i, let M_i denote the number of clauses that contain either x_i or its negation. Then we simply sort the variables so that their M-values are non-increasing. Also, the algorithm has an abort strategy with a parameter θ. At any time of the computation, if the number of remaining nodes after pruning exceeds $2^{\theta n}$, then we stop the computation with assertion that the input formula is satisfiable. Reordering of the variable is essential to our algorithm. We can easily generate numerous instances for which the maximum tree width of the search tree is almost 2^n if the variables are not reordered.

Now we describe the algorithm in more detail. Let F be a formula of n variables. We assume that none of the clauses of F has both the positive form and the negative form of the same variable at the same time. Suppose that the new order of the variables is $x_1, \ldots, x_n$. The algorithm runs with a collection $\mathbf{S}$ of partial assignments, all at the same level of the search tree. At the beginning, $\mathbf{S}$ has only one element, the empty assignment. Then the main loop is executed n times. In the ith iteration of the loop, the partial assignments in $\mathbf{S}$ are extended with respect to x_i and then the partial assignments that already fail to satisfy any clause are removed from $\mathbf{S}$. After that, if $|\mathbf{S}|$ is greater than $2^{\theta n}$, we abort the computation with assertion that F is satisfiable. When the nth iteration of the loop is done, we check whether $\mathbf{S}$ contains a satisfying assignment. The check passes if and only if the algorithm outputs "satisfiable." The main loop at the ith iteration proceeds as follows.

1. The first step of the main loop is to identify partial assignments that enable the use of Rule IV with respect to x_i. We divide $\mathbf{S}$ into three collections $\mathbf{P_0}, \mathbf{P_1}$, and $\mathbf{N}$.
   (a) $\mathbf{P_0}$ consists of the partial assignments in $\mathbf{S}$ that reduce F to formulas in which x_i appears only negatively, and thus, to which we can apply the "negative-form-only" case of Rule IV with respect to x_i.
   (b) $\mathbf{P_1}$ consists of the partial assignments in $\mathbf{S} - \mathbf{P_0}$ that reduce F to formulas in which x_i appears only positively, and thus, to which we can apply the "positive-form-only" case of Rule IV with respect to x_i.
   (c) $\mathbf{N}$ is the remainder; that is, it consists of all the partial assignments in $\mathbf{S}$ that reduce F to formulas to which neither of the two cases of Rule IV with respect to x_i applies.

   In order to split $\mathbf{S}$ into these three collections, let F_{pos} (respectively, F_{neg}) be the formula constructed by collecting all clauses of F containing the literal x_i (respectively, $\overline{x_i}$), then by eliminating all literals with respect to variables that are yet to be covered, i.e., variables $x_j, \dots, x_n$. Note here that, by our assumption that x_i and $\overline{x_i}$ do not appear in the same clause, F_{pos} and F_{neg} are generated from disjoint sets of clauses of F. Now for every partial assignment π in $\mathbf{S}$, it belongs to $\mathbf{P_0}$ if and only if it satisfies F_{pos}. (Note that F_{pos} is not satisfiable if it contains the empty clause. The empty clause is in F_{pos} if and only if F contains a clause in which x_i appears and the other two literals have indices greater than i.) Also, for every partial assignment π in $\mathbf{S} - \mathbf{P_0}$, it belongs to $\mathbf{P_1}$ if and only if it satisfies F_{neg}.
2. The classification having been done, we will join $\mathbf{P_0}, \mathbf{P_1}$, and $\mathbf{N}$ to form $\mathbf{S}$ and then eliminate all the partial assignments that fail to satisfy any clause.
3. The extension having been completed, we will eliminate all partial assignments in $\mathbf{S}$ that fail to satisfy any clause. Since x_i has been just incorporated, we look at only dissatisfaction caused by the incorporation of the variable. Namely, we eliminate all the partial assignments that fail to satisfy any clause C in which x_i or $\overline{x_i}$ appears and the other two literals have indices smaller than i.

One might think of combining several pruning steps into a single one to make the algorithm simpler. However, the steps to be combined should be carefully chosen so as not to increase the total number of DNA strands over the preset threshold for unsatisfiable formulas.

Implementation on DNA Computers.

The algorithm can be implemented on a DNA computer using a method similar to that of Lipton.

We need to determine an encoding scheme. Assignments $x_i = b$ will be denoted by strand $A_{i,b}$. We choose these representations so that they are unique sequences of base (= the number of DNA molecules), e.g., 20. The empty assignment is encoded as a special header H.

In order to collect from a test tube T the strands satisfying a disjunctive clause $C = (l_1 \vee \cdots \vee l_k), k \geq 2$, Lipton [**L**] suggested the following method based on extraction. For $i = 1, \cdots, k$, extract from T all the strands with the pattern corresponding to the assignment that satisfies l_i and put them in the test tube T_i. Then combine the test tubes $x_1, \cdots, x_k$ and replace T by the mixture. In general, for a conjunction of clauses $C_1, \cdots, C_m$, we apply the above method sequentially for $C_1, \cdots, C_m$. By a sequential application of the Lipton's method we can split $\mathbf{S}$ into the three collections $\mathbf{P_0}, \mathbf{P_1}$, and $\mathbf{N}$. Also, the elimination process can be done by Lipton's method. While extraction is error-prone, there is a computationally efficient method to cope with the error with linear overhead on computational time [**KKW**]. Extending of $\mathbf{P_0}$ and $\mathbf{P_1}$ is done by the *append* operation. Each of the strands in the test tube ends with either $A_{i-1,0}$ or $A_{i-1,1}$. We pour a sufficient amount of linker sequences that bind $A_{i,0}$ to each of the two ending patterns for $\mathbf{P_0}$ and those that bind $A_{i,1}$ to the two ending patterns for $\mathbf{P_1}$. Extension of $\mathbf{N}$ is done in the following four phases. (i) We first run a single step of DNA chain synthesis (in one direction) to $\mathbf{N}$. As a result from each molecule of Y of the form $H \cdots A_{i-1,b}$ one molecule of $\overline{Y} = \overline{H \cdots A_{i-1,b}}$ is synthesized without losing the original. (ii) We run four kinds of append steps in parallel: $A_{i,0}$ to $A_{i-1,0}$, $A_{i,0}$ to $A_{i-1,1}$, $\overline{A_{i-1,0}}$ to $\overline{A_{i,1}}$, and $\overline{A_{i-1,1}}$ to $\overline{A_{i,1}}$. (Note that the complementary strand has the reverse direction.) Consequently, to each strand with H at one end is attached $A_{i,0}$ at the other end, and to each strand with $\overline{H}$ at one end (synthesized in the previous phase) is attached $\overline{A_{i,1}}$ at the other end. (iii) We run a single step of DNA chain synthesis again. If we concentrate on the strands with H, those which already had the header prior to this phase remain the same and from each molecule of the form $\overline{H \cdots A_{i-1,b}A_{i,1}}$ one molecule of $H \cdots A_{i-1,b}A_{i,1}$ is synthesized. Therefore, from each molecule of Y of the form $H \cdots A_{i-1,b}$ which existed at the beginning of (i) we have synthesized one molecule of $YA_{i-1,1}$, and we have turned the original into $YA_{i-1,0}$. (iv) We extract all strands with H, thereby completing the two-way extension.

Running Time Analysis.

We analyze the running time by the number of sequential operation steps. Let $C = (x_p, x_q, x_r)$ be a clause with $p < q < r$. The clause can contribute to three activities in the entire computation:

1. Construction of $\mathbf{P_0}$ via evaluation of F_{pos} at the qth iteration, where strands with $A_{q,1}$ are extracted. The test tube containing the extracted strands is processed with more extraction steps if there are more clauses in F_{pos}. Otherwise, it becomes $\mathbf{P_0}$. The remainders of this extraction process should be merged back to $\mathbf{S}$, but it can be postponed until all the clauses of F_{pos} have been processed, all the remainders are merged back to $\mathbf{S}$ in one single step of merger.
2. Construction of $\mathbf{P_0}$ via evaluation of F_{pos} at the rth iteration. Here from a given test tube T, strands with $A_{p,1}$ are extracted into U_1, then strands with $A_{q,1}$ are extracted into U_2 from T, U_1 and U_2 are merged into new T, and T is merged back to $\mathbf{S}$. Again the final merge process can be combined into a single merger step.
3. Elimination of partial assignments that fail to satisfy C. This is at the r the iteration. Here extraction of strands with $A_{p,1}$, extraction of those with $A_{q,1}$, and extraction of those with $A_{r,1}$ are sequentially run on $\mathbf{S}$, and the extracted strands are merged to form a new $\mathbf{S}$.

Therefore, each clause contributes eight operations and there is one merger incurred for constructing $\mathbf{P_0}$ as well as $\mathbf{P_1}$ for each variable (except for the first one). Also, for each variable, one quantitation step is needed for implementing the abort strategy. In regards to the number of steps required for extension, it is one for $\mathbf{P_0}$ as well as $\mathbf{P_1}$, and four for $\mathbf{N}$. Since the extension steps are run on different test tubes,

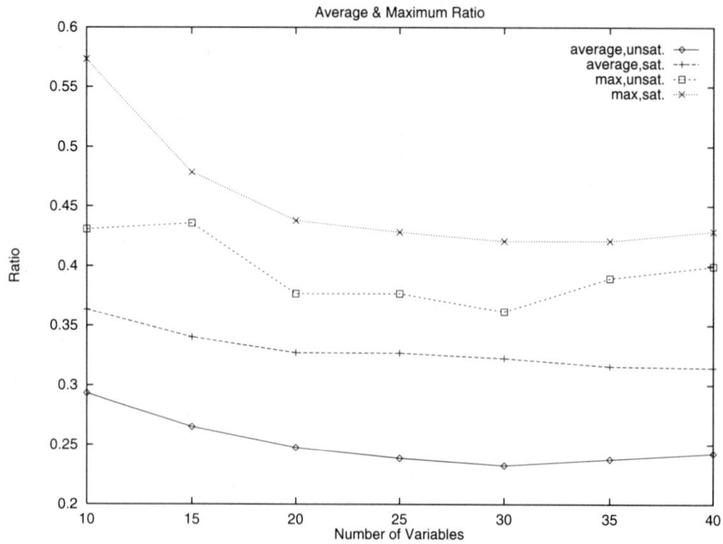

FIGURE 1: SIMULATION RESULT

the number of sequential steps is the maximum of four. Therefore, the additional steps required for a variable is seven (four for the first variable). One remaining operation is detection at the very end of the computation. Hence, the total running time is at most

$$8m + 7(n-1) + 4 + 1 \leq 8m + 7n.$$

Thus, we have proven the following proposition.

PROPOSITION 1. *The running time of the algorithm is at most $7n + 8m$ for a formula of n variables and m clauses.*

Note that the running time is at most twice that of Lipton's method [**L**].

Empirical Results

We have tested the performance of the breadth first search algorithm by computer simulations. On a SPARC workstation we implemented a program that, on an input formula of n variables, simulates our algorithm, computes the maximum number μ of partial assignments that are maintained, and outputs ratio $\rho = (\log_2 \mu)/n$.

In order to generate sample formulas, we wrote a program that, given an integer n, constructs a formula of n variables with the number of clauses between n and $10 \times n$. Here the choice of the range $(0.0, 10.0]$ needs to be explained. It is empirically shown that 3SAT exhibits phase-transition at the clause/variable ratio around 4.3 [**CA**]; i.e., there is a sudden change from satisfiable to unsatisfiable when the ratio passes the transition point. (Weaker theoretical bounds are shown in [**FS,KMPS**].) For practical SAT algorithms, the hardest instances concentrate at the transition point [**SML**]. We believe that sampling the ratio from the interval $(0.0, 10.0]$, will allow us to sample instances that are the hardest and practically the most interesting.

We generated the formulas by first determining the number of clauses uniformly at random, then by picking the desired number of clauses uniformly at random with repetition. In order to pick a clause, we repeatedly selected three literals at a time independently with equal probability, until we picked a clause free from complementary literals or identical literals.

We generated 120 formulas for each n between 10 and 45. Figure 1 shows the average and the maximum value of ρ, where data sets for five consecutive n are combined and the ratio with respect to satisfiable formulas and that of unsatisfiable formulas are plotted separately.

We observe that the average ratio and the maximum ratio stay flat after $n = 20$ and that the maximum ratio is higher for satisfiable formulas. The maximum after $n = 20$ for unsatisfiable formulas is a little less than 0.40, which suggests that we can set θ as small as 0.40.

Simulation by DNA Chemistry

In this simulation by DNA chemistry, variables are represented by unique oligonucleotide sequences, their complements, and adaptors. A total of approximately 500 different oligonucleotides are required to handle formulas with 120 variables. For 120 variables in the case of a satisfiable formula, we will need to discriminate between a complexity of $2.6 \times 10^{14} (= 2^{0.4 \cdot 120})$ and a complexity of 10^{15} to 10^{16} species.

The diagram below explains how the first two iterations of the loop of the algorithm are implemented. Steps 1, 2, and 3 respectively correspond to *append*, *polymerization synthesis*, and *quantitation*. For simplifying the description, here we assume that x_1 appears both positively and negatively in the input formula, so that we need to extend H in two ways by the append operation. We also assume that the two strands from the first iteration belong to $\mathbf{N}$ in the second iteration of the loop and thus we need to extend them in two ways.

The biochemical steps of the program are implemented as follows. Nanomole ($\sim 10^{-14}$ molecules) quantities of the header oligonucleotide H are labeled at the $5'$ end with a long half-life radioactive isotope of phosphorus (^{33}P) by the polynucleotide kinase reaction [**SFM**]. The efficiency of labeling is determined by standard biochemical techniques [**SFM**]. Step 1 for the first variable is implemented by ligating the first variable and its negation (which is not its complementary sequence) to the header with the help of the linker oligonucleotides l_1 and l_1', respectively. The excess headers and linker oligonucleotides are removed by electrophoresis [**OR**]. In step 2, each assignment is hybridized to an appropriate primer to synthesize the complement of the strands made in the previous step 1 of the algorithm. The next variable is then ligated as before, and the complementary strands are synthesized. The synthesis reaction is made by a highly processive version of the DNA polymerase enzyme. Several cycles of denaturation, annealing, and extension drive the reaction to completion. Note that at the end of this step, the total number of DNA strands will only double.

One set of the newly synthesized strands regenerates a new copy of the header, H, which is radioactively labeled. The unreacted headers and the linker oligonucleotides are separated by electrophoresis as before. For implementing step 3, the initial number of DNA strands in each test tube are "counted" by withdrawing a small fraction of the volume, measuring the amount of radioactivity in the sample, and from the sample volume and the specific activity of the radioisotope. Since each individual assignment is present with high redundancy (at least 10^{10} in these

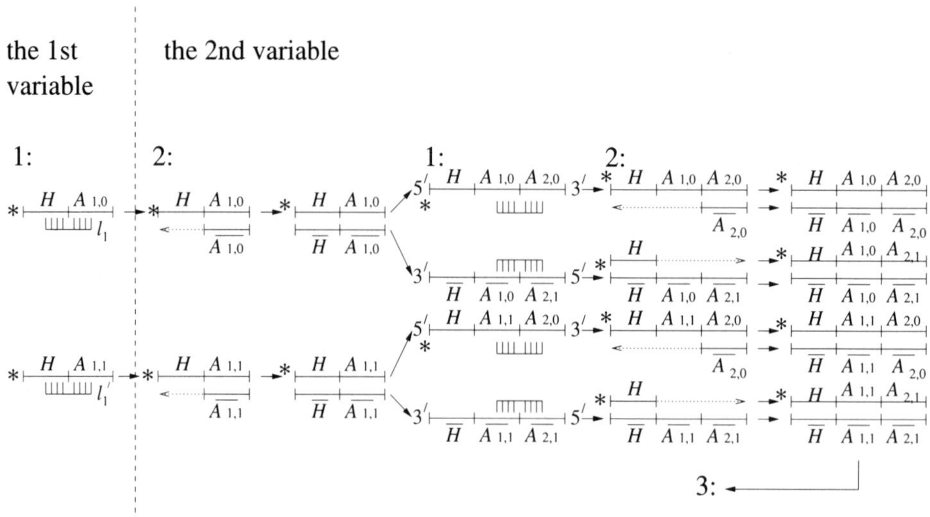

FIGURE 2: DIAGRAM

1. *Here H is always ^{33}P-labeled at the $5'$-end while $\overline{H}$ is not.*
2. *After each synthesis step, unreacted H is removed by size selection. This is done by electrophoresis* [**OR**].
3. *Total radioactivity before and after extraction of all non-satisfiable assignments is determined. The number of molecules is calculated from the measured radioactivity of $^{33}P \times$ (sp.act.)$^{-1}$.*

experiments), withdrawal of the small volume does not significantly deplete any one particular assignment from the total collection. Specific non-satisfiable assignments are extracted out by hybridization to magnetically-labeled oligonucleotide probes as described by Adleman [**A**], but each extraction step is repeated several times on the remainder of each previous step to completely exhaust from the mixture any non-satisfying assignment. The number of such extraction steps needed to accomplish this is empirically determined from controlled experiments with known complexity of sequences (the number of "extractable" sequences not extracted by chance at any one step diminishes exponentially). At the end of this step, the number of DNA strands are again counted by measuring radioactivity.

Although the molecular algorithm described above uses radioactivity as the detection and quantitation methods, it is possible to use faster, alternative techniques. One possibility is to use fluorescent labeled oligonucleotides [**SHH+**]. Hybridization can be done in microcapillaries, followed by fluorescent detection, such that each cycle will take approximately fifteen minutes. The total time for solving a 120 variable CNF SAT instance is 1,800 minutes, which is 2 to 3 order of magnitude slower than the current best digital computer algorithm.

Future Work

The proposed method need to be more rigorously studied theoretically, in particular from the following two angles:

1. estimating how small the parameter θ can be set and

2. designing better reordering methods.

As to the first problem, we note that the number of instances we tried is not very big. We may encounter a ratio μ larger than 0.40 for unsatisfiable formulas if we try more instances, in particular, of larger formulas. We plan to run more intensive experiments to ensure that our observation is correct. One technical obstacle for this would be the memory capacity. If a formula of 60 variables achieves $\mu = 0.45$, the maximum number of partial assignments to be held is $2^{27} = 128 \cdot 10^6$. Since a partial assignment of 60 variables is stored in 2 words, running the algorithm on the formula will require at least 256 words of memory, which may outnumber the memory capacity. Furthermore, sequential processing of such a big amount of partial assignments may be time consuming. We may need to revise the program so that it can handle larger formulas.

Is it possible to prove that μ never exceeds θ? To answer that question, obtaining average case analysis of the maximum number of partial assignments may be a good starting point. As mentioned before, there are a number of papers dealing with average case running time analysis of variants of the Davis-Putnam procedures [**GPB,PB,P,FP,CF,I**]. However, the methods shown in these papers do not seem to offer much direct help. First, the target of their analysis is the running time while ours is the maximum tree width. Second, the order of the variables is always $x_1, \cdots, x_n$ there while we determine it by a greedy method in our algorithm. Third, these papers allow clauses with more than three literals while we don't. So, we will need a different method for theoretical analysis.

As to the second problem, since the extension step preserves the size of $\mathbf{P_0}$ and $\mathbf{P_1}$ while doubles that of $\mathbf{N}$, we want the reordering method to make $\mathbf{N}$ as small as possible. On the other hand, when partial assignments are eliminated after an extension step, $\mathbf{S}$ seems to be significantly decreased if there are many constraints that the partial assignments have to satisfy. The current reordering method surely fulfills the second requirement. However, it is unclear whether the second requirement is met by this algorithm.

References

[A] L. Adleman, *Molecular computation of solutions to combinatorial problems*, Science **266** (1994), 1021–1024.

[BCGT] E. Bach, A. Condon, E. Glaser, and C. Tanguay, *DNA models and algorithms for NP-complete problems*, Proceedings of 11th Conference on Computational Complexity, IEEE Computer Society Press, Los Alamitos, CA, 1996, pp. 290–299.

[BF] R. Beigel and B. Fu, *Molecular computing, bounded nondeterminism, and efficient recursion*, Proceedings of the 24th Conference on Automata, Languages and Programming, Spring-Verlag Lecture Notes in Computer Science, Volume 1256, Berlin, Germany, 1997, pp. 816–826.

[CA] J. Crawford and L. Auton, *Experimental results on the crossover point in random 3-SAT*, Artifcial Intelligence **81** (1996), 31–57.

[CF] M. Chao and J. Franco, *Probabilistic analysis of two heuristics for the 3-satisfiability problem*, SIAM J. Comput. **15** (1986), 1106–1118.

[CS] C. Cantor and R. Schimmel, *Biophysical Chemistry*, vol. 3, W. H. Freeman, New York, NY, 1980.

[DP] M. Davis and H. Putnam, *A computing procedure for quantification theory*, J. Assoc. Comput. Mach. **7** (1960), 201–215.

[FP] J. Franco and M. Paull, *Probabilistic analysis of the Davis-Putnam procedure for solving the satisfiability problem*, Disc. Appl. Math. **5** (1983), 77–87; *Errata* **15** (1987), 295–299.

[FS] A. Frieze and S. Suen, *Analysis of two simple heusistics on a random instance of k-SAT*, J. Algorithms **20** (1996), 312–355.

[GPB] A. Goldberg, P. Purdom, and C. Brown, *Average time analysis of simplified Davis-Putnam procedures*, Inform. Process. Lett. **15** (1982), 72–75; *Corrigendum* **16** (1983), 213.

[H] J. Hartmanis, *The structural complexity column: on the weight of comptuation*, Bull. Europ. Assoc. Theoret. Comput. Sci. **55** (1995), 136–138.

[I] K. Iwama, *CNF satisfiability test by counting and polynomial average time*, SIAM J. Comput. **18** (1989), 385–391.

[K] O. Kullmann, *A systematical approach to 3-SAT-decision, yielding 3-SAT-decision in less than* 1.5045^n *steps*, Manuscript, February (1996).

[KKW] R. Karp, C. Kenyon, and O. Waarts, *Error-resilient DNA computation*, Proceedings of the 7th ACM-SIAM Symposium on Discrete Algorithms, ACM Press/SIAM, New York, NY/Philadelphia, PA, 1996, pp. 458–467.

[KMPS] A. Kamath, R. Motwani, K. Palem, and P. Spirakis, *Tail bounds for occupancy and the satisfiability threshold conjecture*, Proceedings of the 35th Symposium on Foundations of Computer Science, IEEE Computer Society Press, Los Alamitos, CA, 1994, pp. 592–604.

[L] R. Lipton, *DNA solutions of hard computational problems*, Science **268** (1995), 542–545.

[O] M. Ogihara, *Breadth first search 3SAT algorithms for DNA computers*, Technical Report TR 629, Department of Computer Science, University of Rochester, Rochester, NY, July (1996).

[OR] M. Ogihara and A. Ray, *Simulating boolean circuits on DNA computers*, Proceedings of 1st International Conference on Computational Molecular Biology, ACM Press, New York, NY, 1997, pp. 326–331.

[P] P. Purdom, Jr, *Solving satisfiability with less searching*, IEEE Trans. Pattern Anal. Mach. Intelligence **6** (1984), 510–513.

[PB] P. Purdom, Jr. and C. Brown, *The pure literal rule and polynomial average time*, SIAM J. Comput. **14** (1985), 943–953.

[SFM] J. Sambrook, E. F. Fritsch, and T. Maniatis, *Molecular Cloning: a Laboratory Manual*, 2nd edition (1989), Cold Spring Harbor Laboratory Press, Cold Spring Harbor, NY.

[SHH+] D. Simpson, J. Hoijer, W. Hsieh, C. Jou, J. Gordon, T. Theriault, R. Gamble, and J. Baldeschwieler, *Real-time detection of DNA hybridization and melting on oligonucleotide arrays by using optical wave guides*, Proc. Nat'l Acad. Sci. **92** (1995), 6379–6383.

[SML] B. Selman, D. Mitchell, and H. Levesque, *Generating hard satisfiability problem*, Artificial Intelligence **81** (1996), 17–29.

[T] S. Toda, *PP is as hard as the polynomial time hierarchy*, SIAM J. Comput. **20** (1991), 865–877.

[TO] S. Toda and M. Ogiwara, *Counting classes are at least as hard as the polynomial-time hierarchy*, SIAM J. Comput. **21** (1992), 316–328.

[V1] L. Valiant, *The complexity of computing the permanent*, Theoret. Comput. Sci. **8** (1979), 189–201.

[V2] ———, *The complexity of enumeration and reliability problems*, SIAM J. Comput. **8** (19179), 410–421.

Department of Computer Science, University of Rochester, Rochester, NY 14627, USA.
E-mail address: `ogihara@cs.rochester.edu`

Department of Biology University of Rochester, Rochester, NY 14627, USA.
E-mail address: `ray@ar.biology.rochester.edu`

DIMACS Series in Discrete Mathematics
and Theoretical Computer Science
Volume **48**, 1999

Parallel Computation on a DNA Substrate

Andrew J. Blumberg

ABSTRACT. Much work in DNA computing has focused on solution of problems under a passive computational paradigm; all possible solutions are generated and incorrect solutions are filtered away. In this paper, it is argued that such a model of computing is inadequate and a methodology for performing various forms of active computation is presented. Within this framework, I describe how to perform the parallel evaluation of continous functions, how to compute boolean circuits, and I present a direct simulation of a Connection Machine (a massively parallel SIMD computer). There is also a brief presentation of an implementation of a genetic algorithm. Although the algorithms are presented in an error-free regime, I discuss a strategies for ensuring that these algorithms could actually be carried out experimentally.

1. Introduction

In the past few years there has been very exciting research directed at performing computation using DNA and some tools of molecular biology. A significant amount of attention has been directed towards a constraint-based metaphor for solving search problems. That is, many algorithms depend on a generate and test methodology — DNA strands are synthesized such that all possible solutions to the problem in question are presented. Then, this multiset of strands is repeatedly filtered to remove all strands not satisfying some constraint. At the end, the only strands which remain in the tube are those which have satisfied all of the constraints — the solutions to the problem.

There are a number of advantages to this methodology. It is easy to conceptualize. The number of filtering operations performed depends on the complexity of the problem (*e.g.* the depth of the circuit or the size of the graph) but not on

1991 *Mathematics Subject Classification.* Primary 68Q10.

This paper benefited greatly from the direction and guidance of Thomas Knight, Jr. The suggestions of Dr. Knight were invaluable in developing this research. The support and suggestions of John C. Mallery helped enable this research to occur. This paper was improved by comments from Gerald Sussman, Carlos Bustamante, and William A.M. Blumberg, and various anonymous reviewers. This research was conducted at the Artificial Intelligence Laboratory of the Massachusetts Institute of Technology. Support for the Laboratory's artificial intelligence research is provided in part by the Advanced Research Projects Agency of the Department of Defense under Office of Naval Research contract number MDA972-93-1-003N7. The views and conclusions are those of the authors, not the Massachusetts Institute of Technology or the United States Government. Any errors or shortcomings are the responsibility of the author.

the possible number of solutions tested. Furthermore, this model simplifies error handling to some degree.

Unfortunately, for a number of reasons I believe that this conception of DNA computation will be of relatively limited utility. For problems which suffer exponential growth in the space of possible solutions as the size of the problem grows (*e.g.* NP-complete problems), the size of the possible solution space rapidly outstrips the maximum number of DNA strands that can fit in a reasonable amount of solution. For example, let us assume that we are unlikely to have more than 10^{23} strands of DNA in a liter of water (a rather optimistic assessment). In the case of the satisfiability problem (deciding on the existence of a satisfying assignment for a logical expression in CNF form), these strands will suffice to check at most about 76 variables ($2^{76} \approx 10^{23}$), presuming ideal conditions (*i.e.* each strand encodes a single assignment and the process is error-free). Dramatic increases in working volume will not solve this problem; not only will the linear increases in volume be rapidly outstripped by the exponential growth of the problem, but the increased volume will slow the reaction rate because of increased path length between molecules in solution.

Furthermore, this model of DNA computation phrases the computation as a passive process of filtering away incorrect solutions until only the correct one(s) remain. For many problems, it is vastly easier to express the computation as a process of applicative function computation. For example, consider the problem of attempting to minimize a continuous function. It is not clear how to express this in a constraint formalism. However, when phrased in terms of function computation, many solutions are possible on parallel hardware.

It is easy to see that this operation is extremely useful–the arguments are analogous to the arguments for parallel processing computers in general. For example, consider the area of image processing. For concreteness, let us take the example of filtering an 1024 X 1024 image. This is a relatively trivial undertaking given communication between processors–each processor assumes responsibility for a pixel, and the computation can be performed using a small number of local interactions. But in the constraint model, each strand would have to represent a possible solution to the filtering problem–requiring each strand to be at least a million base pairs long (depending on the efficiency of the encoding). Furthermore, an unfeasibly large number of constraints would have to be applied to filter the pool down to the correct filtered solution.

For other examples, consider problems in simulation–*e.g.* air-flow calculations, circuit simulation, *etc.* Given intercommunication between "processing elements", there exist simple algorithms for performing all of these on massively parallel hardware. Furthermore, even were it possible to do these things simply in the constraint model, a great deal of effort would have to be spent developing the algorithms. But why disregard the vast amount of prior research on parallel computing? It seems a much better solution to modify the model to allow us to tap into the rich existing body of parallel algorithms rather than require a whole new discipline of algorithm construction.

As such, it is extremely important to develop a model of DNA computation which allows easy implementation of general purpose parallel computation. There have been efforts in this direction already. In the early phases of research on DNA computing, a number of different efforts demonstrated the universality of computating embodied in DNA by simulation of Turing machines and the like [**4**],

[7]. Reif [17] raised similar concerns to ours and produced an abstract model of DNA computation in which he was able to easily simulate the operation of a CREW PRAM. We distinguish the work presented in this paper for the following reasons. First off, we will present a direct simulation of boolean circuits and a direct simulation of a Connection Machine [11]. The simulations derived in much of the previous work were not designed to be necessarily efficient, but merely to prove computational equivalence. Reif's simulation of a CREW PRAM is efficient, but I believe that there is some utility in discussing an alternative parallel architecture; namely, one which has "virtual routing" rather than a shared memory structure.

2. DNA Computing Model

The base instruction set of the computational model will be grounded on a DNA substrate as follows. A "legal" strand of DNA will correspond to a bit-string using the variable-length word encoding method discussed in Adleman [1] (where the strand consists of *<bit-label-0>* *<bit-value-0>* *<bit-label-0>* *<bit-label-1>* *<bit-value-1>* ...). In general, a strand will correspond to a specific processing element. It will be assumed herein that it is possible to choose labels in such a way so as to prevent overwhelming error associated with cleavage by restriction enzyme and during conversion of single-stranded DNA to double-stranded DNA. This assumption will be examined in more detail when we discuss error handling.

Any tube of DNA manipulated in the following discussions is assumed to contain a multiset populated by legal stands of DNA distributed according to the specific problem. We will assume the following operations can be performed on a tube of DNA :

1. Separate via constraints : In this operation a single tube of DNA is split into two tubes based on the given constraint; all strands satisfying the constraint in one tube, all strands failing to satisfy the constraint in the other tube. These constraints at the physical level are phrased in terms of the presence or absence of base pairs in the strand. However, we will typically consider "higher-order" constraints which refer to bits in the representation encoded on the strand, for conciseness.
2. Merge : Two tubes are mixed together to form one tube.
3. Append (bit or tag): Appends to all strands in a tube the specified object. Typically either bits or tags are appended. A bit is a triple consisting of *<bit-label>* *<bit-value>* *<bit-label>* as detailed above. A tag is simply a unique sequence of base pairs which we use to distinguish different parts of the strand. In discussions to follow, tags will be referred to with names like "NEW". These names are for convenience of discussion only and do not make any implications about the bases which make up the tag.

   These both are higher-order operations built on the basic physical operation of ligating a base.
4. Cut on tag : Cuts all strands in a tube along the specified tag. Actually performed using restriction enzymes.

3. Basic Algorithmic Structure

To describe the algorithms for computation, I need to clarify the exact level at which the algorithm is operating. In the constraint-based model, the instruction stream of the algorithm is embodied wholly in the physical processes going on. The data varies across strands, but the same thing is done to all strands at every step of the computation–the instruction stream is fixed.

Implementing traditional parallel computation (and in particular communication between processing elements) seems to depend on being able to modify the instruction stream. Herein, the modification of the instruction stream will be done by separating the set of strands into a small number of different tubes. For each of these sub-tubes, a distinct activity will be performed and then the tubes are merged together again. This notion turns out to be very powerful–it is never necessary to explicitly examine any strand, but the net result is that each strand appears to have been treated individually.

Essentially, this process takes advantage of localization of computational dependence. Consider data represented as a string of bits. If we can know a priori that the result of some computation on this string of bits only depends on the values of bits n and m, then we can perform this computation in the DNA model easily as follows. We separate all of the strands into four tubes, depending on the pairs of values of bits n and m. Then, for each of the four sub-tubes we perform whatever operation is indicated by the bit values which we know hold over all strands in the given sub-tube. Typically this will involve concatenating some value to the end of the strands. Finally, we merge all of the sub-tubes together again and perform the next operation.

The number of separations performed depends only on the algorithm and the length of the computation, not on any specific details of strands. Of course, the number of sub-tubes required increases exponentially in the number of dependent bits, so it is advantageous to keep the number of bits needed small. Fortunately, this turns out not to be a particularly restrictive constraint.

We will now discuss three applications of the techniques described above.

4. Static Logical Operators

The following operations all take place without communication between strands; each strand is treated as an argument to a function, and then parallel evaluation of the function is computed. This kind of application is somewhere in between the constraint-based paradigm and a general purpose parallel computation.

4.1. Logical Operators. Specify a logical operator, a multiset of n bit strings, and a single n bit string. The following operation will implement the combination (via the logical operator) of the single bit string with each member of the multiset.

Begin with the main tube containing strands representing the multiset of n bit strings.

1. Let *count* $= 0$ (note this is a bookkeeping step - no tube operations are involved).
2. Append a tag ("NEW") to all strands.
3. Separate into two tubes (A and B) based on the value of bit *count* (0 and 1 respectively).
4. (a) Compute a bit by applying the logical operator to 0 and bit *count* of the argument. Append this bit to all strands in tube A.
   (b) Compute a bit by applying the logical operator to 1 and bit *count* of the argument. Append this bit to all strands in tube B.
5. Merge the tubes A and B.
6. Increment *count*.

7. If *count* $< n$, goto step 3. Otherwise, fall through to step 8.
8. Cut all strands at the tag "NEW".
9. Separate into tubes OLD and NEW based on the presence or absence or tag "NEW"

In the tube labelled NEW will be the result of applying the logical operator and the argument to all of the strands in parallel. The tube NEW can now be used as the starting tube (perhaps after amplification using PCR to ensure enough copies of strands) for another operation.

It is possible to convert this algorithm so as to take both arguments from the strands—*e.g.* let each strand consist of the concatenation of the arguments. In this case there are four separation tubes (based on the possible pairings of values at each bit-position) but otherwise the algorithm remains the same. In this case the output will need to have the next argument concatenated to it to continue (although we can apply the uniform operation with global argument to the output at this point with no further processing).

The above algorithm requires 1 separation operation and 2 ligations for each bit in the input strings (although it is independent of the number of strings operated on). Thus, the total number of operations performed by this algorithm is $3n + k$, where k is a constant representing the cost of the initial set-up and the postprocessing.

4.2. Addition. The logical operation algorithm was perhaps the simplest kind of algorithm of the class we are presenting, in that the value of an output bit depended only on the corresponding input bits. The addition algorithm is somewhat more involved, as we must propagate a carry bit throughout the computation. Nonetheless, it is fundamentally similar.

The specification is identical to the specification of the logical operator algorithm (except that the operator here is fixed to be addition). Start with the working tube filled with strands representing the multiset.

1. Let *count* $= 0$ (note this is a bookkeeping step - no tube operations are involved).
2. Append a tag ("NEW") to all strands.
3. Separate into two tubes (C and notC) based on the presence of the tag "CARRY".
4. Separate both C and notC into two tubes based on the value of bit *count* (0 and 1 respectively).
5. (a) To the two tubes from C, remove the tag "CARRY" (this can be done by cutting and separating, since nothing follows this tag). Add 1, bit *count* of the argument, and 0 or 1 (depending on the tube) to get a result bit (and a carry bit). Append this result bit to the tube. Append the tag "CARRY" if the computed carry bit is 1.
   (b) To the two tubes from notC, add bit *count* of the argument and 0 or 1 (depending on the tube) to get a result bit (and a carry bit). Append this result bit to the tube. Append the tag "CARRY" if the computed carry bit is 1.
6. Merge all tubes.
7. Increment *count*.
8. If *count* $< n$, goto step 3. Otherwise, fall through to step 9.

9. Separate into two tubes (C and notC) based on the presence of the tag "CARRY".
10. Remove the tag "CARRY" from tube C (done as described above). Append a 1 to all strands.
11. Merge tubes.
12. Separate into tubes OLD and NEW based on the presence or absence or tag "NEW"

It is clear that this algorithm could also be transformed to perform additions where both arguments come from the bit string (as described in the logical operations algorithm), although this new algorithm will be rather involved. For each bit of the input strings, the above algorithm requires 6 separations, 6 ligations, and 2 cleavages thus implying that the total complexity of the algorithm is $14n+k$, where k again represents external constant costs.

4.3. Multiplication and Polynomials. Multiplication can be carried out in a number of fashions. One possibility is to use the traditional algorithm for multiplication taught in schools - compute terms (appropriately shifted) by multiplying by each individual bit of one of the multiplicands. Concatenate each such term onto the end of the strand and add to an accumulated result. This will require $O(n^2)$ time in the length of the string. A more clever way to do this is to implement a carry-save adder (a classical parallel adder design) [**13**].

Define the squaring operation according to the simple algorithm of concatenating the value of each strand to itself and then performing the multiplication algorithm. Using this squaring operation, it is easy to implement exponentiation. Consider the recursive expansions $b^n = (b^{\frac{n}{2}})^2$ if n is even, $b^n = b(b^{n-1})$ if n is odd [AS85]. Using these relations, we can represent the computation of any exponent of b in terms of two operations; squaring the current value and multiplying the current value by b. This can implemented in the model so long as we store the original argument value as well as a current "working value" on the strand. Then we multiply the working value by the original value or square the working value to form the new working value repeatedly as specified by the compiled recursive decomposition.

Given these operations, we can now evaluate arbitrary polynomials in parallel. Note that via approximation methods this permits us to compute arbitrary continuous functions. To compute a polynomial of degree d requires $O(dm^2 \log d)$ steps, where m is the maximum length in bits of any number appearing in the computation.

4.4. Realistic Implementation. Actually performing this algorithm would require consideration of a number of factors which are mostly overlooked in this discussion. For example, one would probably wish to decide on a floating point format, which would complicate the algorithms described above somewhat. Furthermore, some care in choice of approximation algorithm would be required to ensure that the necessary degree and coefficients can be controlled.

4.5. Observations and Comments. The fact that the format of the data is preserved means we can perform abstraction of operations in this model. The construction of the exponentiation algorithm demonstrates how this abiliity to compose these operations permits the high-level construction of functions from primitives far-removed from the DNA level which can subsequently be compiled down to DNA (although one should keep in mind that such a composition would be sensitive to issues of error propagation).

As an aside, it should be noted that this model of computation is quite well suited for reversible computation—it is possible to encode the process of the computation on the strand as the computation progresses, thereby permitting full reversibility. However, given that the energy requirements of computation using DNA are so favorable compared with silicon, it is not clear to me that there is much utility in this observation.

4.6. Applications to Minimization. One direct application of the algorithms discussed in this paper is to the minimization of an arbitrary continuous function. Via the Stone-Weierstrauss theorem, we can approximate arbitrary continuous functions using polynomials. So to perform the minimization of a function, compute the representation by polynomials (to desired accuracy). Then evaluate this polynomial function over a tube employing the algorithm described above. However, modify the evaluation algorithms slightly so as to preserve a copy of the original argument at the beginning of the string at all times (*e.g.* write a copy of the argument to the "NEW" string before performing the algorithm, etc.) — this is for read-out purposes. Having evaluated the function over the tubes, we now need to extract the minimal string(s).

For this, we need the $< x$ function. This returns all strands representing numbers less than the argument x. To do this, simply separate repeatedly according to the following algorithm.

1. Let *count* = the most significant nonzero bit of the argument.
2. Separate based on all bit-positions larger than *count*. Remove all such strings.
3. Separate the tube based on the value of bit *count*; if *count* is 0 merge into the result tube, if *count* is 1 keep in the working tube.
4. Set *count* to be the next most significant nonzero bit of the argument. If there are no more nonzero bits, return the result tube.
5. Otherwise, goto step 3.

This requires 1 separation and 1 merge for every nonzero bit of x, plus $n - x$ separations, where n is the maximum bitsize we are computing with.

Now using binary search we can find the set of minimum strands rather easily.

1. Set upper-bound = the largest number represented in the current scheme, lower-bound = 0, test = upper-bound.
2. Compute < test into tube A.
3. (a) If tube A is empty, set lower-bound to test. Goto step 4.
   (b) If tube A is not empty, set upper-bound to test. Goto step 4.
4. Set test to $\frac{(upper-bound)-test}{2}$. If upper-bound − lower-bound is within some tolerance level, stop and return < upper-bound as the result.
5. Otherwise goto step 4.

This algorithm requires $n \log n$ steps plus the cost of actually computing the functions, where n is the maximum number of bits in any number considered.

5. Parallel Computation

While interesting and potentially useful, the algorithms described above serve merely to perform in parallel a vast number of computations. We would like parallel computations which involve the sharing of information between strands. A key component of all such algorithms is the method of interstrand communication. One possibility is a shared memory structure, as described in Reif [**17**]. Herein we discuss a model which simulates a message routing architecture.

5.1. The Connection Machine.

5.1.1. *Processors.* The Connection Machine [**11**] consists of a large number of very simple processors. Each processor consists of memory, some flags, and simple logical circuitry. At each step, the processor reads two bits from memory and one flag bit and combines them to produce two bits of output and a flag output.

So let each strand of DNA represent a single processor (for the time being). The basic format for the strands will be as follows—a unique tag specifying the name of the processor, a sequence of bits representing the memory, and a sequence of bits representing the flags. This format will be expanded slightly during the following discussions as needed.

The instruction sequence is implemented at the level of the physical DNA manipulation operations. Given an instruction which specifies two read locations from memory, a flag to read, two write locations in memory, and a flag to write, we perform the following operation :

1. Separate the CM tube into an active tube and an inactive tube depending on the values of certain flag bits (this enables processors to be "turned off" to permit simulation of a wide variety of processor topologies). We will now only work with the active tube.
2. Separate the active tube into eight sub-tubes depending on the values of the read locations and the read flag; there are three bits which control the nature of the operation performed.
3. For each sub-tube, externally perform the logical operation and write the result bits to the write locations and the write flag. That is, the logical operation performed is specified at the instruction (physical) level. So for each sub-tube we compute externally the results and apply this by writing the correct bits and flag value.
4. Merge all of the tubes to reform the general CM tube.

I need to expand on the algorithm for performing the write operation described above. There are a number of ways to write values into the strands. The method we will describe here re-writes each strand, incorporating the new values.

1. Write the tag "NEW" to the end of each strand in the tube.
2. Set $count = 0$ (this is a bookkeeping step external to the tube).
3. If *count* is not equal to one of the two write locations, separate into two tubes based on the value of bit *count*.
4. In this case, we are simply copying the current bit-values.
    (a) For tube 0, append the value 0 to the end of the strand.
    (b) For tube 1, append the value 1 to the end of the strand.
    (c) Merge tubes.
5. If *count* is equal to one of the write locations, append the bit value to be written to all strands.
6. Increment *count*. If $count <$ the size of the processor strand (total number of bits), goto step 3.
7. Cut strands at "NEW" tag.
8. Separate based on presence of "NEW" tag. Throw away old strands.

At the completion of this algorithm, the strands are in the same format as before and so the operation can be iterated (presuming we have a strategy for controlling the inevitable errors which will arise in this computation).

However, there is one more feature of the processor operation in the Connection Machine we have overlooked–the global bit. Each processor can set a global bit, the OR of which is available at the instruction level. This is easily implemented in the DNA model by simply filtering for strands with the proper bit set and performing the detection operation to determine if any such strands exist.

The above simulation requires $O(k)$ DNA operations for every instruction simulated, where k is the length of the Connection Machine strand. With some overhead for various housekeeping tasks, this essentially depends on the size of the memory of the processor. Thus, to simulate an n-instruction program would require time of $O(nk)$ (ignoring the cost of routing presently).

This conception of SIMD computing uses the DNA itself simply as memory—the instruction operations are all carried out at the level of tube operations. This is possible since the power of our operation set enables the strands of DNA to simply contain the data which varies over the processors—there is no need to encode information about how to compute at each processor. This is actually a substantial improvement in space efficiency over the silicon Connection Machine (which had a small general logic unit in each processor).

5.1.2. *Interconnect.* For the most part in the preceding discussion, I have been working with an idealized model of the actual DNA processing in which it is assumed that all of the operations have 100% efficiencies and introduce no errors. This assumption is valid insofar as it is possible to convert these algorithms to error tolerant computations using redundancy and clever composition of operations. However, when discussing the interconnection algorithm we need to consider certain pragmatic issues in order to be assured that our algorithms are correct.

Specifically, we have the following problem. In a realistic implementation of the machines described herein, there will necessarily be redundancy—several thousand identical DNA strands corresponding to a given processor or gate. As long as we ensure that at no time do any processors disappear due to loss of DNA, then we can be assured by subsequently discussed techniques that most of the processors will be in identical and correct states. However, communication between processors raises new and grave dangers. Specifically, any communication algorithm we consider must ensure that each copy of a given processor receives the same information in a given time-step; otherwise we risk divergent behaviors. Although it might be interesting to consider the nondeterministic model of computation supported by having processors which diverge at run-time, for the current applications such divergence will be regarded as a problem.

First, I will present an algorithm whereby a sender can transmit a message to a recipient *which is willing to accept messages from that particular sender*. This algorithm will then be used as the basic building-block of a protocol which will guarantee consistency. In the following algorithm description, the dash will indicate concatenation and the function *WC* will indicate the Watson-Crick complement. Further, assume each processor has been given an unique ID tag.

1. Concatenate to the sending processors the string *"MESSAGE1" (a tag)-WC(ID of recipient)-WC(ID of sender)-WC(data to be sent)-"MESSAGE2" (a tag)*.

2. Cut the strands after the "MESSAGE1" tag.
3. The messages (denoted by the "MESSAGE2" tag) are separated from the tube.
4. (Optional) The messages are amplified.
5. Remove the "MESSAGE2" tag from the messages.
6. Remove the "MESSAGE1" tag from the strands.
7. Concatenate to a given processor its ID tag and the ID tag of the processor it is presently willing to receive data from.
8. The messages are merged with the processor strands. Since the message consists of two ID tags and the data, this will result in the message sticking to the end of a recipient processor strand (with appropriate ID tags) with the message data hanging off the end. Adding polymerase will then result in the copying of the complement of the data onto the processor strand.
9. The messages are melted off the processors and separated out.
10. The message is incorporated into the appropriate location of the strand by a process of cutting and insertion.

Employing this algorithm, I will now describe how we can build hypercubes and thereby implement traditional packet-routing algorithms on our DNA Connection Machine. Let us further expand the representation for each processor such that a given processor incorporates arbitrarily ordered lists of the m processors it can receive from and the m processors it can send to.

Now, perform routing by an $2m$ stage process. During the first m stages, each processor can generate a message to be sent to the mth processor on the list it can send to. Then, in each of the next m stages the processors concatenate with the ID tag of the mth processor they can receive from and then hybridize with appropriate messages. Since at each receiving stage only a single class of message can hybridize to a given type of processor, we are guaranteed that all processors of a specific type will receive the same data (presuming a sufficient quantity of message strands).

Let's clarify this a little bit via an example. Consider process number 11. This processor can receive messages from processors 1, 3, and 6 and can send messages to 1, 2, and 4. So now we will walk through a step of operation focused on this processor. First, the processor performs some computation depending on the global instruction and its flags and memory bits (as described above)–this involves a variety of separations and mixings, but we aren't particularly interested in this process presently. At the conclusion of the computation stage, we enter the communication stage.

Now we go through three sending stages. At the first stage, processor 11 has the option to send a message intended for processor 1. If its program specifies, it will generate a message (as described in the algorithm above) which contains its ID (11), the ID of the recipient processor (1), and some data. At the second stage, 11 can send a message intended for processor 3, and at the last stage it can send a message intended for processor 6. This three-stage event is performed for all processors simultaneously.

Finally we have three receiving stages. At the first stage, processor 11 generates an appropriate tag at its end such that it will hybridize with messages sent from processor 1 to processor 11. All of the messages generated in the sending stage are mixed with the processor strands and allowed to hybridize. If something does hybridize to processor 11, we know this is a message from processor 1 and it can be incorporated into the memory of processor 11 (via the described copying process) appropriately. Next, processor 11 generates the tag to accept messages from processor 2 to processor 11 (having removed the old tag) and similarly incorporates such messages if they arrive. Finally, processor 11 accepts messages from processor 4.

The reason for the incorporation into the recipient of both its own ID and the ID of the processor it is willing to receive from is that this guarantees that at any time, all of the copies of a given processor which receive a message will receive only the same message–even if multiple messages for that particular processor exist. In an implementation in which the processor simply received messages based on its ID, if two different processors sent to it different copies could receive different messages and there would be no local way to determine who had received what.

This scheme is very well-suited for the implementation of a hypercube. At any given routing step a processor can receive information from all of the processors "connected" to it and send information to all of the processors "connected" to it. Expanding to include more processors is trivial—the only consequence is an increase in the size of a given processor strand. The capacity of the lines is dependent on the amount of space on a strand the implementor is willing to devote to message storage. Also, note that since the lists of processors are stored in "memory" on the strand, it should be possible to implement dynamic connection changes depending on software.

The time required for this algorithm to operate is $O(k + \log P)$, where k is the length of a strand (for the copying process) and P is the number of processors. Since it may take up to $\log P$ routing stages for a message to reach its destination, the total time to simulate an n-instruction program is $O(n(k + k\log P + \log^2 P)) = O(nk\log P + n\log^2 P)$.

5.2. Possible Impediments to Realization. There are some potential problems with implementation that should be addressed. First off, the size of the hypercube so simulated may be restricted by the length of the tags of the processors; requiring each processor to keep lists of processors on them increases the length in dramatic and unwholesome ways, as it is unclear what the maximum workable length of a strand of DNA is. In order to ease this problem, it is also possible to restructure the above algorithm so it supports processors which are distributed across multiple strands (for example, lists of processors on one strand, memory on another, *etc.*).

In addition, there are potential problems with solution dynamics; it would be useful to perform some experiments and calculations to understand what the message volume would have to be for there to be a reasonable probability of message receipt in relatively short amounts of time.

6. Boolean Circuits

It should be clear that is now easy to simulate boolean circuits using similar methods. For comparison, consider the work of Ogihara and Ray [**16**]. First, assume the circuits we're concerned with consist of AND and OR gates ordered into levels, and each AND and OR gate has fan-in two. Let each strand represent a gate, with an identifier, a number indicated what level this gate appears at, a tag describing the type of gate, and the pair of identifiers representing the gates which input to this gate. Initialize the algorithm such that the gates at level 0 have the input attached (concatenated on). Then we can evaluate the circuit using the following algorithm.

1. Set *count* to 0.
2. Separate out the gates of stage *count*.
3. Separate the removed gates into two tubes, AND and OR.
4. For each tube, perform the appropriate logical operation (recall that the input is attached to the strand of the gate already), cut off the input, and append the result.
5. Merge the tubes.
6. For each gate, generate a message containing the ID of the gate as well as the result of its computation.
7. Mix this into the main tube (specifically with stage $count + 1$) and perform two-phase recognition; each gate looks for the two gates it receives from. Ligate the received messages onto the strands.
8. Unless $count =$ the depth of the circuit, return to step 2.

This simulation requires d stages (where d is the depth of the circuit). However, in order to count properly we should note that in fact this simulation requires a series of amplifications of the messages. The number of messages we need is clearly bounded by the maximum fan-out F of any gate. Thus, the number of PCR stages needed is $\log_2 F$, since each such stage doubles the number of strands (approximately).

7. Genetic Algorithms

As discussed in the introduction, the utility of using DNA computation for exponential search problems is limited. However, search problems (in particular exponential ones) are of critical importance. Thus, it behooves us to attempt to provide methods for enhancing the solution of such problems even when the solution space is too large for direct attack. The genetic algorithms methodology has proven useful in the context of traditional computers for such problems. Herein I present a simple implementation strategy for genetic algorithms within our model.

The essential loop in a genetic algorithm is as follows.

1. Initialize the population of organisms (typically represented by bit-strings of "genes").
2. Evaluate the fitness of the population according to some fitness function determined by the nature of the problem.
3. Throw out all but the most fit organisms (according to some percentage criterion) and fill the population by combining the "surviving" organisms according to some scheme attempting to mimic sexual reproduction.
4. Goto step 1.

But we can fairly easily implement this in our DNA-based model. We represent each organism as a bit-string representing its genome. We have already described herein how to implement the fitness function and cull out the bottom performers (same as the maximization procedure). Now, given our resulting organisms, we can generate new organisms as follows.

1. Make a copy of the parent tube to work with.
2. Perform arbitrary concatenation of pairs in the tube. This can be performed by adding tags to each strand so that each strand has only one "sticky" end, thus enforcing pairwise (but random) concatenation.

3. Now, perform the following logical operation on the pairs. Iterate over the bits in the pairs. If both bits are the same, append this bit to the end of the strand. If the bits are different, append a random bit to the strand. The random bit can be added by slightly modifying the operation in which a fixed bit is added. In that operation, many copies of an oligonucleotide representing the bit value in question are added to the working tube and ligated. To add a random bit, add to the working tube (in roughly equal proportions) copies of the oligonucleotides for both bit values. Then the specific value ligated to a specific strand is essentially random.
4. Cut off the offspring, and mix with the original parent tube to produce the new working tube to repeat the process on (*i.e.* evaluate the fitness function, cull, *etc.*).

No doubt it is possible (and desirable) to construct algorithms corresponding to more elaborate recombination schemes—but this algorithm should be indicative of the basic strategy and take more as proof of concept than as a recommended genetic algorithm.

8. Implementation Issues

8.1. Selection of Tags. In the preceeding discussion it was tacitly assumed that it is possible to select tags and bit labels such that there would be little interference between different tags and labels. Since the number of tags and bit labels we use is relatively small, one could imagine doing this simply via random search (prior to any computation). In fact, the problem can be solved directly as an exercise in coding theory [**16**].

8.2. Error. There are a wide variety of barriers to an actual implementation of the algorithms presented herein. If we ignore the various issues of solution dynamics mentioned above, the biggest problem is the propagation of error. With the exception of the merge operation, all of the operations discussed above have appreciable error, especially in the solution-based embodiment of DNA computing.

The worst culprit in this regard is without a doubt the operation of separation. Magnetic bead separation is tremendously error-prone, and this is the bane of realizing DNA computing presently. There are a variety of proposed solutions. Liu *et. al.* [**15**] and Cai *et. al.* [**9**] suggest a solution via surface chemistry; the operations embodying separation in this paradigm appear to be considerably more reliable than magnetic bead separation. Other solutions involve the idea of repeated separation; these range considerably in sophistication. Adleman proposes a fairly simple process, aided by the fact that his applications simply require that some good strands survive. Boneh and Lipton [**8**] suggest a methodology for performing repeated separation with amplification. Extremely elegant results of Karp *et. al.* [**13**] provide a demonstration for how one could assemble repeated separations to perform reliable separation without employing amplification. Thus, there is cause to be optimistic about the possibilities for implementing the algorithms we have discussed.

Unfortunately, it is unclear that these results will really help us. Karp's beautiful techniques require separation to be error prone but lossless. In reality, magnetic bead separation can result in enormous loss of DNA (with DNA sticking to the beads) [**12**]. It is possible that the surface chemistry solution will not involve significant losses of DNA; in this case, there is a viable procedure. Otherwise, it is necessary to turn to PCR, as was done in [**8**]. Lamentably, PCR is also error-prone [**12**], although considerably less so than bead separation. It is unclear what the proper answer to this issue is. We can't use separation techniques to verify PCR, so traditional error-checking techniques likely cannot help us, as they would require multiple separations simply to check the result of a PCR. If the rate of separation loss is sufficiently low, it might be possible to use PCR to produce such enormous quantities of DNA that we could in fact afford to use repeated separation to verify the strands and end with a still useable amount of DNA. Another possibility to investigate is to consider running multiple PCR trials in parallel, and then perhaps keeping only DNA which appears in multiple trials. Unfortunately, I am not sure how one would match DNA between trials; strands which differ in only a few base pairs essentially appear the same unless we pull them out individually.

The only other process which might have substantial error is ligation. Since the only ligation that occurs in the above algorithms is uniform across all strands in a tube, we can use separation to check that the ligation has succeeded (presupposing the problem of separation is satisfactorily addressed).

Thus, the fundamental conclusion of this discussion on practical implementation must be one of uncertainty; as the perennial cry goes, we need more experimental experience to know what works and what doesn't.

9. Conclusion

In this paper, it was suggested that the traditional model of DNA computation via constraint-based search is limited. In its stead, I have presented a technique for performing general purpose parallel computation with DNA. Using this technique, we discussed a variety of applications, primarily parallel function evaluation, a Connection machine simulation, and the simulation of boolean circuits. As the simulations are direct, we obtained time bounds somewhat better than what one might obtain via more roundabout simulations. It is hoped that these simulations will help reinforce the point that DNA is a viable medium for active computation as well as the constraint-based search model of computation.

References

[1] L. M. Adleman, Molecular computation of solutions to combinatorial problems. *Science* 266:1021-1024, 1994.

[2] ———, On constructing a molecular computer. Manuscript, Computer Science Department, University of Southern California, 1995.

[3] H. Abelson and G. J. Sussman, *The Structure and Interpretation of Computer Programs*. MIT Press, Cambridge, MA, 1985.

[4] D. Beaver, A universal molecular computer. Penn State University CS Tech Report CSE-95-001, 1995.

[5] A. J. Blumberg, Parallel function application on a DNA substrate. MIT Artificial Intelligence Laboratory Memo 1588, 1996.

[6] ———, General purpose parallel computation on a DNA substrate. MIT Artificial Intelligence Laboratory Memo 1589, 1996.

[7] D. Boneh, C. Dunworth, R. J. Lipton, and J. Sgall, On the computational power of DNA. Princeton University CS Tech Report CS-TR-489-95, 1995.

[8] D. Boneh and R. J. Lipton, Making DNA computers error resistant. Princeton University CS Tech Report CS-TR-499-95.

[9] W. Cai, A. E. Condon, R. M. Corn, E. Glaser, Z. Fei, T. Frutos, Z. Guo, M. Lagally, Q. Liu, L. M. Smith, and A. Thiel, The power of surface-based DNA computation. Manuscript, Chemistry Department, University of Wisconsin, Madison, 1996.

[10] F. Guarnieri, M. Fliss, and C. Bancroft, Making DNA add. *Science* 273:220-223, 1996.

[11] W. D. Hillis, *The* Connection Machine. MIT Press, Cambridge, MA, 1985.
[12] P. Kaplan, G. Cecchi, and A. Libchaber, Molecular computation : Adleman's experiment repeated. NEC Research Institute Technical Report 95-120, 1995.
[13] R. M. Karp, C. Kenyon, and O. Waarts, Error-resilient DNA computation. 7th Annual ACM-SIAM Symposium on Discrete Algorithms, 1996.
[14] R. Lipton, Speeding up computations via molecular biology. Manuscript, Princeton University, 1994.
[15] Q. Liu, Z. Guo, A. E. Condon, R. M. Corn, M. G. Lagally, and L. M. Smith, A surface-based approach to DNA computation. Manuscript, Chemistry Department, University of Wisconsin, Madison, 1996.
[16] M. Ogihara and A. Ray, Simulating boolean circuits on a DNA computer. Rochester University CS Tech Report 631, 1996.
[17] J. Reif, Parallel molecular computation : models and simulations. 7th Annual ACM Symposium on Parallel Algorithms and Architectures, 1995.

MIT ARTIFICIAL INTELLIGENCE LABORATORY, 545 TECH SQUARE, NE43-757, CAMBRIDGE, MA 02139

E-mail address: blumberg@ai.mit.edu

DIMACS Series in Discrete Mathematics
and Theoretical Computer Science
Volume **48**, 1999

Design for a DNA Conformational Processor

Michael Conrad and Klaus-Peter Zauner

ABSTRACT. We describe a conceptual model of DNA-based computing that utilizes conformational dynamics for pattern classification. Input signals are coded into unmethylated and methylated oligonucleotides. Hybridization is used to arrange these input oligonucleotides on a backbone strand that contains complementary sequences. Different input signal patterns are thus re-represented as DNA duplexes with distinctly different conformational dynamics, in particular different equilibria of B and Z DNA that can be read out using circular dichroism. The system could be trained to carry out desired pattern recognition functions (input-output mappings) through evolutionary procedures.

1. Introduction

The recent high interest in DNA computing stems from Adleman's insight that it is possible to use the pattern matching intrinsic to DNA hybridization to support formal computations [**Adl94**]. The conformational polymorphism of DNA ("secondary effects") is either irrelevant to the computation or must be suppressed so as not to interfere. The purpose of the present paper is to exhibit a model that utilizes these secondary effects, which can be substantial [**KS95**], for pattern classification tasks. Pattern matching associated with hybridization is utilized; but the main computational work is done by the conformational dynamics that occur subsequent to the hybridization step.

The model may be very roughly analogized to a neural net which uses the interactions among neurons to combine inputs in space and time and which must be molded for specific function through an adaptive procedure. In the case of our proposed DNA device the input-output transform is mediated by the interactions among the various constituents of the DNA molecule (the various electrons and atomic nuclei). The DNA molecule is in effect used as a processing network that is molded for desired pattern classification tasks through an evolutionary adaptation procedure. The dynamics of DNA secondary structure formation supports the

1991 *Mathematics Subject Classification.* Primary 68Q05; Secondary 68Q10, 68T10, 68T05.

Key words and phrases. DNA computing, conformational processing, pattern recognition, context sensitivity, molecular computing.

This paper is based on research supported by NSF Grant No. ECS-9409780. We thank Svetlana Morosyuk for helpful discussion.

computation in the same general sense that the dynamics of neural networks support computations.

2. Computational Concept

For the purposes of the proposed design we concentrate on the capacity of DNA to switch between right (B) and left (Z) forms in response to methylation of cytosine residues. But before describing the methylation-based model let us briefly consider the operative principle (illustrated in Fig. 1).

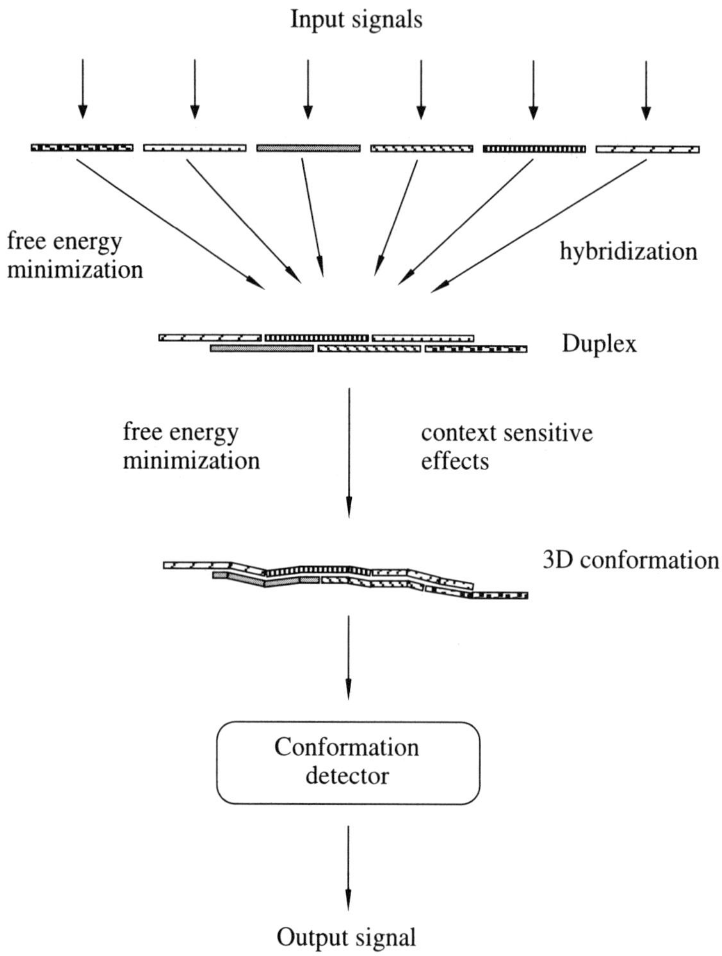

FIGURE 1. General scheme of DNA conformational processing. Input signal patterns are re-represented as different oligonucleotide sequences. These hybridize to form predictable duplexes. The formation of the DNA secondary structures from these duplexes fuses the inputs in a context sensitive manner. The response of the detector to the 3D conformation may then be used to classify the input patterns.

Imagine that input signals arriving along different lines are coded into different base sequences and these are allowed to hybridize to form one or more duplexes. A ligase reaction could be introduced at this point. The duplex in general would be a double helix in the B form. Actually the conformation (to be referred to as the secondary structure) would exhibit slight deviations from a perfect Watson-Crick double helix that depend on base sequence in a context sensitive manner [**GBS90**]. The deviations are determined by the contributing base sequences and therefore by the initial input signals. In effect the input signals are fused into a 3-D conformation by the interactions among the bases. Different features of the conformation would be associated with different groupings of the initial input patterns.

Now imagine that we have a detector that is sufficiently sensitive to measure these slight deviations. This detector would then be sensitive to the input signal pattern through the resulting collective conformational effects. The important point is that input signals (e.g., electric pulses or light signals) have no such collective properties. These are created by the dynamics of secondary structure formation.

The formation of these secondary structures is a free energy minimization process, as is hybridization. The new feature is that the free energy of structure formation is now recruited for solving a computational problem, namely the classification of the initial input signal patterns [**Con92**].

The difficulty with the above scheme is that conformational effects are very small and therefore the sensitivity required of the detector is substantial. On a molecular scale antibodies or enzymes could serve as highly sensitive detectors. However, if methylation is introduced into the scheme it is possible to use macroscopic methods of detection, since transitions between B and Z forms of DNA constitute a major and spectroscopically observable conformational effect.

3. The DNA Conformational Processor

The proposed conformational processor is illustrated in Fig. 2. Input signals arrive along different input lines, each of which is in one of two states. For convenience we will denote these by 0 and 1. A different oligonucleotide is assigned to each input line. An unmethylated version of this sequence is released if the input signal is 0; otherwise a methylated version is released. We will call these the input oligonucleotides. The solution into which these sequences are released contains single stranded DNA molecules with subsequences that are complementary to the released oligonucleotides. We will refer to these as the backbone sequence (a concept introduced in the framework of a formal model by Roweis et al. [**RWB$^+$**]). The released input oligonucleotides hybridize with these backbone molecules, and are therefore lined up according to the arrangement of their complementary sequences on the backbone. A ligase reaction can be used to fix this arrangement.

Different input signal patterns will give rise to different patterns of methylation of the duplex formed during hybridization. The different methylation patterns will in turn give rise to a different equilibrium of B and Z DNA, with the particular equilibrium depending on environmental factors (pH, salt concentration, temperature, solvent properties, and possibly interactions with proteins, cf. [**Zac93**]). The extent to which Z DNA forms is conveniently detected through circular dichroism [**RNW84**]. The circular dichroism (CD) spectrum is then used to produce an output signal that classifies the input pattern.

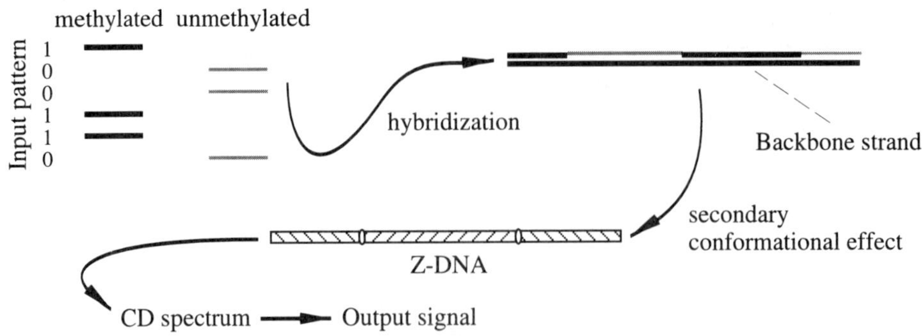

FIGURE 2. B-Z version of DNA conformational processing. Input signals arrive along different input lines, each associated with a specific oligonucleotide sequence. If the input line carries a 1 the corresponding oligonucleotide is methylated; if it carries a 0 it is not methylated. The oligonucleotides are allowed to hybridize with a single stranded (backbone) DNA molecule that contains sequences complementary to those associated with the input lines. The resulting hybrid (which may be fixed through ligation) will exhibit a ratio of Z to B DNA that depends on the input pattern. Circular dichroism (CD) spectroscopy is used to produce the output signal.

The pattern recognition capability of the above setup draws on the cooperativity of the transition from B to Z forms. The B form is strongly favored under common conditions. The junctions between B and Z regions are energetically unfavorable. Whether the B to Z transition occurs in a particular region depends on the context provided by the conformational state of neighboring regions [**SB81, Sas83**]. The input signals are fused in a nonlinear fashion into the final conformational form by these context sensitive interactions.

4. Context Sensitive "Programming"

The context sensitivity of the DNA conformational processor precludes conventional programming. Suppose we desire a particular pattern classification. It is in general not possible to use a set of simple instruction-like rules to determine what combination of backbone and environment would implement this mapping [**Con85, Con88**]. Conformational self-organization is, however, highly amenable to adaptation though directed evolution.

Possible features that can be subject to variation and selection include:

1. The arrangement of input oligonucleotide complements on the backbone. This would alter the strength of the interactions between different input lines and also could be used to increase the weight of particular input lines (by including multiple complements for these lines on the backbone). Simultaneous use of different backbones is also possible.
2. The base sequences of the oligonucleotides used to code the input signals. The backbone would have to be altered accordingly.
3. Environmental factors affecting the B-Z equilibrium. Altering these would allow for fine tuning the B-Z transitions and therefore for fine tuning the pattern grouping functionality.

4. Trigger level of the output signal. The strength of the CD signal used to decide whether the duplex is to be classified as sufficiently Z or sufficiently B is also a variable parameter.

It is possible to modulate the degree of conformation change produced by base sequence variation. Increasing the length of input oligonucleotides by adding neutral nucleotides (i.e., nucleotides that are unmethylated for both 0 and 1 signal states) would weaken B-Z cooperative effects. Including superfluous input lines (represented by both input oligonucleotides and complementary backbone sequences) would also serve to reduce interactions between given input lines. If the superfluous lines were always unmethylated this would increase the tendency of its neighborhood to assume a B state. If it were always methylated it would increase the tendency to assume a Z state. The presence of such redundant elements has a facilitating effect on the evolution of biological systems, since it serves to open up traversable pathways of evolutionary adaptation [**Con72, Con90, CBS86**].

5. Parallelism and Scalability

The conformational self-organization of every single hybridized duplex is a highly parallel process, since numerous electrons and atomic nuclei contribute simultaneously. In effect the search space is explored by the different pathways of secondary structure formation. Note that parallelism here enters subsequent to the hybridization process. It therefore has a different character than in the Adleman scheme, which explores the search space by utilizing the different possible pathways of hybrid formation (cf. [**ZC96**]).

The conformational processor also uses the parallelism of hybrid formation, but does so to sort the input sequences into their arrangement on the backbone. This allows for pattern processors that have an extremely large signal fan-in. The base sequences used to code the input must contain cytosines, since it is these that are methylated. The set of base sequences available for input lines thus grows as $4^L - 3^L$, where L is the length of the sequences and $3L$ is the number of sequences not containing cytosine. Lengths between 20 and 100 are reasonable. Within the range of environmental conditions the sequence should have the potential to form a Z structure. This is most favorable for alternating CG sequences and to a lesser extent for CA sequences [**RNW84**]. A compromise between the potential to form Z structures and the specificity of the hybridization can be reached by utilizing characteristic patterns of purine-pyrimidine alternations. The tendency to form Z DNA could, if necessary, be enhanced by methylating some regions of the backbone. The sequences should have sufficient potential methylation sites to distinguish the two input states. Also it would be desirable to preserve a portion of the sequence space for evolutionary variation. But even with these restrictions the number of possible input lines would be enormous.

Each input line would realistically be represented by a large number of identical molecules. This number could be in the hundreds of thousands without creating an undue burden in terms of the total amount of DNA. Micromolar concentrations are reasonable. For example, suppose that signal patterns arrive along 10^4 input lines. Each of these lines could be associated with 10^{10} DNA molecules. Ideally this would lead to the self-assembly of 10^{10} identical duplexes, each with the conformation representing the solution of the pattern recognition problem. It is not necessary that all of these duplexes form in order to obtain a result as long as a sufficient number of duplexes form to be detectable in the CD spectrum. The numbers assumed in this example would be equivalent to 500*pmol* DNA for all signal sequences in 100μl of reaction solution. These are quantities typical of standard protocols (cf. [**Har96**]).

6. Concluding Remarks

Conformational pattern processors were originally proposed in a framework oriented to protein interactions [**Con86**]. The conformational dynamics of DNA is much more restricted than for proteins, as would be expected from a molecule with information storage functionality. The advantage is that it is easy to predict which complexes will form on the basis of nucleotide sequence, whereas such prediction is generally infeasible on the basis of amino acid sequence. A second advantage is that it is relatively easy, given presently available methods, to produce desired DNA sequences and to vary them for the purposes of evolutionary adaptation. RNA would combine the advantage of predictable hybridization with some of the 3D conformational richness of proteins. Different problem domains would undoubtedly dictate whether DNA, RNA, or protein (or some combination of these) is the material of choice.

Our discussion of conformation-driven computing has been presented in what at first sight might appear to be the restricted domain of pattern recognition. General purpose digital computers can be viewed as built up out of simple pattern recognizing elements, however. The NAND gate, for example, recognizes simple two bit patterns. One can imagine computational architectures, most likely of a highly specialized nature, built up from powerful pattern recognizers operating on the conformational principle.

References

[Adl94] L. M. Adleman, *Molecular computation of solutions to combinatorial problems*, Science **266** (1994), 1021–1024.

[CBS86] M. Conrad, S. Brahmachari, and V. Sasisekharan, *DNA structural variability as a factor in gene expression and evolution*, BioSystems **19** (1986), 123–126.

[Con72] M. Conrad, *Information processing in molecular systems*, Currents in Modern Biology (now BioSystems) **5** (1972), no. 1, 1–14.

[Con85] M. Conrad, *On design priciples for a molecular computer*, Communications of the ACM **28** (1985), 464–479.

[Con86] M. Conrad, *The lure of molecular computing*, IEEE Spectrum **23** (1986), no. 10, 55–60.

[Con88] M. Conrad, *The price of programmability*, The Universal Turing Machine: A Fifty Year Survey (R. Herken, ed.), Oxford University Press, Oxford, 1988, pp. 285–307.

[Con90] M. Conrad, *Molecular computing*, Advances in Computers (M. C. Yovits, ed.), vol. 31, Academic Press, Inc., Boston, 1990, pp. 235–324.

[Con92] M. Conrad, *Molecular computing: The lock-key paradigm*, Computer (IEEE) **25** (1992), no. 11, 11–20.

[GBS90] G. Gupta, M. Bansal, and V. Sasisekharan, *Polymorphism and conformational flexibility of DNA: right and left duplexes*, Int. J. Biol. Macromol. **2** (1990), 368–380.

[Har96] A. J. Harwood, *Basic DNA and RNA protocols*, Methods in Molecular Biology, vol. 58, Humana Press, Totowa, 1996.

[KS95] O. Kennard and S. A. Salisbury, *DNA structure*, Molecular Biology and Biotechnology (R. A. Meyer, ed.), VCH Publishers (Verlag Chemie), Weinheim, 1995, pp. 242–247.

[RNW84] A. Rich, A. Nordheim, and A. H.-J. Wang, *The chemistry and biology of left-handed Z-DNA,*, Annu. Rev. Biochem. **53** (1984), 791–846.

[RWB$^+$] S. Roweis, E. Winfree, R. Burgoyne, N. V. Chelyapov, M. F. Goodman, P. W. K Rothemund, and L. M. Adleman, *A sticker based model for DNA computation*, (presented at the Second Annual Meeting on DNA Based Computing, Princeton University, June, 10-12,1996).

[Sas83] V. Sasisekharan, *Left-handed DNA duplexes*, Cold Spring Harbor Symp. Quant. Biol. **47** (1983), 45–52.

[SB81] V. Sasisekharan and S. K. Brahmachari, *B to Z transitions in DNA fibre: the question of handedness of the duplex*, Curr. Sci. **50** (1981), 10–13.

[Zac93] W. Zacharias, *Methylation of cytosine influences the DNA structure*, DNA Methylation: Molecular Biology and Biological Significance (J. P. Jost and H. P. Saluz, eds.), Birkhäuser Verlag, Basel, 1993, pp. 27–38.

[ZC96] K.-P. Zauner and M. Conrad, *Parallel computing with DNA: toward the anti-universal machine*, Parallel Problem Solving from Nature—PPSN IV (Berlin) (H.-M. Voigt, W. Ebeling, I. Rechenberg, and H.-P. Schwefel, eds.), Lecture Notes in Computer Science, vol. 1141, Springer-Verlag, Berlin, September 1996, pp. 696–705.

Department of Computer Science, Wayne State University, Detroit, Michigan 48202
E-mail address: `biocomputing@cs.wayne.edu`

DIMACS Series in Discrete Mathematics
and Theoretical Computer Science
Volume **48**, 1999

Linear DNA Self-Assembly with Hairpins Generates Linear Context-Free Grammars

Tony L Eng

ABSTRACT. Self-assembly of oligonucleotides in various DNA hybridization models has been shown to be capable of generating sequences equivalent to regular languages and context-free languages. In this paper, we show that the model which generates regular languages can be used to generate linear context-free languages if hairpins are allowed and if certain restrictions are placed on the initial set of DNA complexes used for self-assembly.

1. Introduction

Single-stranded DNA that contain appropriate Watson-Crick complementary base pairs are able to hybridize into a double helical structure. The computational power of hybridization was studied in [**WiYaSe**]. They found that the self-assembly of properly designed oligonucleotides into a linear DNA molecule generated sequences that were equivalent to regular languages. Furthermore, by allowing dendrimer structures, the resulting sequences were equivalent to context-free languages, and that by increasing the dimensions in which these complexes can be formed, the computational power likewise increased.

They also observed that a nonregular language could be generated if hairpins were allowed in the model that generated regular languages, and left open the possibility of exploring further how hairpins might affect computation. Here, we show that the self-assembly of linear duplexes with hairpins generates sequences that are equivalent to linear context-free languages.

2. Background

An alphabet Σ is a set of symbols. A string s is a finite sequence of symbols taken over some alphabet Σ. The set of all such strings is denoted as Σ^*. Let ϵ be the empty string, note that $\epsilon \in \Sigma^*$. Let s^V be the reverse of s. Also for the case when s is a string of nucleotides ($\Sigma = \{A, C, T, G\}$), let $\overline{s}$ represent the Watson-Crick complementary string of s. So, for example, if $s = AGGT$ then $s^V = TGGA$ and $\overline{s} = TCCA$.

1991 *Mathematics Subject Classification.* Primary 92B05, 68Q50; Secondary 03D99.

The author was supported in part by NIH/NHCRI Training Grant in Genomic Science #HG00039 and DARPA Grant #N00014-95-1-1246.

2.1. Context-Free Grammars. The following definition is taken from [**LePa**].

DEFINITION 2.1. A context-free grammar G consists of a quadruple (Σ, T, R, S) where

- Σ is an alphabet,
- T is a set of terminals, $T \subset \Sigma$,
- R is a set of rules, $R \subset (\Sigma - T) \times \Sigma^*$,
- S is the start symbol, $S \in (\Sigma - T)$.

Let $N = (\Sigma - T)$ represent the set of nonterminals. The rules are a set of transformations governing how a nonterminal can be replaced by a string of terminals and nonterminals. A context-free language, $L(G)$, is the set of strings that is generated by a context-free grammar G, i.e. all strings obtainable by transforming S according to its rules R.

2.2. Linear Grammars. A linear grammar is a context-free grammar in which no rule has a right-hand side that contains more than one nonterminal. The rules are of the form:

$$\begin{aligned} \text{type 1:}A &\rightarrow t, \\ \text{type 2:}A &\rightarrow t_1 B t_2, \end{aligned}$$

where $A, B \in N$ and $t, t_1, t_2 \in T^*$. Note that this grammar generates all strings that contain a "center" that is the substring generated by the application of a rule of the first type. This center is flanked by two (possibly empty) strings of terminals generated by the (repeated) application of the second rule. Examples of linear context-free languages are: $\{0^k 1^k | k \geq 0\}$, $\{ss^V | w \in \{0,1\}^*\}$ and $\{s = s^V | s \in \{0,1\}^*\}$.

2.3. Self-Assembly of DNA. A single-stranded DNA(ssDNA) strand has an orientiation; one end is its 5' end and the other, its 3' end. Hybridization will allow two ssDNA strands positioned in opposite orientations to assume a double helix or duplex structure as a result of hydrogen bond formation. The degree to which two strands will hybridize is due in large part to how complementary the two strands are. Duplex structures may have sticky ends (overhangs) on one or both ends, enabling them to hybridize with another complementary DNA strand (which may itself be the sticky end of another duplex) (see Figure 1).

Aside from hybridization, two other operations are possible: denaturing and ligation. Hybridization can be undone by denaturing, which separates the strands of a duplex. Ligation enables certain nicks in DNA duplexes to be repaired by creating a covalent bond in the DNA backbone. In this manner, several DNA strands/duplexes with complementary sticky ends can be hybridized together and ligated, producing a DNA duplex that when denatured, yields two single strands of DNA (rather than several).

3. Linear Self-Assembly with Hairpins

In their proof that linear self-assembly is equivalent to the regular languages, Winfree, et. al. note that the introduction of a hairpin (modulo some change in encoding) allows one to generate the set of all palindromes in which the number of ones is a multiple of four [**WiYaSe**]. We elaborate on this remark and show that

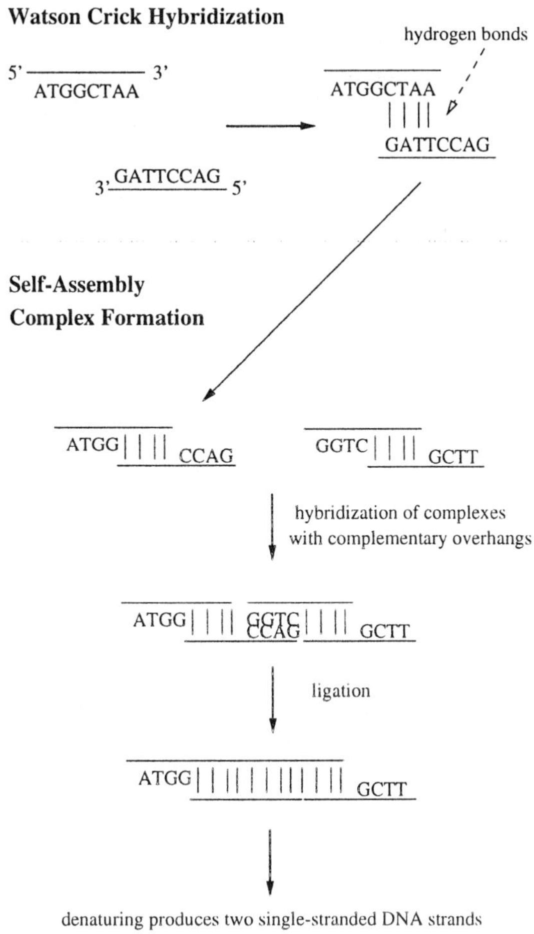

FIGURE 1. Hybridization, Ligation and Denaturing

the introduction of a hairpin in the linear self-assembly process produces complexes analogous to those generated by a linear context-free grammar under conditions specified by our model.

Hairpins or hairpin loops are formed when base pairing between complementary segments of DNA within the same strand occurs. The distance between these segments (and hence the size of the loop) is usually about 6-8 nucleotides. Hairpins may have one sticky end or two sticky ends[1]. Hairpins with one sticky end can terminate elongation at a site, since once the hairpin hybridizes, nothing else can hybridize to it. Hairpins with two sticky ends are a more complex matter. They can be functionally equivalent to a normal duplex with two sticky ends, but also allow complexes to alternate between hybridization with 5' overhangs and hybridization with 3' overhangs. However, two hairpins that are face-to-face may allow dendritic structures to form in which the number of incoming paths and the number of

[1]Hairpins with no sticky ends are uninteresting for our purposes.

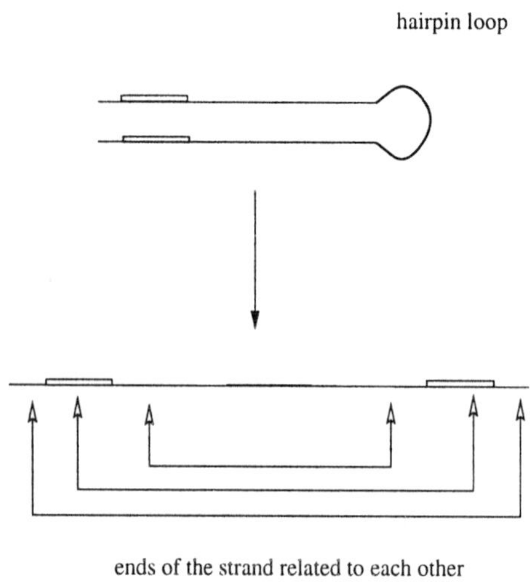

FIGURE 2. Relationships Between Ends of Molecule

outgoing paths are not conserved (one path in, forking into two paths out). In this paper, we only consider hairpins with a single sticky end.

The introduction of a hairpin allows the resulting DNA complex to be denatured into not two single-stranded DNA strands, but *one* single-stranded DNA strand. This doubling over of the molecule in the duplex state enables one end of the denatured single-stranded molecule to be related to the other end of the denatured single-stranded molecule(see Figure 2). Regions that are hybridized to each other in the duplex end up at opposite ends of the single-stranded molecule, enabling the portions that are roughly equidistant from the molecule's center(where the hairpin is) to be related. This is similar to properties exhibited by a linear context-free grammar.

3.1. Self-Assembly Model. Winfree, et.al. created a model for studying the construction of self-assembled DNA complexes. In this model, certain assumptions were made about the self-assembly process including constant temperature during assembly, perfect complementarity and no self interaction of complexes. They defined $\mathcal{L}_{R_1^T,\mathcal{A}}$ to be the set of DNA complexes that could be generated from an initial set $\mathcal{A}$ of complexes according to a set of rules R_1^T which specify which hybridizations were allowed. For the most part, R_1^T represented the constraints governing the conditions for hybridization and the closure properties of the resulting complex. In particular, only overhangs with T or more complementary base pairs could hybridize yielding essentially one-dimensional complexes as a result.

Note that complexes in $\mathcal{L}_{R_1^T,\mathcal{A}}$ can be thought of as "strings of terminal symbols" in the sense that no further hybridizations are possible because no sticky ends are available.

A prefix-free codebook $\mathcal{C}$ was also used such that given a single-stranded DNA sequence s, $C(s)$ translated the bases of s from the 5' end to the 3' end of DNA (ignoring DNA sequences not in the codebook) into another alphabet. This was useful for interpreting the products of the DNA self-assembly process.

The model used in this paper is a more restricted version of Winfree's. Our initial set of complexes $\mathcal{A}$ will consist of duplexes that fall into three classe: terminal duplexes, intermediate duplexes and start duplexes. Terminal complexes are hairpins with a single sticky end. Nonterminal complexes are duplexes with two overhangs, one on each end (see Figures 1 and 3 for examples). And finally, start complexes are duplexes with a single overhang (similar to a terminal complex but without the hairpin). The naming of these complexes will become clear in the context of linear context-free grammars.

In our model, we introduce the notion of a "well-formed" initial set of complexes which place restrictions on the DNA complexes in $\mathcal{A}$.

DEFINITION 3.1. For some constant n, a set of DNA complexes $\mathcal{A}_n$ is said to be well-formed if the following conditions are met:

1. **Overhang Sequence Design** All overhangs are of length n, and overhang sequences are chosen so that if two sequences can hybridize with each other during self-assembly, their sequences are completely complementary; otherwise, they are entirely non-complementary. Furthermore, an overhang sequence should not have long common stretches of similarity or complementarity with a non-overhang sequence.

   (Note that as in Winfree's model, we assume that the temperature remains constant throughout the course of self-assembly, in particular at a temperature that allows the n base pairs of two overhangs to form a stable duplex.)
2. **Overhang Compatibility** Consider the set of all overhangs in $\mathcal{A}$. This set must be partitionable into two disjoint sets I and J such that the overhangs of all terminal complexes belong to one set (e.g. I), the overhangs for all start complexes belong to the other set (e.g. J), and finally, nonterminal complexes contain one overhang from each set. The sticky ends of two DNA duplexes are allowed to anneal only if they belong to two different sets (hence, Watson-Crick complementary overhangs belong to different sets).

The first condition is a simplification that rules out products of partial or incomplete hybridization (e.g. two overhangs that anneal in a fashion that results in bulges or gaps), and allows only simple fully complementary hybridization (or "all or nothing" hybridization). The second restricts hybridization products to only those that contain the start complex, hybridized to zero or more nonterminal complexes, and ending with a terminal complex.

Lastly, we also simply codebook translation by requiring that a codebook be "compatible" with a well-formed initial set $\mathcal{A}_n$.

DEFINITION 3.2. Given a well-formed initial set of complexes $\mathcal{A}_n$, a codebook $\mathcal{C}$ is said to be compatible if no overlap sequence is a proper subsequence of any sequence in $\mathcal{C}$, no prefix of any overlap is the suffix of any sequence in $\mathcal{C}$, and no suffix of any overlap is the prefix of any sequence in $\mathcal{C}$.

This allows for no ambiguity in the translation of DNA sequences by the codebook, as any sequences encoded by the codebook are located entirely in the overlap or entirely in the duplex sequences of the original complexes $\mathcal{A}$.

In the context of grammars, only overhangs may encode sequences representing nonterminal symbols (and this encoding must be completely within the overhang sequence), while duplex regions of initial complexes in $\mathcal{A}$ may contain sequences encoding for terminals.

Note that other formulations of and conditions on $\mathcal{A}$ and $\mathcal{C}$ may be possible and less restrictive.

For the remainder of this paper, we assume that an appropriate n is chosen and fixed so that we can use the simpler notation $\mathcal{A}$ to stand for $\mathcal{A}_n$. Similarly, R will be understood to denote R_1^T.

Thus, given a well-formed set of initial complexes $\mathcal{A}$ and a compatible $\mathcal{C}$, the set of strings corresponding to a language of DNA complexes $\mathcal{L}_{R,\mathcal{A}}$ is given by: $\mathcal{L}_{R,\mathcal{A},\mathcal{C}} = \mathcal{C}(denature(ligate(\mathcal{L}_{R,\mathcal{A}})))$.

3.2. Proofs of Equivalence. Using this model, we argue that any linear context-free grammar can be transformed into our self-assembly model with proper choices of $\mathcal{C}$ and $\mathcal{A}$, and the self-assembly products generated by any well-formed initial set $\mathcal{A}$ produce strings in some linear context-free grammar as translated by any codebook $\mathcal{C}$ for complexes in $\mathcal{A}$.

THEOREM 3.3. *For any language L generated by a linear context-free grammar G, there exists a compatible codebook $\mathcal{C}$ and a choice of well-formed initial complexes $\mathcal{A}$ such that $\mathcal{L}_{R,\mathcal{A},\mathcal{C}} = L(G)$.*

PROOF. Given a linear context-free grammar $G = (\Sigma, T, R, S)$, we exhibit a means for constructing a proper codebook $\mathcal{C}$ and initial complexes $\mathcal{A}$, such that every string generated by G has an analogous DNA complex generated by $\mathcal{L}_{R,\mathcal{A}}$.

The codebook contains entries mapping every symbol in Σ to a distinct oligonucleotide. If $\sigma \in \Sigma$, then there exists an oligo x such that $\mathcal{C}(x) = \sigma$. We will call this x, $\mathcal{C}^{-1}(\sigma)$, which was designed to minimize long stretches of similarity and complementarity with other oligos and with itself, and furthermore chosen so that $\mathcal{C}$ would be compatible with $\mathcal{A}$. For example, we can choose to design this codebook such that the complement of encoding oligos and the reverse of these complements are not present in the codebook (i.e. if $x \in \mathcal{C}$, then $\overline{x}, \overline{x^V} \notin \mathcal{C}$). Finding these actual sequences is not always an easy (or efficient) task, and is an area of current investigation. Here, we assume that these oligos are somehow designed, and that for example, proper oligo lengths are chosen since we have omitted the treatment of oligo length that appears in [**WiYaSe**].

An example of the construction of $\mathcal{A}$ is shown in Figure 3. For the starting nonterminal $S = A$, construct a complex in which a duplex region has a single 5' overhang whose sequence corresponds to that of $\overline{\mathcal{C}^{-1}(A)}$. Although labelled "$\overline{A}$" in Figure 3(a) to convey the semantic intention of the overhang, the sequence really corresponds to $\overline{\mathcal{C}^{-1}(A)}$ – the nucleotide sequence that is complementary to the DNA sequence encoding nonterminal A. Note that the other end of this complex is a blunt end.

The rules R of G are also encoded in $\mathcal{A}$. For each rule of type 1, $A \rightarrow t$, create a hairpin loop in which the overhang is $\mathcal{C}^{-1}(A)$ and $\mathcal{C}^{-1}(t)$ is encoded in the duplex or loop region of the hairpin(see Figure 3(c)). (Note that extraneous nucleotides will be present in the DNA complexes, but should not be sequences that are not present in the codebook and hence noncoding for any symbol in $\mathcal{C}$.)

For each rule of type 2, $A \rightarrow t_1 B t_2$, create a complex with a $\mathcal{C}^{-1}(A)$ overhang, a duplex region encoding $\mathcal{C}^{-1}(t_1)$ and $\mathcal{C}^{-1}(t_2)$, and a $\overline{\mathcal{C}^{-1}(B)}$ overhang(see

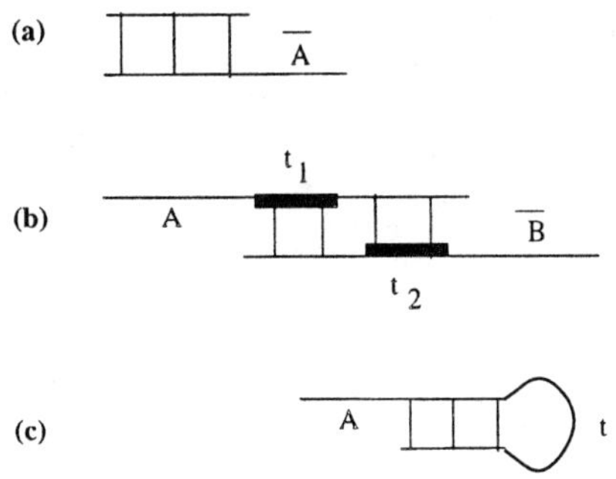

FIGURE 3. Construction of Oligos of $\mathcal{A}$

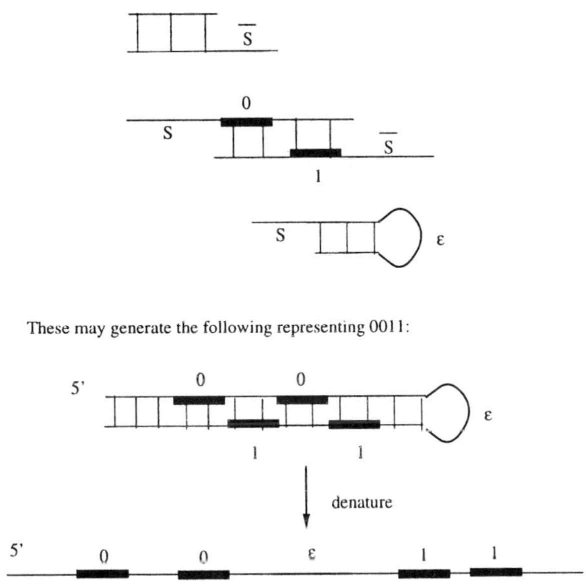

FIGURE 4. An Example for the Language $\{0^k1^k|k \geq 0\}$.

Figure 3(b)). Note that there many be several ways to create the middle duplex region, but again, noncoding sequences will be ignored.

Any string thus generated by G can be similarly generated by $\mathcal{L}_{R,\mathcal{A},\mathcal{C}}$. Figure 4 depicts the set $\mathcal{A}$ for the linear context-free language, $\{0^k1^k|k \geq 0\}$ whose linear context-free grammar is: $S \rightarrow \epsilon, S \rightarrow 0S1$.

□

THEOREM 3.4. *For all well-formed initial sets of complexes $\mathcal{A}$, and all compatible codebooks $\mathcal{C}$, there exists a linear context-free grammar G such that $L(G) = \mathcal{L}_{R,\mathcal{A},\mathcal{C}}$.*

PROOF. Given $\mathcal{A}$ and $\mathcal{C}$, we show how to construct a $G = (\Sigma, T, R, S)$. This is rather straight-forward and is practically the reverse of the previous proof. For each distinct overhang x, $\mathcal{C}(x)$ is a nonterminal in N. The set of terminals are obtained from running $\mathcal{C}$ on the various nucleotide sequences and will become apparent when we consider how to create R.

We create R as follows: Each hairpin with an overhang of x leads to a rule of the form $\mathcal{C}(x) \to \mathcal{C}(y)$ where y is the remainder of the DNA sequence comprising the hairpin when the hairpin is denatured. Similarly, for complexes containing an overhang x_1 and an overhang x_2, and a duplex region consisting of y and $\overline{y}$ (where y is attached to the end of x_1), then the following type 2 rule is created: $\mathcal{C}(x_1) \to \mathcal{C}(y_1)\mathcal{C}(x_2)\mathcal{C}(y_2)$ where $y_1 = y$ and $y_2 = \overline{y}$. (Note that it can also be the case that y_1 is some substring of y and y_2 is some substring of $\overline{y}$ with all other regions of y and $\overline{y}$ are noncoding.) The structures of the rules indicate that G is a linear context-free grammar.

Any complexes generated by $\mathcal{L}_{R,\mathcal{A},\mathcal{C}}$ can also be generated by G. Recall that these complexes are terminal (meaning one end is blunt and the other contains a hairpin loop so that the cannot be elongated anymore).

□

In our construction, only hairpins with single overhangs were considered. Hairpins with two overhangs may be useful for inserting arbitrary sequences into a complex since this sequence is located within the hairpin itself and will not interfere with the self-assembly process. Again, they may also permit junctions to occur if not carefully designed so that dendrimer structures rather than linear structures result. But it may be interesting to consider the non-linear (higher dimensional) use of hairpins.

4. Conclusions

In this paper, we have explored the power of linear self-assembly when augmented to include hairpins with a single sticky end. We show that this generates sequences that are equivalent to strings generated by a linear context-free grammar.

5. Acknowledgments

We would like to thank Professor Michael Sipser for helpful discussions. We are also indebted to Erik Winfree for his suggestions and insights, and for his help in formulating the notion of a "well-formed" set of initial complexes.

References

[LePa] Lewis, H., Papadimitriou, C., Elements of the Theory of Computation, Prentice Hall, USA, 1981.

[Re] Revesz, G., Introduction to Formal Languages, McGraw-Hill, Inc., 1983, USA.

[WiYaSe] Winfree, E., Yang, X., Seeman, N., "Universal Computation via Self-assembly of DNA: Some Theory and Experiments", manuscript, pp. 1-22, May 30, 1996.

LABORATORY FOR COMPUTER SCIENCE, MASSACHUSETTS INSTITUTE OF TECHNOLOGY, CAMBRIDGE, MASSACHUSETTS 02139

Current address: 545 Technology Square, NE43-338, Cambridge, MA 02139

E-mail address: `tleng@theory.lcs.mit.edu`

DIMACS Series in Discrete Mathematics
and Theoretical Computer Science
Volume **48**, 1999

Watson-Crick Finite Automata

R. Freund, Gh. Păun, G. Rozenberg, and A. Salomaa

ABSTRACT. We introduce a new type of automata, working on data structures called Watson-Crick tapes. Such a tape is a double-stranded sequence of symbols related by a complementarity relation (of the sort present in DNA). The automata can scan separately each of the two strands, in a correlated manner, and can have a finite number of states controlling the moves. Moreover, either an automaton moves on both strands in the same direction, or it moves in opposite directions on the strands. Considering different alternatives leads to several types of automata. In many cases, when augmented with squeezing mechanisms like weak codings or deterministic sequential transducers, they characterize the recursively enumerable languages. Techniques and results related to the characterizations of recursively enumerable languages by equality sets and twin-shuffle languages are used.

1. Introduction

We here bring together two important areas of computability, the well developed *automata theory* and the rapidly developing *molecular computing.* The bridging notion is that of *Watson-Crick complementarity*, which (as follows from the results concerning twin-shuffle languages, see, e.g., [**9**]) guarantees universal computation in any model of DNA computers having sufficient capabilities for handling inputs and outputs, [**30**]. In this paper we discuss several models for handling inputs and outputs which yield universality.

The essential difference between our automata and the customary ones lies in the data structures they handle. While the customary automata operate on linear (one-dimensional) strings of symbols, our automata operate on double strands. The double strands resemble DNA molecules in the following sense. The matching letters (nucleotides) are *complementary*, the relation of complementarity being defined for pairs of letters of the basic alphabet, similarly to the Watson-Crick complementarity of the pairs (A, T) and (C, G) of the DNA alphabet. Most importantly, we assume that such double strands are freely available in the sense that we do not have to check anymore that the matching letters are indeed complementary. This assumption can be interpreted to mean, in language-theoretic terms, that our automata have the *twin-shuffle language* as a built-in feature. The formal technicalities will be explained in detail below.

Because of the complementarity, we have chosen to call our automata *Watson-Crick automata.* Our main interest is in the basic variant, where the automaton

1991 *Mathematics Subject Classification.* 68-06, 68Q05.

scans separately each of the two strands in a correlated manner. However, we will briefly investigate also other variants, such as transducers and reverse Watson-Crick finite automata. The essential feature in each case is the complementarity in the data structures handled.

While the real practical significance of *DNA computing* is still unclear, it seems to us that theoretical studies about DNA computing can already use two advantages stemming from the nature of DNA molecules: (i) *Watson-Crick complementarity* which renders the power of the twin-shuffle language available, and (ii) the multitude of DNA molecules which brings *massive parallelism* to the computing scene.

The present paper can be viewed as a contribution to (i). It remains an entirely open area to model (ii), as well as to combine (i) and (ii) into one model of DNA computing.

2. Watson-Crick complementarity and twin-shuffle languages

Watson-Crick complementarity refers to the following. The DNA molecules are double stranded sequences of nucleotides (A, C, G, T). We here ignore the spatial arrangements of these molecules (and the cases where the molecules have loops, branches, etc) and we assume that every DNA molecule is a linear double-stranded sequence, as illustrated in Figure 1.

$$\begin{array}{c} 5' - ACCGACCTTACCGAATCG - 3' \\ 3' - TGGCTGGAATGGCTTAGC - 5' \end{array}$$

FIGURE 1. Double stranded DNA molecule

The nucleotides are always present in the two strands in pairs: A is always paired with T and C with G (in all our considerations we take into account only the standard case, of stable molecules, error-free). This pairing, A – T, C – G, is called the *Watson-Crick complementarity.* When two complementary strings of nucleotides meet each other in a solution, they stick to each other due to the complementarity of the nucleotides, producing a double-stranded DNA molecule. For instance, in this way the single stranded sequences ACCGACCTTACCGAATCG and TGGCTGGAATGGCTTAGC lead to the molecule in Figure 1. Actually, in order that two single-stranded sequences stick together, they should be complementary *and of opposite orientation.* Each single stranded sequence of nucleotides has an orientation indicated by writing $5'$ and $3'$ at the ends of the string (as in Figure 1). A string of the form $5' - x - 3'$ can anneal only to a string of the form $3' - \overline{x} - 5'$, where $\overline{x}$ denotes the complement of x in the Watson-Crick sense.

The fact that the DNA molecules are essentially based on Watson-Crick complementarity has a deep (and, at the first sight, surprising) connection with a powerful result in computability theory: the characterization of recursively enumerable languages by means of the so-called twin-shuffle languages, [**9**], [**4**], [**8**]. Because this connection is very important and illustrative for the computability based on the Watson-Crick complementarity, we present some details here.

Consider an alphabet V and let $\overline{V}$ be its barred version, the set of symbols $\bar{a}$, for each a in V, where we also assume that $V \cap \overline{V} = \emptyset$. The twin-shuffle language over V, denoted by TS_V, consists of the strings obtained by shuffling any string x over V with its barred twin $\bar{x}$ over $\overline{V}$.

For instance, if $x = 01101$, then $\bar{x} = \bar{0}\bar{1}\bar{1}\bar{0}\bar{1}$ and both $01\bar{0}10\bar{1}\bar{1}1\bar{0}\bar{1}$ and $\bar{0}01\bar{1}\bar{1}10\bar{0}\bar{1}1$ are obtained by shuffling x and $\bar{x}$, hence they belong to $TS_{\{0,1\}}$.

As proved in [**9**], for every recursively enumerable language L there is a deterministic gsm mapping g_L such that $L = g_L(TS_{\{0,1\}})$. Therefore, $TS_{\{0,1\}}$ is a generator of the family of recursively enumerable languages; this single language contains, metaphorically speaking, all the recursively enumerable languages, we only have to read them out of it by deterministic gsm's.

Watson-Crick complementarity directly means the twin-shuffle language over two letters: Let us associate, for instance, the symbol 0 to the pair (A, T) and the symbol 1 to the pair (C, G); read a DNA molecule from the left to the right, nucleotide by nucleotide in a nondeterministic way, but always continuously from the left, without omissions; when reading a nucleotide A or T in the upper level we write 0, when reading A or T in the lower level we write $\bar{0}$, when reading a nucleotide C or G in the upper level we write 1, when reading C or G in the lower level we write $\bar{1}$. The string obtained in this way when finishing the reading of the molecule is a string in $TS_{\{0,1\}}$.

For instance, try this procedure with the molecule

$$\begin{matrix} ACCTG \\ TGGAC \end{matrix}$$

whose strands correspond to the strings $x = 01101$ and $\bar{x} = \bar{0}\bar{1}\bar{1}\bar{0}\bar{1}$ mentioned above.

Consequently, the generator $TS_{\{0,1\}}$ of the family of recursively enumerable languages is given by the Watson-Crick complementarity that is inherent to the DNA structure. Hence in order to compute we only need the deterministic gsm associated to each problem.

This observation can directly or indirectly explain many results from the DNA computing area, where models based on operations specific to DNA were proved to be computationally universal, that is to say equal in power to Turing machines; see, e.g., [**3**], [**26**]. However, in the models considered up to now, either a generative strategy has been followed, or direct simulations of Turing machines by DNA techniques were proposed. Here we start an alternative approach, trying to make explicit and full use of the power of Watson-Crick complementarity: we define automata working on tapes which are double-stranded structures observing a relation of the Watson-Crick complementarity type.

3. Some language and automata theory prerequisites

The previously discussed elements of biochemistry, as simple as they are, are sufficient for the reader to understand the remaining sections of this paper. In the following we also specify some notions, notations and results from automata and formal language theory. For details we refer to the many monographs in the area; a comprehensive source of information is [**31**].

For an alphabet V, by V^* we denote the free monoid generated by V under the operation of concatenation; the empty string is denoted by λ. Moreover, $V^+ = V^* - \{\lambda\}$ and $|x|$ is the length of $x \in V^*$. If $U \subseteq V$, then the morphism $pr_U : V^* \longrightarrow U^*$ defined by $pr_U(a) = a$ if $a \in U$, and $pr_U(a) = \lambda$ if $a \in V - U$, is called *a projection.* A morphism $h : V^* \longrightarrow U^*$, for arbitrary alphabets V, U, is called a *coding* if $h(a) \in U$ for each $a \in V$, and a *weak coding* if $h(a) \in U \cup \{\lambda\}$, for each $a \in V$. The Parikh mapping associated to an alphabet V is denoted by Ψ_V.

A Chomsky grammar is written in the form $G = (N, T, S, P)$, where N is the nonterminal alphabet, T is the terminal alphabet, S is the axiom, and P is the set of production rules, written in the form $u \rightarrow v$. The language generated by G is denoted by $L(G)$. By *REG, LIN, CF, CS,* and *RE* we denote the families of regular, linear, context-free, context-sensitive, and recursively enumerable languages, respectively. By MAT^{λ} we denote the family of languages generated by context-free matrix grammars without appearance checking but possibly using λ-rules (see, e.g., [**6**]), and by $ET0L$ the family of languages generated by extended tabled Lindenmayer systems without interaction ([**29**]).

Unless stated otherwise, two languages will be considered to be identical if they differ at most by the empty string.

The families *REG, CF, CS, RE* are characterized not only by classes of grammars but also by classes of automata: finite automata, push-down automata, linear bounded automata, and Turing machines, respectively.

A *finite automaton* will be presented in the form $M = (Q, V, s_0, F, \delta)$, where Q is the set of states, V is the alphabet, s_0 is the initial state, F is the set of final (accepting) states, and $\delta : Q \times V \longrightarrow P(Q)$ is the transition mapping. ($P(X)$ denotes the set of subsets of X; if X is an infinite set, then the set of all *finite* subsets of X is denoted by $P_f(X)$.)

The transitions can also be written in a rewriting manner: $sa \rightarrow s'$ instead of $s' \in \delta(s, a)$, for $s, s' \in Q, a \in V$. Then the mapping δ can be replaced by a set P of rewriting rules of the form above.

A gsm (= generalized sequential machine) is a finite automaton with outputs: $g = (Q, V_I, V_O, s_0, F, \delta)$, where Q is the set of states, V_I is the input alphabet, V_O is the output alphabet, s_0 is the initial state, F is the set of final states, and $\delta : Q \times V_I \longrightarrow P_f(V_O^* \times Q)$; $(x, s') \in \delta(s, a)$ has the following meaning: reading the symbol $a \in V_I$ in state s, one may choose to produce the string $x \in V_O^*$ and go to state s'. For $z, x \in V_O^*, s, s' \in Q, a \in V_I$, and $w \in V_I^*$, we write

$$zsaw \Longrightarrow zxs'w \text{ if and only if } (x, s') \in \delta(s, a).$$

Then, for $w \in V_I^*$ we define the result of the translation of w by g by

$$g(w) = \{y \in V_O^* \mid s_0 w \Longrightarrow^* y s_f,\ s_f \in F\},$$

where $\Longrightarrow^*$ is the reflexive and transitive closure of the relation $\Longrightarrow$. The mapping g is extended in the natural way to languages:

$$g(L) = \bigcup_{x \in L} g(x),$$

for any $L \subseteq V_I^*$.

When for each pair (s, a) we have $card(\delta(s, a)) \leq 1$, then we say that the automaton M (the gsm g, respectively) is *deterministic*.

For any alphabet V, by $\overline{V}$ we denote the alphabet $\{\bar{a} \mid a \in V\}$, of barred variants of symbols in V; $\bar{a}$ is also called the *twin* of a. We assume that $V \cap \overline{V} = \emptyset$. We also use "bar" as an operator: for $x = a_1 \ldots a_n$ with $n \geq 1$ and $a_i \in V$ for $1 \leq i \leq n$, $\bar{x} = \bar{a}_1 \ldots \bar{a}_n$. The *twin-shuffle language over* V is defined by

$$TS_V = \{x \in (V \cup \overline{V})^* \mid \overline{pr_V(x)} = pr_{\overline{V}}(x)\}.$$

By $mi(x)$ we denote the mirror image of $x \in V^*$. The *reverse twin-shuffle* language over V is defined by

$$RTS_V = \{x \in (V \cup \overline{V})^* \mid \overline{pr_V(x)} = mi(pr_{\overline{V}}(x))\}$$

(a string $x \in RTS_V$ is obtained by shuffling a string $y \in V^*$ with the mirror image of its twin $\bar{y}$).

The languages TS_V, RTS_V can also be defined by

$$TS_V = \bigcup_{x \in V^*} (x \sqcup\!\sqcup \bar{x}), \quad RTS_V = \bigcup_{x \in V^*} (x \sqcup\!\sqcup mi(\bar{x})),$$

where $\sqcup\!\sqcup$ is the shuffle operation, defined by

$$x \sqcup\!\sqcup y = \{x_1y_1x_2y_2 \dots x_ny_n \mid n \geq 1, x = x_1x_2 \dots x_n, \; y = y_1y_2 \dots y_n, \text{ for } x_i, y_i \in U^*, 1 \leq i \leq n\},$$

for $x, y \in U^*$ where U is an alphabet.

The following results are fundamental for our study:

THEOREM 1. ([**9**]) *Each language $L \in RE$ can be written in the form $L = g_L(TS_{\{0,1\}})$, where g_L is a deterministic gsm.*

THEOREM 2. ([**4**], [**8**]) *Each language $L \in RE$ can be written in the form $L = g_L(RTS_{\{0,1\}})$, where g_L is a deterministic gsm.*

The proof of Theorem 1 is based on techniques related to equality sets of morphisms, which will also be used below. Theorem 2 is proved directly in [**4**], and in [**8**] it is proved by reduction to Theorem 1.

For two morphisms $h_1, h_2 : V_1^* \longrightarrow V_2^*$, we define the equality set by

$$EQ(h_1, h_2) = \{x \in V_1^* \mid h_1(x) = h_2(x)\}.$$

Proofs of the following representation of RE can be found in [**5**], [**34**].

THEOREM 3. *Each language $L \in RE, L \subseteq T^*$, can be written in the form $L = pr_T(EQ(h_1, h_2) \cap R)$, where $R \subseteq V_1^*$ is a regular language and $h_1, h_2 : V_1^* \longrightarrow V_2^*$ are two λ-free morphisms, $T \subseteq V_1$.*

A variant of this result is proved in [**19**] (Lemma 8 there).

THEOREM 4. *Each language $L \in RE, L \subseteq T^*$, can be written in the form $L = pr_T(h_1(EQ(h_1, h_2)) \cap R)$, where $R \subseteq V_2^*$ is a regular language and $h_1, h_2 : V_1^* \longrightarrow V_2^*$ are two λ-free morphisms, $T \subseteq V_2$.*

A notion which is related to the devices defined in the sequel is that of *two-head finite automata.*

A two-head finite automaton is a construct $M = (Q, V, s_0, F, \delta)$, where Q, V, s_0, F are as in a usual finite automaton and δ is the transition mapping, $\delta : Q \times (V \cup \{\lambda\}) \times (V \cup \{\lambda\}) \longrightarrow P(Q)$. For $x_1, x_2, x_1', x_2' \in V^*$ and $s, s' \in Q$ we write

$$s(x_1, x_2) \Longrightarrow s'(x_1', x_2') \text{ if and only if } x_1 = ux_1', x_2 = vx_2' \text{ and } s' \in \delta(s, u, v), \text{ for } u, v \in V \cup \{\lambda\}.$$

The language recognized by M is defined by

$$L(M) = \{x \in V^* \mid s_0(x, x) \Longrightarrow^* s_f(\lambda, \lambda), s_f \in F\}.$$

By TH we denote the family of languages recognized by such automata.

Some variants of two-head (or, more generally, multi-head) finite automata were thoroughly investigated: deterministic, simple (one head only reads the tape, the others can only distinguish the end markers of the input string), sensing (the heads can sense the case when two of them are placed in the same cell of the tape). Precise definitions, results and further references can be found in [**7**], [**17**], [**18**], [**27**], etc.

4. The operation of sticking

The basic objects with which we work in this paper are double-stranded sequences whose corresponding symbols are paired according to a complementarity relation. In this section we fix some notation and terminology concerning such objects.

In the following we consider an alphabet V and a symmetric relation ρ on V, $\rho \subseteq V \times V$ (ρ is the relation of *complementarity*).

REMARK 1. The property of symmetry is not used in what follows, but we consider it because the complementarity of Watson-Crick pairs is a symmetric one. On the other hand, we impose no further conditions on the relation ρ besides symmetry, although in the Watson-Crick case the relation is also injective: if $(a,b),(c,b) \in \rho$, then $a = c$. It is a topic for further investigations to consider automata as those introduced below, involving a relation ρ with other properties than symmetry alone.

Besides the monoid V^*, of strings over V, to V we also associate the monoid of pairs, $V^* \times V^*$. In accordance with the way of representing DNA molecules, where one considers the two strands as placed one over the other, we write the elements (x_1, x_2) of $V^* \times V^*$ in the form $\begin{pmatrix} x_1 \\ x_2 \end{pmatrix}$. Therefore, the concatenation of two pairs $\begin{pmatrix} x_1 \\ x_2 \end{pmatrix}, \begin{pmatrix} y_1 \\ y_2 \end{pmatrix}$ is $\begin{pmatrix} x_1y_1 \\ x_2y_2 \end{pmatrix}$. In this way, we identify $\begin{pmatrix} x_1 \\ x_2 \end{pmatrix}\begin{pmatrix} y_1 \\ y_2 \end{pmatrix}$ with $\begin{pmatrix} x_1y_1 \\ x_2y_2 \end{pmatrix}$. We also write $\begin{pmatrix} V^* \\ V^* \end{pmatrix}$ instead of $V^* \times V^*$.

We also denote

$$\begin{bmatrix} V \\ V \end{bmatrix}_\rho = \{\begin{bmatrix} a \\ b \end{bmatrix} \mid a, b \in V, (a,b) \in \rho\},$$

$$WK_\rho(V) = \begin{bmatrix} V \\ V \end{bmatrix}_\rho^*.$$

The set $WK_\rho(V)$ is called the *Watson-Crick domain* associated to V and ρ. The elements $\begin{bmatrix} a_1 \\ b_1 \end{bmatrix}\begin{bmatrix} a_2 \\ b_2 \end{bmatrix}\dots\begin{bmatrix} a_n \\ b_n \end{bmatrix} \in WK_\rho(V)$ are also written in the form $\begin{bmatrix} w_1 \\ w_2 \end{bmatrix}$, for $w_1 = a_1a_2\dots a_n$, and $w_2 = b_1b_2\dots b_n$. We call such elements $\begin{bmatrix} w_1 \\ w_2 \end{bmatrix} \in WK_\rho(V)$ well-formed double-stranded sequences, or simply *double-stranded sequences*, or *molecules*, in order to remind the reality they are modeling. When used as input or output sequences for automata, such objects are also called *Watson-Crick tapes*. The two component strings, w_1, w_2, are also called strands, the *upper strand* and the *lower strand*, respectively.

Note that also $\begin{bmatrix} \lambda \\ \lambda \end{bmatrix}$ is a molecule, although it has no biochemical counterpart.

We emphasize the two properties characterizing the elements $\begin{bmatrix} w_1 \\ w_2 \end{bmatrix}$ in $WK_\rho(V)$:

- the two strands w_1, w_2 are of the same length,
- the corresponding symbols in the two strands are complementary in the sense of the relation ρ.

In the automata we shall introduce in this paper we make an essential use of these properties which have a significant influence on the power of our automata. However, this power is provided to us "for free": the DNA molecules *are* Watson-Crick tapes, and when computing with DNA molecules we do not have to care about these properties as they are inherently present.

We shall also consider a bigger set of molecules that includes "incomplete molecules" (or "molecules with overhangings"):

$$W_\rho(V) = \begin{bmatrix} V \\ V \end{bmatrix}_\rho^* \left(\begin{pmatrix} \lambda \\ V^* \end{pmatrix} \cup \begin{pmatrix} V^* \\ \lambda \end{pmatrix} \right).$$

Thus, the elements of $W_\rho(V)$ will be written in the form $\begin{bmatrix} w_1 \\ w_2 \end{bmatrix} \begin{pmatrix} x_1 \\ x_2 \end{pmatrix}$, where $\begin{bmatrix} w_1 \\ w_2 \end{bmatrix} \in WK_\rho(V)$ and $\begin{pmatrix} x_1 \\ x_2 \end{pmatrix} \in \begin{pmatrix} V^* \\ \lambda \end{pmatrix} \cup \begin{pmatrix} \lambda \\ V^* \end{pmatrix}$. For this reason we also call the elements of $W_\rho(V)$ *well-started double stranded* sequences. Clearly, $WK_\rho(V) \subset W_\rho(V)$.

The concatenation of two elements of $\begin{pmatrix} V^* \\ V^* \end{pmatrix}$ is always defined and the result is an element of $\begin{pmatrix} V^* \\ V^* \end{pmatrix}$. Also for elements $\begin{bmatrix} x_1 \\ x_2 \end{bmatrix}, \begin{bmatrix} y_1 \\ y_2 \end{bmatrix}$ in $WK_\rho(V)$ we always have $\begin{bmatrix} x_1 y_1 \\ x_2 y_2 \end{bmatrix} \in WK_\rho(V)$. On the other hand, a similar componentwise concatenation is not possible with the elements of $W_\rho(V)$ and can only partially be defined as an operation between elements $\begin{bmatrix} w_1 \\ w_2 \end{bmatrix} \begin{pmatrix} x_1 \\ x_2 \end{pmatrix} \in W_\rho(V)$ and $\begin{pmatrix} y_1 \\ y_2 \end{pmatrix} \in \begin{pmatrix} V^* \\ V^* \end{pmatrix}$. Thus formally, the *sticking* of $\begin{bmatrix} w_1 \\ w_2 \end{bmatrix} \begin{pmatrix} x_1 \\ x_2 \end{pmatrix}$ and $\begin{pmatrix} y_1 \\ y_2 \end{pmatrix}$ as above is defined by:

$$\mu\left(\begin{bmatrix} w_1 \\ w_2 \end{bmatrix} \begin{pmatrix} x_1 \\ x_2 \end{pmatrix}, \begin{pmatrix} y_1 \\ y_2 \end{pmatrix} \right) = \begin{cases} \begin{bmatrix} z_1 \\ z_2 \end{bmatrix} \begin{pmatrix} u_1 \\ u_2 \end{pmatrix}, & \text{if } z_1 u_1 = w_1 x_1 y_1, \\ & z_2 u_2 = w_2 x_2 y_2, \\ & \begin{bmatrix} z_1 \\ z_2 \end{bmatrix} \in WK_\rho(V), \\ & \begin{pmatrix} u_1 \\ u_2 \end{pmatrix} \in \\ & \begin{pmatrix} V^* \\ \lambda \end{pmatrix} \cup \begin{pmatrix} \lambda \\ V^* \end{pmatrix} \\ \text{undefined} & \text{otherwise.} \end{cases}$$

5. Simple sticker systems

Based on the idea of Watson-Crick complementarity and the operations defined above, a language generating mechanism has been defined in [**19**]. Because the Watson-Crick automata we shall define and investigate in the following sections are automata counterparts of these grammar-like devices, we recall here some definitions and results from [**19**].

Actually, the devices introduced in [**19**] use arbitrary sequences which are not necessarily well started. Moreover, in [**10**] bidirectional systems are considered, where the sequences are grown in both directions, to the left and to the right. Here we need only the so-called simple sticker systems from [**19**], which prolong sequences only to the right, using blocks of the form $\begin{pmatrix} \lambda \\ y_2 \end{pmatrix}$ or $\begin{pmatrix} y_1 \\ \lambda \end{pmatrix}$ in the way described by the restricted sticking defined in the previous section.

A *simple sticker system* is a construct

$$\gamma = (V, \rho, A, B_d, B_u),$$

where V is an alphabet, $\rho \subseteq V \times V$ is a symmetric relation, A is a finite subset of $W_\rho(V)$, and B_d, B_u are finite sets, $B_d \subseteq \begin{pmatrix} \lambda \\ V^+ \end{pmatrix}, B_u \subseteq \begin{pmatrix} V^+ \\ \lambda \end{pmatrix}$. The elements of A are called axioms, those of B_d, B_u are called lower level (*down*) and upper level (*up*) blocks.

For $x, z \in W_\rho(V)$ we write

$$x \Longrightarrow z \text{ if and only if } z = \mu(x, y), \text{ for some } y \in B_d \cup B_u.$$

We denote by $\Longrightarrow^*$ the reflexive and transitive closure of the relation $\Longrightarrow$. A sequence $\sigma : x_1 \Longrightarrow x_2 \Longrightarrow \ldots \Longrightarrow x_k$ such that $x_1 \in A$, is called a *computation* in γ. Then σ is said to be *complete* if $x_k \in WK_\rho(V)$ (a molecule is completed).

The language generated by γ is defined by

$$L(\gamma) = \{w \in WK_\rho(V) \mid x \Longrightarrow^* w,\ x \in A\}.$$

A complete computation $\sigma : x_1 \Longrightarrow x_2 \Longrightarrow \ldots \Longrightarrow x_k$ is said to be *primitive* if no proper initial part of it is complete ($x_i \notin WK_\rho(V)$ for $1 \le i \le k-1$). Associating labels from a set *Lab* to the elements in the sets B_d, B_u, we can consider the control words, $c_d(\sigma), c_u(\sigma)$, of the labels of blocks in B_d, B_u, respectively, used in a given computation σ. Then, σ is *fair* if $|c_d(\sigma)| = |c_u(\sigma)|$, and *coherent* (with respect to mappings c_d and c_u) when $c_d(\sigma) = c_u(\sigma)$. By $L_p(\gamma)$, $L_f(\gamma)$, $L_c(\gamma)$, $L_{pf}(\gamma)$, $L_{pc}(\gamma)$ we denote the language of all strings in $L(\gamma)$ generated by primitive, fair, coherent, primitive and fair, primitive and coherent computations, respectively. By *SL, PSL, FSL, CSL, PFSL, PCSL* we denote the families of languages of the form $L(\gamma), L_p(\gamma), L_f(\gamma), L_c(\gamma), L_{pf}(\gamma), L_{pc}(\gamma)$, defined as above.

Because building a double-stranded sequence in $WK_\rho(V)$ by adding lower or upper level blocks in B_d, B_u can be controlled by a finite memory in such a way that the delay (the length difference) in growing the two strands is bounded, the families SL, PSL contain only regular languages. Moreover, we have

THEOREM 5. ([**19**]) *Each regular language is a coding of a language in the family SL or PSL.*

Because the fairness condition is of a non-regular type, the families $FSL, PFSL$ contain non-regular languages, but they do not characterize RE, even when using squeezing tools stronger than codings, for instance morphisms, intersections with regular languages, or gsm mappings (it is proved in [**19**] that $FSL \subseteq MAT^{\lambda}$ and MAT^{λ} is closed under the operations mentioned above, [**6**]). For the coherent restriction the following result has been proved in [**19**]:

THEOREM 6. *Each recursively enumerable language is the weak coding of a language in the family CSL or PCSL.*

Several results in the subsequent sections are based on this result.

6. Watson-Crick finite automata

We are now going to define the main class of automata we have announced in the Introduction. They are a counterpart of finite automata (they use states which control the transitions, as usual in automata theory), but work on Watson-Crick tapes.

A *Watson-Crick finite automaton* is a construct

$$M = (V, \rho, Q, s_0, F, \delta),$$

where V and Q are disjoint alphabets, $\rho \subseteq V \times V$ is a symmetric relation, $s_0 \in Q$, $F \subseteq Q$, and $\delta : Q \times \begin{pmatrix} V^* \\ V^* \end{pmatrix} \longrightarrow P(Q)$ is a mapping such that $\delta(s, \begin{pmatrix} x \\ y \end{pmatrix}) \neq \emptyset$ only for finitely many triples $(s, x, y) \in Q \times V^* \times V^*$.

The elements of Q are called *states*, V is the (input) alphabet, ρ is a complementarity relation on V, s_0 is the initial state, F is the set of final states, and δ is the transition mapping. The interpretation of $s' \in \delta(s, \begin{pmatrix} x_1 \\ x_2 \end{pmatrix})$ is: in state s, the automaton passes over x_1 in the upper level strand and over x_2 in the lower level strand of a double-stranded sequence, and enters the state s'.

REMARK 2. In contrast to the case of finite automata, in Watson-Crick finite automata we have written first the alphabet V and the complementarity relation ρ, and after that the set Q of states, in order to stress the fact that the pair (V, ρ) plays a fundamental role in our machines. Working with double-stranded sequences is the crucial difference between traditional finite automata and Watson-Crick finite automata.

For $\begin{pmatrix} x_1 \\ x_2 \end{pmatrix}, \begin{pmatrix} u_1 \\ u_2 \end{pmatrix}, \begin{pmatrix} w_1 \\ w_2 \end{pmatrix} \in \begin{pmatrix} V^* \\ V^* \end{pmatrix}$ such that $\begin{bmatrix} x_1 u_1 w_1 \\ x_2 u_2 w_2 \end{bmatrix} \in WK_{\rho}(V)$ and $s, s' \in K$, we write

$$\begin{pmatrix} x_1 \\ x_2 \end{pmatrix} s \begin{pmatrix} u_1 \\ u_2 \end{pmatrix} \begin{pmatrix} w_1 \\ w_2 \end{pmatrix} \Longrightarrow \begin{pmatrix} x_1 \\ x_2 \end{pmatrix} \begin{pmatrix} u_1 \\ u_2 \end{pmatrix} s' \begin{pmatrix} w_1 \\ w_2 \end{pmatrix}$$

$$\text{if and only if } s' \in \delta(s, \begin{pmatrix} u_1 \\ u_2 \end{pmatrix}).$$

By $\Longrightarrow^*$ we denote the reflexive and transitive closure of the relation $\Longrightarrow$.

Three languages can be associated to a Watson-Crick finite automaton in the following way:

$$L_{ud}(M) = \{w \in WK_\rho(V) \mid s_0 w \Longrightarrow^* s_f, s_f \in F\},$$

$$L_u(M) = \{w_1 \in V^* \mid \left[\begin{array}{c} w_1 \\ w_2 \end{array} \right] \in L_{ud}(M), \text{ for some } w_2 \in V^*\},$$

$$L_d(M) = \{w_2 \in V^* \mid \left[\begin{array}{c} w_1 \\ w_2 \end{array} \right] \in L_{ud}(M), \text{ for some } w_1 \in V^*\}.$$

REMARK 3. Depending on the complementarity relation ρ, these languages can be very different from one another: consider the case when the relation ρ is not injective and on the lower strand we use one symbol only; for every context-free $L_u(M)$ we have $L_d(M)$ regular and $L_{ud}(M)$ identical with $L_u(M)$ modulo the identification of two level ssymbol with the symbol on their upper level. In this paper we are mainly interested in the possibilities of characterizing the recursively enumerable languages by means of the languages recognized by our automata. Because $L_d(M)$ and $L_u(M)$ are codings of $L_{ud}(M)$, the same results hold true for $L_{ud}(M)$ as for the other two languages. Because, in most cases, we shall use an injective complementarity relation, a coding will map $L_u(M)$ to $L_d(M)$ (and conversely). Moreover, for an automaton M we can consider the up-down reversed automaton M' for which we obviously have $L_u(M') = L_d(M), L_d(M') = L_u(M)$. Therefore, from now on we shall not investigate all languages $L_d(M), L_u(M), L_{ud}(M)$ defined above, but only the language $L_u(M)$.

Another important language associated to a Watson-Crick automaton can be defined taking into account the transitions, not the recognized sequence.

For a Watson-Crick finite automaton $M = (V, \rho, Q, s_0, F, \delta)$, let us define the set of transition rules

$$P(\delta) = \{(s, \left(\begin{array}{c} x_1 \\ x_2 \end{array} \right), s') \mid s' \in \delta(s, \left(\begin{array}{c} x_1 \\ x_2 \end{array} \right)), \ s, s' \in Q, x_1, x_2 \in V^*\}.$$

By the definition of the mapping δ, the set $P(\delta)$ is a finite set. Consider a labeling $e : P(\delta) \longrightarrow Lab$, of rules in $P(\delta)$ with elements in a set Lab. For a computation $\sigma : s_0 w \Longrightarrow^* s_f$, $w \in WK_\rho(V)$, $s_f \in F$, denote by $e(\sigma)$ the control word of σ, that is the sequence of labels of transition rules used in σ. In this way a fourth language is obtained:

$$L_{ctr}(M) = \{e(\sigma) \mid \sigma : s_0 w \Longrightarrow^* s_f, w \in WK_\rho(V), s_f \in F\}.$$

REMARK 4. The control word $e(\sigma)$ associated to a computation σ in a Watson-Crick finite automaton can be particularly useful in DNA computing, where we work with words over a prescribed reduced (four letter) alphabet, hence we need codifications of symbols of larger alphabets arising from the problems we want to solve. Consider, for instance, the very initial experiment in DNA computing, that in [**1**]. Associate a Watson-Crick automaton to a graph by using the codes of nodes on the upper level and the codes of the edges on the lower level of the double-stranded sequences as transitions. Let each transition parse either a node or an edge. Label each transition with the name of the corresponding node or edge. Then the control word of a computation will be a shuffle of the description of the path associated to our computation, written as a sequence of nodes and simultaneously as a sequence of vertices. By a weak code, we can select from the control word the path description we want. Thus, in this case the control word of

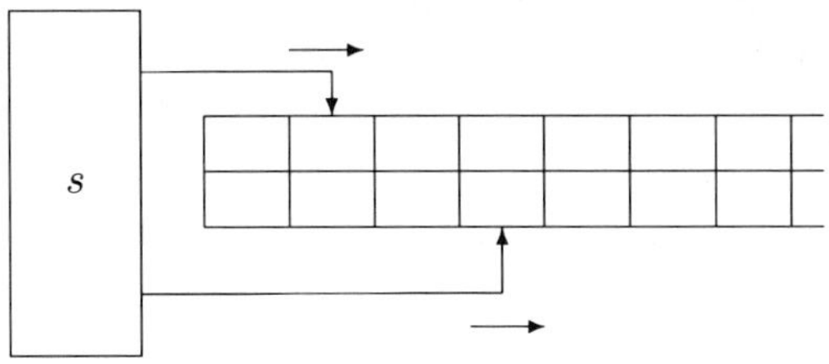

FIGURE 2. Watson-Crick finite automaton

a computation is more explicit than the recognized sequence. In particular, like in Adleman's experiment, we can let the automaton work on nondeterministically chosen sequences, selecting the control words of interest.

We say that all languages $L_\alpha(M), \alpha \in \{u, d, ud, ctr\}$, are *recognized* by the Watson-Crick finite automaton M. As we have said above, here we explicitly investigate mainly the languages $L_u(M)$ and $L_{ctr}(M)$, but the results also hold true for the other languages.

We note again that the work of Watson-Crick automata is defined for elements of $WK_\rho(V)$ only, that is for double-stranded sequences of elements in V paired according to the complementarity relation ρ. We can represent such a machine as consisting of a double tape on which an element of $WK_\rho(V)$ is written, a finite memory, able to store a state from a finite set of states, and two read-only heads, one of them scanning the upper level and the other one scanning the lower level of the tape. One starts with the two heads placed before the first symbol of each level, in state s_0. The two heads are moved to the right, according to the current state of the machine, as indicated by the transition mapping (the transition rules). Here a transition means to move the two heads across blocks defined by the transition rules. Figure 2 illustrates this representation.

One stops and accepts the starting sequence when both heads reach the right end of the sequence written on the tape, entering a final state.

Several variants of Watson-Crick finite automata are worth to be defined. We say that $M = (V, \rho, Q, s_0, F, \delta)$ is:

- *stateless*, if $Q = F = \{s_0\}$;
- *all-final*, if $F = Q$,
- *simple*, if for all $(s, \begin{pmatrix} x_1 \\ x_2 \end{pmatrix}, s') \in P(\delta)$ we have either $x_1 = \lambda$ or $x_2 = \lambda$.

In the specification of a stateless automaton, the components Q, s_0, F can be omitted, hence the transition rules can be written in the form $\begin{pmatrix} x_1 \\ x_2 \end{pmatrix}$. Then, the mapping δ is replaced by a set P of such pairs $\begin{pmatrix} x_1 \\ x_2 \end{pmatrix}$ and the automaton is written in the form $M = (V, \rho, P)$.

By $UWK_\alpha, NWK_\alpha, SWK_\alpha, AWK_\alpha, NSWK_\alpha, ASWK_\alpha$, we denote the families of languages of the form $L_\alpha(M)$, $\alpha \in \{u, ctr\}$, recognized by Watson-Crick finite automata which are unrestricted (U), stateless (N, from "no state"), simple (S), all-final (A), stateless simple (NS), and all-final simple (AS), respectively. If necessary, we can also consider α above equal to d or to ud. (The basic abbreviation, WK, is obtained by selecting the beginning and the end symbols of the single stranded sequence W A T S O N C R I C K.) We will use the generic term "WK families" to refer to all these families of languages.

7. The relationships between the WK families

In this section we investigate the relationships between the families of languages introduced in the previous section as well as the relationship of these families to the families in the Chomsky hierarchy.

Directly from the definitions we obtain:

LEMMA 1. $XWK_\alpha \subseteq UWK_\alpha$, $\alpha \in \{u, ctr\}, X \in \{N, S, A, NS, AS\}$.

LEMMA 2. $NWK_\alpha \subseteq AWK_\alpha$, $NSWK_\alpha \subseteq ASWK_\alpha$, $\alpha \in \{u, ctr\}$.

LEMMA 3. $XSWK_\alpha \subseteq SWK_\alpha$, $XSWK_\alpha \subseteq XWK_\alpha$, $\alpha \in \{u, ctr\}, X \in \{N, A\}$.

Moreover, it is easy to see that we also have the following relations:

LEMMA 4. $REG \subseteq SWK_\alpha$, $\alpha \in \{u, ctr\}$.

LEMMA 5. $UWK_\alpha \subseteq CS$, $\alpha \in \{u, ctr\}$.

LEMMA 6. *Each language in a family* XWK_u *is a coding of a language in the family* $XWK_{ctr}, X \in \{U, N, S, A, NS, AS\}$.

It is also obvious that a two-head finite automaton is a particular case of an unrestricted Watson-Crick finite automaton – the complementarity relation is the identity: $(a, b) \in \rho$ if and only if $a = b$.

On the other hand, a Watson-Crick finite automaton can be simulated by a two-head finite automaton: one head parses the input string simulating the upper head of the Watson-Crick automaton, the second one parses this string acting as the lower head of the Watson-Crick automaton: it guesses a complement of the current symbol and it crosses a symbol a only if the lower head of the Watson-Crick automaton can cross – in the same state - a symbol b which is complementary to a.

Consequently, we get the next equality:

LEMMA 7. $TH = UWK_u$.

REMARK 5. A Watson-Crick finite automaton can also be viewed as a two-tape, two-head finite automaton (see, e.g, [**7**], [**17**], [**18**], [**27**]), but of a very particular type: the two tapes are interrelated through the complementarity relation; if this relation is one-to-one, then one tape precisely defines the other one – this is the case of the DNA molecules.

The use of states is powerful, in the sense that arbitrary transition rules can be replaced by simple transition rules without decreasing the power. The following lemma can also be viewed as a "normal form" result, customary in automata theory.

LEMMA 8. $UWK_u \subseteq SWK_u$.

PROOF. Consider an unrestricted Watson-Crick finite automaton $M = (V, \rho, Q, s_0, F, \delta)$ and construct the simple Watson-Crick automaton

$$M' = (V, \rho, Q', s_0, F, \delta'),$$

with

$$Q' = Q \cup \{[s', v] \mid s' \in \delta(s, \begin{pmatrix} u \\ v \end{pmatrix}), s \in Q, u, v \in V^*\},$$

$$\delta'(s, \begin{pmatrix} u \\ \lambda \end{pmatrix}) = \{[s', v] \mid s' \in \delta(s, \begin{pmatrix} u \\ v \end{pmatrix}), v \in V^*\},\ s \in Q, u \in V^*,$$

$$\delta'([s', v], \begin{pmatrix} \lambda \\ v \end{pmatrix}) = \{s'\},\ [s', v] \in Q' - Q, v \in V^*,$$

$$\delta'(q, \begin{pmatrix} x \\ y \end{pmatrix}) = \emptyset, \text{ otherwise.}$$

Because $\delta(s, \begin{pmatrix} u \\ v \end{pmatrix})$ is non-empty only for finitely many triples $(s, u, v) \in Q \times V^* \times V^*$, it follows that δ' is non-empty only for finitely many triples in $Q' \times V^* \times V^*$. No parsing can stop in a state in $Q' - Q$ and the passing from s to $[s', v]$ and then to s' corresponds to the use of a transition $s' \in \delta(s, \begin{pmatrix} u \\ v \end{pmatrix})$ in M. Consequently, $L_u(M') = L_u(M)$. □

COROLLARY 1. $SWK_u = UWK_u$.

The construction given above modifies the language $L_{ctr}(M)$; we do not know whether or not the inclusion in Lemma 8 also holds for the families SWK_{ctr}, UWK_{ctr}.

For an easy reference, we summarize the relations from the previous lemmas in the diagrams in Figure 3 (for *ctr*) and in Figure 4 (for u); the arrows indicate inclusions which are not necessarily proper.

In the simple stateless case, the parsing of a sequence in $WK_\rho(V)$ can be controlled by examining a subsequence of length at most the length of the longest string w_1, w_2 in transition rules $\begin{pmatrix} w_1 \\ w_2 \end{pmatrix}$: because one of w_1, w_2 is always empty, we can continue with the level whose reading head is behind, thus bounding the distance (delay) between the two heads. Consequently, we obtain:

LEMMA 9. $NSWK_u \subseteq REG$.

On the other hand, we have the following strengthening of Lemma 4:

LEMMA 10. $REG \subseteq ASWK_u$.

PROOF. Consider a finite automaton $M = (Q, V, s_0, F, \delta)$ and construct the all-final simple Watson-Crick finite automaton

$$M' = (V, \rho, Q', s_0, Q', \delta'),$$

with

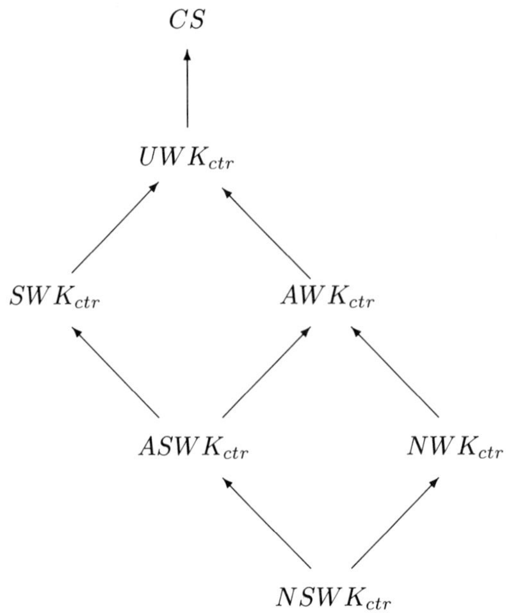

FIGURE 3. Hierarchy of *ctr* families of languages

$$\rho = \{(a, a)) \mid a \in V\},$$
$$Q' = Q \cup \{s_f\},\ s_f \notin Q,$$
$$\delta'(s, \begin{pmatrix} a \\ \lambda \end{pmatrix}) = \delta(s, a) \cup F(s, a),\ s \in Q, a \in V,$$
$$\text{where } F(s, a) = \begin{cases} \{s_f\}, & \text{if } \delta(s, a) \cap F \neq \emptyset, \\ \emptyset, & \text{otherwise,} \end{cases}$$
$$\delta'(s_f, \begin{pmatrix} \lambda \\ a \end{pmatrix}) = \{s_f\},\ a \in V,$$
$$\delta'(s, \begin{pmatrix} u \\ v \end{pmatrix}) = \emptyset, \text{ in all other cases.}$$

The recognition of a sequence $\begin{bmatrix} w \\ w \end{bmatrix}$ proceeds as follows: one first parses the first strand, from left to right, exactly as in M except for the last step, when M reaches a final state; then M' switches to s_f. Then one can also parse the second strand, making possible the completion of the recognition. Therefore, $L_u(M') = L(M)$. □

In some sense, all the families considered above are "small": the languages in these families satisfy very strong conditions. The following lemmas give two such necessary conditions.

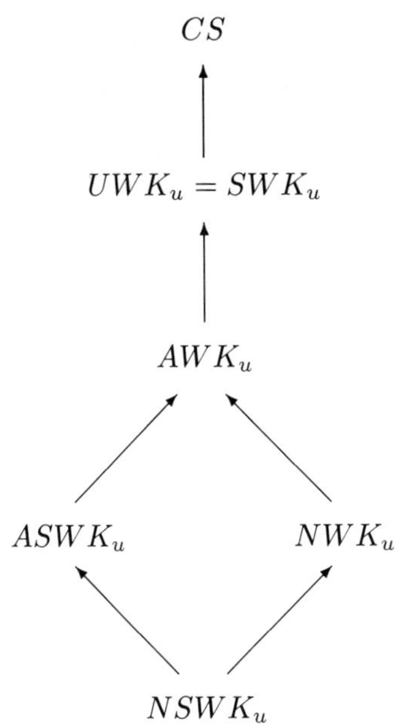

FIGURE 4. Hierarchy of u families of languages

LEMMA 11. *If $L \in NWK_\alpha$ for $\alpha \in \{u, ctr\}$, then $L = L^+$.*

PROOF. Consider a stateless Watson-Crick finite automaton $M = (V, \rho, P)$. If $w_1, w_2 \in WK_\rho(V)$ can be parsed by M, then also $w_1 w_2$ can be parsed, using the same elements of P. Hence, $L_\alpha(M)^+ \subseteq L_\alpha(M), \alpha \in \{u, crt\}$. The opposite inclusion, $L_\alpha(M) \subseteq L_\alpha(M)^+$, is obvious. □

COROLLARY 2. *$REG - XWK_\alpha \neq \emptyset$, $\alpha \in \{u, ctr\}, X \in \{NS, N\}$.*

COROLLARY 3. *Corollary 3 $ASWK_u - NWK_u \neq \emptyset$.*

PROOF. By Lemma 10, the language ab^* is in $ASWK_u$. This language does not have the property stated in Lemma refLemma11, so $ab^* \notin NWK_u$. □

COROLLARY 4. *The inclusions $NWK_u \subset AWK_u$ and $NSWK_u \subset ASWK_u$ are proper.*

LEMMA 12. *Every one-letter language in UWK_u is regular.*

PROOF. Consider an unrestricted Watson-Crick finite automaton $M = (V, \rho, Q, s_0, F, \delta)$. If there is a transition $s' \in \delta(s, \begin{pmatrix} a^i \\ w \end{pmatrix})$ such that w contains a symbol b and $(a, b) \notin \rho$, then this transition can never be used when producing strings in $L_u(M)$. Thus, all such transitions can be ignored, that is we may assume that for all b as above we have $(a, b) \in \rho$.

We construct the linear grammar

$$G = (Q, \{a, b\}, s_0, P),$$

with

$$\begin{aligned} P &= \{s \to a^i s' b^j \mid s' \in \delta(s, \begin{pmatrix} a^i \\ w \end{pmatrix}), s, s' \in Q, i \geq 0, j = |w|\} \\ &\cup \{s \to a^i b^j \mid \delta(s, \begin{pmatrix} a^i \\ w \end{pmatrix}) \cap F \neq \emptyset, s \in Q, i \geq 0, j = |w|\}. \end{aligned}$$

Consider also the linear language

$$L = \{a^n b^n \mid n \geq 1\}.$$

According to [**12**], Corollary 5.3.1, $L(G) \cap L$ is a linear language. (We have $L(G) \cap L = \{a^n b^m \mid (n, m) \in \Psi_{\{a,b\}} \cap \{(p, p) \mid p \in \mathbf{N}\}\}$, and the intersection of two semilinear sets is a semilinear set, hence $\Psi_{\{a,b\}}(L(G) \cap L)$ is a semilinear set. Together with the result from [**12**] mentioned above, this implies that $L(G) \cap L$ is a linear language.) For the weak coding h defined by

$$h(a) = a, \; h(b) = \lambda,$$

we have

$$L_u(M) = h(L(G) \cap L),$$

which implies that $L_u(M)$ is regular. □

COROLLARY 5. *$UWK_u \subset CS$ is a strict inclusion.*

However, the families considered above (with the exception of $NSWK_u$) contain languages of a very high complexity: in the next section we shall see that characterizations of recursively enumerable languages can be obtained starting from them and using AFL operations (in fact, weak codings and deterministic gsm's). We close this section with a result proving directly that stateless Watson-Crick finite automata can recognize complex languages; this is, of course, due to the work with double-stranded sequences.

LEMMA 13. *NWK_u contains non-semilinear languages.*

PROOF. Consider the stateless Watson-Crick finite automaton

$$M = (\{a, b, c, d, e, f\}, \rho, P),$$

with

$$\rho = \{(a, a), (b, c), (c, b), (a, d), (d, a), (e, f), (f, e)\},$$

$$P = \{\begin{pmatrix} c \\ \lambda \end{pmatrix}, \begin{pmatrix} d \\ \lambda \end{pmatrix}, \begin{pmatrix} a \\ a \end{pmatrix}, \begin{pmatrix} b \\ b \end{pmatrix}, \begin{pmatrix} b \\ c \end{pmatrix}, \begin{pmatrix} e \\ c \end{pmatrix}, \begin{pmatrix} e \\ d \end{pmatrix}, \begin{pmatrix} \lambda \\ f \end{pmatrix}\}.$$

Consider also the regular language described by the following regular expression

$$R = c(dd^+ b)(aa^+ b)^+ a^+ e^+,$$

and the weak coding h defined by

$$h(a) = a, \; h(b) = h(c) = h(d) = h(e) = \lambda.$$

The molecules recognized by M and having a string in R in their upper strand must be of the form

$$\begin{bmatrix} cd^{n_1} ba^{n_2} b \ldots ba^{n_{m-1}} ba^{n_m} e^{n_{m+1}} \\ b x_1 \, c x_2 \, c \ldots c x_{m-1} \, c x_m \, f^{n_{m+1}} \end{bmatrix},$$

with $m \geq 3, n_i \geq 2, 1 \leq i \leq m, n_{m+1} \geq 1$. Because of the complementarity, we also have $x_1 = a^{n_1}$, $x_i \in \{a, d\}^*, |x_i| = n_i$, for $2 \leq i \leq m$.

Because of the form of the pairs in P, each b in the upper strand is paired with an occurrence of b or c in the lower strand, as suggested by the subscripts of these symbols b, c in the following writing:

$$\left[\begin{array}{c} c_0 d^{n_1} b_1 a^{n_2} b_2 \dots b_{m-2} a^{n_{m-1}} b_{m-1} a^{n_m} e_m e^{n_{m+1}-1} \\ b_1\, x_1\, c_2\, x_2\, c_3\, \dots\, c_{m-1}\, x_{m-1}\, c_m\, x_m\, f\, f^{n_{m+1}-1} \end{array}\right].$$

We have also indicated the fact that the last occurrence of c in the lower strand is paired with the first occurrence of e in the upper strand.

Now, because of this precise pairing and because (1) the symbol a apears only in pairs $\left(\begin{array}{c} a \\ a \end{array}\right)$ of P, and (2) the symbol d can be introduced in the lower strand only by the pair $\left(\begin{array}{c} e \\ d \end{array}\right)$, it follows that:

1. $x_i \in a^+, 2 \leq i \leq m-1$,
2. $x_m = d^{n_{m+1}-1}$,
3. $|x_i| = n_{i+1}, 1 \leq i \leq m-1$.

From $|x_i| = n_i$ and $|x_i| = n_{i+1}$ we get $n_i = n_{i+1}, 1 \leq i \leq m-1$.

Therefore, the recognized molecule has the form

$$\left[\begin{array}{c} cd^n ba^n b \dots ba^n ba^n e^{n+1} \\ ba^n ca^n c \dots ca^n cd^n f^{n+1} \end{array}\right],$$

with at least two blocks ba^n and $n \geq 2$.

Consequently, we obtain

$$\begin{aligned} h(L_u(M) \cap R) &= \{a^{nm} \mid n, m \geq 2\} \\ &= \{a^p \mid p \text{ is a composite number}\}. \end{aligned}$$

This language is not semilinear. □

COROLLARY 6. *$NWK_u - MAT^\lambda \neq \emptyset$.*

PROOF. The language $\{a^p \mid p$ is a composite number$\}$ from the previous proof is not regular. Each one-letter language in MAT^λ is regular, [**13**]. The family MAT^λ is closed under intersection with regular languages and arbitrary morphisms. Consequently, for the automaton M constructed above we have $L_u(M) \notin MAT^\lambda$. □

COROLLARY 7. *The inclusion $NSWK_u \subset NWK_u$ is proper.*

8. Characterizations of recursively enumerable languages

We now shall give a series of representation results for recursively enumerable languages starting from WK languages. Because RE is closed under the used operations, we have, in fact, characterizations of RE.

The proof of the following lemma, although technically simple, captures the essence of Watson-Crick tapes and the interconnection to the twin-shuffle language.

LEMMA 14. *For every alphabet V, we have $TS_V \in NSWK_{ctr}$.*

PROOF. Consider the simple stateless Watson-Crick finite automaton

$$M = (V, \rho, P),$$

with

$$\rho = \{(a,a) \mid a \in V\},$$
$$P = \{\begin{pmatrix} a \\ \lambda \end{pmatrix}, \begin{pmatrix} \lambda \\ a \end{pmatrix} \mid a \in V\},$$

with the labeling

$$e(\begin{pmatrix} a \\ \lambda \end{pmatrix}) = a, \ e(\begin{pmatrix} \lambda \\ a \end{pmatrix}) = \bar{a}, \ a \in V.$$

If $w = \begin{bmatrix} x \\ x \end{bmatrix} \in WK_\rho(V)$ and $\sigma : \begin{bmatrix} x \\ x \end{bmatrix} \Longrightarrow^* \begin{bmatrix} \lambda \\ \lambda \end{bmatrix}$ is a computation in M, then $e(\sigma) \in x \sqcup\!\sqcup \bar{x}$, hence $e(\sigma) \in TS_V$.

Because x above can be any string in V^* and each element of $x \sqcup\!\sqcup \bar{x}$ describes a correct computation in M, we have the equality $L_{ctr}(M) = TS_V$. □

THEOREM 7. *Each language in the family RE is the image of a deterministic gsm mapping of a language in any family $XWK_{ctr}, X \in \{U, S, A, N, AS, NS\}$.*

PROOF. This is a direct consequence of the previous lemma and Theorem 1. □

LEMMA 15. $CSL \subseteq AWK_{ud}$.

PROOF. Consider a simple sticker system $\gamma = (V, \rho, A, B_d, B_u)$, with $A \subseteq W_\rho(V)$, $B_d \subseteq \begin{pmatrix} \lambda \\ V^* \end{pmatrix}$, $B_u \subseteq \begin{pmatrix} V^* \\ \lambda \end{pmatrix}$, with the elements of B_l, B_u labeled by a mapping $e : B_l \cup B_u \longrightarrow Lab$.

We construct the all-final Watson-Crick finite automaton

$$M = (V, \rho, Q, s_0, Q, \delta),$$

with

$$Q = \{s_0, s_1\},$$
$$\delta(s_0, \begin{bmatrix} u_1 \\ u_2 \end{bmatrix}\begin{pmatrix} v_1 \\ v_2 \end{pmatrix}) = \{s_1\}, \text{ for } \begin{bmatrix} u_1 \\ u_2 \end{bmatrix}\begin{pmatrix} v_1 \\ v_2 \end{pmatrix} \in A,$$
$$\delta(s_1, \begin{pmatrix} x \\ y \end{pmatrix}) = \{s_1\}, \text{ for } \begin{pmatrix} \lambda \\ y \end{pmatrix} \in B_l, \begin{pmatrix} x \\ \lambda \end{pmatrix} \in B_u$$
$$\text{such that } e(\begin{pmatrix} \lambda \\ y \end{pmatrix}) = e(\begin{pmatrix} x \\ \lambda \end{pmatrix}),$$
$$\delta(s, \begin{pmatrix} u \\ v \end{pmatrix}) = \emptyset, \text{ otherwise.}$$

The coherent correct computations in γ can be simulated by correct computations in M and, conversely, the sequences recognized by M are in $L_c(\gamma)$. Indeed, the fact that we always have to start from an axiom is ensured by the initial state, s_0, while the coherence is ensured by the transition mapping, which is defined only for pairs $\begin{pmatrix} u \\ v \end{pmatrix}$ such that $\begin{pmatrix} u \\ \lambda \end{pmatrix} \in B_u$ and $\begin{pmatrix} \lambda \\ v \end{pmatrix} \in B_d$ and these blocks have the same label. Consequently, $L_c(\gamma) = L_{ud}(M)$. □

THEOREM 8. *Each recursively enumerable language is the weak coding of a language in any family $XWK_u, X \in \{S, A, U\}$.*

PROOF. As we have pointed out in Section 6 (Remark 3), each language in a family XWK_u is a coding of a language in the corresponding family XWK_{ud}. Now the theorem follows from Lemma 15 and Theorem 5 (the composition of two weak codings is a weak coding). □

LEMMA 16. *If $h_1, h_2 : V_1^* \longrightarrow V_2^*$ are two morphisms, then $h_1(EQ(h_1, h_2)) \in NWK_u$.*

PROOF. For h_1, h_2 given, we construct the stateless Watson-Crick finite automaton

$$M = (V_2, \rho, P),$$

with

$$\rho = \{(a, a) \mid a \in V_2\},$$
$$P = \{\begin{pmatrix} h_1(b) \\ h_2(b) \end{pmatrix} \mid b \in V_1\}.$$

A sequence $w = \begin{bmatrix} w_1 \\ w_2 \end{bmatrix} \in WK_\rho(V_2)$ is successfully parsed by M if and only if $w_1 = w_2$ (due to the relation ρ) and $w_1 = h_1(x), w_2 = h_2(x)$, for some $x \in V_1^*$. Consequently, $x \in EQ(h_1, h_2)$ and $w \in h_1(EQ(h_1, h_2))$, which implies $L_u(M) = h_1(EQ(h_1, h_2))$. □

THEOREM 9. *Each language $L \in RE$ can be written in the form $L = h(L' \cap R)$, where $L' \in NWK_u, R \in REG$, and h is a projection.*

PROOF. This is a direct consequence of the previous lemma and of Theorem 4. □

Although a result like that in Lemma 16 seems not to be true for the family $ASWK_u$, a result like that in Theorem 9 holds true for this family, too.

THEOREM 10. *Each language $L \in RE$ can be written in the form $L = h(L' \cap R)$, where $L' \in ASWK_u, R \in REG$, and h is a projection.*

PROOF. Using Theorem 4, we can write L as $L = h(h_1(EQ(h_1, h_2)) \cap R_0)$, where h_1, h_2 are λ-free morphisms, h is a projection, and R_0 is a regular language. For $h_1, h_2 : V_1^* \longrightarrow V_2^*$, we construct the all-final simple Watson-Crick automaton

$$M = (V, \rho, Q, s_0, Q, \delta),$$

with

$$\begin{aligned}
&V = V_2 \cup \{c\},\\
&\rho = \{(a,a) \mid a \in V_2\} \cup \{(c,c)\},\\
&Q = \{s_0, s_1\} \cup \{s_a \mid a \in V_1\},\\
&\delta(s_0, \begin{pmatrix} h_1(a) \\ \lambda \end{pmatrix}) = \{s_a\},\\
&\delta(s_a, \begin{pmatrix} \lambda \\ h_2(a) \end{pmatrix}) = \{s_0\},\ a \in V_1,\\
&\delta(s_0, \begin{pmatrix} c \\ \lambda \end{pmatrix}) = \{s_1\},\\
&\delta(s_1, \begin{pmatrix} \lambda \\ c \end{pmatrix}) = \{s_1\},\\
&\delta(s, \begin{pmatrix} u \\ v \end{pmatrix}) = \emptyset \text{ in all other cases.}
\end{aligned}$$

For the regular language

$$R_1 = V_2^*\{c\}$$

we obviously have

$$L_u(M) \cap R_1 = h_1(EQ(h_1,h_2))\{c\}.$$

Considering $R = R_0\{c\}$ and extending the projection h by $h(c) = \lambda$, we obtain $L = h(L_u(M) \cap R)$. □

Obviously, the projection h and the intersection with the regular language R in Theorems 9 and 10 can be done by a deterministic gsm, hence a result like Theorem 7 holds true for the families NWK_u and $ASWK_u$, too.

Consequently, we may say that modulo a deterministic gsm – and in some cases modulo a weak coding only – all families XWK_α, with the exception of $NSWK_u$ (Lemma 9), are equal to RE, and, in this sense, equal to each other. This also leads to the following more general result:

COROLLARY 8. *For each family of languages FL such that $FL \subset RE$ and FL is closed under intersection with regular languages and arbitrary morphisms, we have $XWK_u - FL \neq \emptyset$, $X \in \{U,S,A,N,AS\}$.*

Among the important language families FL satisfying the premise of the statement of Corollary 7 are MAT^λ and $ET0L$.

Note that in the results stated above the final states play no role, they do not increase the power of Watson-Crick finite automata modulo a deterministic gsm or, in certain cases, modulo a weak coding – and that simple Watson-Crick finite automata (with only three states) suffice to characterize RE modulo a weak coding. Automata without states, using transition pairs $\begin{pmatrix} u \\ v \end{pmatrix}$ with non-restricted strings u, v, are also very powerful. These observations illustrate again the power of Watson-Crick complementarity; they will also be confirmed by the machines considered in the sequel of this paper.

It is also worth mentioning that in most of the constructions given above (this is the case for the proofs of Theorem 10 and of Lemmas 14 and 16) the complementarity relation is the identity relation, $(a,b) \in \rho$ if and only if $a = b$. This is not the case for the proof of Theorem 6, hence for Theorem 8, too. As we have already mentioned several times, in the DNA case the complementarity relation is one-to-one. When trying to be closer to the "DNA reality", we have to take a symmetric one-to-one relation as a basic complementarity relation for our models. This can raise some problems, because the proof given above of Theorem 6 uses a relation which is not even injective: $(s,s) \in \rho$ and $(s,[s,k]) \in \rho$, too.

9. Watson-Crick finite transducers

An output can be associated to a Watson-Crick finite automaton in the same way as an output is associated to a finite automaton to form a gsm. This output is written on a usual tape rather than on a Watson-Crick double-stranded tape. Figure 5 illustrates this idea.

A *Watson-Crick gsm* is a construct

$$g = (V_I, \rho_I, V_O, Q, s_0, F, \delta),$$

where V_I is the input alphabet, $\rho_I \subseteq V_I \times V_I$ is a symmetric relation (of complementarity), V_O is the output alphabet, Q is the finite set of states, $s_0 \in Q$ is the initial state, $F \subseteq Q$ is the set of final states, and $\delta : Q \times \begin{pmatrix} V_I^* \\ V_I^* \end{pmatrix} \longrightarrow P_f(V_O^* \times Q)$ is a mapping such that $\delta(s, \begin{pmatrix} u \\ v \end{pmatrix}) \neq \emptyset$ only for a finite number of triples $(s,u,v) \in Q \times V_I^* \times V_I^*$. The interpretation of $(x,s') \in \delta(s, \begin{pmatrix} u \\ v \end{pmatrix})$ is: in state s, the transducer passes over u in the upper level and over v in the lower level of a double-stranded sequence, produces the output x, and enters the state s'.

Formally, for $w_1, w_2, w_1', w_2' \in V_I^*, z \in V_O^*$, and $s, s' \in Q$, we write

$$zs\begin{pmatrix} w_1 \\ w_2 \end{pmatrix} \Longrightarrow zxs'\begin{pmatrix} w_1' \\ w_2' \end{pmatrix} \quad \text{if and only if} \quad w_1 = x_1w_1', w_2 = x_2w_2',$$
$$\text{for } (x,s') \in \delta(s, x_1, x_2), x_1, x_2 \in V_I^*.$$

For a sequence $w = \begin{bmatrix} w_1 \\ w_2 \end{bmatrix} \in WK_{\rho_I}(V_I)$ we define

$$g(w) = \{z \in V_O^* \mid s_0 \begin{bmatrix} w_1 \\ w_2 \end{bmatrix} \Longrightarrow^* zs_f \begin{bmatrix} \lambda \\ \lambda \end{bmatrix}, s_f \in F\}.$$

We extend this definition to languages in $WK_{\rho_I}(V_I)$ in the natural way.

As for Watson-Crick finite automata, we consider stateless, simple, all-final, stateless simple, and all-final simple Watson-Crick gsm's.

Because the labeling of transitions in a stateless simple Watson-Crick finite automaton can be expressed as the output function of a Watson-Crick gsm, from Lemma 14 we get the following result.

COROLLARY 9. *For every alphabet V, there is a stateless simple Watson-Crick gsm g_V such that $TS_V = g_V(WK_\rho(V))$, for $\rho = \{(a,a) \mid a \in V\}$.*

This implies a result like that in Theorem 7: each language $L \in RE$ can be written in the form $L = g(g_V(WK_\rho(V)))$, where g is a usual gsm (deterministic) and g_V is a stateless simple Watson-Crick gsm.

In the same way, from Theorem 10 we obtain:

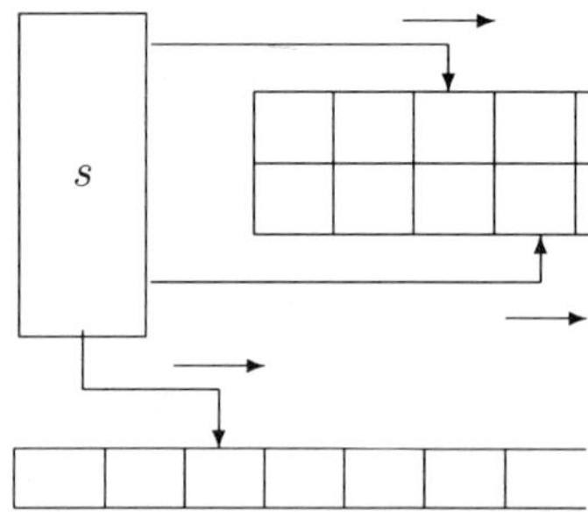

FIGURE 5. Watson-Crick gsm

THEOREM 11. *For each language $L \in RE, L \subseteq V^*$, there is a simple all-final Watson-Crick gsm g_L such that $L = g_L(WK_\rho(V'))$, for some Watson-Crick domain $WK_\rho(V')$.*

In the same way as each two-head finite automaton can be considered to be an unrestricted Watson-Crick finite automaton, a two-head finite transducer can be considered to be a special case of an unrestricted Watson-Crick gsm. From the results above we get

COROLLARY 10. *Each language $L \in RE$ can be written in the form $L = g(V^*)$, for a two-head finite transducer g, and an alphabet V.*

Hence, Watson-Crick finite transducers (and two-head finite transducers) not only do not preserve the families in the Chomsky hierarchy, but they map (very simple) regular languages in such a way that they cover the whole family RE.

10. Further variants of Watson-Crick finite automata

Among the Watson-Crick finite automata considered in Section 6, the most intimately related to the DNA structure are the stateless automata which are using only Watson-Crick complementarity, but no additional automata features. These automata deserve further investigations.

10.1. Initial stateless Watson-Crick finite automata. Because $h(w_1w_2) = h(w_1)h(w_2)$, where h is a morphism, from Lemma 11 it follows that we cannot represent every regular language as the morphic image of a language in $NWK_\alpha, \alpha \in \{u, ctr\}$. Therefore, on the one hand, starting from $NSWK_u$ we cannot even obtain represetations of REG using weak codings (as in Theorem 8), while on the other hand, the results in Theorems 7 and 10 cannot be improved by replacing the deterministic gsm mapping by a morphism, or not using an intersection with a regular language.

The "weak" point of stateless Watson-Crick finite automata is that they cannot control the first step (or the last step) of a computation. This suggests the following definition.

An *initial stateless Watson-Crick finite automaton* is a quadruple

$$M = (V, \rho, P_0, P),$$

where V is an alphabet, $\rho \subseteq V \times V$ is a symmetric relation, P_0 and P are finite subsets of $\begin{pmatrix} V^* \\ V^* \end{pmatrix}$.

A double-stranded sequence $\begin{bmatrix} w_1 \\ w_2 \end{bmatrix} \in WK_\rho(V)$ is recognized by M if and only if

$$w_1 = x_0 x_1 \ldots x_n,\ w_2 = y_0 y_1 \ldots y_n,\ n \geq 0,$$
$$\text{where } \begin{pmatrix} x_0 \\ y_0 \end{pmatrix} \in P_0, \begin{pmatrix} x_i \\ y_i \end{pmatrix} \in P, 1 \leq i \leq n.$$

Therefore, at the beginning of the parsing of $\begin{bmatrix} w_1 \\ w_2 \end{bmatrix}$ we have to use an element of P_0 and after that no element of P_0 is used again – unless $P_0 \cap P \neq \emptyset$.

As usual, by $L_\alpha(M), \alpha \in \{u, d, ud, ctr\}$, we denote the languages associated to correct computations in M. As in the previous sections, only the cases $\alpha \in \{u, ctr\}$ will be considered explicitly. By $INWK_\alpha, INSWK_\alpha$ we denote the families of languages recognized by initial stateless and by initial stateless simple Watson-Crick finite automata.

As expected, controlling the first step of a recognition process increases the power of our machines. Still, a result like that in Lemma 9 is true:

LEMMA 17. *$INSWK_u \subseteq REG$.*

On the other hand, we have the following result, which does not hold for non-initial automata.

THEOREM 12. *Each regular language is a coding of a language in the family $INSWK_u$.*

PROOF. Consider a deterministic finite automaton $M = (Q, V, s_0, F, \delta)$ and construct the initial simple stateless Watson-Crick finite automaton

$$M' = (U, \rho, P_0, P),$$

where

$$\begin{aligned}
U &= \{[s,a,s']_i \mid s,s' \in Q, a \in V, 1 \le i \le 4\},\\
\rho &= \{([s,a,s']_i,[s,a,s']_i) \mid s,s' \in Q, a \in V, 1 \le i \le 4\},\\
P_0 &= \{\begin{pmatrix} [s_0,a,s']_1 \\ \lambda \end{pmatrix} \mid s' = \delta(s_0,a), a \in V\},\\
&\cup \{\begin{pmatrix} [s_0,a,s']_3 \\ \lambda \end{pmatrix} \mid s' = \delta(s_0,a) \in F, a \in V\},\\
P &= \{\begin{pmatrix} [s,a_1,s']_2[s',a_2,s'']_1 \\ \lambda \end{pmatrix} \mid s' = \delta(s,a_1), s'' = \delta(s',a_2),\\
&\quad s,s',s'' \in Q, a_1,a_2 \in V\}\\
&\cup \{\begin{pmatrix} [s,a_1,s']_2[s',a_2,s'']_3 \\ \lambda \end{pmatrix} \mid s' = \delta(s,a_1), s'' = \delta(s',a_2),\\
&\quad s,s' \in Q, s'' \in F, a_1,a_2 \in V\}\\
&\cup \{\begin{pmatrix} [s,a,s']_4 \\ \lambda \end{pmatrix} \mid s' = \delta(s,a) \in F, s \in Q, a \in V\}\\
&\cup \{\begin{pmatrix} \lambda \\ [s,a_1,s']_1[s',a_2,s'']_2 \end{pmatrix} \mid s' = \delta(s,a_1), s'' = \delta(s',a_2),\\
&\quad s,s',s'' \in Q, a_1,a_2 \in V\}\\
&\cup \{\begin{pmatrix} \lambda \\ [s,a_1,s']_1[s',a_2,s'']_4 \end{pmatrix} \mid s' = \delta(s,a_1), s'' = \delta(s',a_2) \in F,\\
&\quad s,s' \in Q, s'' \in F, a_1,a_2 \in V\}\\
&\cup \{\begin{pmatrix} \lambda \\ [s,a,s']_3 \end{pmatrix} \mid s' = \delta(s,a) \in F, s \in Q, a \in V\}.
\end{aligned}$$

Each recognition of a sequence must start with $\begin{pmatrix} [s_0,a,s_1]_1 \\ \lambda \end{pmatrix}$ in P_0, with the exception of the recognition of sequences of length one, which starts with $\begin{pmatrix} [s_0,a,s_1]_3 \\ \lambda \end{pmatrix}$ in P_0, and ends immediately by using $\begin{pmatrix} \lambda \\ [s_0,a,s_1]_3 \end{pmatrix}$ in P. All elements of P of the form $\begin{pmatrix} x \\ \lambda \end{pmatrix}$ have the strand x starting with a symbol $[s,a,s']_2$ or $[s,a,s']_4$; all elements of P of the form $\begin{pmatrix} \lambda \\ y \end{pmatrix}$ have y starting with a symbol $[s,a,s']_1$ or $[s,a,s']_3$. Consequently, after completing the parsing of an element of $WK_\rho(U)$ no further step can be done, because the pairing imposed by ρ cannot be observed.

This means that we can successfully parse only sequences $w \in WK_\rho(U)$ composed of two identical strands of the form

$$[s_0,a_1,s_1]_1[s_1,a_2,s_2]_2[s_2,a_3,s_3]_1[s_3,a_4,s_4]_2 \dots [s_{k-1},a_k,s_k]_i[s_k,a_{k+1},s_{k+1}]_j,$$

with $i = 2, j = 3$ when k is even, and $i = 1, j = 4$ when k is odd; moreover, $s_0 s_1 \dots s_{k+1}$ is a state sequence corresponding to the recognition of the string $a_1 a_2 \dots a_{k+1}$ in M.

Let h be the coding that maps triples $[s,a,s']_i$ to a. Then we obtain $L(M) = h(L_u(M'))$. □

Because each usual stateless Watson-Crick finite automaton can be considered to be an initial one, by taking $P_0 = P$, we obtain the inclusions $XWK_\alpha \subseteq IXWK_\alpha, X \in \{N, NS\}, \alpha \in \{u, ctr\}$. Consequently, the characterizations of the family RE in Theorems 7 and 9 hold true also for families $INSWK_\alpha, INWK_\alpha, \alpha \in \{u, crt\}$.

10.2. Reverse Watson-Crick finite automata. As we have mentioned in Section 3, the two strands of a DNA molecule have opposite $5' - 3'$ orientations (see Figure 1). This suggests to consider a variant of Watson-Crick finite automata that parse the two strands of a Watson-Crick tape in opposite directions. Figure 6 illustrates the initial configuration of such an automaton.

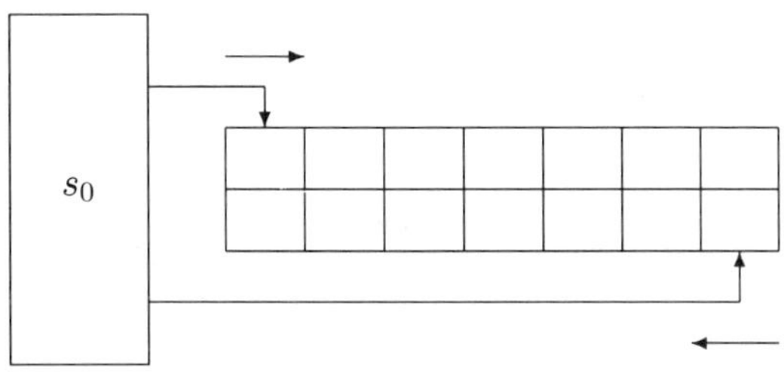

FIGURE 6. Reverse Watson-Crick finite automaton

Formally, a *reverse Watson-Crick finite automaton* is a construct

$$M = (V, \rho, Q, s_0, F, \delta),$$

with the components defined exactly as for Watson-Crick finite automata, but with the relation $\Longrightarrow$ defined as follows.

For $w_1, w_2, x, y \in V^*, s, s' \in Q$, we write

$$\begin{pmatrix} \lambda \\ w_2 y \end{pmatrix} s \begin{pmatrix} x w_1 \\ \lambda \end{pmatrix} \Longrightarrow \begin{pmatrix} \lambda \\ w_2 \end{pmatrix} s' \begin{pmatrix} w_1 \\ \lambda \end{pmatrix} \text{ if and only if } s' \in \delta(s, \begin{pmatrix} x \\ y \end{pmatrix}).$$

Then $\begin{bmatrix} w_1 \\ w_2 \end{bmatrix} \in WK_\rho(V)$ is recognized by M if and only if

$$\begin{pmatrix} \lambda \\ w_2 \end{pmatrix} s_0 \begin{pmatrix} w_1 \\ \lambda \end{pmatrix} \Longrightarrow^* s_f, \text{ for } s_f \in F.$$

As in Section 6 we can associate four languages $L_\alpha(M), \alpha \in \{u, d, ud, ctr\}$, with a reverse Watson-Crick finite automaton M. These families of languages associated with reverse Watson-Crick finite automata corresponding to families XWK_α are denoted by $XRWK_\alpha, X \in \{U, N, S, A, NS, AS\}$.

Clearly, the diagrams as those in Figures 3 and 4 hold true for these families, too.

Because in simple stateless automata the direction of moving a head is not crucial, we obtain:

LEMMA 18. $NSWK_u = NSRWK_u$.

In view of Lemma 9, we obtain

COROLLARY 11. *$NSRWK_u \subseteq REG$.*

Also the counterpart of Theorem 7, viz. Theorem 13, remains true in the new framework.

LEMMA 19. *For every alphabet V, we have $RTS_V \in NSRWK_{ctr}$.*

PROOF. For the reverse simple stateless Watson-Crick automaton M in the proof of Lemma 14 we have $L_{ctr}(M) = RTS_V$. □

THEOREM 13. *Each language in RE is the image by a deterministic gsm mapping of a language in any family $XRWK_{ctr}, X \in \{U, N, A, S, NS, AS\}$.*

We do not know which of the other results in Section refCharRE are valid also for reverse Watson-Crick finite automata. Anyway, we have

THEOREM 14. *$NRWK_u - CF \neq \emptyset$.*

PROOF. Consider the reverse stateless Watson-Crick finite automaton

$$M = (\{a, b, c\}, \rho, P),$$

with

$$\rho = \{(a, a), (b, b), (c, c)\},$$
$$P = \{\begin{pmatrix} a \\ c \end{pmatrix}, \begin{pmatrix} b \\ \lambda \end{pmatrix}, \begin{pmatrix} c \\ b \end{pmatrix}, \begin{pmatrix} \lambda \\ a \end{pmatrix}\},$$

and the regular language $R = a^+b^+c^+$.

We are interested in the intersection $L_u(M) \cap R$. Because of the form of ρ, only sequences $\begin{bmatrix} w \\ w \end{bmatrix}$ with $w = a^n b^m c^p$, $n, m, p \geq 1$, can be considered as inputs for M which can lead to strings in $L_u(M) \cap R$.

Because a in the upper strand and c in the lower one can be parsed only by $\begin{pmatrix} a \\ c \end{pmatrix} \in P$, we must have $n = p$; similarly, $m = p$, because the only way to parse c in the upper strand and b in the lower one is by using $\begin{pmatrix} c \\ b \end{pmatrix} \in P$. Therefore,$n = m = p$, that is $L_u(M) \cap R \subseteq \{a^n b^n c^n \mid n \geq 1\}$.

This is an equality, because any $\begin{bmatrix} w \\ w \end{bmatrix}, w = a^n b^n c^n, n \geq 1$, can be parsed as follows:

$$\begin{aligned}
\begin{pmatrix} \lambda \\ a^n b^n c^n \end{pmatrix} s_0 \begin{pmatrix} a^n b^n c^n \\ \lambda \end{pmatrix} &\Longrightarrow^* \begin{pmatrix} \lambda \\ a^n b^n \end{pmatrix} s_0 \begin{pmatrix} b^n c^n \\ \lambda \end{pmatrix}, \\
&\quad \text{by the rule } \begin{pmatrix} a \\ c \end{pmatrix} \in P, \\
&\Longrightarrow^* \begin{pmatrix} \lambda \\ a^n b^n \end{pmatrix} s_0 \begin{pmatrix} c^n \\ \lambda \end{pmatrix}, \\
&\quad \text{by the rule } \begin{pmatrix} b \\ \lambda \end{pmatrix} \in P, \\
&\Longrightarrow^* \begin{pmatrix} \lambda \\ a^n \end{pmatrix} s_0 \begin{pmatrix} \lambda \\ \lambda \end{pmatrix}, \\
&\quad \text{by the rule } \begin{pmatrix} c \\ b \end{pmatrix} \in P, \\
&\Longrightarrow^* s_0, \\
&\quad \text{by the rule } \begin{pmatrix} \lambda \\ a \end{pmatrix} \in P.
\end{aligned}$$

Consequently,

$$L_u(M) \cap R = \{a^n b^n c^n \mid n \geq 1\},$$

which is not a context-free language. □

THEOREM 15. $ASRWK_u - CF \neq \emptyset$.

PROOF. Consider the all-final simple reverse Watson-Crick finite automaton

$$M = (\{a, b, c\}, \rho, Q, s_0, Q, \delta),$$

with

$$\begin{aligned}
&\rho - \{(a, a), (b, b), (c, c)\}, \\
&Q = \{s_0, s_1, s_2, s_3\}, \\
&\delta(s_0, \begin{pmatrix} a \\ \lambda \end{pmatrix}) = \{s_1\}, \\
&\delta(s_1, \begin{pmatrix} \lambda \\ c \end{pmatrix}) = \delta(s_0, \begin{pmatrix} b \\ \lambda \end{pmatrix}) = \delta(s_2, \begin{pmatrix} \lambda \\ b \end{pmatrix}) = \{s_0\}, \\
&\delta(s_0, \begin{pmatrix} c \\ \lambda \end{pmatrix}) = \{s_2\}, \\
&\delta(s_0, \begin{pmatrix} \lambda \\ a \end{pmatrix}) = \delta(s_3, \begin{pmatrix} \lambda \\ a \end{pmatrix}) = \{s_3\}, \\
&\delta(s, \begin{pmatrix} u \\ v \end{pmatrix}) = \emptyset, \text{ otherwise.}
\end{aligned}$$

For the regular language $R = a^+ b^+ c^+$ we obtain

$$L_u(M) \cap R = \{a^n b^n c^n \mid n \geq 1\}$$

(the states in Q control the work of M when parsing a sequence $\left[\begin{array}{c} w \\ w \end{array} \right]$ with $w = a^n b^m c^p$ in the same way as the pairs $\left(\begin{array}{c} u \\ v \end{array} \right) \in P$ do in the proof of Theorem 14). □

COROLLARY 12. *The inclusions* $NSRWK_u \subset ASRWK_u, NSRWK_u \subset NRWK_u$ *are proper.*

A natural generalization of Watson-Crick finite automata, suggested also by the idea of reverse automata, are the *Watson-Crick two-way finite automata*, where the two read heads can move on the two strands of a Watson-Crick tape in both directions. More formally, such a machine is a construct

$$M = (V, \rho, Q, s_0, F, \delta),$$

with all the components as in a Watson-Crick finite automaton, except that $\delta : Q \times (V^* \times \{R, L\}) \times (V^* \times \{R, L\}) \longrightarrow P(Q)$; if $s' \in \delta(s, (x_1, d_1), (x_2, d_2))$, then one passes from state s to state s', scanning x_1 in direction d_1 in the upper strand and x_2 in the direction d_2 in the lower strand of a Watson-Crick tape, starting from the current position of the two heads; of course, R stands for "right" and L stands for "left"; the work of M starts in state s_0, with the read heads placed at the left hand end of the tape, and one ends in a final state with both heads at the right hand end of the tape.

We do not formally define the transition relation $\Longrightarrow$ for such machines.

All characterizations of recursively enumerable languages in Section 8 remain true for the corresponding variants of two-way Watson-Crick automata, too. Moreover, also Lemma 9 remains true: the moves on the two strands are independent of each other in the case of simple stateless automata; two-way finite automata recognize regular languages, [**16**]; checking the correctness of the pairing of the symbols on the two strands according to the complementarity relation can be done by a literal shuffle followed by a gsm, hence this is an operation which preserves the regularity.

If we supplement the model also with end markers of the input, then a two-way Watson-Crick finite automaton (with states) can also simulate a reverse Watson-Crick finite automaton.

We do not enter further studies of these variants of Watson-Crick finite automata in this paper.

11. Directions for further research

In this paper we have only investigated some basic questions about Watson-Crick automata which are automata very essentially using a new data structure, the Watson-Crick tapes, which are based on the complementarity relation modeling the complementarity relation as present in DNA sequences. We here have considered especially one class of problems (important from the point of view of the universality of DNA computations): representations of recursively enumerable languages. The results are encouraging: most of our machines characterize RE modulo a simple squeezing device – a weak coding in some cases and a deterministic gsm in other cases. Such squeezing mechanisms are inherent to DNA computing because of the necessity of encoding the information we deal with using the alphabet of the four DNA letters A, C, G, T.

Many other (classes of) problems remain to be investigated. We mention some of them now.

1. Investigate the "pure" families of languages associated with Watson-Crick automata (where "pure" means: without using squeezing mechanisms). What are their mutual relationships and what are their relationships with the families in the Chomsky hierarchy, or in any other standard hierarchy of languages ?
2. Improve, if possible, the characterizations of RE presented here, by using simpler Watson-Crick automata and/or simpler squeezing mechanisms.
3. Consider deterministic Watson-Crick automata of various types. Are they strictly weaker than the non-deterministic ones ? The determinism here can also be defined in a dynamic manner, as the possibility of branching during a computation (remember that we use transition rules of the form $s' \in \delta(s, x_1, x_2)$ with x_1, x_2 being strings, possibly empty, but not necessarily symbols).
4. Define and investigate complexity classes based on Watson-Crick automata. Because an essential part of the information necessary during a computation is embedded in the data structure we use, the Watson-Crick tape, and this is considered as provided for free, by DNA strands "automatically" checked for the Watson-Crick complementarity, it is expected that usual complexity classes, based on Turing machines and string-like data structures, will be different from the Watson-Crick complexity classes. This research topic (based on answers to (2)) is of a particular interest for the DNA computing area, because it can prove (or disprove) the usefulness of using DNA molecules - or associated theoretical models – for computing.
5. Define and investigate descriptional complexity measures for Watson-Crick automata and for their languages. The number of states is the basic measure to be considered, as well as the number of transition rules (the cardinality of the set $P(\delta)$, as defined in Section 6). This latter measure is particularly interesting for stateless automata.
6. All of these classes of problems deal with the automata introduced in this paper and, thus, are concerned with the DNA-associated advantage (i), mentioned in the Introduction. The problem area concerning the advantage (ii) remains entirely open. In our estimation, the full usage of (ii) may render drastic changes to some basic ideas of complexity such as deterministic and nondeterministic polynomial time.

As a general "background" research topic there remains the question of implementability – if not implementation – of a Watson-Crick automaton of any of the variants considered here, or of other types which might be introduced. However, this is a challenge which should be approached in an interdisciplinary manner/team.

Acknowledgements

The authors are indebted to Hendrik Jan Hoogeboom and Niké van Vugt for useful comments on a previous version of this paper.

References

[1] L. M. Adleman, Molecular computation of solutions to combinatorial problems, *Science*, 226 (Nov. 1994), 1021 – 1024.

[2] L. M. Adleman, On constructing a molecular computer, in [**22**], 1 – 22.
[3] E. Baum, D. Boneh, P. Kaplan, R. Lipton, J. Reif, N. Seeman (eds.), *DNA Based Computers*, Proc. of the Second DIMACS Meeting, Princeton Univ., 1996.
[4] F. J. Brandenburg, Representations of language families by homomorphic equality operations and generalized equality sets, *Theoretical Computer Science*, 55 (1987), 183 – 263.
[5] K. Culik II, A purely homomorphic characterization of recursively enumerable sets, *Journal of the ACM*, 26 (1979), 345-350.
[6] J. Dassow, Gh. Păun, *Regulated Rewriting in Formal Language Theory*, Springer-Verlag, Berlin, Heidelberg, 1989.
[7] P. Duris, J. Hromkovic, One-way simple multihead finite automata are not closed under concatenation, *Theoretical Computer Science*, 27 (1983), 121 – 125.
[8] J. Engelfriet, Reverse twin-shuffles, *Bulletin of the EATCS*, 60 (1996), 144 (see also `http://www.wi.leidenuniv.nl/`).
[9] J. Engelfriet, G. Rozenberg, Fixed point languages, equality languages, and representations of recursively enumerable languages, *Journal of the ACM*, 27 (1980), 499 – 518.
[10] R. Freund, Gh. Păun, G. Rozenberg, Salomaa, Bidirectional sticker systems, *Third Annual Pacific Conf. on Biocomputing*, Hawaii, 1998 (R. B. Altman, A. K. Dunker, L. Hunter, T. E. Klein, eds.), World Sci., Singapore, 1998, 535 – 546.
[11] R. Freund, Gh. Păun, G. Rozenberg, A. Salomaa, Watson-Crick automata, *Techn. Report 97-13*, Dept. of Computer Sci., Leiden Univ., 1997.
[12] S. Ginsburg, *The Mathematical Theory of Context-free Languages*, McGraw Hill Book Comp., New York, 1966.
[13] D. Hauschild, M. Jantzen, Petri nets algorithms in the theory of matrix grammars, *Acta Informatica*, 31 (1994), 719 – 728.
[14] T. Head, Formal language theory and DNA: an analysis of the generative capacity of specific recombinant behaviors, *Bulletin of Mathematical Biology*, 49 (1987), 737 – 759.
[15] T. Head, Gh. Păun, D. Pixton, Language theory and molecular genetics. Generative mechanisms suggested by DNA recombination, chapter 7 in vol. 2 of *Handbook of Formal Languages* (G. Rozenberg, A. Salomaa, eds.), Springer-Verlag, Heidelberg, 1997, 295 – 360.
[16] J. E. Hopcroft, J. D. Ullman, *Introduction to Automata Theory, Langages and Computation*, Addison-Wesley, Reading, Mass., 1979.
[17] J. Hromkovic, One-way multihead deterministic finite automata, *Acta Informatica*, 19, 4 (1983), 377 – 384.
[18] O. H. Ibarra, C. E. Kim, On 3-head versus 2-head finite automata, *Inform. Control*, 4 (1975), 193 – 200.
[19] L. Kari, Gh. Păun, G. Rozenberg, A. Salomaa, S. Yu. DNA computing, sticker systems, and universality, *Acta Informatica*, in press.
[20] R. J. Lipton, Using DNA to solve NP-complete problems, *Science*, 268 (Apr. 1995), 542 – 545.
[21] R. J. Lipton, Speeding up computations via molecular biology, in [**22**], 67 – 74.
[22] R. J. Lipton, E. B. Baum (eds.), *DNA Based Computers*, Proc. of DIMACS Workshop, Princeton Univ., 1995, American Math. Soc., 1996.
[23] Gh. Păun, Splicing. A challenge to formal language theorists, *Bulletin of the EATCS*, 57 (1995), 183 – 194.
[24] Gh. Păun, G. Rozenberg, A. Salomaa, Computing by splicing, *Theoretical Computer Science*, 168, 2 (1996), 321 – 336.
[25] Gh. Păun, G. Rozenberg, A. Salomaa, *DNA Computing. New Computing Paradigms*, Springer-Verlag, Heidelberg, 1998.
[26] Gh. Păun, A. Salomaa, DNA computing based on the splicing operation, *Mathematica Japonica*, 43, 3 (1996), 607 – 632.
[27] A. L. Rosenberg, On multihead finite automata, *IBM J. R. and D.*, 10 (1966), 388 – 394.
[28] S. Roweis, E. Winfree, R. Burgoyne, N. Chelyapov, M. Goodman, P. Rothemund, L. Adleman, A sticker based architecture for DNA computation, in [**3**], 1 – 27.
[29] G. Rozenberg, A. Salomaa, *The Mathematical Theory of L Systems*, Academic Press, New York, 1980.
[30] G. Rozenberg, A. Salomaa, Watson-Crick complementarity, universal computations and genetic engineering, *Techn. Report* 96-28, Dept. of Computer Science, Leiden Univ., Oct. 1996.

[31] G. Rozenberg, A. Salomaa (eds.) *Handbook of Formal Languages*, Springer-Verlag, Heidelberg, 1997, 3 volumes.
[32] A. Salomaa, *Formal Languages*, Academic Press, New York, London, 1973.
[33] A. Salomaa, Equality sets for homomorphisms of free monoids, *Acta Cybernetica*, 4 (1978), 127 – 139.
[34] A. Salomaa, *Jewels of Formal Language Theory*, Computer Science Press, Rockville, Maryland, 1981.
[35] E. Winfree, X. Yang, N. Seeman, Universal computation via self-assembly of DNA; some theory and experiments, in [**3**], 172 – 190.

INSTITUTE OF COMPUTER LANGUAGES, VIENNA UNIVERSITY OF TECHNOLOGY, KARLSPLATZ 13, A-1040 WIEN, AUSTRIA
E-mail address: rudi@logic.at

INSTITUTE OF MATHEMATICS OF THE ROMANIAN ACADEMY, PO BOX 1 – 764, 70700 BUCUREŞTI, ROMANIA
E-mail address: gpaun@imar.ro

DEPARTMENT OF COMPUTER SCIENCE, LEIDEN UNIVERSITY, PO BOX 9512, 2300 RA LEIDEN, THE NETHERLANDS
E-mail address: rozenb@wi.leidenuniv.nl

ACADEMY OF FINLAND AND TURKU UNIVERSITY, DEPARTMENT OF MATHEMATICS, 20014 TURKU, FINLAND
E-mail address: asalomaa@sara.utu.fi

DIMACS Series in Discrete Mathematics
and Theoretical Computer Science
Volume **48**, 1999

At the Crossroads of DNA Computing and Formal Languages: Characterizing Recursively Enumerable Languages Using Insertion-Deletion Systems*

Lila Kari[a], Gheorghe Păun[b], Gabriel Thierrin[a], Sheng Yu[a]
[a]Department of Computer Science, University of Western Ontario
N6A 5B7 London, Ontario, Canada
E-mails: lila@csd.uwo.ca/ gab@csd.uwo.ca/ syu@csd.uwo.ca
[b]Institute of Mathematics of the Romanian Academy
PO Box 1-764, 70700 Bucureşti, Romania
E-mail: gpaun@imar.ro

Abstract

Several characterizations of recursively enumerable (RE) languages are presented, using insertion-deletion systems. Such a system generates the elements of a language by inserting and deleting words, according to their contexts (the insertion-deletion rules are triples (u, z, v), with the meaning that z can be inserted or deleted in/from the context (u, v)). Grammars based on insertion rules have already been considered in [10] with linguistic motivation. Insertion/deletion operations are also basic in DNA and RNA processing, [5]. Our results show that these operations, even with strong restrictions on the length of the contexts and/or on the length or on the form of the inserted/deleted words are computationally complete, that is, they can simulate the work of any Turing machine. A problem formulated in [30] is solved in this context.

1 Introduction

The insertion-deletion operation is fundamental in many areas dealing with symbol processing (in spite of the fact that the formal language theory is mainly

*Research supported by Grants OGP0007877 and OGP0041630 of the Natural Sciences and Engineering Research Council of Canada and by Project 11281 of the Academy of Finland

1991 *Mathematics Subject Classification.* Primary 68Q05, 68Q42, 68Q45, 03D10.

devoted to the study of *rewriting*). Natural and fruitful from a mathematical point of view (see, e.g., [14]), the insertion (adjoining) of a string or of a pair of strings in (to) a given string, depending on certain local conditions, is a basic operation in linguistics. The string-context interplay is probably one of the most important ingredients of all descriptive linguistic theories ([17]). Several generative mechanisms based on this operation have already been considered. In contextual grammars introduced in [18] (up-to-date details can be found in the forthcoming monograph [24]) pairs of strings are inserted, depending on the bracketed string. A similar effect is accomplished by the tree adjoining operation in tree adjunct grammars, [13]. A sort of dual model has been considered in [10]: strings are inserted in given associated contexts. In all these cases only context-sensitive languages are obtained.

A natural extension of these generative mechanisms is to consider also *erasing operations*. For contextual grammars this has been done in [25]. Insertion-deletion systems, corresponding to Galiukschov grammars with deletion, were investigated in [15].

Because the families of languages generated by types of grammars with insertion rules only seem to be "small" in comparison with families in the Chomsky hierarchy (often, they do not cover the family of regular or of linear languages), it is rather surprising to find characterizations of recursively enumerable languages starting from "insertion languages" and using certain operations with languages. For instance, in [8] it is shown that each RE language can be written as the quotient by a regular language of a language generated by a contextual grammar with finitely many selection strings associated with each context (this is the most restricted class of contextual grammars). A similar result is obtained in [19] for Galiukschov grammars.

The "explanation" of such results is, roughly speaking, the following one: in general, the ability of sensing contexts, supplemented by erasing possibilities can do everything that a type-0 grammar (a Turing machine) can do. In the results mentioned above, the context sensing is ensured by the context-dependent insertion, whereas the erasing is provided by the quotient operation.

However, as we have pointed out above, the erasing feature can be introduced in the generative mechanism, aiming in this way at obtaining direct characterizations of RE languages. This has been done in [15], where a way to simulate Turing machines by insertion-deletion systems is provided. Another proof of this characterization of RE is given in [19].

In the above mentioned papers no care has been paid to the length of the contexts used by the insertion-deletion operation, or to the length or to the shape of the inserted-deleted strings. We address here this problem, proving that quite restricted insertion-deletion grammars (we call them *systems*) are still able to characterize RE.

Besides being fundamental in formal language theory, the operations of insertion and deletion have recently become of interest in connection with the topic of molecular computing. The area of molecular computing was born in 1994 when Adleman, [1], succeeded to solve an instance of the Directed Hamiltonian Path Problem solely by manipulating DNA strands. This marked the first instance where a mathematical problem could be solved by biological means and, besides further research [16], gave rise to a couple of interesting problems: a) can *any* algorithm be simulated by means of DNA manipulation, and b) is it possible, at least in theory, to design a programmable molecular computer? To answer these questions, various models of molecular computation have been proposed, and for some of these models it has been shown that the bio-operations involved can simulate the actions of a Turing machine (see, for example, [2], [4], [12], [3], [9], [27], [33], [23]).

Besides their being theoretically interesting, one of the motivations for studying insertions and deletions is that these operations can be used as the sole primitives needed for modeling DNA computation, [15], and moreover, they are already implementable in the laboratory. Indeed, by using available reagents and a standard technique called *PCR site-specific oligonucleotide mutagenesis* [7] one can perform insertions and deletions of nucleotide sequences. (A similar operation, substitution, has been proposed in [3] as a bio-operation necessary to simulate a universal Turing machine.)

The results obtained here address a related problem: how to obtain full computational power by insertions and deletions of a single nucleotide in an RNA sequence. Indeed, it is known that the U nucleotide can be relatively easily inserted and deleted in/from RNA sequences, [5]. This has led to the open problem whether or not each Turing machine can be simulated, in a convenient encoding, just by inserting/deleting strings consisting of one symbol only, U (see [30]). We prove here that the answer is affirmative. The proof that insertions and deletions of a single nucleotide are enough to simulate the actions of a Turing machine opens thus another possible way for designing a molecular computer.

2 Insertion-deletion systems

In this section we introduce the basic model investigated in this paper, the *insdel systems*. We denote by V^* the free monoid generated by an alphabet V under the operation of concatenation; the empty string is denoted by λ and the length of $x \in V^*$ is denoted by $|x|$. By *FIN, REG, LIN, CF, CS, RE* we denote the families of finite, regular, linear, context-free, context-sensitive, and of recursively enumerable languages, respectively. For basic elements of formal language theory, we refer to [28], [29].

An *insertion-deletion* (shortly, *insdel*) *system*, [15], is a construct

$$\gamma = (V, T, A, I, D),$$

where V is an alphabet, $T \subseteq V$, A is a finite subset of V^*, and I, D are finite subsets of $V^* \times V^* \times V^*$.

The alphabet T is the terminal alphabet of γ, A is the set of axioms, I is the set of insertion rules, and D is the set of deletion rules. An insertion/deletion rule is given in the form (u, z, v); in order to increase the readability, we sometimes write $(u, z, v)_{ins}$ in order to indicate that (u, z, v) is an insertion rule and $(u, z, v)_{del}$ for $(u, z, v) \in D$. In an insertion/deletion rule (u, z, v), u, v represent the context of insertion/deletion (i.e. the insertion/deletion will take place between u and v), whereas z represents the string to be inserted/deleted.

For $x, y \in V^*$ we write $x \Longrightarrow y$ iff one of the following two cases holds:

(i). $x = x_1uvx_2$, $y = x_1uzvx_2$, for some $x_1, x_2 \in V^*$ and $(u, z, v) \in I$ (an insertion step);

(ii). $x = x_1uzvx_2$, $y = x_1uvx_2$, for some $x_1, x_2 \in V^*$ and $(u, z, v) \in D$ (a deletion step).

Denoting by $\Longrightarrow^*$ the reflexive and transitive closure of the relation $\Longrightarrow$, the language generated by γ is defined by

$$L(\gamma) = \{w \in T^* \mid x \Longrightarrow^* w, \text{ for some } x \in A\}.$$

An insdel system $\gamma = (V, T, A, I, D)$ is said to be of *weight* (n, m, p, q) if

$$\begin{aligned}
&\max\{|z| \mid (u, z, v) \in I\} = n,\\
&\max\{|u| \mid (u, z, v) \in I \text{ or } (v, z, u) \in I\} = m,\\
&\max\{|z| \mid (u, z, v) \in D\} = p,\\
&\max\{|u| \mid (u, z, v) \in D \text{ or } (v, z, u) \in D\} = q.
\end{aligned}$$

Intuitively, n is an upper bound for the length of the inserted string, while m is an upper bound for the length of the left and right context. The constants p and q represent the analogous upper bounds for deletion.

We denote by $INS_n^m DEL_p^q$, $n, m, p, q \geq 0$, the family of languages $L(\gamma)$ generated by insdel systems of weight (n', m', p', q') such that $n' \leq n$, $m' \leq m$, $p' \leq p$, $q' \leq q$. Because insertion/deletion of empty strings changes nothing, we ignore such rules; namely, when $n = 0$ we also assume that $m = 0$, and when $p = 0$ we also assume that $q = 0$. The meaning of INS_0^0 is that no insertion rule is used, and the meaning of DEL_0^0 is that no deletion rule is used. When one of the parameters n, m, p, q is not bounded, we replace it by ∞. Thus, the family of all insdel languages is $INS_\infty^\infty DEL_\infty^\infty$.

3 Preliminary results

The families $INS_\infty^m DEL_0^0$ are those generated by Galiukschov grammars. Proofs of the following results can be found in [10], [20], [21], [31]:

(i). $FIN \subset INS_\infty^0 DEL_0^0 \subset INS_\infty^1 DEL_0^0 \subset INS_\infty^2 DEL_0^0 \subset \ldots \subset INS_\infty^\infty DEL_0^0 \subset CS$.

(ii). REG is incomparable with all families $INS_\infty^m DEL_0^0, m \geq 0$, and $REG \subset INS_\infty^\infty DEL_0^0$.

(iii). $INS_\infty^1 DEL_0^0 \subset CF$, but CF is incomparable with all families $INS_\infty^m DEL_0^0, m \geq 2$, and with $INS_\infty^\infty DEL_0^0$; $INS_\infty^2 DEL_0^0$ contains non-semilinear languages.

(iv). LIN is incomparable with all families $INS_\infty^m DEL_0^0, m \geq 0$, and with $INS_\infty^\infty DEL_0^0$.

(v). All families $INS_\infty^m DEL_0^0, m \geq 0$, are anti-AFL's (that is, they are closed under none of the following operations: union, concatenation, Kleene closure, direct and inverse morphisms, intersection with regular languages).

(vi). Each regular language is the morphic image of a language in the family $INS_{\infty}^{1}DEL_{0}^{0}$.

Moreover, in [19] it is proved that

(vii). Each language $L \in RE$ can be written in the form $L = g(h^{-1}(L'))$, for a morphism h, a weak coding g, and $L \in INS_{\infty}^{7}DEL_{0}^{0}$. (In fact, from the proof in [19], we can see that $L' \in IND_{4}^{7}DEL_{0}^{0}$.) As a corollary, we get the fact that L can be also written as $L = R\backslash L'$, for $R \in REG$ and $L' \in INS_{\infty}^{7}DEL_{0}^{0}$ (in fact, from the proof we get $L' \in INS_{5}^{7}DEL_{0}^{0}$).

From the definitions, we obviously have

Theorem 1. *$INS_{n}^{m}DEL_{p}^{q} \subseteq INS_{n'}^{m'}DEL_{p'}^{q'}$ for all $0 \leq n \leq n'$, $0 \leq m \leq m'$, $0 \leq p \leq p'$, $0 \leq q \leq q'$.*

A proof of the following result can be found in [15].

Theorem 2. *$RE = INS_{\infty}^{\infty}DEL_{\infty}^{\infty}$.*

In fact, from the proof in [15] we have $RE = INS_{3}^{6}DEL_{2}^{7}$. Another proof of the result in Theorem 2 is given in [19]. In order to see the power of insertion/deletion operations, we recall here the construction in [19], making also explicit the weight of the obtained insdel system.

Theorem 3. *$RE = INS_{3}^{2}DEL_{3}^{0}$.*

Proof. Take a language $L \subseteq T^*$ generated by a grammar $G = (N, T, S, P)$ in Kuroda normal form, hence with P containing context-free rules of the form $X \to x, |x| \leq 2$, and non-context-free rules of the form $XY \to UZ$, for $X, Y, U, Z \in N$. We construct the insdel system

$$\begin{aligned}
\gamma &= (N \cup T \cup \{E, K_1, K_2\}, T, \{SEE\}, I, D),\\
I &= \{(X, K_1x, \alpha_1\alpha_2) \mid X \to x \in P, \alpha_1, \alpha_2 \in N \cup T \cup \{E\}\}\\
&\cup \{(XY, K_2UZ, \alpha_1\alpha_2) \mid XY \to UZ \in P, \alpha_1, \alpha_2 \in N \cup T \cup \{E\}\},\\
D &= \{(\lambda, XK_1, \lambda) \mid X \in N\}\\
&\cup \{(\lambda, XYK_2, \lambda) \mid X, Y \in N\}\\
&\cup \{(\lambda, EE, \lambda)\}.
\end{aligned}$$

The symbol E is a dummy symbol used when checking the context $\alpha_1\alpha_2$ at the end of the string. The symbols K_1, K_2 are "killers": K_1 removes one symbol, namely the one placed immediately to its left, and K_2 removes two symbols, those placed immediately to its left. Making use of these symbols, the rules in I simulate the rules in P. Symbols already marked by the "killers" K_1, K_2 cannot be used as contexts of rules in I. Consequently we get $L(G) = L(\gamma)$. □

4 Restricting the weight

A natural problem in this framework is to look for characterization results $RE = INS_{n}^{m}DEL_{p}^{q}$ with as small as possible values for n, m, p, q. We give here three results proving the surprising power of "small" insdel systems: systems of weight (1, 2, 1, 1), or (2, 1, 2, 0), or (1, 2, 2, 0) are able to simulate any type-0 grammar.

These results do not settle the above formulated problem. For instance, we do not know the power of insdel systems of weight (n, m, p, q) strictly smaller than (1, 2, 1, 1) and than (2, 1, 2, 0) and than (1, 2, 2, 0), or of weight (n, m, p, q) incomparable to these quadruples. Particularly interesting are the systems of weights (1, 1, 1, 1), (1, 1, 2, 0), (2, 1, 1, 1), and (1, 2, 1, 0). Are such systems generating non-context-free languages ? We *conjecture* that the answer is negative.

Theorem 1. $RE = INS_1^2DEL_1^1$.

Proof. According to the Church-Turing thesis, we only have to prove the inclusion $\subseteq$.

Consider a language $L \subseteq T^*, L \in RE$, generated by a grammar $G = (N, T, S, P)$ in the Penttonen normal form, [26], that is containing context-free rules $X \to x$ with $|x| \leq 2$, and non-context-free rules of the form $XY \to XZ$, for $X, Y, Z \in N$.

Without loss of generality we may assume that in each rule $X \to \alpha_1\alpha_2 \in P$ we have $X \neq \alpha_1$, $X \neq \alpha_2$, $\alpha_1 \neq \alpha_2$. (If necessary, we replace $X \to \alpha_1\alpha_2$ with $X \to X'$, $X' \to \alpha_1\alpha_2'$, $\alpha_2' \to \alpha_2$, where X', α_2' are new symbols.) Similarly, we may assume that for each rule $XY \to XZ \in P$ we have $X \neq Y$, $X \neq Z$, $Y \neq Z$. Moreover, by replacing each rule $X \to \alpha \in P$, $\alpha \in N \cup T$, by $X \to \alpha Z, Z \to \lambda$, we obtain an equivalent grammar. Hence, we may assume that the rules in P are of the following three forms:

1. $X \to \alpha_1\alpha_2,\ \alpha_1, \alpha_2 \in N \cup T, X \neq \alpha_1, X \neq \alpha_2, \alpha_1 \neq \alpha_2,$
2. $X \to \lambda,$
3. $XY \to XZ,\ X, Y, Z \in N, X \neq Y, X \neq Z, Y \neq Z.$

Moreover, we assume the rules of P labelled in an one-to-one manner.

We construct the insdel system

$$\gamma = (V, T, A, I, D),$$

where

$$\begin{aligned} V &= N \cup T \cup \{[r], (r) \mid r \text{ is the label of a rule in } P\} \\ &\cup \{B, E\}, \\ A &= \{BSE\}, \end{aligned}$$

and the sets I, D are constructed as follows.

(i). For each rule $r : X \to \alpha_1\alpha_2 \in P, \alpha_1, \alpha_2 \in N \cup T$, of type 1, we consider the following insertion/deletion rules:

(r.1.) $(\beta_1, [r], X\beta_2)_{ins}, \beta_1 \in N \cup T \cup \{B\}, \beta_2 \in N \cup T \cup \{E\},$
(r.2.) $([r]X, (r), \beta)_{ins}, \beta \in N \cup T \cup \{E\},$
(r.3.) $([r], X, (r))_{del},$

$$\begin{aligned}
&(r.4.) \quad ([r], \alpha_1, (r))_{ins},\\
&(r.5.) \quad (\alpha_1, \alpha_2, (r))_{ins},\\
&(r.6.) \quad (\lambda, [r], \alpha_1)_{del},\\
&(r.7.) \quad (\alpha_2, (r), \lambda)_{del}.
\end{aligned}$$

(ii). For each rule $r : X \to \lambda \in P$ of type 2, we introduce the deletion rule

$$(r.1.) \quad (\beta_1, X, \beta_2)_{del}, \beta_1 \in N \cup T \cup \{B\}, \beta_2 \in N \cup T \cup \{E\}.$$

(iii). For each rule $r : XY \to XZ \in P$, $X, Y, Z \in N$, of type 3, we consider the following insertion/deletion rules:

$$\begin{aligned}
&(r.1.) \quad (\beta_1 X, [r], Y\beta_2)_{ins}, \beta_1 \in N \cup T \cup \{B\}, \beta_2 \in N \cup T \cup \{E\},\\
&(r.2.) \quad ([r]Y, (r), \beta)_{ins}, \beta \in N \cup T \cup \{E\},\\
&(r.3.) \quad ([r], Y, (r))_{del},\\
&(r.4.) \quad ([r], Z, (r))_{ins},\\
&(r.5.) \quad (X, [r], Z)_{del},\\
&(r.6.) \quad (Z, (r), \lambda)_{del}.
\end{aligned}$$

(iv). We also consider the deletion rules

$$\begin{aligned}
&(\lambda, B, \lambda),\\
&(\lambda, E, \lambda).
\end{aligned}$$

We claim that the equality $L(G) = L(\gamma)$ holds.

($\subseteq$) Each derivation step $w \Longrightarrow w'$ in G is simulated in γ by a derivation $BwE \Longrightarrow^* Bw'E$, using the rules $(r.1.) - (r.i.)$ associated as above to the rule in P used in $w \Longrightarrow w'$. For instance, assume that $w = w_1 X w_2$, $w' = w_1 \alpha_1 \alpha_2 w_2$, for $r : X \to \alpha_1 \alpha_2 \in P$. Then we successively obtain:

$$\begin{aligned}
Bw_1Xw_2E &\Longrightarrow Bw_1[r]Xw_2E && \text{by the rule } (r.1.)\\
&\Longrightarrow Bw_1[r]X(r)w_2E && \text{by the rule } (r.2.)\\
&\Longrightarrow Bw_1[r](r)w_2E && \text{by the rule } (r.3.)\\
&\Longrightarrow Bw_1[r]\alpha_1(r)w_2E && \text{by the rule } (r.4.)\\
&\Longrightarrow Bw_1[r]\alpha_1\alpha_2(r)w_2E && \text{by the rule } (r.5.)\\
&\Longrightarrow Bw_1\alpha_1\alpha_2(r)w_2E && \text{by the rule } (r.6.)\\
&\Longrightarrow Bw_1\alpha_1\alpha_2w_2E && \text{by the rule } (r.7.)\\
&= Bw'E.
\end{aligned}$$

We proceed in a similar way when $w \Longrightarrow w'$ is obtained by using a rule $r : XY \to XZ$. The details are left to the reader.

A derivation starts from BSE; at any moment, the markers B, E can be removed. Thus, any terminal string generated by G is in $L(\gamma)$.

($\supseteq$) Consider a string BwE; initially we have $w = S$. We can apply to it a rule $(r.1.)$ from group 1, or a deletion rule (β_1, X, β_2), or a rule $(r.1.)$ from group 3. Assume that we apply $(\beta_1, [r], X\beta_2)_{ins}$ for some $r : X \to \alpha_1\alpha_2 \in P$. We have

$$Bw_1Xw_2E \Longrightarrow Bw_1[r]Xw_2E.$$

Because the rules in P are labelled in an one-to-one way, because $X \neq \alpha_1$, and because rules of the form of (r.1.) in groups 1 and 3 have a left context checking the symbol placed immediately to the left of X (the same for the deletion rules in group 2), the only rule which can use the symbol X is (r.2.). Eventually this rule must be applied, otherwise the derivation cannot lead to a terminal string. Thus, the substring $[r]X$ of $Bw_1[r]Xw_2$ leads to $[r]X(r)$. Again there is only one possible continuation, by the rule (r.3.), which erases the symbol X. Only after inserting α_1 in-between $[r]$ and (r) we can remove the symbol $[r]$. In the presence of α_1 and of (r) we can introduce α_2, too, by the rule (r.5.). Because (r) is introduced after $[r]$, and $X \neq \alpha_1$, the symbol α_1 used by this rule (r.5.) as a left context should be introduced at a previous step, by the corresponding rule (r.4.). After introducing α_2, which is different from both α_1 and X, we can delete (r), by the rule (r.7.). Due to the contexts, no other rule can use the mentioned symbols as contexts or can delete any of them. Thus, after using (r.1.), we have to use all rules $(r.i.), 2 \leq i \leq 7$, associated with $r : X \to \alpha_1\alpha_2$, simulating the use of $X \to \alpha_1\alpha_2$.

In the same way, after using a rule $(\beta_1 X, [r], Y\beta_2)$ associated with $r : XY \to XZ \in P$, we have to continue with $(r.i.), 2 \leq i \leq 6$ (possibly not immediately or at consecutive steps, but using the same symbols of the current string), hence we have to simulate the rule $XY \to XZ$.

The deletion rules (β_1, X, β_2) directly correspond to erasing rules in P. The markers B, E can be deleted at any step. Consequently, γ can generate only strings in $L(G)$. □

Theorem 2. $RE = INS_2^1DEL_2^0$.

Proof. For every $L \subseteq T^*, L \in RE$, there is a grammar $G = (N, T, S, P)$ such that $L = L(G)$ and P contains two types of rules: context-free rules and rules of the form $XY \to \lambda, X, Y \in N$ (see [11], [22], [32]). Without loss of generality, we may assume that the context-free rules in P are of the form $X \to u$ with $|u| \leq 2$. Moreover, from the proofs in [11], [22], [32] we see that the symbols X, Y in rules $XY \to \lambda$ do not appear in the left hand members of the context-free rules in P.

Consider such a grammar G, where the rules in P of the form $X \to \alpha_1\alpha_2$ have been labelled in an one-to-one manner, and construct the insdel system $\gamma = (V, T, A, I, D)$ with

$$\begin{aligned}
V &= N \cup T \cup \{F_r, F'_r \mid r : X \to \alpha_1\alpha_2 \in P, \alpha_1, \alpha_2 \in N \cup T\} \\
&\cup \{B, K\}, \\
A &= \{BS\}, \\
I &= \{(\beta, \alpha_1 F_r, X),\ (\beta, \alpha_2 K, F_r),\ (K, F'_r, F_r) \mid \\
&\qquad r : X \to \alpha_1\alpha_2 \in P, \alpha_1, \alpha_2 \in N \cup T, \beta \in N \cup T \cup \{B\}\} \\
&\cup \{(\beta, \alpha K, X) \mid X \to \alpha \in P, \alpha \in N \cup T, \beta \in N \cup T \cup \{B\}\}, \\
D &= \{(\lambda, F'_r F_r, \lambda) \mid r : X \to \alpha_1\alpha_2, \alpha_1, \alpha_2 \in N \cup T\} \\
&\cup \{(\lambda, KX, \lambda) \mid X \to u \in P\}
\end{aligned}$$

$$\cup \quad \{(\lambda, XY, \lambda) \mid XY \to \lambda \in P\}$$
$$\cup \quad \{(\lambda, B, \lambda)\}.$$

We obtain the equality $L(G) = L(\gamma)$.

($\subseteq$) Consider a derivation step in G, $w \Longrightarrow w'$. If the used rule is of the form $XY \to \lambda$, then clearly $Bw \Longrightarrow Bw'$ in γ, by using the rule $(\lambda, XY, \lambda) \in D$. If the used rule is $X \to \alpha, \alpha \in N \cup T$, then

$$\begin{aligned} Bw = Bw_1Xw_2 &\Longrightarrow Bw_1\alpha KXw_2 && \text{by the rule } (\beta, \alpha K, X)_{ins} \\ &\Longrightarrow Bw_1\alpha w_2 && \text{by the rule } (\lambda, KX, \lambda)_{del} \\ &= Bw'. \end{aligned}$$

If the used rule is $r : X \to \alpha_1\alpha_2$, then we proceed as follows

$$\begin{aligned} Bw = Bw_1Xw_2 &\Longrightarrow Bw_1\alpha_1F_rXw_2 && \text{by the rule } (\beta, \alpha_1F_r, X)_{ins} \\ &\Longrightarrow Bw_1\alpha_1\alpha_2KF_rXw_2 && \text{by the rule } (\beta, \alpha_2K, F_r)_{ins} \\ &\Longrightarrow Bw_1\alpha_1\alpha_2KF'_rF_rXw_2 && \text{by the rule } (K, F'_r, F_r)_{ins} \\ &\Longrightarrow Bw_1\alpha_1\alpha_2KXw_2 && \text{by the rule } (\lambda, F'_rF_r, \lambda)_{del} \\ &\Longrightarrow Bw_1\alpha_1\alpha_2w_2 && \text{by the rule } (\lambda, KX, \lambda)_{del} \\ &= Bw'. \end{aligned}$$

Consequently, each derivation $S \Longrightarrow^* x$ in G can be reproduced in γ as $BS \Longrightarrow^* Bx$; then B can be removed by the rule $(\lambda, B, \lambda) \in D$. If $x \in T^*$, then $x \in L(\gamma)$.

($\supseteq$) Consider a string of the form Bw with $w \in (N \cup T)^*$. (Initially we have $w = S$.) If a rule $(\lambda, XY, \lambda) \in D$ can be applied to w, then it corresponds to the use of the rule $XY \to \lambda$ for rewriting w. No other deletion rule can be used.

If we use an insertion rule $(\beta, \alpha K, X)$, then $w = w_1Xw_2$ (w_1 can be λ, and then $\beta = B$). We get the string $Bw_1\alpha KXw_2$. The symbol X in the right hand of K cannot be used by any other rule of γ excepting $(\lambda, KX, \lambda) \in D$ (the symbol K marks the symbols which cannot be used in further insertion rules). Eventually, this deletion rule should be used, hence we simulate in this way the rule $X \to \alpha \in P$.

Another possibility is to use the insertion rule (β, α_1F_r, X), for some $r : X \to \alpha_1\alpha_2 \in P$. We get a string of the form $Bw_1\alpha_1F_rXw_2$. No rule in γ can use the symbols F_r and X, excepting $(\beta, \alpha_2K, F_r) \in I$. Eventually, this rule is used, hence α_1F_rX is replaced by $\alpha_1\alpha_2KF_rX$. The only way to remove the nonterminals K, F_r, X is by first introducing F'_r, by the rule $(K, F'_r, F_r) \in I$ (hence obtaining $\alpha_1\alpha_2KF'_rF_rX$), then erasing F'_rF_r and KX, by using the rules $(\lambda, F'_rF_r, \lambda)_{del}$ and $(\lambda, KX, \lambda)_{del}$, respectively. This corresponds to the use of the rule $X \to \alpha_1\alpha_2$.

The symbol B ensures the fact that we can always check the left context β of insertion rules.

Therefore, $L(\gamma) = L(G)$. □

Combining the ideas of the previous proofs, we also get

Theorem 3. $RE = INS_1^2DEL_2^0$.

Proof. Start as in the proof of Theorem 2 from a grammar $G = (N, T, S, P)$ with P containing context-free rules of the form $X \to u$, with $|u| = 2$ or $u = \lambda$, and non-context-free rules of the form $XY \to \lambda$. Without loss of generality we may assume that if $X \to \alpha_1\alpha_2 \in P$, then $\alpha_1 \neq X$, $\alpha_2 \neq X$, and that X, Y in rules of the form $XY \to \lambda$ do not appear as left hand members of context-free rules.

Then we construct the insdel system $\gamma = (V, T, A, I, D)$, with

$$\begin{aligned} V &= N \cup T \cup \{[r], (r) \mid r : X \to \alpha_1\alpha_2 \in P\} \\ &\cup \{B, E, K\}, \\ A &= \{BSE\}, \end{aligned}$$

and the sets I, D are constructed as follows.

(i). For each rule $r : X \to \alpha_1\alpha_2 \in P$ we consider the following rules:

$$\begin{aligned} &(\beta_1, [r], X\beta_2)_{ins}, \ \beta_1 \in N \cup T \cup \{B\}, \beta_2 \in N \cup T \cup \{E\}, \\ &([r]X, (r), \beta)_{ins}, \ \beta \in N \cup T \cup \{E\}, \\ &([r]X, K, (r))_{ins}, \\ &(\lambda, XK, \lambda)_{del}, \\ &([r], \alpha_1, (r))_{ins}, \\ &([r]\alpha_1, \alpha_2, (r))_{ins}, \\ &([r], K, \alpha_1\alpha_2)_{ins}, \\ &(\lambda, [r]K, \lambda)_{del}, \\ &(\alpha_1\alpha_2, K, (r))_{ins}, \\ &(\lambda, K(r), \lambda)_{del}. \end{aligned}$$

(They simulate the rule r in the same way as the rules $(r.1.) - (r.7.)$ in the proof of Theorem 1, with the difference that the control symbols $[r], (r)$ are deleted in the presence of the "killer" K, which is introduced only after replacing X by $\alpha_1\alpha_2$.)

(ii). For each rule $r : X \to \lambda \in P$, we introduce the rules

$$\begin{aligned} &(X, K, \beta)_{ins}, \ \beta \in N \cup T \cup \{E\}, \\ &(\lambda, XK, \lambda)_{del}. \end{aligned}$$

(We cannot introduce two or more occurrences of K for the same X, due to the presence of the symbol β.)

(iii). For each rule $r : XY \to \lambda \in P$ we introduce the rule

$$(\lambda, XY, \lambda)_{del}.$$

As in the previous proofs, one can see that $L(G) = L(\gamma)$. □.

5 Restricting the form of the inserted-deleted strings

In the molecular computing framework we deal with a restricted number of symbols (four in the case of DNA: A, C, G, T, denoting the four nucleotides, adenine, cytosine, guanine, thymine; similarly for RNA). Moreover, the insertion/deletion of some of these symbols is easier than for other symbols. Thus it is of practical interest to consider insdel systems with rules of the form (u, z, v) with $z \in c^*$ for a specified symbol c. In [30] one explicitly asks whether or not such systems can simulate any given Turing machine. Of course, because we can no longer introduce or erase symbols different from c, we must start with them introduced from the very beginning as a sort of workspace, whereas the result of a rewriting (we may also call it a *computation*) should be also embedded in a specified workspace. In such conditions, we prove below that the question posed in [30] has a positive answer. Moreover, we can work with only one symbol different from c.

Specifically, a *restricted-insdel system* is a construct

$$\gamma = (V, \{a, c\}, A, I, D, h),$$

where V is an alphabet, a, c are specified symbols (not necessarily from V), A is a finite subset of $\{a, c\}^*$, I, D are finite subsets of $\{a, c\}^* \times c^* \times \{a, c\}^*$, and $h : V^* \longrightarrow \{a, c\}^*$ is a morphism. (Note that all insertion-deletion rules are of the form $(u, c^i, v), u, v \in \{a, c\}^*, i \geq 0$.) The relation $\Longrightarrow$ is defined in the usual way, over $\{a, c\}^*$. Then, the language generated by γ is

$$L(\gamma) = h^{-1}(\{w \in \{a, c\}^* \mid z(aca)^n \Longrightarrow^* (aca)^m w, \text{ for some } n, m \geq 0, z \in A\}.$$

In words, we start from an axiom $z \in A$, prolonged with an arbitrary number of "empty spaces" aca, we use arbitrarily many insertion/deletion rules, we discard the "spaces" aca placed to the left hand end of the obtained string, and we map by h^{-1} the remaining string into a string in V^*. In this way, strings w for which $h^{-1}(w)$ is not defined are removed, hence we can ensure the termination of the derivation in the same way as when using a specified terminal alphabet.

A restricted-insdel system

$$\gamma = (V, \{a, c\}, A, I, D, h)$$

is said to be *faithful* if the morphism h is an injective morphism. Let $RINSDEL$ and $FRINSDEL$ denote the family of languages generated respectively by restricted-insdel and faithful restricted-insdel systems of arbitrary weight; because we work here with a codification of strings over V as strings over $\{a, c\}$, we cannot keep bounded (independent of the cardinality of V, for instance) the weight of the used systems.

Expected from the point of view of Theorems 1, 2, 3 and encouraging from DNA/RNA computing point of view, we have the following result.

Theorem 4. $RE = FRINSDEL$.

Proof. Consider a language $L \subseteq T^*, L \in RE$, and consider a grammar $G = (N, T, S, P)$ in the Kuroda normal form generating L. Assume that

$$N \cup T = \{\alpha_1, \ldots, \alpha_n\},$$

with $\alpha_1 = S$.

The rules in P can be of the following forms:

$$\begin{aligned} &1. \quad \alpha_i \to \alpha_j, \\ &2. \quad \alpha_i \to \lambda, \\ &3. \quad \alpha_i \to \alpha_j\alpha_k, \\ &4. \quad \alpha_i\alpha_j \to \alpha_p\alpha_q. \end{aligned}$$

Consider one new special symbol, $\alpha_0 = \#$, denoting an empty space, and assume that we start not from S but from $S\#^s$, for some $s \geq 0$. Then we may replace all rules of forms 1, 2, 3 in P with rules of the form 4, as follows:

$$\begin{aligned} &1. \quad \alpha_i\alpha_0 \to \alpha_j\alpha_0, \text{ for } \alpha_i \to \alpha_j \in P, \\ &2. \quad \alpha_i\alpha_0 \to \alpha_0\alpha_0, \text{ for } \alpha_i \to \lambda \in P, \\ &3. \quad \alpha_i\alpha_0 \to \alpha_j\alpha_k, \text{ for } \alpha_i \to \alpha_j\alpha_k \in P, \end{aligned}$$

providing that we also add the rules of the form

$$5. \quad \alpha_i\alpha_0 \to \alpha_0\alpha_i, \ 1 \leq i \leq n.$$

(They move the symbol $\alpha_0 = \#$ to the left.)

Let us denote by G' the grammar obtained in this way.

Then, for every derivation $S \Longrightarrow^* w$ in the grammar G we can find a derivation $S\#^s \Longrightarrow^* \#^t w$ in G'.

Starting from G', we construct a faithful restricted-insdel system $\gamma = (N \cup T \cup \{\#\}, \{a, c\}, A, I, D, h)$ as follows.

Consider the morphism h defined by $h(\alpha_i) = ac^{i+1}a, \ 0 \leq i \leq n$. (Hence the space $\#$ is encoded by aca.) This morphism h is injective. Indeed suppose that h is not injective. Then there exist $u, v \in V^+$ such that $h(u) = h(v)$. Let

$$u = \beta_1 \cdots \beta_r, v = \gamma_1 \cdots \gamma_s, \beta_p, \gamma_q \in V, r, s \geq 1$$

This implies

$$h(u) = h(\beta_1) \cdots h(\beta_r) = h(v) = h(\gamma_1) \cdots h(\gamma_s)$$

The set

$$h(V) = \{h(\alpha_i) | 1 \leq i \leq n\} = \{ac^{i+1}a | 1 \leq i \leq n\}$$

is a prefix code. Indeed, suppose that $h(V)x \cap h(V) \neq \emptyset$, i.e. $ac^{i+1}ax = ac^{j+1}a$. This implies $i = j$ and hence $x = 1$. Hence $h(V)$ is a prefix code. Therefore, since $h(u) = h(v)$, we have $r = s$ and $h(\beta_p) = h(\gamma_p)$ for $p = 1, \cdots, r$. This implies $\beta_p = \gamma_p$ and $u = v$.

We have

$$A = \{ac^2a\},$$
$$h(\alpha_i) = ac^{i+1}a, \text{ for } \alpha_i \text{ a terminal symbol of } G,$$

and the sets I, D contain the following rules:

For the q-th rule of G', $r_q : \alpha_i\alpha_j \to \alpha_k\alpha_p$, we consider the "codified" rule

$$ac^{i+1}aac^{j+1}a \to ac^{k+1}aac^{p+1}a$$

and we introduce in I, D the rules

$$\begin{array}{ll}(r_q.1.) & (ac^{i+1}aa, c^{(2q-1)(n+1)}, c^{j+1}aac^{s+1}a)_{ins}, 0 \le s \le n,\\ (r_q.2.) & (ac^{i+1}, c^{2(q-1)(n+1)}, aac^{(2q-1)(n+1)+j+1}a)_{ins},\\ (r_q.3.) & (ac^{(2q-1)(n+1)+i+1}aa, c^{(2q-1)(n+1)+j-p}, c^{p+1}a)_{del},\\ (r_q.4.) & (ac^{k+1}, c^{2(q-1)(n+1)+i-k}, aac^{p+1}a)_{del}.\end{array}$$

Note that:

- no rule in G' is of the form $\#\alpha_j \to \alpha_k\alpha_p$, hence in the previous rules we have $i \ge 2$;
- however, $j, k, p \ge 1$ (and they can be equal to 1).

We obtain $L(G) = L(\gamma)$. According to the previous discussion, it is enough to prove that

$$\{w \in T^* \mid S\#^s \Longrightarrow^* \#^t w \text{ in } G', s, t \ge 0\} = L(\gamma).$$

($\subseteq$) Consider a derivation step $x \Longrightarrow y$ in G', using a rule $r_q : \alpha_i\alpha_j \to \alpha_k\alpha_p$. Therefore,

$$x = x_1\alpha_i\alpha_jx_2 \Longrightarrow x_1\alpha_k\alpha_px_2 = y.$$

The "codified" string corresponding to x is $g(x)$ such that

$$h(x) = h(x_1)ac^{i+1}aac^{j+1}ah(x_2).$$

Because we start from $S\#^s$ with large enough s, we may assume that $h(x_2) \neq \lambda$. Thus, we can use the associated rules $(r_q.1.)$ – $(r_q.4.)$ and we successively get

$$\begin{array}{rcl}h(x) & = & h(x_1)ac^{i+1}aac^{j+1}ah(x_2)\\ & \Longrightarrow & h(x_1)ac^{i+1}aa^{(2q-1)(n+1)+j+1}ah(x_2)\\ & \Longrightarrow & h(x_1)ac^{2(q-1)(n+1)+i+1}aac^{2(q-1)(n+1)+j+1}ah(x_2)\\ & \Longrightarrow & h(x_1)ac^{2(q-1)(n+1)+i+1}aac^{p+1}ah(x_2)\\ & \Longrightarrow & h(x_1)ac^{k+1}aac^{p+1}ah(x_2)\\ & & = h(y).\end{array}$$

Therefore, each derivation in G' can be simulated as above by a derivation in γ. Using rules of the form $\alpha_i\alpha_0 \to \alpha_0\alpha_i$, for each terminal derivation in G', producing a string w, we can find a derivation in γ producing a string $\#^t h(w)$. Therefore, $h^{-1}(h(w)) = w$, that is $L(G) \subseteq L(\gamma)$.

($\supseteq$) Consider a string

$$z = ac^{i_1}bac^{i_2}b \dots ac^{i_k}b,$$

where $1 \leq i_j \leq n+1, 1 \leq j \leq k$.

A block $ac^i a$, with $1 \leq i \leq n+1$, is said to be *low*; in contrast, $ac^i a$ with $i > n+1$ is said to be *high*.

To a string z as above (with all blocks being *low*) we can only apply a rule in I of the form (r_q.1.), hence corresponding to a rule r_q in P. Let the "codified" rule associated to r_q be of the form

$$ac^{i_s}ac^{i_{s+1}}a \to ac^{p_1}aac^{p_2}a.$$

Using the rule (r_q.1.) we get

$$z \Longrightarrow ac^{i_1}a \dots ac^{i_{s-1}}aac^{i_s}aac^{(2q-1)(n+1)+i_{s+1}}aac^{i_{s+2}}a \dots ac^{i_k}a = z'.$$

Note that a *high* block has been introduced, $ac^{(2q-1)(n+1)+i_{s+1}}a$. When using a rule of type (r_q.1.), a *low* block, surrounded by two *low* blocks, is replaced by a *high* block. Specifically, $(2q-1)(n+1)$ new occurrences of the symbol c are introduced; because $2q-1$ is odd, we say that we have an *odd high* block.

No rule in $I \cup D$ can be used in such a way to replace a *low* block adjacent to the *high* block in z' by a *high* block; no rule in $I \cup D$ can use the *high* block in z' as a context, excepting the rule (r_q.2.) for the same q as above. The inverse morphism h^{-1} is not defined on *high* blocks, hence eventually the rule (r_q.2.) should be applied. This means that we replace the substring

$$ac^{i_s}aac^{(2q-1)(n+1)+i_{s+1}}a$$

of z' by

$$ac^{2(q-1)(n+1)+i_s}aac^{(2q-1)(n+1)+i_{s+1}}a. \qquad (*)$$

Therefore, the *low* block $ac^{i_s}a$ has been replaced by the *high* block $ac^{2(q-1)(n+1)+i_s}a$. Because $2(q-1)$ is even, we say that we have an *even high* block.

Thus, we have obtained a string containing a pair of *high* blocks, one *even* and one *odd*, precisely identified by the index q of the rule r_q in G'.

None of the rules in $I \cup D$ can use these high blocks, excepting (r_q.3.), which replaces the substring $(*)$ by

$$ac^{2(q-1)(n+1)+i_s}aac^{p_2}a.$$

Thus, the previous *odd high* block is replaced by a *low* block. The only possible way to replace the remaining *even high* block by a *low* one is by using the rule (r_q.4.), which leads to

$$ac^{p_1}aac^{p_2}a.$$

Thus, the rule $ac^{i_s}aac^{i_{s+1}}a \to ac^{p_1}aac^{p_2}a$ has been simulated and this is the only way to proceed towards a successful derivation (that is a derivation which produces a string in $(aca)^*h(T^*)$). Consequently, when $ac^2a(aca)^s \Longrightarrow^* (aca)^t h(w)$, for some $w \in T^*$ (only for such strings w we have $h^{-1}(w)$ defined), we have $S\#^s \Longrightarrow^* \#^t w$ in G', $w \in T^*$. This implies $w \in L(G)$, completing the proof. □

Corollary. $RE = RINSDEL$.

6 Concluding remarks

We have proved that the insdel systems of weights (1, 2, 1, 1), (2, 1, 2, 0) and (1, 2, 2, 0) have the same power as Turing machines, they are *computationally complete*. The same result holds for restricted-insdel systems, which insert and delete only strings over an one letter alphabet. Moreover, from the proofs we can obtain universality results: there are insdel systems $\gamma_u = (V_u, T, -, I_u, D_u)$ of weight (1, 2, 1, 1), or (2, 1, 2, 0), or (1, 2, 2, 0) (similarly for restricted-insdel systems) such that for any insdel system $\gamma = (V_0, T, A_0, I_0, D_0)$ we can construct a set $A(\gamma)$ over V_u such that $L(\Gamma'_u) = L(\gamma)$, for $\gamma'_u = (V_u, T, A(\gamma), I_u, D_u)$. Therefore, the universal insdel system emulates the given insdel system γ, the particular system γ can be run on γ_u as a program. This can be obtain as follows. Take a universal type-0 grammar, that is a grammar with all components but the axiom fixed, $G_u = (N_u, T, -, P_u)$, and construct the equivalent insdel system as in the proof of Theorems 1, 2, 3. Because G_u has no axiom, the obtained system, γ_u, will have no axiom set. However, γ_u is the universal system we look for. If we consider an insdel system γ, with the same terminal alphabet, then there is a type-0 grammar, $G = (N, T, S, P)$, equivalent with γ. Construct (for instance, as in [6]) the code $w(G)$ of this grammar. In the axiom strings BSE, BS, BSE used in the proofs of Theorems 1, 2, 3, respectively, replace S by $w(G)S$. Then, because $L(G'_u) = L(G)$, for $G'_u = (N_u, T, w(G)S, P_u)$ and $L(G) = L(\gamma)$, $L(G'_u) = L(\gamma'_u)$, for γ'_u obtained from γ_u as above, we get $L(\gamma'_u) = L(\gamma)$, hence the universality property holds. One sees that, in fact, the "program" of γ to be run on γ_u consists of one string only. The same result holds true for restricted-insdel systems, with the difference that the unique program-axiom should be supplemented with arbitrarily many empty spaces (aca, in the codification in the proof of Theorem 4). Thus, via the existence of universal Turing machines and of universal Chomsky type-0 grammars, we find a proof of the theoretical possibility of designing universal (programmable) molecular computers based on the insertion-deletion operations.

The proof of Theorem 4 can suggests another interesting speculation, concerning the so-called "junk DNA". It is known that a large part of the human genome, about 97% of it, consists of short repeated sequences, thus unable to encode much information, and having no known function. A "computer science explanation" of this situation is given in [30]. We do not enter into details, but we only mention the fact that the basic assumption is that the higher life forms have to have complete computational power in order to possess an efficient immune system. But the computational completeness is fragile, hence dangerous for the life itself. Hence it must be kept under control. Thus, a "replicon police" ("replicon killers") should exist. Conclusion: "junk DNA may be in large part composed of the corpses of former replicons", [30], page 14.

The proof above provides a much more "peaceful" explanation: if we need computational complexity, then we need an arbitrarily large workspace, which at the beginning of the computation is given as a sequence of the encoding of the empty space (aca, here); during the computation the empty space is shuffled with the current meaningful string, the spaces are consumed and reintroduced in the string, such that at the end of the computation we again have a sequence of repeated strings, which can be arbitrarily long. In short, the junk DNA might be the working space of the cell "computation device". This explains both its abundance and the fact that it consists of repeats of the same, short, string. In the terms of [30], these speculations show that computer science approaches to DNA/RNA might be at least as interesting to biology as it is for DNA/RNA computing.

References

[1] L.Adleman, Molecular computation of solutions to combinatorial problems. *Science*, 266 (1994), 1021 – 1024.

[2] L.Adleman. On constructing a molecular computer. *1st DIMACS workshop on DNA based computers*, Princeton, 1995. In *DIMACS Series*, 27 (1996), 1 – 21.

[3] D. Beaver, A universal molecular computer. *1st DIMACS workshop on DNA based computers*, Princeton, 1995. In *DIMACS Series*, 27 (1996), 29 – 36.

[4] D.Boneh, R.Lipton, C.Dunworth, J.Sgall. On the computational power of DNA. *http://www.cs.princeton.edu/~ dabo.*

[5] R. Benne (ed.), *RNA-Editing: The Alteration of Protein Coding Sequences of RNA*, Ellis Horwood, 1993.

[6] C. Calude, Gh. Păun, Global syntax and semantics for recursively enumerable languages, *Fundamenta Informaticae*, 4, 2 (1981), 245 – 254.

[7] C.W.Dieffenbach, G.S.Dveksler, Eds. *PCR primer: a laboratory manual*, Cold Spring Harbor Laboratory Press, 1995, 581-621.

[8] A. Ehrenfeucht, Gh. Păun, G. Rozenberg, On representing recursively enumerable languages by internal contextual languages, *Theor. Computer Sci.*, 1996.

[9] R. Freund, L. Kari, Gh. Păun, DNA computing based on splicing: the existence of universal computers. To appear.

[10] B. S. Galiukschov, Semicontextual grammars (in Russian), *Mat. logica i mat. ling.*, Kalinin Univ., 1981, 38 – 50.

[11] V. Geffert, Normal forms for phrase-structure grammars, *RAIRO. Th. Inform. and Appl.*, 25 (1991), 473 – 496.

[12] T.Head. Formal language theory and DNA: an analysis of the generative capacity of recombinant behaviors. *Bulletin of Mathematical Biology*, 49(1987), 737-759.

[13] A. K. Joshi, L. S. Levi, M. Takahashi, Tree adjunct grammars, *J. Computer Syst. Sci.*, 19, 1 (1975), 136 – 163.

[14] L. Kari. *On Insertion and Deletion in Formal Languages*, Ph.D. Thesis, University of Turku, 1991.

[15] L. Kari, G. Thierrin, Contextual insertion/deletion and computability, *Information and Computation*, 131, no.1(1996), 47 – 61.

[16] R.Lipton, DNA solution of hard computational problems. *Science*, 268 (1995), 542 – 545.

[17] S. Marcus, *Algebraic Linguistics. Analytical Models*, Academic Press, New York, 1967.

[18] S. Marcus, Contextual grammars, *Rev. Roum. Math. Pures Appl.*, 14 (1969), 1525 – 1534.

[19] C. Martin-Vide, Gh. Păun, A. Salomaa, Characterizations of recursively enumerable languages by means of insertion grammars, submitted, 1996.

[20] Gh. Păun, On semicontextual grammars, *Bull. Math. Soc. Sci. Math. Roumanie*, 28(76) (1984), 63 – 68.

[21] Gh. Păun, Two theorems about Galiukschov semicontextual grammars, *Kybernetika*, 21 (1985), 360 – 365.

[22] Gh. Păun, A characterization of recursively enumerable languages, *Bulletin of the EATCS*, 45 (1991), 218 – 222.

[23] Gh. Păun, Five (plus two) universal DNA computing models based on the splicing operation, *Second Annual Meeting on DNA Based Computers*, Princeton, 1996, 67 – 86.

[24] Gh. Păun, *Contextual Grammars. From Natural Languages to Formal Languages and Back*, in preparation, 1996.

[25] Gh. Păun, G. Rozenberg, A. Salomaa, Contextual grammars: erasing, determinism, one-sided contexts, in *Developments in Language Theory* (G. Rozenberg, A. Salomaa, eds.), World Sci. Publ., Singapore, 1994, 370 – 388.

[26] M. Penttonen, One-sided and two-sided contexts in phrase structure grammars, *Inform. Control*, 25, 4(1974), 371–392.

[27] P.Rothemund. A DNA and restriction enzyme implementation of Turing machines. *1st DIMACS workshop on DNA based computers*, Princeton, 1995. In *DIMACS Series*, 27 (1996), 75 – 119.

[28] G. Rozenberg, A. Salomaa (eds.), *Handbook of Formal Languages*, Springer-Verlag, Heidelberg, 1996.

[29] A. Salomaa. *Formal Languages*, Academic Press, New York, 1973.

[30] W. Smith, DNA computers in vitro and in vivo, *1st DIMACS workshop on DNA based computers*, Princeton, 1995. In *DIMACS Series*, 27 (1996), 121 – 187.

[31] C. C. Squier, Semicontextual grammars: an example, *Bull. Math. Soc. Sci. Math. Roumanie*, 32(80) (1988), 167 – 170.

[32] P. Turakainen, A unified approach to characterizations of recursively enumerable languages, *Bulletin of the EATCS*, 45 (1991), 223 – 228.

[33] E. Winfree, On the computational power of DNA annealing and ligation, *1st DIMACS workshop on DNA based computers*, Princeton, 1995. In *DIMACS Series*, 27 (1996), 199 – 221.

DIMACS Series in Discrete Mathematics
and Theoretical Computer Science
Volume **48**, 1999

DNA-EC : A Model of DNA-Computing Based on Equality Checking

Takashi Yokomori and Satoshi Kobayashi

ABSTRACT. This paper proposes new models for DNA computation based on a simple principle called *equality checking*. The advantages of the proposed models may include (*i*) the universal computability of the general models, (*ii*) the clarity and simplicity of molecular biological operations employed, and therefore (*iii*) the potential of high feasibility in molecular biological implementation of the models.

1. Introduction

Since Adleman's ground-breaking work on the DNA implementation of computing a small instance of directed Hamiltonian path problem ([**Adl94**]), a numerous number of literature on this new computation paradigm have been published. In fact, Adleman's model has been extensively studied by many researches, for generalizing his technique to solve larger class of problems ([**Lip95a**], [**Lip95b**],[**Bea95**]), for providing abstract DNA computer models with Turing computability ([**Adl95**],[**Bea95**], [**Rot95**],[**WYS96**]), and so forth. In spite of those efforts on pursuing possible implementation methods for DNA computers with Turing computability using a finite set of molecular biological operations, there is no commonly acceptable models of DNA computation yet.

In this paper, we propose **DNA-EC**, new models of DNA computation based on a simple computational principle called *equality checking*. It will be shown that our models have the universal computability of Turing machines. The equality checking principle (EC principle) is simple enough for human to understand and for machines of any kind to implement. In particular, it seems very much suitable for biomolecular machines, because this EC principle only requires the task of equality checking of two memories,

---

1991 *Mathematics Subject Classification.* Primary 68Q05, 68Q10; Secondary 03D10, 92C40.

This work was supported in part by "Research for the Future" Program No. JSPS-RFTF 96I00101 from the Japan Society for the Promotion of Science.

represented as strings of symbols, just like a completely hybridized *double stranded* DNA *sequence.*

It would be common that the Turing machine or the grammatical equivalent (like phrase-structure grammar) is the computation model at the starting point in order to prove the universal computability of the new model of computation at issue. In contract to this, our model of DNA-computing has its unique origin : an Equality Machine (EM), a simplified variant of Turing machine with EC principle where "concatenation" is only performed on the working tapes.

This preliminary research is motivated by the work of characterization results for the universal computability of an EM based on EC principle in [**ER80**] where it is shown that, roughly speaking, the Turing computability can be achieved by the finite control transformation and the checking ability of two memory tapes' equality. The finite control transformation is a common concept (e.g., a generalized sequential machine mapping : GSM mapping), and seems so simple as to be realized by the state-of-the-art technology of molecular biology. Further, checking equality may be one of just the right tasks for molecular manipulation. Thus, we think that our models bear several advantages over the other models of DNA-computing discussed in the literature.

The objective of this paper is to propose new models **DNA-EC** as the abstract model of DNA-computing and to show their universal computabilities of Turing machines, while the issues of biologically implementing the models together with experimental study will be reported in a separate paper.

2. Abstract Model of DNA Computation

In this section, we introduce the abstract model of our molecular computation based on a simple and powerful characterization result on the computational power in formal language theory.

For a given finite alphabet Σ, let $\overline{\Sigma} = \{\overline{a} | a \in \Sigma\}$ be the alphabet of barred symbols of Σ. In analogy of DNA (or RNA) molecules, $\overline{\Sigma}$ can be regarded as the *complementary alphabet* of Σ, so that we assume that $\overline{(\overline{a})} = a$ for all $a \in \Sigma$. For representing a double stranded string, we prepare the alphabet $D = D_0 \cup D_1 \cup D_2$, where $D_0 = [\Sigma/\overline{\Sigma}] = \{[a/\overline{a}] | a \in \Sigma\}$, $D_1 = [\Sigma/\emptyset] = \{[a/\ ] | a \in \Sigma\}$ and $D_2 = [\emptyset/\overline{\Sigma}] = \{[\ /\overline{a}] | a \in \Sigma\}$. A string $u \in D^*$ is said to be *double stranded* iff it is in $(D_1^* \cup D_2^*)D_0^*(D_1^* \cup D_2^*)$, and it is called *complete* double stranded string iff it is in D_0^*. (A completely hybridized double stranded DNA sequence is such an example.) When a string is in either D_1^* or D_2^*, we regard it as a single stranded string (i.e., ordinary string in the usual sense) rather than a double stranded string.

We assume standard notions and notations from the theory of formal languages. (See, e.g., [**HU79**].)

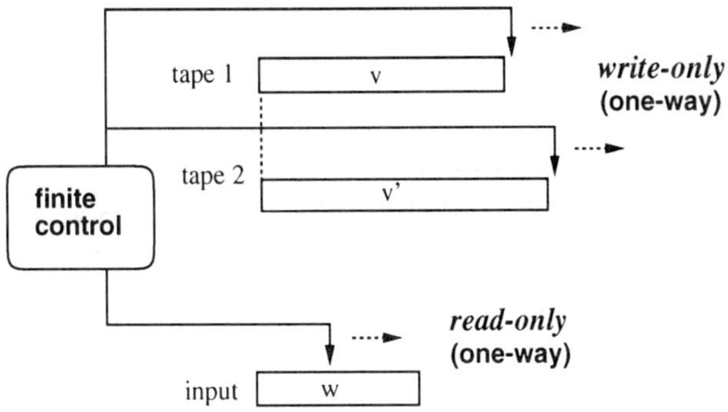

FIGURE 1. Equality Machine : EM

2.1. Equality Machines. The *equality machine*([**ER80**]) is an abstract acceptor shown in Figure 1 where it has a one-way *input* tape, a *finite control*, and two *write-only* tapes for its memory. With the initial empty memory tapes, the machine starts to read the input from the left and changes its state, then extends one of the two memory tapes by concatenating an output string, according to the transition rules specified. At the end of computation where the input string has been fully read, the machine accepts it only if the contents of the two memory tapes are identical.

Formally, an equality machine (EM) M is a structure $(Q, \Sigma, \Delta, \delta, q_0, F)$, where Q is the finite set of states, $q_0 \in Q$ is the initial state, $F \subseteq Q$ is the set of final states, Σ is the input alphabet, Δ is the output alphabet, and δ is a mapping from $Q \times \Sigma^*$ into the finite subsets of $Q \times (\Delta \times \{1,2\})^*$.

Let $M = (Q, \Sigma, \Delta, \delta, q_0, F)$ be an EM. Then,

1. for $u \in (\Delta \times \{1,2\})^*$, we define $[u] : \Delta^* \times \Delta^* \to \Delta^* \times \Delta^*$ as follows: For $v_1, v_2 \in \Delta^*$ and $a \in \Delta$, $[(a,1)](v_1, v_2) = (v_1 a, v_2)$, and $[(a,2)](v_1,v_2) = (v_1, v_2 a)$. For $u = u_1 \cdots u_n$ $(u_i \in \Delta \times \{1,2\}, n \geq 1))$, $[u](v_1, v_2) = [u_n](\cdots [u_1](v_1,v_2)\cdots)$. Further, for $u = \lambda$, $[u](v_1, v_2) = (v_1, v_2)$, where λ is an empty string.
2. A configuration of M is of the form $(p, w, (v_1, v_2))$, where $p \in Q$, $w \in \Sigma^*$, and $v_1, v_2 \in \Delta^*$. As a notation, $(p, w) \to (q, u) \in \delta$ stands for $(q, u) \in \delta(p, w)$. (In this sense, δ is called the set of *transition rules*.) If $(p, w) \to (q, u) \in \delta$, then we write $(p, ww', (v_1, v_2)) \vdash (q, w', [u](v_1, v_2))$. The reflexive and transitive closure of $\vdash$ is denoted by $\vdash^*$.
3. The *language accepted* by M is $L(M) = \{w \in \Sigma^* | (q_0, w, (\lambda, \lambda)) \vdash^* (q, \lambda, (v, v))$ for some $q \in F$ and $v \in \Delta^*\}$.
4. An EM M is *deterministic* if δ is a mapping from $Q \times \Sigma$ into $Q \times (\Delta \times \{1,2\})^*$.

The following normal form theorem for EMs is easily proved:

Proposition 1. *Given an EM M, one can find an equivalent EM M′ such that each transition rule of M′ is of the form :* $(p, a) \to (q, u)$, *where* $p, q \in Q$, $a \in \Sigma \cup \{\lambda\}$ *and* $u \in (\Delta \times \{1\})^* \cup (\Delta \times \{2\})^*$.

This implies that we can denote each transition rule by $(p, a) \to (q, u, i)$, where $p, q \in Q$, $a \in \Sigma \cup \{\lambda\}$, $u \in \Delta^*$, and $i \in \{1, 2\}$.

The classes of languages accepted by nondeterministic and deterministic EMs are denoted by $\mathcal{L}_N(EM)$ and $\mathcal{L}_D(EM)$, respectively. Further, $\mathcal{RG}, \mathcal{CF}, \mathcal{CS}$ and $\mathcal{RE}$ denotes the four language classes of the Chomsky hierarchy.

Some of very interesting results from [**ER80**] are summed up here which motivated the current research in this paper.

Proposition 2. ([**ER80**]) *The following relations hold true :*
(*i*) $\mathcal{L}_N(EM) = \mathcal{RE}$
(*ii*) $\mathcal{L}_D(EM) \subseteq$ **DSPACE**$(\log n)$, *and*
$\mathcal{RG} \subset \mathcal{L}_D(EM) \subset$ **NSPACE**$(n) = \mathcal{CS}$
(*iii*) $\mathcal{L}_D(EM)$ *is incomparable to* $\mathcal{CF}$ *but not disjoint.*

2.2. DNA-EC : A Model of DNA-Computing. Following the notation in [**RW96**], we describe our DNA computation model called **DNA-EC** which is based on the characterization results given in the previous subsection.

In the *test tube computation* discussed below, T denotes a test tube containing a set of strings over some fixed alphabet Σ, and by $I(T)$ we denote its contents. Further, we assume that $I(T) \subseteq \Sigma^*$ unless stated otherwise.

First, we employ a basic operation :

- AM (*Amplify*): Given a test tube T, produce two tubes T' and T'' such that $I(T) = I(T') = I(T'')$.

Then, we assume the following two *set test* operations :

- EM (*Emptiness Test*): Given a test tube T, return "*yes*" if $I(T)$ contains a string (i.e., $I(T) \neq \emptyset$) and "*no*" if it contains none.
- EQ (*Equivalence Test*) : Given a test tube T (where $I(T) \subseteq D^*$) that may contain double stranded strings, return "*yes*" if $I(T)$ contains (at least) one complete double stranded string, and "*no*" otherwise.

Note that :

(*i*) Applying AM enables us to use set variables as many times as desired at any time in the programming via **DNA-EC**.
(*ii*) An operation EM is often called *Detect* in some literature (e.g., [**Adl95**]). Thus, EM in this paper answers "yes" when the test tube is *non*-empty, while the one in [**RW96**] answers "yes" when it is empty.
(*iii*) EQ is a "double stranded version" of EM to detect whether or not there is a completely hybridized double stranded string in the test tube.

Let $k \geq 0$ and let T, T_1, T_2 be set variables for test tubes used in **DNA-EC**.

The following table shows a collection of set operations we consider in **DNA-EC**, where $a, b, c, d \in \Sigma$, $u, v, w \in \Sigma^*$:

Table 1. Set operations in **DNA-EC**

Name	Syntax	$I(T)$ after the Operation
UN(union)	$T := T_1 \cup T_2$	$I(T_1) \cup I(T_2)$
LC(left cut)	$T := a \backslash T_1$	$\{u \mid au \in I(T_1)\}$
LA(left adding)	$T := a \cdot T_1$	$\{a\} \cdot I(T_1)$
EX(extraction)	$T := \mathrm{Ex}(T_1, w)$	$I(T_1) \cap \Sigma^*\{w\}\Sigma^*$
DE(deletion)	$T := \mathrm{De}(T_1, w)$	$\{uv \mid uwv \in I(T_1)\}$
RB(2 bit replacement)	$T := \mathrm{Rb}(T_1, ab, cd)$	$\{xcdy \mid xaby \in I(T_1)\}$
RE(replacement)	$T := \mathrm{Re}(T_1, u, v)$	$\{xvy \mid xuy \in I(T_1)\}$
RV(reversal)	$T := T_1^R$	$\{u^R \mid u \in I(T)\}$
RC(right cut)	$T := T_1 / a$	$\{u \mid ua \in I(T_1)\}$
RA(right adding)	$T := T_1 \cdot a$	$I(T_1) \cdot \{a\}$
CU(cut)	$T := \mathrm{Cu}(T_1, a)$	$\{u, v \mid uav \in I(T_1)\}$ $(I(T_1) \subseteq D^*)$
IN(initialization)	$T := \mathrm{In}(\Sigma, k)$	Σ^k

Notes. EX is essentially the same as the one often referred to as *Separation* in the literature. u^R is the reversal of u.

Definition 1. Let $\mathcal{O}$ and $\mathcal{T}$ be collections of set operations and set test operations introduced above. Then, DNA$(\mathcal{O}, \mathcal{T})$-EC is the class of languages recognized by **DNA-EC** which, besides AM, allows the use of set operations from $\mathcal{O}$ and of set test operations from $\mathcal{T}$.

3. Computation Power of DNA-EC Model

3.1. Universal Computability. We now prove that DNA-EC model has the universal computability of Turing machines. To this end, from Proposition 1 and (i) of Proposition 2, it suffices to show that given an EM $M = (Q, \Sigma, \Delta, \delta, q_0, F)$ in normal form and $w \in \Sigma^*$, one can simulate the computation process of M with w within DNA-EC model.

We may assume that each element of δ is of the form : $(p, a) \rightarrow (q, u, i)$, where $p, q \in Q$, $a \in \Sigma \cup \{\lambda\}$, $u \in \Delta^*$ and $i \in \{1, 2\}$.

Let $c = (p, w, (v, v'))$ be a configuration of M. We encode c as a string over $Q \cup \Sigma \cup \Delta \cup \{\#, 1, 2\}$ as follows :

- $\langle c \rangle_0 = q_0\, w \# 1$ and $\langle c \rangle_0' = q_0\, w \# 2$ (if $c = (q_0, w, (\lambda, \lambda))$),
- $\langle c \rangle_1 = p\, w \# u\, 1$ (if $v = v'u$),
- $\langle c \rangle_2 = p\, w \# u\, 2$ (if $v' = vu$),
- $\langle c \rangle$ is undefined, (otherwise).

(0) *Initialization*: Given an input w, we starts with the encoded initial configurations $\langle c \rangle_0$ and $\langle c \rangle_0'$, and set the initial tube T_0 so that

$$I(T_0) = \{q_0\, w \# 1,\ q_0\, w \# 2\}.$$

It is clear that starting with $\langle c\rangle_0$ or $\langle c\rangle'_0$, we have only to keep the encoded configurations of type of either $\langle c\rangle_1$ or $\langle c\rangle_2$ during the simulation, because those are only qualified to lead to successful computations of accepting w in M.

Let $m = |\delta|$ and let T be the current tube.

(1) *Parallel Simulation of Rules*: Using *Amplify*, construct m copies $T_1, ..., T_m$ of T, i.e.,

for all $j = 1, ..., m, \quad T_j := T.$

Suppose that for each $1 \le j \le m$ we have an encoded configuration : $\langle c\rangle_i = p\,x\,\#\,v\,i$ ($i = 1$ or 2) in T_j. This corresponds to the result of an original configuration sequence : $(q_0, w, (\lambda, \lambda)) \vdash^* (p, x, (uv, u))$ or $(p, x, (u, uv))$ for some $u \in \Delta^*$. (Thus, an idea is that an encoded expression keeps only the difference of w and w', provided that w is a prefix of w' or vice versa in the memory tapes of the original configuration $(p, x, (w, w'))$.)

Let $r_j : (p, a) \to (q, y, i)$ be the j-th rule in δ. Then,

for all $j = 1, ..., m$, do the following procedure :

(See (ii) of Figure 2.)

[(1)-1] $i = 1$ in r_j and a successful application of r_j to the configuration $(p, x, (uv, u))$ produces $(q, x', (uvy, u))$, where $x = ax'$: This case is simulated in **DNA-EC** by a series of operations

(a) $T_j := \mathrm{Ex}(T_j, pa)$;
(b) $T_j := p\backslash T_j$;
(c) $T_j := a\backslash T_j$;
(d) $T_j := q \cdot T_j$;
(e) $T_j := \mathrm{Re}(T_j, 1, y1)$;

[(1)-2] $i = 2$ in r_j and a successful application of r_j to the configuration $(p, x, (u, uv))$ produces $(q, x', (u, uvy))$, where $x = ax'$: This case is simulated in **DNA-EC** in the same manner as above.

[(1)-3] $i = 1$ in r_j and a successful application of r_j to the configuration $(p, x, (u, uv))$ produces either $(q, x', (uy, uyv'))$ with $v = yv'$ or $(q, x', (uvy', uv))$ with $y = vy'$, where $x = ax'$: The former case is simulated in **DNA-EC** as follows. (See (ii-2) of Figure 2.)

(a) $T_j := \mathrm{Ex}(T_j, pa)$;
(b) $T_j := p\backslash T_j$;
(c) $T_j := a\backslash T_j$;
(d) $T_j := q \cdot T_j$;
(e) $T_j := \mathrm{Re}(T_j, \#y, \#)$;

The latter can be done in a similar manner by replacing (e) with : $T_j := \mathrm{Re}(T_j, \#v2, \#y'1)$, where $y = vy'$. In particular, to simulate the case of $v' = \lambda$(i.e., $v = y$), we replace (e) with a series of operations :

(i) Make a copy T'_j of T_j using *Amplify* ;

(ii) $T_j' := \mathrm{Re}(T_j', \#y1, \#1)$;
(iii) $T_j := \mathrm{Re}(T_j, \#y1, \#2)$;
(iv) $T_j := T_j \cup T_j'$,

so that T_j may contain two encoded strings of both "$q\,x'\,\#\,1$" and "$q\,x'\,\#\,2$".

[(1)-4] In the case when $i = 2$ in r_j, a successful application of r_j to the configuration $(p, x, (uv, u))$ can be simulated in a symmetric fashion.

Thus, in each case, we will be able to have a new encoded configuration : $\langle c' \rangle_i = q\,x'\,\#\,v'\,i$ (for some v'; $i = 1$ or 2) in T_j which corresponds to the original configuration $(q, x', (u'v', u'))$ or $(q, x', (u', u'v'))$, for some u', obtained in M after successfully applying r_j to $(p, x, (uv, u))$ or $(p, x, (u, uv))$.

We proceed to the task of merging all test tubes :

$$T := T_1 \cup \cdots \cup T_m.$$

(2) *Checking Solution*: In order to complete a successful simulation, we have only to check whether or not the current T contains a string "$q\,\#i$" with $q \in F$ and $i \in \{1, 2\}$, which is performed as follows :

Make $2|F|$ more copies of T using *Amplify* and keep the original T ;
For all $q \in F$;
$T^q := \mathrm{Ex}(T, q\#1) \cup \mathrm{Ex}(T, q\#2)$;
$T := \bigcup_{q \in F} T^q$;
do *Emptiness Test* to T ;
If "*yes*" is returned, then $w \in L$ and halts ;
else go back to (1) and work on the original T kept above ;

The whole procedural schema is shown in (i) of Figure 2.

Theorem 1. *The class* DNA({UN, LC, LA, EX, RE}, {EM})-EC *coincides with* $\mathcal{RE}$.

We observe that

(1). LC can be generally simulated by LA and DE.
(2). DE is a special version of RE in the sense that $\mathrm{De}(T, w) = \mathrm{Re}(T, w, \lambda)$.
(3). EX is a special case of RE in the sense that $\mathrm{Ex}(T, w) = \mathrm{Re}(T, w, w)$.
(4). We take the position that DE (therefore, RE) *subsumes* EX as a part of its task.
(5). In fact, a series of operations of (a)-(d) above can be done by using a single RE.

Hence, we have :

Corollary 1. *The class* DNA({UN, RE}, {EM})-EC *coincides with* $\mathcal{RE}$.

Remarks. Thus, RE plays a crucial role in achieving the universal computability in **DNA-EC**. Further, it should be noted, however, that RE always replaces very specific patterns (like $\#y$, where y is a bounded-length memory string of a given rule) with another (like $\#$) at a unique fixed site. This feature seems a great advantage of this model from the actual molecular biological implementation point of view. (We will discuss more about this in Section 4.)

3.2. Another Model of DNA-EC — Double Stranded Approach. Admitted that the model **DNA-EC** described in the previous subsection only requires very restricted manner of applying RE in the computation

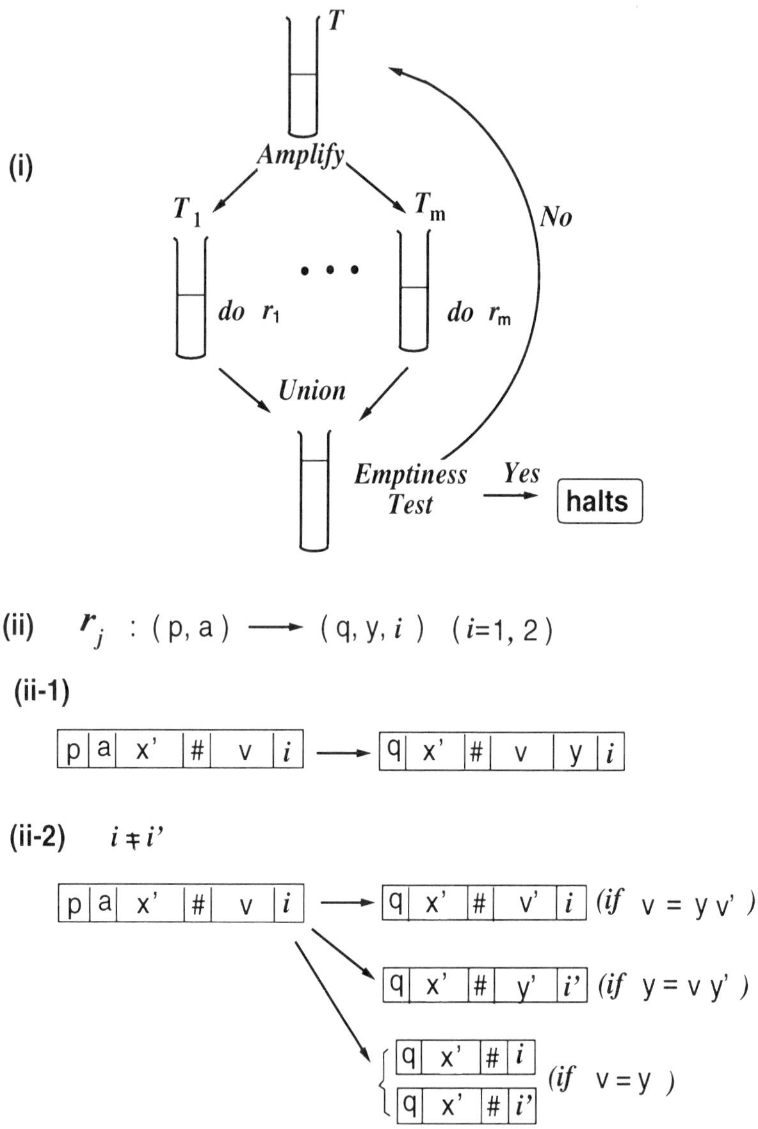

FIGURE 2. Test Tube Computation in **DNA-EC**

process, it would be better if we can replace it with other operation that is easier to be implemented. We will show that this is possible if a set test operation EQ is allowed to use instead of EM. That is, we use the property of double stranded molecular strings.

We may assume that each element of δ is of the form : $(p, a) \to (q, u, i)$, where $p, q \in Q$, $a \in \Sigma \cup \{\lambda\}$, $u \in \Delta^*$ and $i \in \{1, 2\}$.

Given an input w, suppose that $c = (p, w', (v, v'))$ be a configuration of an EM M such that the prefix of w of length k has already been read, where $w = xw'$ and $lg(x) = k$. We use the following encoding of c as a string over $Q \cup \Sigma \cup \Delta \cup \overline{\Delta} \cup \{\#, \$, 1, 2\}$:

$$\langle c \rangle \; = \; 2\,\overline{v'}^R\, p\, \overbrace{\$ \cdots \$}^{k}\, w' \,\#\, v\, 1$$

(Note that the encoded initial configuration is : $\langle c \rangle_0 = 2\, p_0\, w \,\#\, 1$ for $c = (q_0, w, (\lambda, \lambda))$.)

Thus, given an input w, we start with $\langle c \rangle_0$ and set up the initial tube T_0 so that $I(T_0) = \{2\, p_0\, w \,\#\, 1\}$.

Again, let $m = |\delta|$ and T be the current test tube. Then,

(1) using *Amplify*, make m copies of T, i.e.,
for all $j = 1, \ldots, m, \quad T_j := T$.

Suppose that for each $1 \le j \le m$ we have an encoded configuration : $\langle c \rangle$ (= $2\,\overline{v'}^R p\, \$^k\, a\, x' \,\#\, v\, 1$) in T_j.

The idea is that starting with $\langle c \rangle_0$, we simulate the original configuration of M in such a naive manner that for each c the encoded configuration $\langle c \rangle$ is simply kept during the simulation.

The way of simulating one transition step by a rule of δ is now straightforward. (See (1) of Figure 3.) Let $r_j =$"$(p, a) \to (q, y, i)$" be a rule to be applied to a configuration c. Then, it is only needed to update an encoded configuration $\langle c \rangle$ as either

$$\langle c \rangle' \; = \; 2\,\overline{v'}^R q\, \$^{k+1}\, x' \,\#\, v\, y\, 1$$

or

$$\langle c \rangle' \; = \; 2\,\overline{y}^R\, \overline{v'}^R q\, \$^{k+1}\, x' \,\#\, v\, 1$$

depending upon $i = 1$ or $i = 2$. It is clear that this task requires RB, LA, LC, RV, and EX. (Or one could carry it out using only RE, if it is allowed to use.)

(2) In order to finish the simulation, we have to check whether or not the current T contains $\langle c \rangle = 2\,\overline{v'}^R q\, \$^k \,\#\, v\, 1$ such that q is in F and $v = v'$. (See (2) of Figure 3.) We first make a duplication of the current test tube T.

(2.1) One of the procedures for this purpose makes use of the *hybridization mechanism* of DNA molecular sequences. That is, after extracting only strings that contain both substrings "$\$\#$" and "$q\$$"(where $q \in F$), making them hybridized and cutting down a *loop* part by CU(e.g., in terms of an appropriate restriction enzyme), one can produce a test tube T which contains possible double stranded strings corresponding to the correct strings satisfying the conditions above, where we design $\overline{\$} = \#$. Then, the *Equivalence Test* can be applied to the current test tube. If "*yes*" is returned, then $w \in L$ and it halts. Otherwise, go back to (1) and work on the original T duplicated above.

(2.2) The other way of checking equivalence of two strings of v and v' in the intermediate string $\langle c\rangle$ shown in (2) of Figure 3 is to use RE (or more exactly DE) operation : a series of applications of k times of De($T, \$$), followed by De($T, q\#$), De($T, \overline{a}a$) and Ex(T, 21) leads to a test tube which contains "21" iff $v = v'$ in $\langle c\rangle$ holds. (Remember that applying De(T, w) includes the task of Ex(T, w).) It should be also noted that we could have even more efficient starting string $\langle c\rangle$ which contains no symbol \$ if instead of RB (Rb($T, p\$, q\$$) and Rb($T, \$a, \$\$$)), we use RE (Re(T, pa, q)) in the simulation process of (1). These considerations again lead us to the universal computability result presented in Corollary 1.

Summing up the discussion above, we have :

Theorem 2. *The class* DNA({UN, RB, LA, LC, RV, EX, CU}, {EQ})-EC *coincides with* $\mathcal{RE}$.

3.3. Yet Another Model — A Naive Thought. As yet another approach to DNA-computing within **DNA-EC**, one may think of the following naive procedure where we assume set operations LC, LA and EX that are counterparts of LC, LA and EX, acting on double stranded strings.

The idea is quite simple: Using the double stranded string property of complementary pairs of symbols, we simulate the computation process of an Equality Machine $M = (Q, \Sigma, \Delta, \delta, p_0, F)$ in the manner shown in (3) of Figure 3. It is almost obvious that to do this, all we need are operations LC, LA and EX that are for replacing "$[p/\overline{p}][a/\overline{a}]$" with "$[q/\overline{q}]$". In order to append y or $\overline{y}$ to the memory tape specified by a rule from δ, we assume the hybridization property of DNA molecules. Finally, EQ checks whether or not there is a complete double stranded string in the current T.

Thus, we have :

Theorem 3. *The class* DNA({UN, LA, LC, EX}, {EQ})-EC *coincides with* $\mathcal{RE}$.

3.4. Polynomial-Time Computation in DNA-EC Model. We now briefly examine the computation power of **DNA-EC** where any computation is done in polynomial-time. To begin with this,

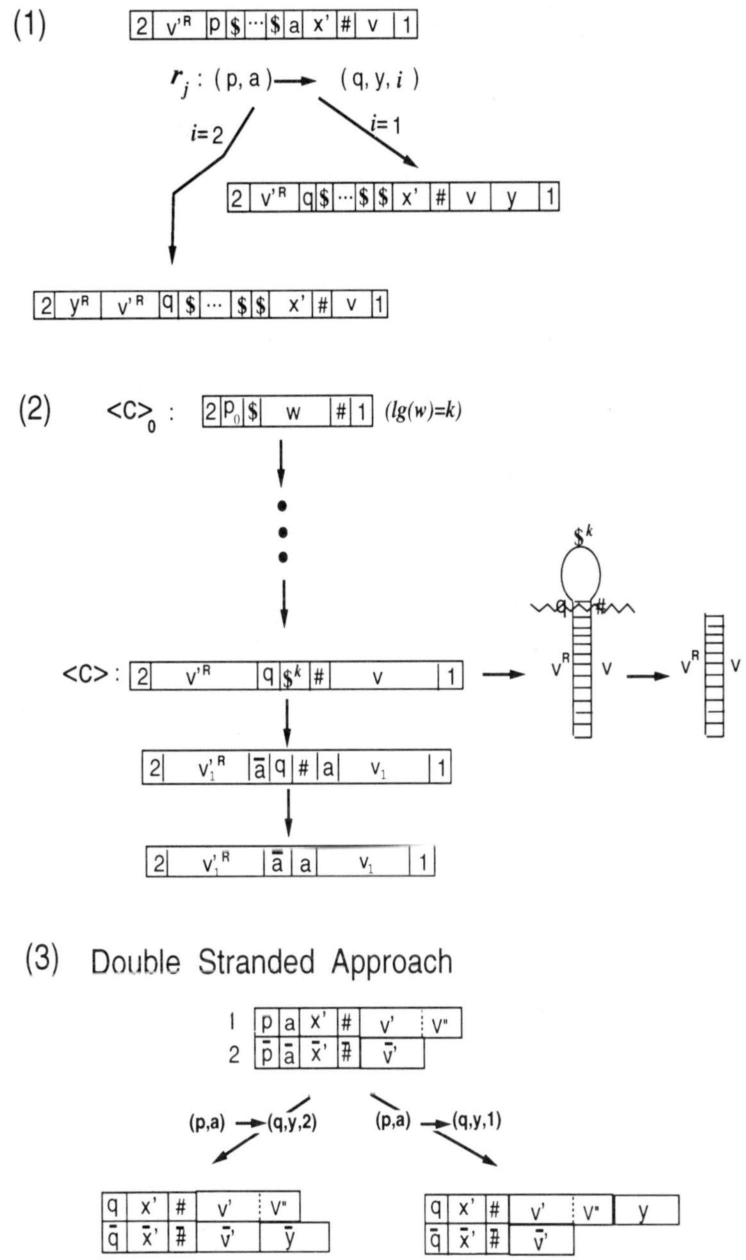

FIGURE 3. Double Stranded Approach to DNA Computation in **DNA-EC**

Definition 2. Let $\mathcal{O}$ and $\mathcal{T}$ be collections of set operations and set test operations introduced above. Then, DNA($\mathcal{O}, \mathcal{T}$)-PEC is the class of languages recognized in *polynomial-time* by **DNA-EC** which, besides AM, allows the use of set operations from $\mathcal{O}$ and of set test operations from $\mathcal{T}$.

We note that almost all of the known works on the power of DNA computation assume (take advantage of) the use of the operation IN (in Table 1) which offers us the ability to create the total set of strings of a fixed length over the given alphabet Σ. That is, given $k \geq 0$, by using IN, we have a test tube T such as $I(T) = \Sigma^k$ within a (fixed) constant time.

In [**RW96**] the following result is shown : DNA({UN,IN,BX},{EM})-P=$\Delta_2^p(= \mathbf{P^{NP}})$, where BX is the bit-wise counterpart of our EX in that it has one more parameter in its syntax to designate the position of target bit.

It is not so hard to see that these two operations BX and EX have actually the same effects within the polynomial time computation (by using a coding technique). Therefore, we have :

Theorem 4. DNA({UN,IN,EX},{EM})-PEC=Δ_2^p

4. Discussion

In this paper, we have proposed a new type of abstract models for DNA-computing that is based on a simple computation principle of checking string equivalence introduced and intensively studied by Engelfriet and Rozenberg in [**ER80**]. We have investigated some of the fundamental models with Turing universal computability within the framework of **DNA-EC**, i.e., DNA($\mathcal{O}, \mathcal{T}$)-EC. It is almost clear that one may make many proposals of DNA-computing models of other types, varing parameters $\mathcal{O}$ and $\mathcal{T}$ from one to another.

On the other hand, there have already been made many attempts to discover grammatical DNA computation models with Turing computability ([**FKP95**],[**Pau95**],[**YKF97**], etc.) based on the *splicing* operation (originally proposed by Tom Head in [**Hea87**]) and its extensions. Rather recently, Kari and Thierrin ([**KT96**]) propose and investigate another type of grammatical model for DNA computation based on insertion/deletion schemes. Contextual insertion/deletion in their work can be regarded as a special type of RE, while the use of RE in our models only requires special types of contexts. It seems interesting to study trade-off relations of biological operations between these two models. It may be summarized, however, that primary interests of most existing works lie in the *generative* capacity of the proposed grammatical models and the formal language theoretic characterizations, while a clear difference of our models from others is that we use a collection of set (test tube) operations suitable for molecular biological implementation and our models benefit from a great efficiency of "molecular parallelism" in more natural manner.

Recently Reif ([**Rei95**]) has proposed a parallel molecular computation model PAM which can *theoretically* efficiently simulate a nondeterministic Turing machine and a PRAM, where a PA-Match operation, equivalently the join operation in relational database, is employed in an effective manner. Due to the complexity of the molecular biological transformation procedure to realize a PRAM program, it seems need more study to fill in the technological gaps between two computation procedures in different levels.

More recently, Rooß and Wagner ([**RW96**]) propose a formal framework to discuss the computational power of various abstract models for DNA computation and show several interesting results of relationships between proposed models and known computational complexity classes within a polynomial time DNA computation. In their paper a variety of set operations concerning *Extraction* and *Replacement* operations, (like Bit Extraction, Bit Replacement) is introduced and used in the analysis of their computation models. Although those operations have theoretically the same power as our counterparts discussed in this paper within the polynomial-time computation using Initialization operation, that would make a great difference as far as the molecular implementation of those operations is concerned. That is, due to the simplicity, our operations seem to have more molecular biological feasibility.

References

[Adl94] L.Adleman. Molecular Computation of Solutions to Combinatorial Problems. *Science* **266**, pp.1021-1024, 1994.

[Adl95] L.Adleman. On Constructing A Molecular Computer. *DNA Based Computers* (Selected papers from Proc. of DIMACS Workshop on DNA Based Computers'95), DIMACS Series in Discrete Mathematics and Theoretical Computer Science, **27**, pp.1-21, 1996.

[BB96] E.B.Baum and D.Boneh. Running Dynamic Programming Algorithms on a DNA Computer. in *Proc. of DIMACS Workshop on DNA Based Computers'96*, pp.141-147, 1996.

[Bea95] D.Beaver. A Universal Molecular Computer. *DNA Based Computers* (Selected papers from Proc. of DIMACS Workshop on DNA Based Computers'95), DIMACS Series in Discrete Mathematics and Theoretical Computer Science, **27**, pp.29-36, 1996.

[BDLS95] D.Boneh, C.Dunworth, R.Lipton and J.Sgall. On Computational Power of DNA. Princeton CS Tech-Report CS-TR-499-95, 1995.

[ER80] J.Engelfriet and G.Rozenberg. Fixed Point Languages, Equality Languages, and Representation of Recursively Enumerable Languages, *J. ACM*, **27**, pp.499-518, 1980.

[FKP95] R.Freund, L.Kari and Gh.Paun. DNA Computing Based on Splicing: The Existence of Universal Computers. Tech. Rep., Fachgruppe Informatik, Tech. Univ. Wien., 1995,(to appear in *J. ACM*).

[Hea87] T.Head. Formal Language Theory and DNA : An Analysis of the Generative Capacity of Specific Recombinant Behaviors. *Bulletin of Mathematical Biology*, **49**, pp.737-759, 1987.

[HU79] J.Hopcroft and J.Ullman. *Introduction to Automata Theory, Languages and Computation.* Addison-Wesley, 1979.

[KT96] L.Kari and G.Thierrin. Contextual Insertions/Deletions and Computability. *Information and Computation*, **131**, no.1, pp.47-61, 1996.

[KYSM97] S.Kobayashi, T.Yokomori, G.Sampei and K.Mizobuchi. DNA Implementation of Simple Horn Clause Computation. *Proc. of IEEE Intern. Conf. on Evol. Comput.* (ICEC'97), pp.213-217, 1997.

[Lip95a] R.J.Lipton. DNA Solution of Hard Computational Problems. *Science* **268**, pp.542-545, 1995.

[Lip95b] R.J.Lipton. Speeding Up Computations via Molecular Biology. *DNA Based Computers* (Selected papers from Proc. of DIMACS Workshop on DNA Based Computers'95), DIMACS Series in Discrete Mathematics and Theoretical Computer Science, **27**, pp.67-74, 1996.

[Pau95] Gh.Paun. Regular Extended H Systems are Computationally Universal. *J. Automata, Languages and Combinatorics* (formerly *EIK*), **1**, pp.27-36, 1996.

[Rei95] J.H.Reif. Parallel Molecular Computation : Models and Simulations. *Proc. of Seventh Annual ACM Symposium on Parallel Algorithms and Architectures*(SPAA'95), pp.213-223, 1995.

[RW96] D.Rooß and K.W.Wagner. On the Power of DNA-Computing, *Information and Computation*, **131**, pp.95-109, 1996.

[Rot95] P.W.K.Rothemund. A DNA and Restriction Enzyme Implementation of Turing Machines. *DNA Based Computers* (Selected papers from Proc. of DIMACS Workshop on DNA Based Computers'95), DIMACS Series in Discrete Mathematics and Theoretical Computer Science, **27**, pp.75-119, 1996.

[Win95] E.Winfree. Complexity of Restricted and Unrestricted Models of Molecular Computation. *DNA Based Computers* (Selected Papaers from Proc. of DIMACS Workshop on DNA Based Computers'95), DIMACS Series in Discrete Mathematics and Theoretical Computer Science, **27**, pp.187-198, 1996.

[WYS96] E.Winfree, X.Yang and N.C.Seeman. Universal Computation via Self-assembly of DNA : Some Theory and Experiments. *Proc. of DIMACS Workshop on DNA Based Computers'96*, pp.172-190, 1996.

[YKF97] T.Yokomori, S.Kobayashi and C.Ferretti. On the Power of Circular Splicing Systems and DNA Computability. *Proc. of IEEE Intern. Conf. on Evol. Comput.* (ICEC'97), pp.219-224, 1997.

Department of Computer Science and Information Mathematics, The University of Electro-Communications, 1-5-1, Chofugaoka, Chofu, Tokyo 182, Japan

E-mail address: {yokomori, satoshi}@cs.uec.ac.jp

DIMACS Series in Discrete Mathematics
and Theoretical Computer Science
Volume **48**, 1999

Splicing on Tree-like Structures

Yasubumi Sakakibara and Claudio Ferretti

ABSTRACT. In this paper, we provide a method to increase the power of splicing systems. We introduce the splicing systems on trees to be built as partially annealed single strands, that is a quite similar notion and a natural extension of splicing systems on strings. Trees are a common and useful data structure in computer science and have a biological counterpart such as molecular sequences with secondary structures, which are typical structures in RNA sequences. Splicing on trees involves (1) complete subtrees as axioms, (2) restriction operated on the annealed subsequences, (3) rules to substitute a complete subtree with another. We show that splicing systems on trees with finite sets of axioms and finite sets of rules can generate the class of context-free languages without the need of imposing multiplicity constraints.

1. Introduction

We extend the notion of splicing systems, usually defined on strings, to splicing systems on trees.

Splicing systems (also called H systems) have been developed as a new computational tool for generating formal languages, and are inspired by molecular biology [**3**]. Splicing operations were originally defined as those that work on linear strings like DNA sequences. Main theoretical result which has been established in splicing systems is that extended H systems with finite sets of axioms and finite sets of rules can exactly generate the class of regular languages [**6**]. They can characterize the class of recursively enumerable languages, if we allow constraints on the number of copies of some molecules in the system, instead of assuming that they all have infinitely many duplicates [**2**]. However, such a multiplicity constraint is very unnatural from the biological point of view, and further it destroys the advantage of massive parallelism in molecular computation [**1**].

In this paper, we provide a method to increase the power of splicing systems. We introduce the splicing systems on trees, to be built as partially annealed single strands, and this is a notion quite similar to that of splicing systems on strings, and a natural extension of it. Trees are a common and useful data structure in computer science and have a biological counterpart such as molecular sequences with secondary structures, which are typical structures in RNA sequences. This

1991 *Mathematics Subject Classification.* Primary: 68Q45; Secondary: 92C40, 92D20.

This work is supported in part by "Research for the Future" Program No. JSPS-RFTF 96I00101 from the Japan Society for the Promotion of Science.

extension is motivated by ideas from two fields of research: genetic programming, [**4**] and stochastic context-free grammars for modeling RNA [**8**].

Genetic programming is the extension of the genetic algorithm into the space of tree-like structures. That is, the objects that constitute the population are not fixed-length character strings that encode possible solutions, they are programs expressed in tree-like structures. This extension succeeds to provide the substantial increase of the power and applicability of genetic algorithm.

Stochastic context-free grammars have successfully been applied to modeling families of RNA sequences and their effectiveness has been shown in computational biology problems such as folding, aligning and discriminating RNA sequences. This research seems to imply the possibility of splicing on secondary structures which are typical in RNA families to get a generative power for context-free languages.

1.1. Introduction to splicing. *Splicing* is a model of the recombinant behavior of double stranded molecules of DNA under the action of restriction enzymes and lygases. A single strand of DNA is an oriented sequence of nucleotides `A`, `C`, `G` and `T`, but since `A` can bind to `T` and `G` to `C`, two strands of DNA bind together to form a doubled stranded DNA molecule, if they have matching pairs of nucleotides when reading the second one along the reverse orientation. For instance, the strands `AAGC` and `GCTT` bind together to form a double stranded DNA molecule. Of course, given the fixed pairings between nucleotides, we can model such a molecule simply as $aagc$, moving toward the realm of formal languages.

Restriction enzymes operate on double stranded DNA molecules recognizing specific subsequences of nucleotides, and *cut* each strand at fixed points inside that matching site, generally leaving short single stranded overhangs. We can represent the action of a restriction enzyme on a string as a triple (a, x, b), where axb is the recognized site, and x is the central single stranded segment between the two cuts operated on each strand.

Lygases are able to join together two complementary single stranded overhangs, always according to the `A-T` and `G-C` pairings of nucleotides. This means that two strings $saxbt$ and $pcxdq$, first cut by two restriction enzymes (a, x, b) and (c, x, d), can either recombine in the original form or they can build two strings $saxdq$ and $pcxbt$, since lygases can operate on the common complementary overhang x.

In this way we can model a test tube, containing DNA molecules together with a set of restriction enzymes and lygases, as a generating mechanism of formal languages: the starting set of strings is continuously cut and pasted to generate new strings. Then we can rise a theoretical question: which language will be generated?

In the scientific literature, characterizations in terms of splicing systems have already been given for language classes such as locally testable languages, regular languages, and recursively enumerable languages. These results are obtained using variations on the definition of the splicing systems. In all these results restriction enzymes are modeled in pairs described by quadruples, instead than by a triple for each enzyme as we have seen above. A pair of restriction enzymes is represented by a *rule* of the form:

$$r = u_1 \# u_2 \$ u_3 \# u_4$$

Such a rule directly transforms the strings su_1u_2t and pu_3u_4q into su_1u_4q and pu_3u_2q, applying cut and paste between them. Making further assumptions on the behavior of the splicing system leads to the characterization of various language classes:

- the class of locally testable languages can be characterized by splicing systems where the splicing of two strings on a restriction site containing as a substring another site, does not destroy the latter (*persistent* systems [**3**])
- a way to generate all and only the regular languages is to use splicing systems where, among all the generated strings, we can select only those without some special symbols (*extended* splicing systems [**2**])
- splicing systems where most molecules are present in am unlimited number of copies, but where some strings have only a finite number of duplicates, characterize the class of recursively enumerable languages (*multiplicity constraints* [**2**]).

In this paper we define another variation of the splicing operation, with the goal of obtaining results concerning context free languages. This class is strangely almost absent in the literature of this field. For instance, in [**6**] there is a characterization of context free languages, provided that we can use any language from that same class as starting set of strings for the splicing system. But when we consider to use only finite sets of axioms, as it is also done in the present paper, known characterizations go suddenly from the class of regular language to the universal class.

2. Preliminaries

A *tree* over an alphabet V is a rooted, directed, connected acyclic finite graph in which the direct successors of any node are linearly-ordered from left to right. The formal definition of the tree is as follows. Let $\mathbb{N}$ be the set of natural numbers. A *ranked alphabet* V is a finite set of symbols associated with a finite relation called the *rank* relation $r_V \subseteq V \times \mathbb{N}$. V_n denotes the subset $\{f \in V \mid (f,n) \in r_V\}$ of V. In many cases the symbols in V_n are considered as *function symbols*. We say that a function symbol f has an *arity* n if $f \in V_n$ and a symbol of arity 0 is called a *constant symbol* which corresponds to a terminal symbol in the grammar.

A *tree* over V is a mapping t from Dom_t into V where (1) the domain Dom_t is a finite nonempty subset of $\mathbb{N}^*$; (2) if $x \in Dom_t$ and $y < x$, then $y \in Dom_t$; (3) if $y \cdot i \in Dom_t$ and $i \in \mathbb{N}$, then $y \cdot j \in Dom_t$ for $0 \leq j \leq i$, $j \in \mathbb{N}$; (4) $t(x) \in V_n$, whenever for $i \in \mathbb{N}$, $x \cdot i \in Dom_t$ if and only if $0 \leq i \leq n-1$. An element of the tree domain Dom_t is called a *node* of t. If $t(x) = A$, then we say that A is the *label* of the node x of t. If we consider V as a set of function symbols, the trees over V can be identified with *well-formed terms* over V and written linearly with commas and parentheses. Let V^T denote the set of all trees over V. Let λ denote the null tree in V^T.

For the notational convention, we often use the form $u(u_1, \dots, u_n)$ to denote the tree defined by

$$u(u_1, \dots, u_n)\ (x) = \begin{cases} u(x) & \text{if } x \in Dom_u \text{ and } u(x) \neq \lambda_i, \\ u_i(y) & \text{if } x = z \cdot y,\ u(z) = \lambda_i \text{ and } y \in Dom_{u_i} \\ & \text{for } 1 \leq i \leq n \end{cases}$$

for the tree $u(\lambda_1, \dots, \lambda_n) \in (V \cup \{\lambda\})^T$ (where λ_i denotes the null tree) and $u_i \in V^T$ $(1 \leq i \leq n)$. That is, $u(u_1, \dots, u_n)$ is the tree obtained from $u(\lambda_1, \dots, \lambda_n)$ by replacing each null tree λ_i with a tree u_i.

Formally, a *grammar* is composed of three parts. The first is a finite alphabet Σ of *terminal symbols*. For DNA sequences, this alphabet comprises the four nucleotides `A`, `C`, `G` and `T`, and for RNA sequences, the nucleotides `A`, `C`, `G` and `U` instead of `T`. The second is a finite set N of *nonterminal symbols* and a special *start symbol* S. The third is a set P of rewrite rules (or *production rules*) that specify how sequences containing nonterminals may be rewritten by expanding those embedded nonterminals to new subsequences. The language generated by the grammar is the set of all sequences of terminal symbols that can be derived from the start symbol S by repeatedly applying productions from P.

A *context-free grammar* (CFG) is defined by a quadruple $G = (N, \Sigma, P, S)$. Each production rule in P has the form $X \to \alpha$ for $X \in N$ and $\alpha \in (N \cup \Sigma)^*$, indicating that the nonterminal X can be replaced by the sequence α. The *language generated* by a CFG G is denoted $L(G)$.

A *derivation* is a rewriting of a sequence using the rules of the grammar. It begins with a sequence that consists only of the start symbol S. In each step of the derivation, a nonterminal from the current sequence is chosen and replaced with the right-hand side of a production rule for that nonterminal. This replacement process is repeated until the sequence consists of terminal symbols only. A simple derivation is shown in Figure 1.

A derivation can be arranged in a tree structure called a *derivation tree*. Let $D(G)$ denote the set of derivation trees of a CFG G.

2.1. Splicing systems on tree structures. In the following definitions we basically move from splicing on linear strings, to splicing on tree-like structures that still resemble some natural molecular assemblies. The same generative mechanism, based on starting structures and rules operating by cut and paste on them, will be kept. We use finite set of splicing rules described by quadruples, instead of the triples of [**3**], but we still use finite set of starting strings and of rules. We do not make assumptions such as that of having persistent splicing systems, and we neither impose multiplicity constraints.

We define the splicing operation on tree structures. A *splicing rule on tree structures* over V is of the form

$$r \;=\; u_1(\lambda, \ldots, \#(u_2(\lambda, \ldots, \lambda)), \ldots, \lambda) \;\$\; u_3(\lambda, \ldots, \#(u_4(\lambda, \ldots, \lambda)), \ldots, \lambda)$$

for $u_i \in V^T$ $(1 \le i \le 4)$ and for two special symbols $\#$, $\$$ not in V. $\#$ is used as a marker for specifying a single node in a tree. (See Figure 2.)
For such a rule r, suppose two trees x and y in V^T which contain $u_1(\lambda, \ldots, u_2(\lambda, \ldots, \lambda), \ldots, \lambda)$ and $u_3(\lambda, \ldots, u_4(\lambda, \ldots, \lambda), \ldots, \lambda)$ as a part of the tree respectively. That is, x contains a subtree constructed from $u_1(\lambda, \ldots, u_2(\lambda, \ldots, \lambda), \ldots, \lambda)$ by replacing each null tree λ by some tree in V^T. For such a rule r, such trees $x, y \in V^T$, and a tree $z \in V^T$, we define the operation $\vdash_r$ as follows:

$$\begin{aligned} (x, y) \vdash_r z \quad \text{iff} \quad & x = x_0(u_1(x_1, \ldots, u_2(x'_1, \ldots, x'_{m_2}), \ldots, x_{m_1})) \\ & y = y_0(u_3(y_1, \ldots, u_4(y'_1, \ldots, y'_{n_2}), \ldots, y_{n_1})) \\ & z = x_0(u_1(x_1, \ldots, u_4(y'_1, \ldots, y'_{n_2}), \ldots, x_{m_1})) \\ \text{for some} \quad & x_0, \ldots, x_{m_1}, x'_1, \ldots, x'_{m_2}, y_0, \ldots, y_{n_1}, y'_1, \ldots, y'_{n_2} \in V^T. \end{aligned}$$

(See Figure 3.)

a. Production rules

$$P = \{\ S_0 \to S_1,\quad S_1 \to \mathtt{C}\ S_2\ \mathtt{G},\quad S_1 \to \mathtt{A}\ S_2\ \mathtt{U},\quad S_2 \to \mathtt{A}\ S_3\ \mathtt{U},\quad S_3 \to S_4\ S_9,\quad S_4 \to \mathtt{U}\ S_5\ \mathtt{A},\quad S_5 \to \mathtt{C}\ S_6\ \mathtt{G},\quad S_6 \to \mathtt{A}\ S_7,\quad S_7 \to \mathtt{U}\ S_7,$$
$$S_7 \to \mathtt{G}\ S_8,\quad S_8 \to \mathtt{G},\quad S_8 \to \mathtt{U},\quad S_9 \to \mathtt{A}\ S_{10}\ \mathtt{U},\quad S_{10} \to \mathtt{C}\ S_{10}\ \mathtt{G},\quad S_{10} \to \mathtt{G}\ S_{11}\ \mathtt{C},\quad S_{11} \to \mathtt{A}\ S_{12}\ \mathtt{U},\quad S_{12} \to \mathtt{U}\ S_{13},\quad S_{13} \to \mathtt{C}\ \}$$

b. Derivation

$$S_0 \Rightarrow S_1 \Rightarrow \mathtt{C}S_2\mathtt{G} \Rightarrow \mathtt{CA}S_3\mathtt{UG} \Rightarrow \mathtt{CA}S_4S_9\mathtt{UG}$$
$$\Rightarrow \mathtt{CAU}S_5\mathtt{A}S_9\mathtt{UG} \Rightarrow \mathtt{CAUC}S_6\mathtt{GA}S_9\mathtt{UG}$$
$$\Rightarrow \mathtt{CAUCA}S_7\mathtt{GA}S_9\mathtt{UG} \Rightarrow \mathtt{CAUCAG}S_8\mathtt{GA}S_9\mathtt{UG}$$
$$\Rightarrow \mathtt{CAUCAGGGA}S_9\mathtt{UG} \Rightarrow \mathtt{CAUCAGGGAA}S_{10}\mathtt{UUG}$$
$$\Rightarrow \mathtt{CAUCAGGGAAG}S_{11}\mathtt{CUUG}$$
$$\Rightarrow \mathtt{CAUCAGGGAAGA}S_{12}\mathtt{UCUUG}$$
$$\Rightarrow \mathtt{CAUCAGGGAAGAU}S_{13}\mathtt{UCUUG}$$
$$\Rightarrow \mathtt{CAUCAGGGAAGAUCUCUUG}.$$

c. Parse tree

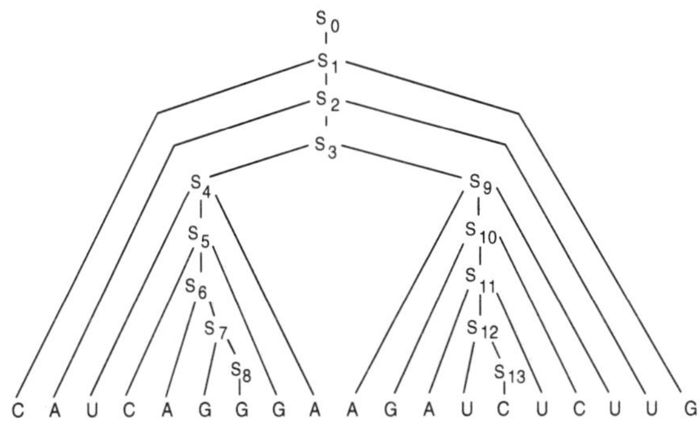

FIGURE 1. (*a*) A simple CFG which may be used to derive a set of RNA sequences including the specific example illustrated here, CAUCAGGGAAGAUCUCUUG. S_0 (start symbol), $S_1, \dots, S_{13}$ are nonterminal symbols; A, U, G and C are terminal symbols representing the four nucleotides. (*b*) Application of the production rules P could generate the example sequence by the derivation indicated. (*c*) The derivation process in *b* may be arranged in a tree structure called a *derivation tree*.

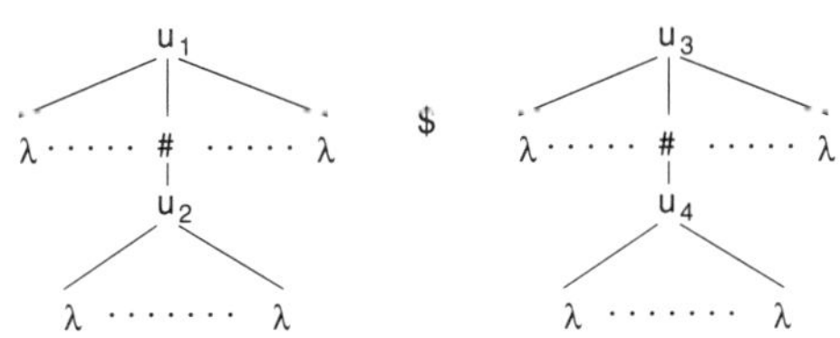

FIGURE 2. Splicing rule on tree structures.

A *splicing scheme on tree structures* is a pair $\gamma = (V, R)$, where V is an alphabet and R is a set of splicing rules on tree structures over V. A pair $\mathcal{S} = (\gamma, L)$ for a splicing scheme γ and a tree language $L \subseteq V^T$ is called *splicing system on tree structures* (or HT system).

For an HT system $\mathcal{S} = (\gamma, L)$, we define

$$\gamma(L) = \{z \in V^T \mid (x, y) \vdash_r z \text{ for some } x, y \in L, r \in R\}.$$

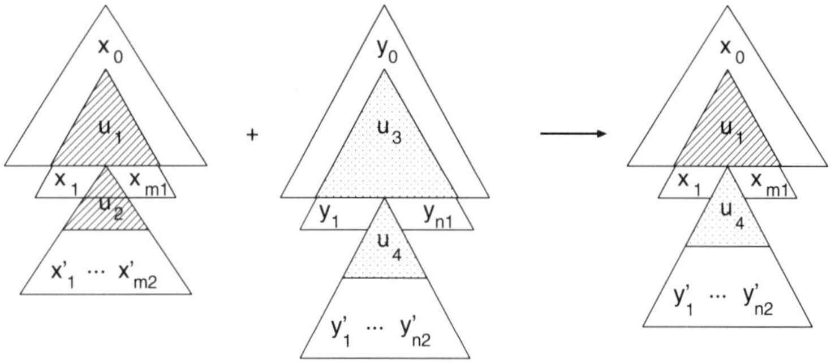

FIGURE 3. Splicing on tree structures.

We also define the *iterated splicing* as follows:

$$\begin{aligned}\gamma^0(L) &= L,\\ \gamma^{i+1}(L) &= \gamma^i(L) \cup \gamma(\gamma^i(L)),\ i \geq 0,\\ \gamma^*(L) &= \bigcup_{i\geq 0} \gamma^i(L).\end{aligned}$$

For a class of tree languages F_1 and a class of sets of splicing rules F_2, we denote by $HT(F_1, F_2)$ the set of all tree languages $\gamma^*(L)$, for $L \in F_1$ and $\gamma = (V, R)$, $R \in F_2$. Especially, we denote by FIN_T the class of finite tree languages, by FIN_R the class of finite sets of splicing rules, and by $HT(FIN_T, FIN_R)$ the set of all tree languages $\gamma^*(L)$ generated from a finite tree language $L \in FIN_T$ and a finite set of splicing rules on tree structures in FIN_R.

3. RNA secondary structures and tree structures

We demonstrate that a special kind of trees, called skeletons, have a biological counterpart such as RNA sequences with secondary structures and splicing operations on skeletons correspond to splicing on secondary structures of molecular sequences.

Let σ denote a special symbol used for labeling internal nodes of trees. A tree defined over $(\{\sigma\} \cup \Sigma)$ is called a *skeleton*. A skeleton is a tree in which all the internal nodes are labeled by σ and other nodes are by symbols in Σ. A skeleton describes only the shape and terminal nodes of the derivation tree.

On the other hand, in RNA, the nucleotides `A`, `C`, `G` and `U` interact in specific ways to form characteristic secondary-structure motifs such as helices, loops and bulges. In general, the folding of an RNA chain into a functional molecule is largely governed by the formation of intramolecular `A-U` and `G-C` Watson-Crick pairs. Such base pairs constitute the so-called *biological palindromes* in the genome.

We consider that such a physical secondary structure of molecular sequence is a reflection of a skeleton. A skeleton represents the syntactic structure of an RNA sequence (Figure 4, right), and this syntactic structure corresponds to the physical secondary structure (Figure 4, left). Therefore, splicing on secondary structures of RNA sequences (like shown in Figure 5) corresponds to splicing on skeletons. We call the HT system on skeletons the *HTS system*. Thus HTS system is a formalization for splicing on secondary structures.

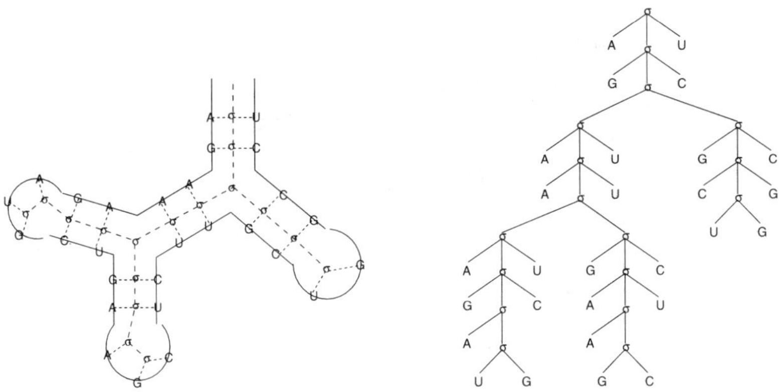

FIGURE 4. A physical secondary structure of RNA sequence and a derivation tree (skeleton) which corresponds to the secondary structure.

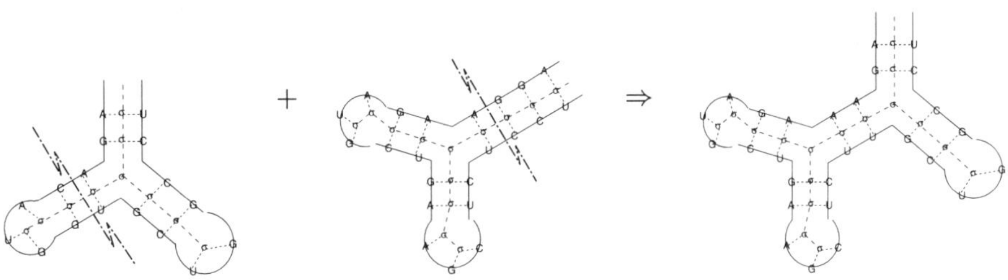

FIGURE 5. Splicing operation on secondary structures.

4. The power of HT systems

First we show that splicing on tree structures generates exactly the class of context-free languages. Next we show our main result that splicing on skeletons is almost equal to the power of context-free grammars. Both proofs are natural extensions of the one (Theorem 9) in Paun [**5**].

THEOREM 1. *For any context-free grammars G, the set of derivation trees of G is a tree language in the class $HT(FIN_T, FIN_R)$ over $V = N \cup \Sigma$.*

PROOF. Let $D(G)$ be the set of derivation trees generated by a CFG $G = (N, \Sigma, P, S)$ with the rules in P of the forms $X \to YZ$, $X \to aY$, and $X \to a$, for $X, Y, Z \in N$, $a \in \Sigma$ (it is clear that these rules are enough for generating every CFG because it includes the Chomsky normal form).

Consider the alphabet

$$V \;=\; N \cup \Sigma,$$

the splicing scheme

$$\begin{aligned}\gamma \;=\; (V,\; & \{X(\#(Y), Z)\ \$\ \lambda(\#(Y)) \mid X, Y, Z \in N\} \cup \\ & \{X(Y, \#(Z))\ \$\ \lambda(\#(Z)) \mid X, Y, Z \in N\} \cup \\ & \{X(a, \#(Y))\ \$\ \lambda(\#(Y)) \mid X, Y \in N, a \in \Sigma\}),\end{aligned}$$

and the finite tree language $L_0 \subseteq D(G)$ defined by

$$L_0 = \{t \in D(G) \mid \begin{array}{l}\text{every path in } t \text{ has at most two occurrences of the}\\ \text{same rule in } P\end{array}\}.$$

We show the relation

$$D(G) = \gamma^*(L_0).$$

First, each tree in L_0 is obviously a derivation tree of G. Next, if x and y are trees in $\gamma^i(L_0)$ and derivation trees of G, and of the forms $x_1(X(U(x_2), Y(x_3)))$ and $y_1(Y(y_2))$ respectively, then $z = x_1(X(U(x_2), Y(y_2))) \in \gamma^{i+1}(L_0)$ and obviously z corresponds to a derivation tree of G, too. Hence $\gamma^*(L_0) \subseteq D(G)$.

Conversely, consider the derivation trees of G. Such derivation trees t of depth less than two are in L_0, hence in $\gamma^*(L_0)$. Assume that all such derivation trees of depth less than some $n \geq 2$ are in $\gamma^*(L_0)$ and consider a derivation tree t of the smallest depth greater than n in $D(G)$. If $t \notin L_0$, it follows that a path in t contains a production rule $X \to UY$ with $X, Y \in N$ and $U \in N \cup \Sigma$ on three different positions:

$$t = x_0(\ X(U(u_1), Y(x_1(\ X(U(u_2), Y(x_2(\ X(U(u_3), Y(x_3)))\ ))\ )))\ ).$$

Then

$$\begin{aligned} t' &= x_0(X(U(u_1), Y(x_1(X(U(u_2), Y(x_3)))))), \\ t'' &= x_0(X(U(u_1), Y(x_2(X(U(u_3), Y(x_3)))))). \end{aligned}$$

are correct derivation trees of G and the depths of both trees t' and t'' are less than the depth of t and hence less than n. Therefore $t', t'' \in \gamma^*(L_0)$. From the form of t', t'' and of splicing rules of γ, it is obvious that $t \in \gamma^*(L_0)$. Hence $D(G) \subseteq \gamma^*(L_0)$. □

Now we show that the class $HTS(FIN_T, FIN_R)$ is almost equal to the class of derivation trees of context-free grammars. We consider a mapping g from skeletons to derivation trees. For a set of skeletons L, let $g(L)$ denote the set of derivation trees $\{g(s) \mid s \in L\}$ and we call it the *coding* of L.

THEOREM 2. *For a mapping g from skeletons to derivation trees, the set of derivation trees of a context-free grammars G is the coding of a tree language in the class $HTS(FIN_T, FIN_R)$.*

PROOF. Let $D(G)$ be the set of derivation trees generated by a CFG $G = (N, \Sigma, P, S)$ with the rules in P of the forms $X \to YZ$, $X \to aY$, and $X \to a$, for $X, Y, Z \in N$, $a \in \Sigma$.

Consider the alphabet

$$\begin{aligned} V = \ &\{\sigma([X,*,*],\ \sigma(\sigma([*,*,Y],\lambda,[*,*,Y]),\sigma([*,*,Z],\lambda,[*,*,Z])),\ [X,*,*]) \\ &\qquad \mid X \to YZ \in P,\ X,Y,Z \in N, \text{a special symbol } * \text{ not in } N,\ \Sigma\} \cup \\ &\{\sigma([X,a,Y],\lambda) \mid X \to aY \in P,\ X,Y \in N, a \in \Sigma\} \cup \\ &\{\sigma([X,a,*]) \mid X \to a \in P,\ X \in N, a \in \Sigma\}. \end{aligned}$$

The symbols in the alphabet V represent the production rules in P in the forms of skeletons (See Figure 6). Note that any tree over V is a skeleton over $\{\sigma\} \cup \Sigma'$

$X \to YZ,$

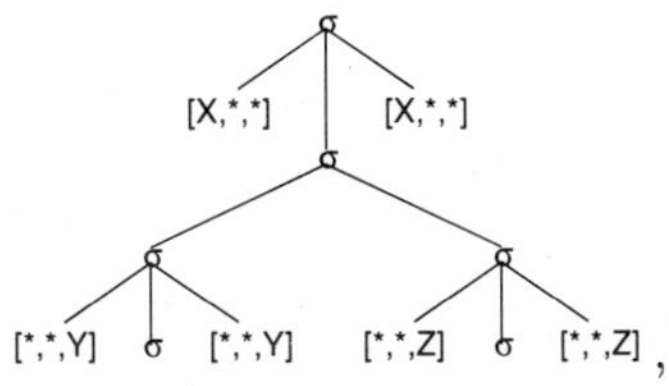

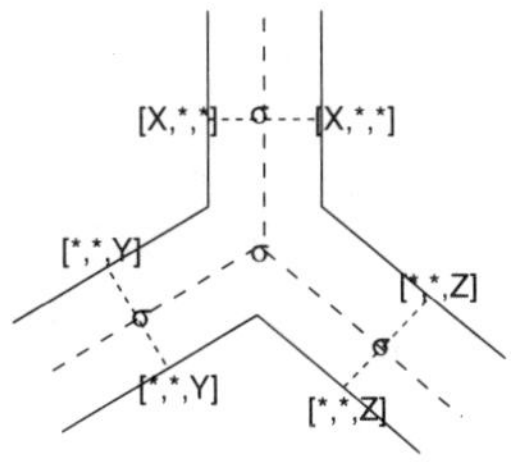

FIGURE 6. A production rule of the form $X \to YZ$, its tree representation on skeletons, and the molecular implementation.

for $\Sigma' = \{[X,*,*], [*,*,X], [X,a,Y], [X,a,*] \mid X, Y \in N, a \in \Sigma\}$. Consider also the splicing scheme

$$\begin{aligned}\gamma \;=\; (V,\; &\{\sigma([*,*,X], \#(\lambda), [*,*,X]) \,\$\, \lambda(\#(\sigma([X,*,*], \lambda, [X,*,*]))) \mid X \in N\} \cup \\ &\{\sigma([*,*,X], \#(\lambda), [*,*,X]) \,\$\, \lambda(\#(\sigma([X,a,Y], \lambda))) \mid X, Y \in N, a \in \Sigma\} \cup \\ &\{\sigma([*,*,X], \#(\lambda), [*,*,X]) \,\$\, \lambda(\#(\sigma([X,a,*]))) \mid X \in N, a \in \Sigma\} \cup \\ &\{\sigma([X,a,Y], \#(\lambda)) \,\$\, \lambda(\#(\sigma([Y,*,*], \lambda, [Y,*,*]))) \mid X, Y \in N, a \in \Sigma\} \cup \\ &\{\sigma([X,a,Y], \#(\lambda)) \,\$\, \lambda(\#(\sigma([Y,b,Z], \lambda))) \mid X, Y \in N, a, b \in \Sigma\} \cup \\ &\{\sigma([X,a,Y], \#(\lambda)) \,\$\, \lambda(\#(\sigma([Y,b,*]))) \mid X, Y \in N, a, b \in \Sigma\}),\end{aligned}$$

and the finite tree language L_0 defined by

$$L_0 \;=\; g(L_0'),$$

where

$$L_0' \;=\; \{t \in D(G) \mid \begin{array}{l}\text{every path in } t \text{ has at most two occurrences of the}\\ \text{same rule in } P\end{array}\},$$

and g is a mapping from $(N \cup \Sigma)^T$ to V^T defined recursively by

$$\begin{aligned}g(X(Y(t_1), Z(t_2))) \;&=\; \sigma([X,*,*],\\ &\quad \sigma(\sigma([*,*,Y], g(Y(t_1)), [*,*,Y]), \sigma([*,*,Z], g(Z(t_2)), [*,*,Z])), [X,*,*]),\\ g(X(a, Y(t_1))) \;&=\; \sigma([X,a,Y], g(Y(t_1))),\\ g(X(a)) \;&=\; \sigma([X,a,*]).\end{aligned}$$

Note that $g(L_0')$ is a set of skeletons.

Consider also the mapping $h : V^T \to (N \cup \Sigma)^T$ defined recursively by

$$\begin{aligned}&h(\sigma([X,*,*], \sigma(\sigma([*,*,Y], t_1, [*,*,Y]), \sigma([*,*,Z], t_2, [*,*,Z])), [X,*,*]))\\ &\quad = X(h(t_1), h(t_2)),\\ &h(\sigma([X,a,Y], t_1)) = X(a, h(t_1)),\\ &h(\sigma([X,a,*])) = X(a).\end{aligned}$$

Obviously, a pair $\mathcal{S} = (\gamma, L_0)$ is an HTS system. We show the relation

$$D(G) \;=\; h(\gamma^*(L_0)).$$

First, from the definitions of the mappings g and h, it is obvious that each skeleton in L_0 describes a derivation tree of G, that is, $h(g(t)) \in D(G)$ for $t \in L_0'$.

Next, assume that x and y are trees in $\gamma^i(L_0)$ describing derivation trees of G. That is, $h(x), h(y) \in D(G)$. If x and y are of the forms:

$$x_1(\sigma([*,*,X], x_2, [*,*,X])) \text{ and } y_1(\sigma([X,*,*], y_2, [X,*,*]))$$

respectively, then

$$z = x_1(\sigma([*,*,X], \sigma([X,*,*], y_2, [X,*,*]), [*,*,X])) \in \gamma^{i+1}(L_0)$$

and obviously z describes a derivation tree of G, too. That is, $h(z) \in D(G)$. If x and y are of the forms $x_1(\sigma([*,*,X], x_2, [*,*,X]))$ and $y_1(\sigma([X,a,Y], y_2))$ respectively, then $z = x_1(\sigma([*,*,X], \sigma([X,a,Y], y_2), [*,*,X])) \in \gamma^{i+1}(L_0)$ and $h(z) \in D(G)$. If x and y are of the forms $x_1(\sigma([X,a,Y], x_2))$ and $y_1(\sigma([Y,*,*], y_2, [Y,*,*]))$ respectively, then $z = x_1(\sigma([X,a,Y], \sigma([Y,*,*], y_2, [Y,*,*]))) \in \gamma^{i+1}(L_0)$ and $h(z) \in D(G)$. If x and y are of the forms $x_1(\sigma([X,a,Y], x_2))$ and $y_1(\sigma([Y,b,Z], y_2))$ respectively, then $z = x_1(\sigma([X,a,Y], \sigma([Y,b,Z], y_2))) \in \gamma^{i+1}(L_0)$ and $h(z) \in D(G)$. Consequently, $h(\gamma^*(L_0)) \subseteq D(G)$.

Conversely, consider the skeletons s over V describing derivation trees of G. That is, $h(s) \in D(G)$. Such skeletons s which describe derivation trees of G of depth less than two are in L_0, hence in $\gamma^*(L_0)$. Assume that all such skeletons which describe derivation trees of G of depth less than some $n \geq 2$ are in $\gamma^*(L_0)$ and consider a skeleton s which describe a derivation tree of G of the smallest depth greater than n, that is, $h(s)$ is of the smallest depth greater than n in $D(G)$. If $s \notin L_0$, it follows that a path in s contains a symbol α in V on three different positions. If α is of the form $\alpha(\lambda,\lambda) = \sigma([X,*,*], \sigma(\sigma([*,*,Y], \lambda, [*,*,Y]), \sigma([*,*,Z], \lambda, [*,*,Z])), [X,*,*])$, assume

$$s = x_0(\alpha(y_1, x_1(\alpha(y_2, x_2(\alpha(y_3, x_3)))))).$$

Then

$$\begin{aligned} s' &= x_0(\alpha(y_1, x_1(\alpha(y_2, x_3)))), \\ s'' &= x_0(\alpha(y_1, x_2(\alpha(y_3, x_3)))). \end{aligned}$$

describe correct derivation trees of G, that is, $h(s'), h(s'') \in D(G)$, and the depths of both trees s' and s'' are less than the depth of s. Hence the depths of $h(s')$ and $h(s'')$ are less than n. Therefore $s', s'' \in \gamma^*(L_0)$. From the form of s', s'' and of splicing rules of γ, it is obvious that $s \in \gamma^*(L_0)$. If α is of the form $\alpha(\lambda) = \sigma([X,a,Y], \lambda)$, assume

$$s = x_0(\alpha(x_1(\alpha(x_2(\alpha(x_3)))))).$$

Then

$$\begin{aligned} s' &= x_0(\alpha(x_1(\alpha(x_3)))), \\ s'' &= x_0(\alpha(x_2(\alpha(x_3)))). \end{aligned}$$

describe correct derivation trees of G and the depths of both trees s' and s'' are less than the depth of s. Hence the depths of $h(s')$ and $h(s'')$ are less than n. Therefore $s', s'' \in \gamma^*(L_0)$. From the form of s', s'' and of splicing rules of γ, it is obvious that $s \in \gamma^*(L_0)$.

Consequently, for each derivation tree t of G, we find a skeleton $s \in \gamma^*(L_0)$ such that $h(s)$ is exactly the derivation tree t, i.e., $h(s) = t$. Hence $D(G) \subseteq h(\gamma^*(L_0))$. □

Conversely, we show that the generative power of splicing on tree structures is not greater than that of context-free grammars. That is, a tree language generated by splicing systems with an initial set of derivation trees is a set of derivation trees generated by a context-free grammar.

THEOREM 3. *Let RT be the class of skeleton languages generated by context-free grammars. A tree language in the class* $HTS(RT, FIN_R)$ *is a set of skeletons generated by a context-free grammar G.*

PROOF. The proof is based on Pixton's proof [**7**] for showing the regularity of linear splicing systems and uses the same type of induction on a non-linear notion of *complexity* for derivation trees (we will explain it later). His result is shown for splicing rules of the form $r = (\alpha, \alpha'; \beta)$ and is easily applied to our type of splicing rules using the form $(u_1u_2, u_3u_4; u_1u_4)$.

Let us define the mapping s from trees to skeletons. For a tree t,

$$s(t)(x) = \begin{cases} t(x) & \text{if } x \text{ is a terminal node,} \\ \sigma & \text{if } x \text{ is an internal node.} \end{cases}$$

Thus the skeleton tells us about the shape and terminal nodes of the tree. For a set of trees T, let us define $s(T) = \{s(t) \mid t \in T\}$.

Let $\gamma = (V, R)$ be a splicing scheme on skeletons for $R \in FIN_R$ and $V = (\{\sigma\} \cup \Sigma)$, and $L_0 \in RT$ be an initial skeleton language.

We first construct a set of production rules to capture the splicing scheme. For each splicing rule

$$r = u_1(\lambda, \ldots, \#(u_2(\lambda, \ldots, \lambda)), \ldots, \lambda) \ \$ \ u_3(\lambda, \ldots, \#(u_4(\lambda, \ldots, \lambda)), \ldots, \lambda)$$

in R, we construct a CFG $B^{(r)} = (N^{(r)}, \Sigma, P^{(r)}, S^{(r)})$ which only generates the skeleton

$$\mathbf{u_1}(\lambda_1, \ldots, \mathbf{u_4}(\lambda'_1, \ldots, \lambda'_n), \ldots, \lambda_m),$$

that is,

$$D(B^{(r)}) = \{b_1^{(r)}(\lambda_1, \ldots, b_4^{(r)}(\lambda'_1, \ldots, \lambda'_n), \ldots, \lambda_m)\}$$

and

$$s(b_1^{(r)}(b_4^{(r)})) = u_1(u_4),$$

where λ_i and λ'_j denote null trees and we denote by $x(y)$ the tree $x(\ \lambda, \ldots, y(\lambda, \ldots, \lambda), \ldots, \lambda)$ for short. We refer to the production rule of the form $A \to \lambda_i$ in $P^{(r)}$ as $Z_i^{(r)} \to \lambda_i$ for $1 \leq i \leq m$ and the production rule of the form $A \to \lambda'_j$ as $\hat{Z}_j^{(r)} \to \lambda'_j$ for $1 \leq j \leq n$. We construct these CFGs $B^{(r)}$ $(r = 1, \ldots, |R|)$ to be pairwise disjoint and we let B be their union $B^{(1)} \cup B^{(2)} \cup \cdots \cup B^{(R)}$.

Let $G_0 = (N_0, \Sigma, P_0, S)$ be a CFG which generates L_0, that is, $s(D(G_0)) = L_0$. We assume G_0 is disjoint from B. We will construct a sequence of CFGs $G_k = (N_k, \Sigma \cup \{[\![,]\!]\}, P_k, S)$ which generates $s(D(G_k))$ as follows, where we will use two specific symbols $[\![$ and $]\!]$ to mark the production rules that connect G_k to $B^{(r)}$.

We consider each splicing rule $r = u_1(\#(u_2))\$u_3(\#(u_4))$ for which both $u_1(u_2)$ and $u_3(u_4)$ occur as subtrees of elements of $s(D(G_k))$. For each such rule we consider all derivation trees of G_k which contain $u_1(u_2)$ or $u_3(u_4)$ as subtrees. If $\alpha_0(\phi_1(\alpha_1, \ldots, \phi_2(\alpha'_1, \ldots, \alpha'_{m_2}), \ldots, \alpha_{m_1}))$ is a derivation tree of G_k such that

$$s(\phi_1(\lambda_1, \ldots, \phi_2(\lambda'_1, \ldots, \lambda'_{m_2}), \ldots, \lambda_{m_1})) = u_1(\lambda_1, \ldots, u_2(\lambda'_1, \ldots, \lambda'_{m_2}), \ldots, \lambda_{m_1})$$

(where λ_i and λ'_j are null trees) and the root of ϕ_1 is a nonterminal X and the root of α_i is a nonterminal Y_i for $1 \leq i \leq m_1$, then we add a new production rule $X \to [\![S^{(r)}$ and new production rules $Z_i^{(r)} \to]\!]Y_i$ where $S^{(r)}, Z_i^{(r)} \in N^{(r)}$ $(1 \leq i \leq m_1)$. Similarly, if there is a derivation tree $\beta_0(\phi_3(\beta_1, \ldots, \phi_4(\beta'_1, \ldots, \beta'_{n_2}), \ldots, \beta_{n_1}))$ of G_k such that $s(\phi_3(\phi_4)) = u_3(u_4)$ and the root of β'_i is a nonterminal Y'_i for $1 \leq i \leq n_2$, then we add new production rules $\hat{Z}_i^{(r)} \to]\!]Y'_i$ $(1 \leq i \leq n_2)$. A production rule of the form $X \to [\![S^{(r)}$ is called an *initial marker-production of level* $k+1$ if it is not included in G_k, and a production rule of the form $Z \to]\!]Y$ is called an *terminal marker-production of level* $k+1$ if it is not included in G_k. If the set of production rules $P^{(r)}$ is not included in G_k, then we say that $B^{(r)}$ has level $k+1$.

We obtain G_{k+1} from G_k by adding all marker-productions and CFGs $B^{(r)}$ of level $k+1$ except production rules $Z_i^{(r)} \to \lambda_i$ and $\hat{Z}_j^{(r)} \to \lambda'_j$ (that is, $P^{(r)} - \{Z_i^{(r)} \to \lambda_i, \hat{Z}_j^{(r)} \to \lambda'_j \mid 1 \leq i \leq m_1, 1 \leq j \leq n_2\}$).

Since the number of possible marker-productions is finite, eventually there is a level n for which $G_{n+1} = G_n$. Now we will show that the tree language $s(D(G_n))$ is closed under splicing and that $s(D(G_n))$ is included in $\gamma^*(L_0)$. We extend the mapping s to $s(X[\![t) = s(t)$ and $s(Z]\!]t) = s(t)$ for dealing with marker-productions where X and Z are nonterminals.

To see closure under splicing, suppose that two skeletons

$$x_0(u_1(x_1, \ldots, u_2(x'_1, \ldots, x'_{m_2}), \ldots, x_{m_1})) \text{ and}$$

$$y_0(u_3(y_1, \ldots, u_4(y'_1, \ldots, y'_{n_2}), \ldots, y_{n_1}))$$

are in $s(D(G_n))$ and that a splicing rule r

$$\begin{aligned} r &= u_1(\lambda_1, \ldots, \#(u_2(\lambda'_1, \ldots, \lambda'_{m_2})), \ldots, \lambda_{m_1})\ \$ \\ &\quad u_3(\lambda_1, \ldots, \#(u_4(\lambda'_1, \ldots, \lambda'_{n_2})), \ldots, \lambda_{n_1}) \end{aligned}$$

is in R. Note that $b_1^{(r)}(b_4^{(r)}) \in D(B^{(r)})$ and $s(b_1^{(r)}(b_4^{(r)})) = u_1(u_4)$. Let

$$\alpha_0(\phi_1(\alpha_1, \ldots, \phi_2(\alpha'_1, \ldots, \alpha'_{m_2}), \ldots, \alpha_{m_1})) \text{ and}$$

$$\beta_0(\phi_3(\beta_1, \ldots, \phi_4(\beta'_1, \ldots, \beta'_{n_2}), \ldots, \beta_{n_1}))$$

be derivation trees of G_n such that $s(\alpha_0) = x_0$, $s(\phi_1(\phi_2)) = u_1(u_2)$, $s(\alpha_i) = x_i$ $(1 \leq i \leq m_1)$ and $s(\alpha'_j) = x'_j$ $(1 \leq j \leq m_2)$, and similarly for β_0, ϕ_3, ϕ_4, β_i, β'_j. If the root of ϕ_1 is a nonterminal X then there is an initial marker-production $X \to [\![S^{(r)}$, and if the root of α_i is a nonterminal Y_i then there are terminal marker-productions $Z_i^{(r)} \to]\!]Y_i$ $(1 \leq i \leq m_1)$, and if the root of β'_i is a nonterminal Y'_i then there are terminal marker-productions $\hat{Z}_i^{(r)} \to]\!]Y'_i$ $(1 \leq i \leq n_2)$. Hence

$$\alpha_0(X[\![b_1^{(r)}(Z_1^{(r)}]\!]\alpha_1, \ldots, b_4^{(r)}(\hat{Z}_1^{(r)}]\!]\beta'_1, \ldots, \hat{Z}_{n_2}^{(r)}]\!]\beta'_{n_2}), \ldots, Z_{m_1}^{(r)}]\!]\alpha_{m_1}))$$

is a derivation tree of G_n and so

$$\begin{aligned} & x_0(u_1(x_1, \ldots, u_4(y'_1, \ldots, y'_{n_2}), \ldots, x_{m_1})) \\ &= s(\alpha_0(X[\![b_1^{(r)}(Z_1^{(r)}]\!]\alpha_1, \ldots, b_4^{(r)}(\hat{Z}_1^{(r)}]\!]\beta'_1, \ldots, \hat{Z}_{n_2}^{(r)}]\!]\beta'_{n_2}), \ldots, Z_{m_1}^{(r)}]\!]\alpha_{m_1}))) \end{aligned}$$

is in $s(D(G_n))$.

To show that $s(D(G_n))$ is included in $\gamma^*(L_0)$, we will show how to obtain $s(t) \in s(D(G_n))$ by a finite number of splicings starting with elements of L_0.

We use the following notion defined by Pixton [**7**]. The *complexity* of a derivation tree t of G_n is the n-tuple of natural numbers $c = \langle c_1, \dots, c_n \rangle \in \mathbb{N}^n$, where c_j is the number of marker-productions used in t of level j. We order the set of these tuples lexicographically. The proof is constructed based on induction on the partial order of this complexity.

First, a derivation tree t of complexity $0 = \langle 0, \dots, 0 \rangle$ is generated by G_0, so $s(t)$ is in L_0.

Suppose that $c \in \mathbb{N}^n$ with $c > 0$ and suppose that for all derivation trees t' of G_n of complexity less than c, $s(t')$ is in $\gamma^*(L_0)$. Suppose that a derivation tree t of G_n has complexity c. Then t starts with the start symbol S, and t contains at least one marker-production and furthermore it contains at least one initial marker-production. We select an initial marker-production $X \to [\![S'$ such that

$$\xi(\ X[\![S'(\ w\ )\ ) = t$$

and w contains no initial marker-production. Since w contains some terminal marker-productions, we select terminal marker-productions $Z_i \to]\!] Y_i$ such that

$$\theta_1(\ Z_1]\!]Y_1(\eta_1), \dots, \theta_2(\ Z_1']\!]Y_1'(\eta_1'), \dots, Z_{q_2}']\!]Y_{q_2}'(\eta_{q_2}')\ ), \dots, Z_{q_1}]\!]Y_{q_1}(\eta_{q_1})\ ) = w$$

and $\theta_1(\theta_2)$ contains no terminal marker-production. Note that

$$t = \xi(X[\![S'(\theta_1(Z_1]\!]Y_1(\eta_1), \dots, \theta_2(Z_1']\!]Y_1'(\eta_1'), \dots, Z_{q_2}']\!]Y_{q_2}'(\eta_{q_2}')), \dots, Z_{q_1}]\!]Y_{q_1}(\eta_{q_1})))).$$

Now we can see that there is a splicing rule $r = u_1(\#(u_2))\$u_3(\#(u_4))$ such that

$$b_1^{(r)}(\lambda, \dots, b_4^{(r)}(\lambda, \dots, \lambda), \dots, \lambda) =$$

$$S'(\theta_1(Z_1(\lambda), \dots, \theta_2(Z_1'(\lambda), \dots, Z_{q_2}'(\lambda)), \dots, Z_{q_1}(\lambda)))$$

and $s(b_1^{(r)}(b_4^{(r)})) = u_1(u_4)$.

Suppose that $X \to [\![S'$ has level k and $S' = S^{(r)}$, and each $Z_i \to]\!] Y_i$ has level at most k' and $Z_i = Z_i^{(r)}$ $(1 \leq i \leq q_1)$, and each $Z_j' \to]\!] Y_j'$ has level at most k' and $Z_j' = \hat{Z}_j^{(r)}$ $(1 \leq j \leq q_2)$. Then there is a derivation tree

$$\alpha_0(\phi_1(\alpha_1, \dots, \phi_2(\alpha_1', \dots, \alpha_l'), \dots, \alpha_{q_1}))$$

in G_{k-1} such that the root of ϕ_1 is a nonterminal X, the root of α_i is a nonterminal Y_i $(1 \leq i \leq q_1)$, and $s(\phi_1(\phi_2)) = u_1(u_2)$, and there is a derivation tree

$$\beta_0(\phi_3(\beta_1, \dots, \phi_4(\beta_1', \dots, \beta_{q_2}'), \dots, \beta_m))$$

in $G_{k'-1}$ such that the root of β_j' is a nonterminal Y_j' $(1 \leq j \leq q_2)$ and $s(\phi_3(\phi_4)) = u_3(u_4)$. It follows that

$$\xi(\phi_1(Y_1(\eta_1), \dots, \phi_2(\alpha_1', \dots, \alpha_l'), \dots, Y_{q_1}(\eta_{q_1}))) = t_1$$

and

$$\beta_0(\phi_3(\beta_1, \dots, \phi_4(Y_1'(\eta_1'), \dots, Y_{q_2}'(\eta_{q_2}')), \dots, \beta_m)) = t_2$$

are derivation trees in G_n, and that

$$s(t_1) = s(\xi)(u_1(s(Y_1(\eta_1)), \dots, u_2(s(\alpha_1'), \dots, s(\alpha_l')), \dots, s(Y_{q_1}(\eta_{q_1}))))$$

and

$$s(t_2) = s(\beta_0)(u_3(s(\beta_1), \dots, u_4(s(Y_1'(\eta_1')), \dots, s(Y_{q_2}'(\eta_{q_2}'))), \dots, s(\beta_m)))$$

splice together using the rule r to give

$$s(\xi)(u_1(s(Y_1(\eta_1)), \dots, u_4(s(Y_1'(\eta_1')), \dots, s(Y_{q_2}'(\eta_{q_2}'))), \dots, s(Y_{q_1}(\eta_{q_1}))))$$

$$= s(\xi(b_1^{(r)}(Y_1(\eta_1), \dots, b_4^{(r)}(Y_1'(\eta_1'), \dots, Y_{q_2}'(\eta_{q_2}')), \dots, Y_{q_1}(\eta_{q_1})))$$

$$= s(t).$$

We can see that t_1 and t_2 have lower complexities than t. By the induction hypothesis, both $s(t_1)$ and $s(t_2)$ are in $\gamma^*(L_0)$, and hence $s(t)$ is also in $\gamma^*(L_0)$. This concludes the proof. □

COROLLARY 4. *A skeleton language in the class $HTS(FIN_T, FIN_R)$ is generated by a context-free grammar.*

5. Conclusions

It is known that splicing systems, with finite set of rules, and finite set of starting strings, can generate only regular languages. As soon as we can use multiplicity constraints on the same class of splicing systems, we are able to generate any recursively enumerable languages. In this paper we looked for variations, other than multiplicity, to the basic definition of splicing system, with the goal of going beyond the generation of regular languages.

We defined a class of splicing systems able to generate context-free languages, or structures closely related to them, like derivation trees or skeletons. An interesting feature is that our model keeps good molecular feasibility, since it generates folded strings similar to the representation in terms of secondary structures of RNA molecules.

We could even consider directly the implementation of our splicing systems by DNA, since it has been proved that it can be assembled in unusual spatial structures, using strands partially annealed as we also propose. See for instance the solid-support based methodology to build complex DNA structures defined in [**11**], but also the suggestive comments on the experimental pitfalls of this constructions reported in [**9**].

Recently, Winfree, Yang and Seeman discussed in [**10**] a way of generating context-free languages similar to ours: it is based on the iterated joining of short double stranded sequences, reproducing in the finally resulting molecule all the derivation of a string. With respect to that paper, the main advantage of our proposal is that while they encode in the molecule all the terminals *and* the nonterminals involved in the derivation, here we can work with cleaner structures, like those representing the skeletons. We can simulate a correct derivation of trees by splicing even without keeping record of nonterminals, since the restriction enzymes recognize specific substructures in the molecules, and not directly the nonterminals, enough to allow them to splice two trees and so to continue the derivation process.

References

[1] L. Adleman. Molecular computation of solutions to combinatorial problems. *Science*, 266:1021–1024, 1994.

[2] R. Freund, L. Kari, and Gh. Paun. DNA computing based on splicing : The existence of universal computers. Technical report, Fachgruppe Informatik, Tech. Univ. Wien.

[3] T. Head. Formal language theory and DNA : An analysis of the generative capacity of specific recombinant behaviors. *Bulletin of Mathematical Biology*, 49:737–759, 1987.

[4] J.R. Koza. *Genetic Programming.* MIT Press, 1992.

[5] G. Paun. On the splicing operation. *Discrete Applied Mathematics*, 70:57–79, 1996.

[6] Gh. Paun, G. Rozenberg, and A. Salomaa. Computing by splicing. *Theoretical Computer Science*, 168(2):321–336, 1996.

[7] D. Pixton. Regularity of splicing languages. *Discrete Applied Mathematics*, 69:101-124, 1996.

[8] Y. Sakakibara, M. Brown, R. Hughey, I. S. Mian, K. Sjolander, R. C. Underwood, and D. Haussler. Stochastic context-free grammars for tRNA modeling. *Nucleic Acids Research*, 22:5112–5120, 1994.

[9] N.C. Seeman, H. Wang, B. Liu, J. Qi, X. Li, X. Yang, F. Liu, W. Sun, Z. Shen, R. Sha, C. Mao, Y. Wang, S. Zhang, T-J. Fu, S. Du, J.E. Mueller, Y. Zhang, J. Chen, The Perils of Polynucleotides: The Experimental Gap Between the Design and Assembly of Unusual DNA Structures. In *Proceedings of Second Meeting on DNA Based Computers*, Princeton 1996, to appear in AMS - DIMACS Series in Discr. Math. and Theor. Comp. Sci.

[10] E. Winfree, X. Yang, N.C. Seeman. Universal computation via self-assembly of DNA: some theory and experiments. In *Proceedings of Second Meeting on DNA Based Computers*, Princeton 1996, to appear in AMS - DIMACS Series in Discr. Math. and Theor. Comp. Sci.

[11] Y. Zhang, N.C. Seeman, A Solid-Support Methodology for the Construction of Geometrical Objects from DNA. *J. Am. Chem. Soc.*, 114:2656–2663, 1992.

Department of Information Sciences, Tokyo Denki University, Hatoyama, Hiki-gun, Saitama 350-03, Japan

E-mail address: `yasu@j.dendai.ac.jp`

Dipartimento di Scienze dell'Informazione, Università di Milano, via Comelico 39, 20135 Milano, Italy

E-mail address: `ferretti@dsi.unimi.it`

Selected Titles in This Series

(Continued from the front of this publication)

19 **Ingemar J. Cox, Pierre Hansen, and Bela Julesz, Editors,** Partitioning Data Sets

18 **Guy E. Blelloch, K. Mani Chandy, and Suresh Jagannathan, Editors,** Specification of Parallel Algorithms

17 **Eric Sven Ristad, Editor,** Language Computations

16 **Panos M. Pardalos and Henry Wolkowicz, Editors,** Quadratic Assignment and Related Problems

15 **Nathaniel Dean and Gregory E. Shannon, Editors,** Computational Support for Discrete Mathematics

14 **Robert Calderbank, G. David Forney, Jr., and Nader Moayeri, Editors,** Coding and Quantization: DIMACS/IEEE Workshop

13 **Jin-Yi Cai, Editor,** Advances in Computational Complexity Theory

12 **David S. Johnson and Catherine C. McGeoch, Editors,** Network Flows and Matching: First DIMACS Implementation Challenge

11 **Larry Finkelstein and William M. Kantor, Editors,** Groups and Computation

10 **Joel Friedman, Editor,** Expanding Graphs

9 **William T. Trotter, Editor,** Planar Graphs

8 **Simon Gindikin, Editor,** Mathematical Methods of Analysis of Biopolymer Sequences

7 **Lyle A. McGeoch and Daniel D. Sleator, Editors,** On-Line Algorithms

6 **Jacob E. Goodman, Richard Pollack, and William Steiger, Editors,** Discrete and Computational Geometry: Papers from the DIMACS Special Year

5 **Frank Hwang, Fred Roberts, and Clyde Monma, Editors,** Reliability of Computer and Communication Networks

4 **Peter Gritzmann and Bernd Sturmfels, Editors,** Applied Geometry and Discrete Mathematics, The Victor Klee Festschrift

3 **E. M. Clarke and R. P. Kurshan, Editors,** Computer-Aided Verification '90

2 **Joan Feigenbaum and Michael Merritt, Editors,** Distributed Computing and Cryptography

1 **William Cook and Paul D. Seymour, Editors,** Polyhedral Combinatorics